Costa Rican Mammals

THE NATURAL HISTORY OF
Costa Rican Mammals

Text & Illustrations by
Mark Wainwright

Preface by Oscar Arias

A Zona Tropical Publication

ISBN 0-9705678-1-2

Editor: David Featherstone
Book design: Servigráficos S.A.
Designer: Adrián Soto Piedra

Published by Distribuidores Zona Tropical, S.A.
S.J.O. 1948
P.O. Box 025216
Miami, FL 33102-5216
www.zonatropical.net

CONTENTS

PREFACE

Costa Ricans are tremendously proud of their national parks and private reserves, which together protect nearly a third of our country's territory. The biodiversity of our small country, which is exceedingly rich, entices scientists and ecotourists alike to choose our tropical forests for study and for vacations. During the years of my presidency, we were able to convert a significant amount of territory into protected land through debt-for-nature swaps, a procedure by which a nation's debt is forgiven or restructured in exchange for an active commitment to setting aside land for environmental protection. Our success has become a model for other developing countries under threat of deforestation.

One of the most important aspects of conservation work is to attend to the local human population within an ecosystem that is to be protected. While some conservation efforts seem to pit the local communities against the environment, it should be remembered that people do not destroy the rainforest or poach endangered animals for fun; they do it to secure their own livelihood. Costa Rica is proud to have made great strides in training local guides and encouraging ecotourism in a way that provides for sustainable human communities within conservation projects that safeguard our flora and fauna. It is now common knowledge among our people that the giant turtles that nest on our beaches and the pacas that scurry about on the forest floor are worth more to the local community alive than dead. The work residents have found as park rangers and guides has created a consciousness about the delicacy of nature at the same time that it produces steady incomes for families in poor rural zones who would otherwise find it necessary to clear land for farming or hunt and trap species that are officially protected.

Both the enjoyment of nature and the efforts to conserve its richness involve a process of education. *The Natural History of Costa Rican Mammals* will be invaluable in this process. More than a simple pictorial guide to help visitors identify one species or another, the book is an in-depth and fascinating look at the natural history of those mammals; it not only describes their habits and habitats, but also how and when they became residents of Costa Rica's forests, mountains, and coastal waters.

This book is for the observer who wants to learn and not just see, and for the guide who strives to be truly knowledgeable about his or her subject matter. It will surely be found in my knapsack on my next visit to one of our national environmental treasures.

Oscar Arias
Former President of Costa Rica
1987 Nobel Peace Laureate

INTRODUCTION

In Costa Rica, there is an unusual intensity of interest in the natural world, thanks to the extraordinary combination of vast biological wealth, an extensive system of protected areas, and political stability. Numerous biologists have chosen to do their research in Costa Rica, and the wildlife here may consequently be better understood than that of any other Latin American country. There is also a huge audience for the discoveries made by these researchers—from ecotourists and naturalist guides to local and international students, park guards, and other interested local residents.

Given this perhaps unrivalled potential for environmental education, it is a shame that most of the information that has been accrued about Costa Rica's wildlife is not readily accessible to the general public. The majority is published only in scientific journals, many of which are difficult to acquire in Costa Rica; and it is presented, necessarily, in a rigorous fashion that can be overwhelming for the nonspecialist. At the other end of the spectrum, the relatively few popular publications on Costa Rican wildlife, while having their own value, seldom provide more than superficial treatment of the fascinating complexities revealed by scientific research. An additional hurdle for Costa Ricans is that the vast majority of both scientific and popular publications is available only in English.

This book is an attempt to bridge that gap—at least for one group of organisms, Costa Rican mammals—by bringing together detailed information from hundreds of different scientific articles and presenting it in a single, more user-friendly format, both in this English volume and, in a future publication, in Spanish. This book should be a useful complement to the various field guides that are now available for the region's mammals because it focuses less on physical description and species identification and more on natural history and conservation. A book like this inevitably runs the risk of appalling biologists with simplifications of complex issues, or of boring nonbiologists with excessive details. While this book probably manages to do both, hopefully it will foster a greater awareness of what is known (and, just as importantly, all that is not known) about Costa Rica's mammals.

In general, the more one knows about something, the more one cares about it. Ultimately, this book is intended to make readers understand and care more about the environment in general, and strive harder to slow the terrible, unsustainable, and irreversible destruction that we are causing.

This book provides extensive information about the natural history and conservation of all Costa Rica's readily identifiable terrestrial and freshwater mammals. Although physical descriptions are not the focus of this book, the key distinguishing features of each species are noted in the plates. When different species could be confused, more detailed descriptions are included in the species accounts. Included in the descriptions are the head-and-body length (the distance between the tip of the snout and the base of the tail) and weight of each mammal. These measurements are approximate averages for adult animals; obviously, the exact size and weight varies between individuals.

Bats and small rodents are treated by family or subfamily rather than by species; descriptions and illustrations are provided for the most common, distinctive, or representative species within each group. Identification of these small mammals is, for the most part, the realm of the specialist, for few species can be distinguished unless they are trapped and then closely analyzed with the guidance of a scientific key (and, generally, a scientist). Covering all the small mammals individually would require extensive physical descriptions, and that would draw the book away from its natural history and conservation focus. Moreover, there is already an excellent, fully illustrated field guide to Costa Rica's small (and large) mammals in Fiona Reid's *A Field Guide to the Mammals of Central America and Southeast Mexico.* The bat and small-rodent accounts should be sufficiently rich, however, to convey the diversity, complexity, and importance of these fascinating mammals.

With the exception of the bats and small rodents, this book covers all terrestrial and freshwater species currently known to occur in Costa Rica. In general, each chapter treats a mammal order and is comprised of a broad introduction followed by species or family accounts. However, the order Carnivora includes such a large portion of the species covered in this book that it has been split into four chapters, each of which covers a family. Every account comes with an extensive list of references. The lengthy bibliography is provided as a potentially valuable resource for students and other readers who wish to pursue more detailed information.

Mammal Classification

The principal characteristics that set mammals apart from other vertebrates are a unique bone arrangement where the jaw meets the skull (see p. 14), the presence of a diaphragm (the muscular wall separating the intestines from the rib cage), body hair (albeit very little in a few species), and, of course, mammary glands, which produce milk to feed offspring.

To understand mammal taxonomy, it is helpful to think of the class Mammalia as a tree. The mammal tree starts where mammals first diverge from their reptilian ancestors, some 190 million years ago. The first branch occurs where the ancestors of most mammals (those included in the subclass Theria) diverge from egg-laying mammals (subclass Prototheria). The Therian branch forks again about 75 million years ago into placental mammals (infraclass Eutheria) and marsupials (infraclass Metatheria). Subsequently, the tree branches into orders, then families, then genera (the plural of genus), and so on, each category being a subset of the category that precedes it. At many levels, there are additional intermediate categories such as suborders or subfamilies. The tree terminates with the roughly 4,000 species of mammals alive today.

The last two major categories on the tree—the genus and species—form an organism's scientific name, such as, for example, *Homo sapiens.* This binomial system was first invented and used in the eighteenth century by Swedish biologist Carl Linnaeus. Scientific names stem from Latin and

Greek, and each is unique. Thus, although a species may be referred to by dozens of different local names across its range, it also has a universally understood name.

There are no absolute criteria for defining a species. Generally, organisms are thought to belong to the same species if they are capable of breeding with each other and producing fertile offspring. This definition doesn't always hold true, however. Domestic dogs and coyotes, for example, breed readily and produce fertile offspring, yet they are considered different species. The criteria by which species are defined can also be hard to prove. Many species have large geographic ranges, where pockets of individuals are isolated from one another by, for example, mountains or water, and thus can breed with each other only in theory. Members of the same species from different geographic areas may look or behave quite differently. Scientists sometimes recognize such differences by splitting species into subspecies. The allocation of subspecies is sometimes rather arbitrary, but can be useful in indicating distinct populations that, perhaps because they are only partially isolated or because they have become isolated only recently, are still reproductively compatible. Such subspecies may be on the road to becoming separate species. They are the still-growing twig tips of the taxonomic tree.

Mammal Evolution

The principal process responsible for the growth of our taxonomic tree was first described by Charles Darwin in 1859. Different genetic combinations ensure that the members of a given species are seldom identical. Inevitably, some individuals are slightly better adapted to their environments than others, and those individuals have a slightly better chance of surviving and passing their genes to other generations. Thus, features that increase a species' chance of survival tend to prevail, through a now famous process that Darwin termed natural selection.

Around 280 million years ago, a series of mammal-like features was being selected in a group of reptiles known as the synapsids. These features, which may have helped synapsids take advantage of an abundance of insects at the time, included a smaller body size and, partly as a result of that, both greater efficiency in controlling body temperature and greater mobility. One such synapsid was *Dimetrodon*, a ten-foot-long animal with a tall fan of skin on its back. The fan must have acted as a solar panel or radiator that helped *Dimetrodon* heat up or cool down quickly. By around 240 million years ago, an even more mammal-like line of reptiles known as the therapsids had appeared. One of these was *Thrinaxodon*, a doglike reptile that probably hunted small vertebrates.

Dimetrodon (rear) and *Thrinaxodon*.

Thrinaxodon had even evolved hair, which may have enabled it to stay warm enough to hunt by night.

Just as these reptiles were on the verge of evolving into the first true mammals, another line of reptiles—the dinosaurs—was entering its era of glory. By the time the first-known mammals appeared, about 200 million years ago, the period of relative abundance that the synapsids and therapsids enjoyed was over. For their first 135 million years, mammals were to live in the shadows of the dinosaurs, playing but a minor role in the world's ecosystems.

The earliest known mammal—a shrew-like creature called *Morganucodon*—was just 3 cm (an inch) long. It hunted insects in the undergrowth by night, when the sun-dependent dinosaurs were inactive. A coat of fur kept it warm, and large eyes and long, sensitive whiskers helped it navigate in the dark. Fossils show that its young had milk teeth, so they were probably suckled. Another feature that defines *Morganucodon* as the earliest mammal is the lack of certain bones in its lower jaw. Reptiles' jaws contain five bones, but mammals' jaws contain only two. By the time *Morganucodon* appeared, the three "missing" bones had shifted to inside the middle ear, where they formed part of the apparatus that directs the vibrations of the eardrum to the inner ear. This feature, a characteristic of the Mammalia today, is one reason we mammals can hear so well. The ear bones are illustrated on p. 24.

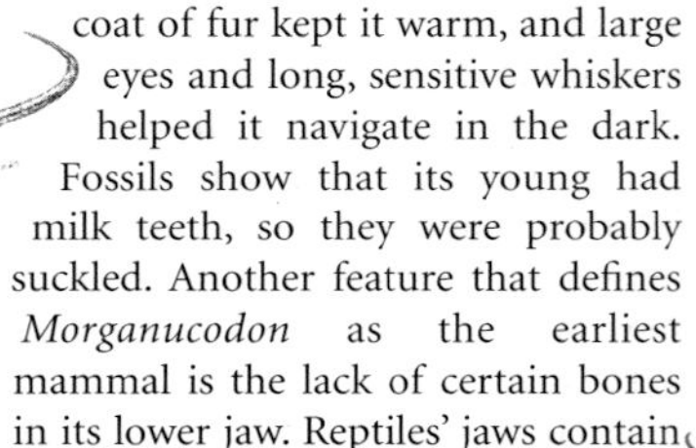

Morganucodon.

APPROXIMATE ARRANGEMENT OF THE WORLD, 200 MILLION YEARS AGO TO PRESENT.

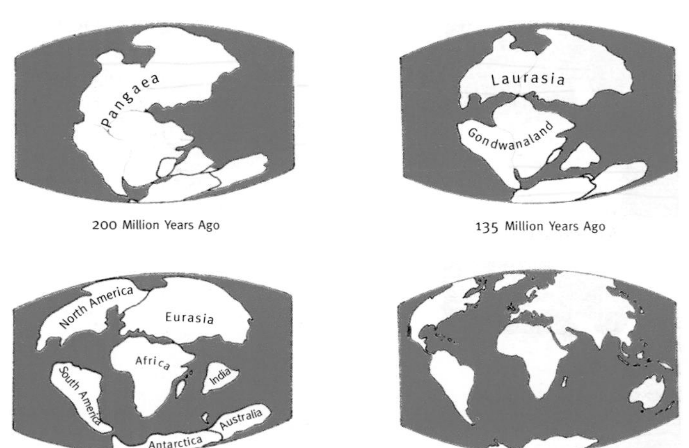

200 Million Years Ago

135 Million Years Ago

65 Million Years Ago

Today

When the first mammals appeared, the world's continents were locked together in a single, huge landmass known as Pangaea. Shortly thereafter (in relative terms), however, Pangaea started to break apart. By the beginning of the Cretaceous period, about 135 million years ago, there were two supercontinents—Laurasia to the north, and Gondwanaland to the south. By the end of the Cretaceous, some 65 million years ago, Gondwanaland itself had split. The fragmentation of the world's continents separated mammal populations into isolated groups that subsequently evolved in different ways, so although the long reign of the dinosaurs hindered mammalian progress, by the end of the Cretaceous, mammals had quietly become quite diverse. About 65 million years ago, most dinosaurs suddenly disappeared, possibly as the result of a meteor strike and/or a worldwide change in climate. Whatever the reason, the dinosaurs' disappearance opened the door for mammals to enter their own age of glory.

For the next 60 million years or so, North and South America remained isolated from one another, and became home to completely different mammal faunas. Members of the Carnivora—dogs, cats, bears, weasels, and raccoons—were restricted to the north, where they preyed upon other northern groups such as ungulates, rabbits, and sciuromorph and myomorph rodents. In South America, large carnivorous opossums preyed upon smaller opossums, sloths, anteaters, armadillos, monkeys, and caviomorph rodents.

Perhaps around 35 million years ago, land that was to become southern Central America started to emerge from the sea, lifted above the water level through the action of plate tectonics. What is now Costa Rica's Talamanca mountain range—the tallest in southern Central America—was among the first islands to appear. As these islands were raised ever so

THE FORMATION OF THE PANAMANIAN LAND BRIDGE.

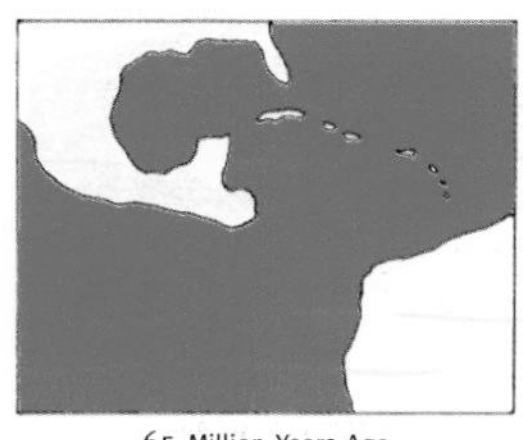

65 Million Years Ago

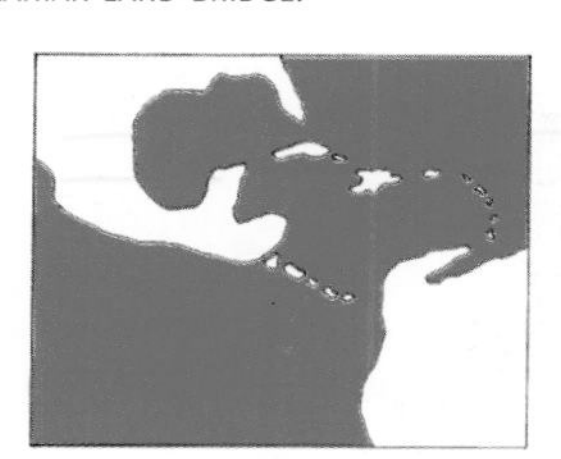

35 Million Years Ago

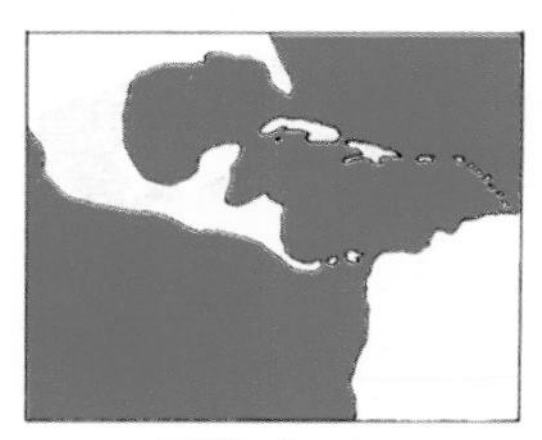

15 Million Years Ago

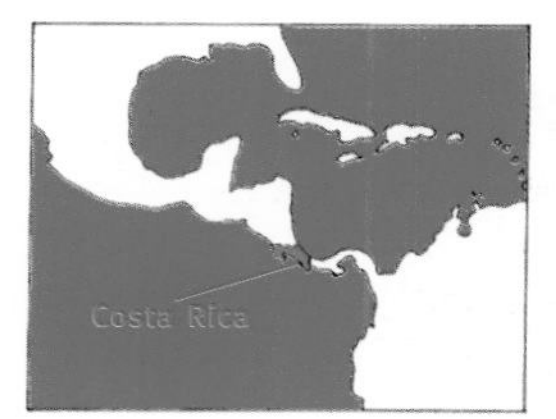

Today

gradually out of the water, they started to fill in the gap between North and South America.

By around seven to two million years ago, this process had formed a continuous land bridge, commonly referred to as the "Panamanian land bridge." As species crossed the bridge in both directions, the distribution of mammals within the Americas underwent profound and rapid change. Situated right on the bridge, Costa Rica has been center stage for one of the most dramatic episodes in mammal history.

The fascinating history of each mammal group is described in the order and family accounts.

Observing Mammals

Compared to animals such as birds or insects, mammals are hard to observe in the field, for several reasons. First, larger mammals, especially carnivores, tend to be less numerous than smaller animals because they are situated higher up the food pyramid. For example, a jaguar needs to eat many medium-sized animals, and each of those medium-sized animals needs to eat many smaller animals or other organisms, and so on down the food chain, such that the jaguar is necessarily much rarer than smaller animals. Second, while we naturalists are predominantly diurnal and ground-dwelling, most mammals are neither. Of Costa Rica's roughly 225 terrestrial and freshwater species, less than 30 are consistently active during the daytime. Even if one excludes the country's 116 or so bat species, about three quarters of the remaining species are active mostly at night. Likewise, more than three quarters of Costa Rica's mammals either fly or spend a significant portion of their time up trees, under the ground, or in water. Finally, most mammals tend to be shy. They can use their keen senses to detect naturalists, along with their generally cryptic coloration and superior speed and agility to evade them.

Given all this, it is helpful to bear in mind a few simple tips that will increase the chances of finding mammals. One necessity is to walk quietly and listen. Although mammals tend not to be as vociferous as many other animals, a few do make conspicuous and frequent vocalizations. Mammals such as monkeys, olingos, kinkajous, coatis, or squirrels are often located thanks entirely to their calls. Even mammals that call rarely can still be heard as they move through the vegetation. Canopy dwellers in particular are apt to shake branches as they move around; they may also draw attention by dropping pieces of food to the ground.

Another helpful strategy is to locate feeding sites and wait quietly nearby or check the sites frequently. A large fig tree or other fruiting species can attract a variety of mammal species in the course of a day or night, as can a watering hole in northwestern Costa Rica during that portion of the country's severe dry season.

Whatever habitat one is in, going out at night with a strong flashlight and a pair of binoculars is well worthwhile. A good way to locate mammals at night is to scan the area with a flashlight held at eye level and look for eyeshine. Eyeshine is produced by a reflective layer in the back of most nocturnal mammals' eyes known as the tapetum lucidum. The tapetum lucidum reflects incoming light back through an eye's retina, such that the light hits the retina twice. By maximizing available light within the eye, the tapetum lucidum improves a mammal's ability to see in the dark. Since mammals usually have large, dilated pupils at night, the beam of a flashlight held at eye level reflects back as eyeshine. Once a mammal has been located in this way, a pair of binoculars can be used to get a closer look (ideally while someone else aims the flashlight).

Mammal Signs

Of course, one can learn a lot about mammals in the field without ever seeing them. Signs such as scats (fecal droppings), tracks, dens, runways, or indications of feeding areas can reveal the presence of elusive species and may provide clues about these mammals' natural history as well. The study of scats alone has provided extensive information on the diets of several Costa Rican mammals, such as jaguars, pumas, and ocelots (Chinchilla 1997), Neotropical river otters (Spínola 1995b), coatis (Sáenz 1994a), tapirs (Naranjo 1995b), and collared peccaries (McCoy 1985). For the amateur also, identifying and interpreting, or at least speculating about, mammal signs can be a fun exercise. For these reasons, this book includes illustrations or descriptions of a wide variety of signs left by Costa Rican mammals.

The scats of many mammals are illustrated in the species accounts. It should be noted that just as a species' diet can vary considerably, so can the shape and size of its scats; it is often impossible to deduce the identity of the defecator with certainty. Nonetheless, the illustrations should enable the reader to identify some scats and to make educated guesses about who might be the authors of others. An analysis of the contents of the scat, the habitat in which it was found, and any tracks in the vicinity can provide further clues.

The most frequently observed mammal sign is tracks, and Costa Rica's muddy trails, sandy beaches, and riverbeds are excellent track substrates. This book includes illustrations of the tracks of all species likely to leave them. Information about tracks is provided alongside the illustrations and under the body-design heading in the family and order accounts. To identify tracks, one should try to distinguish forefeet from hindfeet, count the number of toes on each, and determine whether the mammal is plantigrade, digitigrade, or unguligrade.

Plantigrade mammals press their toes and heels to the ground when standing. Examples include primates, members of the raccoon family, opossums, skunks, and porcupines. These mammals tend to be the most dexterous, but they are also relatively slow moving. They take food that does not require fast pursuit but which might require some manipulation, and they rely on tactics other than running to escape predators. Many are excellent tree-climbers, and some, such as skunks or porcupines, use other defenses. Digitigrade mammals stand on their toes and never touch their heels to the ground. They can move faster than plantigrades, but are less dexterous. They include mammals such as dogs, cats, and carnivorous members of the weasel family that require speed to catch prey. Unguligrades are hooved mammals, such as deer, peccaries, and tapirs, which walk on the tips of their toes. They have great speed, which they use to escape predators, but they have very little dexterity and feed by pulling or biting off vegetation with their mouths.

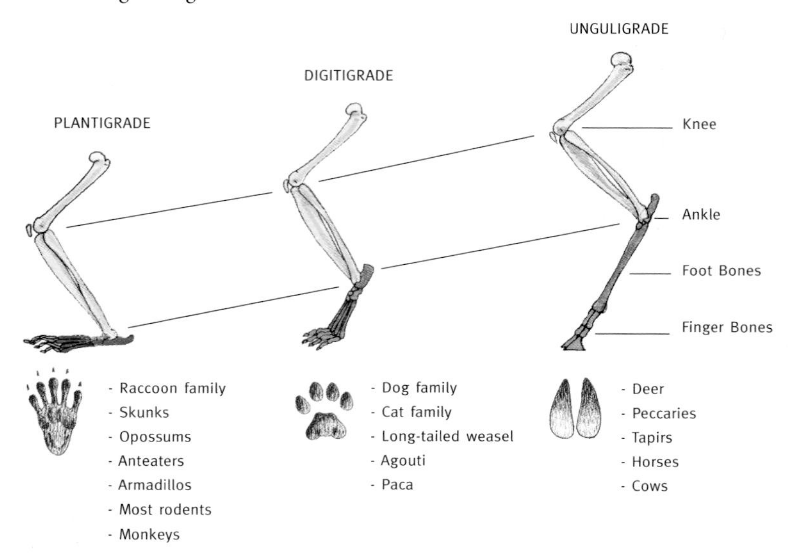

Collecting Tracks

One of the nice things about tracks is that they can be "collected" as plaster of paris molds and then studied in more detail at home. All that is needed to make such molds is plaster of paris, water, and a container in which to mix the two together. It is well worth bringing these simple ingredients along on any field trip.

To make a mold, one should first check the lie of the track. If the track is on a slope, a small wall of mud or sand must be constructed along the downhill end to prevent the mix from running out of the print. Upon mixing, plaster and water should have the consistency of thin pancake batter. If too much water is added, the mix will never harden; if the mix has too much plaster, the cast will turn out chalky and fragile. Once combined, plaster and water thicken quickly, becoming solid within about 15 minutes. The time during which the plaster mix is perfect for pouring lasts no more than 30 seconds. It should not be poured into tracks immediately, since while it is still very thin it tends to seep into the ground or run out at the sides. Nor should it be left to thicken too much, for it becomes difficult to pour and may not penetrate fine details such as claw marks. Extra water cannot be added after the mix has started to thicken, for this only produces a viscous goop that never hardens. It is worth making sure that a little mix will be left over after the initial pouring. Once this excess becomes stiff enough that it will not run, it can be applied over the cast's thin points—such as between the toes and the pad in plantigrade animals. This helps ensure that the cast does not break apart once removed.

Casts should be hard enough to remove after 20 to 30 minutes. To make sure the cast doesn't break upon removal, one should cut around the cast with a penknife and then place the tool underneath the cast to lever it out. In dry sand or sloppy mud, a penknife may not be necessary, but the cast should still be levered out from below. Dirt clinging to the cast can be carefully pulled or washed away. More thorough cleaning can be done with an old toothbrush.

Collecting Skeletons

Along trails and roadsides, one occasionally comes across the carcasses of wild mammals. For the nonsqueamish, reasonably intact specimens that have not been pancaked are worth rescuing, since mammal skeletons tell fascinating stories. When possible, such finds should be donated directly to museums, where they can be prepared properly and used for reference. When lugging around a carcass is not feasible, as is often the case, there is a simple way for the amateur to preserve a skeleton, and that is to bury it.

The carcass should be wrapped either in one of the woven-plastic feed sacks that are abundant in Costa Rica or in a thin wire mesh. If neither is available, a plastic bag will do. There should be a few holes in the wrapping so that beetles and other cleaners can get in, but the carcass should be carefully enclosed on the bottom and on the sides so that none of the bones will fall out as the carcass decomposes. The wrapped carcass should be buried in a two- to three-foot-deep hole and the site covered with rocks or other heavy objects to keep away scavengers. Place a stake with the name of the animal and date of burial next to the site to avoid confusion at the time of retrieval.

How long a carcass should be left underground varies with the size of the carcass and the climate—the smaller the corpse and the wetter the climate, the faster the decomposition. In Costa Rica, a medium-sized mammal such as an olingo or an armadillo should be ready to exhume after about six months, and a large mammal such as a tapir after about nine months. As with a cake in an oven, early is better than late since an "underdone" specimen can always be reburied. If a skeleton is exhumed too early, pieces of flesh will still be clinging to the bones; if it is left too long, the bones themselves will start to disintegrate. In wet, tropical climates, bones will become irretrievably decomposed within a year and a half.

When picking a skeleton out of its burial site, or upon finding an already clean skeleton in the wild, one should try to locate and collect the teeth and the smallest bones—the toes and the smallest tail vertebrae—before moving anything, lest they be lost forever. The teeth are particularly important for identifying and learning about a mammal. Note that teeth will often fall out of a clean skull as soon as it is moved, even if they still appear to be lodged in their sockets. Once exhumed, the bones should be washed thoroughly with water and then bleached by soaking either in a 2% hydrogen peroxide solution (available at many stores) or in a solution of bleach (one part) and water (10 parts). If the latter is used, the bones must then be washed thoroughly again, since traces of bleach will corode the bones.

Mounting skeletons for presentation requires practice and patience. The most common technique is to pass a stiff but bendable wire through the tail and back vertebrae and into the back of skull, and then mount this on two vertical support poles attached to a wooden stand. The wire should be bent first to match whatever natural curvature the skeleton would have in a live animal. All remaining bones are attached from the backbone outward. Strong bones can be nailed or screwed in place, while more fragile bones can be tied on with thin but strong wire such as surgical steel, or glued. Holes for nailing, screwing, and tying can be made with a tiny drill such as those used by jewelry-makers.

The Mammal Skeleton

A mammal's skeleton can be comprised of more than 200 bones, the most important of which are identified in the following illustration.

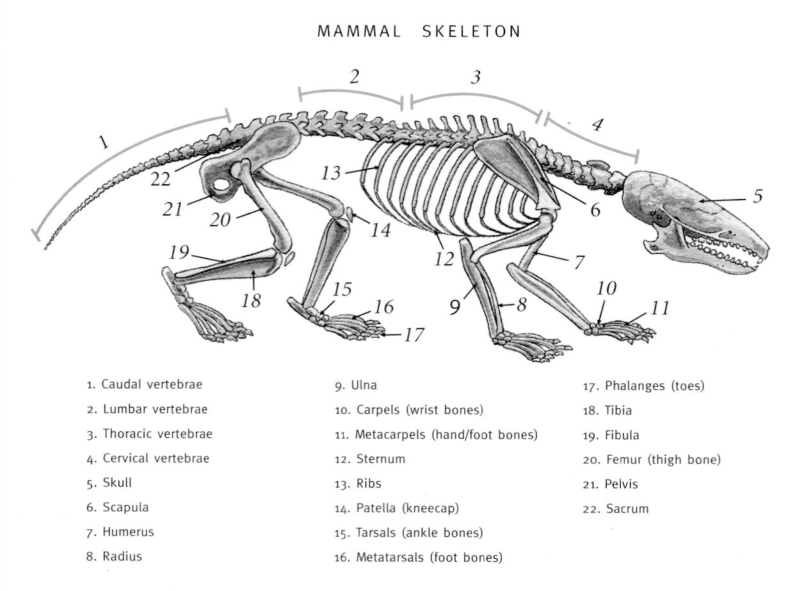

This skeleton is fictional, but approximates that of the earliest, shrewlike mammals. When observing modern mammals or their skeletons, it can be instructive to compare their design to this prototype. Features that diverge from this primitive design have evolved through millions of years of selective pressure, and are thus excellent indicators of a mammal's ecology. For example, selective forces have favored shorter limbs in mammals that dig underground (gophers, for example) and longer limbs in mammals that need to run fast in open areas, such as white-tailed deer or coyotes. Likewise, while some mammals such as shrews and members of the weasel family have retained the five digits on each limb that were present in the earliest mammals, loss of digits has improved the climbing ability of sloths and spider monkeys, and the running ability of horses.

One could fill entire books (and some specialists have done so) speculating about the evolutionary factors that have shaped the bones of a single mammal species. This book describes and illustrates some of the most important skeletal adaptations of Costa Rica's mammals. It includes a number of full skeleton illustrations and, for almost all the mammals described, an illustration of the single most revealing part of the skeleton—the skull.

The Mammal Skull

The following illustration identifies the mammal skull's most prominent features. The function of those features, and what they can tell us about a mammal's ecology, are discussed below.

MAMMAL SKULL

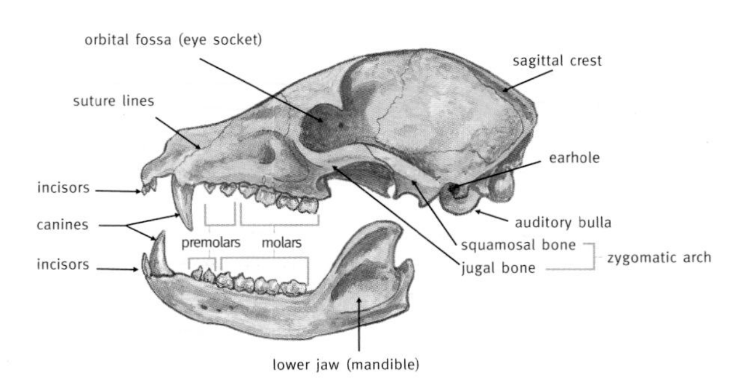

Most of the wiggly cracks that traverse skulls are not caused by random wear and tear; they are **suture lines**, where the different bones that comprise a mammal's skull join. As a mammal gets older, the bones fuse, and eventually (at the age of 30 to 40 in humans, for example), the cracks start to disappear. The cracks can thus help experts determine a mammal's age at its time of death.

The skull's most revealing feature is its dentition. With the exception of the New World anteaters, the similar but unrelated echidnas of Australia and New Guinea, and the baleen whales, all mammals have teeth. Teeth are among the most durable components of a mammal's skeleton, and they are often all that is left in the fossilized remains of ancient species. Partly because of this, teeth are important in classification. Where possible, a mammal's dentition is described with a formula similar to that of the coati, which is: I 3/3, C 1/1, P 4/4, and M 2/2 = 40. The letters stand for the four tooth categories—incisors, canines, premolars, and molars. Such formulas list the teeth in one side of the mouth only, with teeth in the upper jaw written above those of the lower jaw, like the numbers of a fraction. The number at the end of the formula is the total number of teeth. In the coati, the sum of the numbers in the fractions—the teeth on one side of the mouth—is 20, so the total for the whole mouth is 40.

Incisors are situated at the front of the mouth. They tend to be small, although elephants' incisors—the tusks—are a dramatic exception. Incisors are generally used for gnawing, snipping, or pulling and are particularly important to rodents, rabbits, and shrews. Immediately behind the incisors are the **canines**, which are typically large and daggerlike. They are prominent in carnivores (meat-eaters), which use them for stabbing and holding on to prey. Canines also form the tusks of pigs and peccaries, which use them in a variety of other ways (see p. 304). Surprisingly, the largest canines in the mammal world belong to a toothed whale, the narwhal, whose upper left canine forms a three-meter-long lancelike tusk that projects out in front of its head. Behind the canines are the cheek teeth—the **premolars** and the **molars**—which are typically more bulky in shape; most mammals use them for crushing and grinding. They are best-developed in herbivores (plant-eaters) such as tapirs, deer, and peccaries.

The shape of the **lower jaw** or **mandible** can provide further clues about a mammal's feeding habits. A short or curved mandible allows more pressure to be exerted at the front of the mouth, where the incisors and canines are located. Thus, short curved mandibles tend to be prevalent in mammals that need power at the front of their mouths, such as carnivores (that stab, grip, and tear at victims with their incisors and canines) and rodents (that gnaw at seeds with their incisors). A long, straight mandible allows greatest pressure to be exerted in the middle and back of the mouth. This design is common in mammals such as deer, peccaries, and tapir, which crush and grind food with their premolars and molars. Omnivorous mammals (those that eat a bit of everything) tend to have intermediate designs. Dogs, for example, have mandibles that are slightly curved but more elongate than those of their more strictly carnivorous relatives, the cats, while we primates have short but straight jaws. Chewing style is not necessarily the only factor that influences jaw shape. In primates and cats, for example, a longer mandible and snout would reduce the eyes' field of vision and reduce the depth perception that is so important for moving about in trees or pouncing on small, restless prey.

Just in front of the point where the mandible hinges to the rest of the skull is the **zygomatic arch**, or cheek bone. The zygomatic arch really consists of two bones—the **jugal bone** and the **squamosal bone**. The crack that divides these two bones is clearly visible in most skulls. These bones protect the eye and anchor jaw muscles. They were two of the bones that changed shape and position as the mammallike reptiles, over the course of millions of years, evolved into mammals. As these bones aligned to form the zygomatic arch, they provided space and anchorage for more powerful muscles and permitted mammals a greater diversity of chewing styles. Their realignment was also one the key prerequisites that enabled mammals to evolve much more diverse tooth designs and feeding habits than those found in reptiles.

Like the teeth and mandible designs, the shape and relative size of the zygomatic arch can provide clues about a mammal's diet. The size of the gap beneath the arch generally reflects the size of the temporalis muscle, which passes beneath the arch from the back of the lower jaw to the cranium. One of two major jaw muscles, the temporalis is especially important in providing power to the front of the mouth when the jaws are agape. Thus, the zygomatic arches and temporalis muscles tend to be proportionately largest in mammals that need biting power at the front of their mouths, such as, again, carnivores and (to a lesser degree) rodents.

Another feature that reveals the presence of a large temporalis muscle is the **sagittal crest** along the cranium at the top, rear end of the skull. The presence of a crest indicates that the temporalis reaches all the way to the top of the skull, where the crest provides the muscle firm anchorage. On skulls without crests, the temporalis is shorter, anchoring on the side rather than at the top of the cranium. Most carnivores have prominent sagittal crests, while most other mammals lack them.

The second major jaw muscle is known as the masseter muscle. It passes from the rear of the lower jaw to the side of the zygomatic arch. It is especially important in providing power to the rear of the mouth when the jaws are nearly closed, and tends to be proportionately largest in mammals that grind food with their molars, such as, again, deer, peccaries, and tapirs.

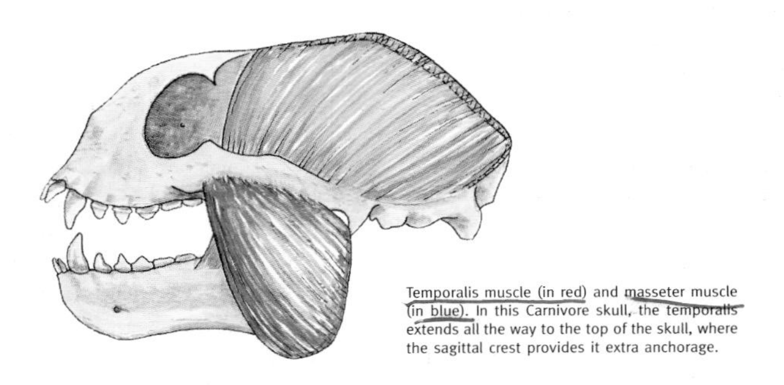

Temporalis muscle (in red) and masseter muscle (in blue). In this Carnivore skull, the temporalis extends all the way to the top of the skull, where the sagittal crest provides it extra anchorage.

Just behind the base of the zygomatic arch is the **auditory bulla** and **earhole.** The earhole tends to be large in mammals with good hearing. Note, for example, how relatively large the earholes of small, nocturnal rodents or many bats (see plates) are compared to most other mammals.

In live mammals, there are three tiny bones just inside the earhole that are responsible for directing sound vibrations to the ear drum. These bones, illustrated below, typically fall out of a clean skull, however.

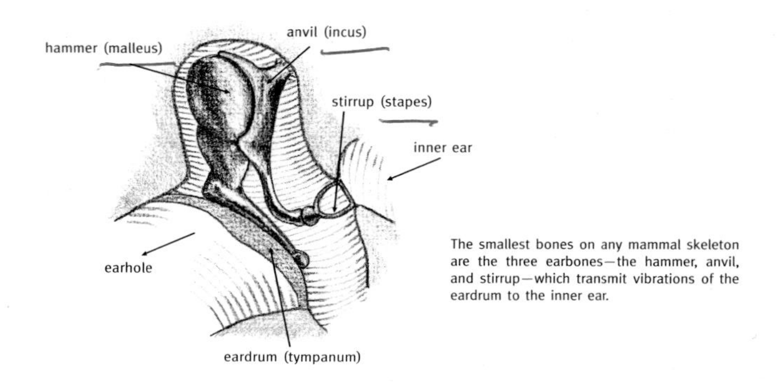

The smallest bones on any mammal skeleton are the three earbones—the hammer, anvil, and stirrup—which transmit vibrations of the eardrum to the inner ear.

Likewise, nocturnal and keen-sighted diurnal mammals tend to have large **orbital fossae**, or **eye sockets**. Each socket houses the eyeball, a set of muscles that control eye movement, a tear gland, and an optic nerve that connects to the brain. One can infer aspects of the ecology of a mammal by looking at the position of the sockets. Mammals whose sockets face forward probably have good depth perception because their eyes' fields of view overlap to create a stereoscopic image. To understand the importance of stereoscopic vision in depth perception, try hammering a nail with one eye shut. Forward-facing sockets prevail in arboreal creatures, such as monkeys, which need good depth perception to jump between branches, and in carnivores, which need it to close in on prey. Mammals such as rabbits or deer with laterally oriented sockets have poor depth perception, but their broader field of view helps them detect predators.

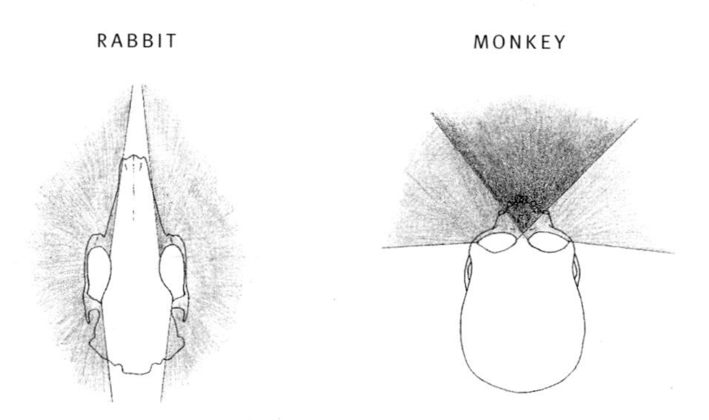

The laterally-positioned eyes of the rabbit have a broader total field of view (shaded in gray) than the forward-facing eyes of the monkey. Forward-facing eyes, however, have overlapping fields of view (shaded in dark gray). This stereoscopic vision provides much better depth perception.

Mammal Diversity in Costa Rica

Most people are aware that Costa Rica has more than its fair share of the world's flora and fauna—about 5%, to use a crude but often-quoted figure, despite the fact that Costa Rica represents only 0.03% of the world's land surface. Mammals are no exception. Thanks in large part to its extraordinary array of bats, Costa Rica is home to about 6% of the world's mammals—some 240 species. At least seven of these, including at least one shrew (*Cryptotis jacksoni*), a spiny pocket mouse (*Heteromys oresterus*), a deer mouse (*Reithrodontomys rodriguezi*), four pocket gophers (*Orthogeomys* spp.), and the woolly olingo (*Bassaricyon lasius*), are endemic to Costa Rica; that is to say, they are found nowhere else in the world.

Several factors may contribute to this diversity. The tropical climate has helped nurture an enormous wealth of plants, insects, and other foods for mammals. Four mountain ranges and varied weather systems blowing in from both oceans at different times of the year react together in diverse ways to create a dense and varied mosaic of habitats. Twelve different life zones exist in Costa Rica. While the tropical wet forests of the Atlantic and southern Pacific lowlands are wet almost year round and receive more than 5 meters (17 feet) of rain annually, the tropical dry forests of the northwest have a severe and prolonged dry season and receive just 1.5 meters (5 feet) of rain each year. And while temperatures in this latter region can exceed 30° C (90° F), the temperature sometimes drops below freezing on the tropical subalpine rain paramos on the high peaks of the Talamanca mountains.

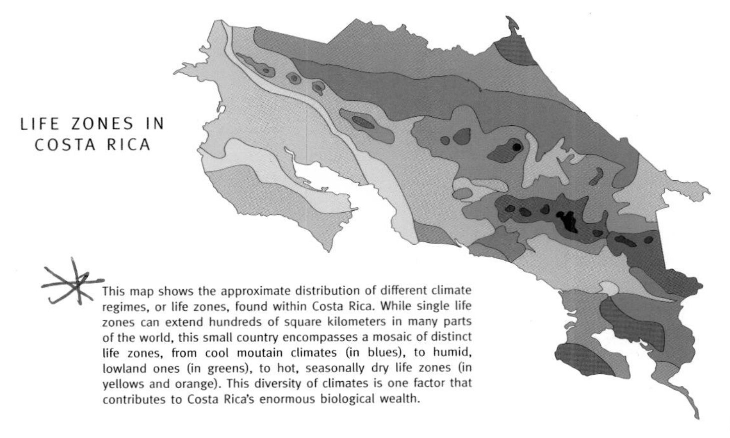

LIFE ZONES IN COSTA RICA

This map shows the approximate distribution of different climate regimes, or life zones, found within Costa Rica. While single life zones can extend hundreds of square kilometers in many parts of the world, this small country encompasses a mosaic of distinct life zones, from cool moutain climates (in blues), to humid, lowland ones (in greens), to hot, seasonally dry life zones (in yellows and orange). This diversity of climates is one factor that contributes to Costa Rica's enormous biological wealth.

Costa Rica's diversity is also influenced by its location right between North and South America. The two major land masses of the New World are very different biologically, not only because of current differences in climate, but also because they were isolated from one another for close to 60 million years. Most groups of mammals were restricted to one or the other until North and South became connected by the Panamanian land bridge. Located right on the bridge, Costa Rica harbors a blend of species typical of both regions. Not including the endemics, 21 mammal species reach the southern limits of their ranges, and 27 species the northern limits, in Costa Rica.

Mammal Conservation in Costa Rica

As we can infer from the graphic below, the second half of this century has been a dismal period for many Costa Rican mammals. Loss of habitat due to deforestation has not been mammals' only problem. Due to lack of awareness as well as to weak or poorly enforced laws, industrial, agricultural, and domestic pollution has destroyed many of Costa Rica's freshwater habitats. Hunting has also decimated numerous mammal populations and put some species, such as the giant anteater, jaguar, manatee, and white-lipped peccary, on the verge of extinction.

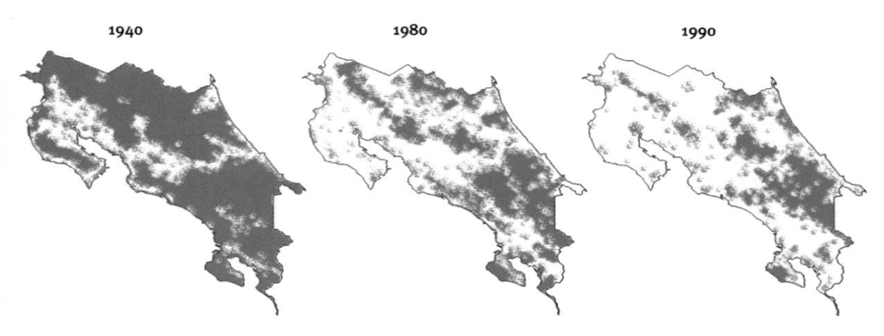

Costa Rica's shrinking forests.

In its latest publication, the Convention on International Trade of Endangered Species (CITES) places 14 of Costa Rica's mammals on their Appendix I, which means that they are threatened with extinction. Another 11 species found in Costa Rica are on the CITES Appendix II, which means that they are not necessarily threatened with extinction now, but may become so unless practices are strictly regulated. And an additional 14 species, not including those on the CITES lists, are deemed scarce by the Costa Rican government and protected by law within the country. As Costa Rica's current population of four million doubles over the next 30 years, environmental problems have the potential to get much worse.

There are three interrelated factors, however, that provide some hope for the future. First, efforts begun in 1970 to create a national park system, as well as the work of a number of private organizations, has led to the protection in some form or another of an impressive 28% of the country. Second, and partly as a result, ecotourism has grown enormously over the last decade, symbolically overtaking coffee, bananas, and cattle products—the major causes of deforestation—as the country's top industry. For many land owners, protecting forest and opening it up to tourism has become the most lucrative option. And third, ecotourism has in turn helped to heighten the general public's environmental awareness. Hopefully, this book will help promote such awareness a little more.

References

Aranda 1981; Evans 1999; Macdonald 1995; Murie 1974; Radinsky 1987; Reid 1997; Rodríguez & Chinchilla 1996; Schmitt 1966; Searfoss 1995; Valerio 1999; Vaughan 1972.

Opossums (order Didelphimorphia)

Distribution and Classification

There are about 270 species of marsupials in the world. Almost 200 of these occur in Australia and on nearby islands. The rest are found in the New World, principally in South America. Only nine species occur in Central America, and only one north of Mexico.

Marsupials are fundamentally different from other mammals. Mammals are divided into two groups, or subclasses, based on how they reproduce. The subclass Prototheria contains the three mammals that lay eggs (the duck-billed platypus and spiny anteater of Australia and the giant echidna of New Guinea). All the rest of the world's mammals—those that do not lay eggs—belong to the subclass Theria. The Theria is divided in turn into the infraclasses Metatheria, the marsupials, and Eutheria, the placental mammals. The world's 3,800 or so species of placental mammals are split into 19 different orders while, until recently, the roughly 270 species of marsupials were placed in a single order, the Marsupialia. Even though marsupials are far less numerous than placental mammals, however, some marsupial groups differ from one another as much as certain placental mammals that are placed in different orders. Consequently, the marsupials are now split into seven orders. Almost all New World marsupials, including the eight or nine species that occur in Costa Rica, belong to the family Didelphidae, in the order Didelphimorphia.

Evolution

The two centers of marsupial diversity—America and Australia—are widely separated today, but when marsupials first appeared, perhaps 120 million or more years ago, these two areas were connected by Antarctica in the large land mass known as Gondwanaland. At that time, Antarctica was forested, not frozen, and available as a bridge between the two areas, as suggested by the recent discovery of a marsupial fossil there.

Where the first marsupials appeared is not known. The oldest fossils that are definitely marsupial are from North America and date back about 75 million years. By about 30 million years ago, North American marsupials had spread to Europe and to parts of Africa and Asia. Over time, however, these populations were replaced by placental mammals; they disappeared in Europe about 25 million years ago, and in North America 15 to 20 million years ago.

Marsupials became most diverse in the southern hemisphere. The marsupials found in Costa Rica today came up from South America following the completion of the Panamanian land bridge some seven to two million years ago. During the 60 million years that South America was an isolated continent, marsupials had limited competition from placental mammals and filled feeding niches that are occupied only by placental mammals in the New World today. Large, carnivorous marsupials such as *Borhyaena*, which resembled a hyena, and *Thylacosmilus*, which resembled a saber-toothed cat, roamed in South America until the appearance of the land bridge. With the subsequent invasion of placental mammals from the north, these large carnivorous species and a number of others became extinct.

Thylacosmilus (left) and *Borhyaena.*

Teeth and Feeding

The dental formula for all didelphids is I 5/4, C1/1, P 3/3, M 4/4 = 50. This large number of teeth, and especially, of incisors, distinguishes didelphid skulls from those of most other mammals, since few placental mammals have more than 44 teeth. Didelphids have small, cone-shaped incisors, large canines, and sharp, pointed premolars and molars. The molars are tricuspidate, and have a distinctive triangular shape when viewed from above. Most didelphids are omnivorous.

Part of the upper left jaw of a common opossum, showing triangular molars—the four rearmost teeth.

Body Design

Compared to the up-to-30 kg (65 lbs) kangaroos and other large marsupials that inhabit Australia, American opossums are relatively small. Didelphid body weights range from 3 kg (6 ½ lbs) down to just 10 g (⅓ oz). Most didelphids are excellent climbers. Most species have naked, strongly prehensile tails, and all species have opposable thumbs on the hindfeet. These thumbs angle backwards such that they fold in the opposite direction to the other fingers, enabling opossums to keep a firm grip on branches. These hindthumbs, as well as the presence of five toes on all feet, make opossum tracks unmistakable.

Reproduction

The name Didelphidae means double-wombed: unlike other mammals, female didelphids have two uteri and two vaginas. The young are born through a third canal, which forms only temporarily for each birth. To match the double vagina of the females, many males have a bifid (two-pronged) penis.

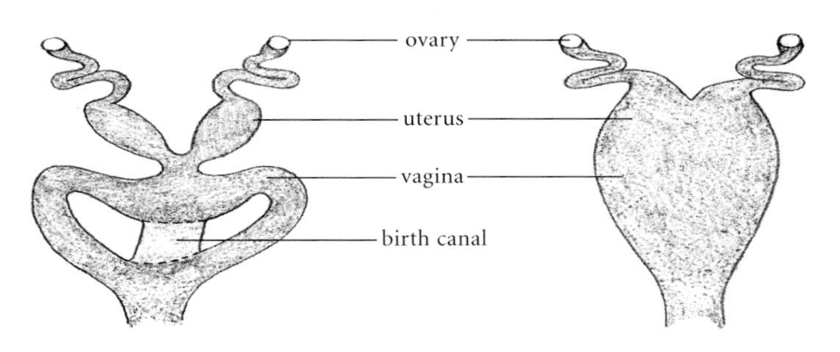

The drawing on the left is not a happy alien, but, rather, a simplified diagram of the reproductive system of marsupials, showing the double vagina and uterus. In most marsupials, the birth canal forms only temporarily for each birth.

The eggs that female didelphids produce are not firmly connected to the uterus wall with a placenta, as in most mammals. Instead, the embryos feed on nutrients that are transferred across the wall of the yolk sac. (This is why eutherians—all mammals other than marsupials, echidnas, and the platypus—are called placental mammals.) Young are born less than two weeks after fertilization, at which time they are so poorly developed that they are referred to in some technical literature as larvae. The larvae do have relatively strong forelimbs and a keen sense of smell, however, which they use to pull themselves through the mother's belly fur and to locate her nipples. The mother was previously thought to flatten a trail for the larvae by licking her fur, but it is now known that she licks her fur only after the larvae have passed to clean off fluids they leave behind. Her habit of licking her fur around birth time, the minute size of the young, and the presence of the bifid penis in males gave rise in times past to the fascinating myth that female opossums mate and give birth through their noses.

Once a larva has attached itself to a nipple, the nipple swells, filling the larva's mouth and ensuring that the larva stays firmly lodged while it continues to develop. In many, but not all, marsupials, the nipples and larvae are housed in a pouch. The Greek word for pouch, *marsypos*, gave rise to the term *marsupial.* Until their jaw muscles are sufficiently developed, some four to seven weeks later, the young are physically incapable of detaching themselves from the nipples. Once the young dislodge, however, the female deposits them in a nest where she continues to nurse them for another three to seven weeks. In all didelphids that have been studied, the female makes the nest of dried leaves that she collects with her coiled tail.

Conservation

Opossums are not highly sought after for their meat or fur. Some species survive or even do well in disturbed habitats, but many others have been affected by loss of habitat. Species that eat some fruit and frequent open areas (see common opossum and common gray four-eyed opossum) may play an important role in forest regeneration.

References
Eisenberg & Golani 1977; Hunsaker 1977a,b; Kirsch 1977; Martin 1977; McManus 1967 & 1970; Medellín 1994; Nowak 1991; O'Connell 1984; Sharman 1970; Woodburne & Zinsmeister 1982, 1984.

SPECIES ACCOUNTS

O: Didelphimorphia F: Didelphidae

COMMON OPOSSUM
Didelphis marsupialis
(plate 1)

Names: Also known as the southern opossum. Scientific name from the Greek words *dis*, meaning double, *delphys*, meaning womb, and *marsypos*, or purse. There are three other species in this genus: the Virginia opossum (p. 35), the white-eared opossum (*D. albiventris*), and the southeastern common opossum (*D. aurita*), the latter two occurring in South America. **Spanish:** *Zorro pelón, zarigüeya.*

Range: Mexico to Bolivia, Paraguay, and Argentina; from sea level to about 2,200 m (7,200 ft), but mostly below about 1,500 m (5,000 ft); in both pristine and disturbed habitats, often in secondary forest and around rural towns.

Size: 40 cm, 1.5 kg (16 in., 3 ½ lbs); males larger than females.

Similar species: See Virginia opossum (pl. 1, p. 35).

Natural history: The common opossum forages by night, usually alone. Its varied diet includes insects, small vertebrates, snails, land crabs, carrion, fruits, nuts, flowers, and nectar. Sometimes it irks biologists by preying on rodents and bats caught in traps and mist nets.

Common opossums studied at La Pacífica in Guanacaste spent most of their foraging time on the ground. Although they occasionally reached the canopy in search of fruit or flowers, they used terminal branches infrequently, preferring thicker, sturdier limbs. Such use of trees contrasts with that of the more arboreal Central American woolly opossum (see p. 38), and may minimize competition between the two species in the numerous areas where they coexist. When startled, the common opossum tends to come down from trees, while the woolly opossum tends to climb upward.

Common opossum scat. Each dropping is 1 to 3 cm long.

Like the Virginia opossum, the common opossum may play dead when cornered (see pp. 35-36), but it does so only rarely. If it can't flee, it may rock back and forth on its forefeet, bite, hiss, or lunge; if picked up, it squirts urine, feces, and other foul-smelling secretions over its aggressor.

Studies done in Venezuela and Panama suggest that, in contrast to the Virginia opossum, the common opossum tends to stay within a fairly stable home range. Although didelphids are generally nonterritorial, female common opossums in Venezuela were found to have exclusive ranges of about 12 ha (30 A) for at least part of the year. Males had larger ranges of about 122 ha (300 A), and showed the more typical didelphid habit of sharing their ranges with other males and females. Males probably compete for females and frequently show battle scars, especially during breeding season. Common opossums den in cavities in trees; under rocks, logs, or piles of brush; or in burrows abandoned by other creatures. Dens are seldom used for long—females studied in Panama seldom spent more than a few days in a den, except when they were nursing young too old to be carried around, and males were found to change dens almost daily.

After a gestation period of 12 ½ to 13 days, females give birth to as many as 20 or more 1 cm (⅓ in.) long, 0.2 g (1/130 oz) larvae. Females have an average of only nine functional nipples, however, and many larvae, unable to latch on to a nipple, die shortly after birth. More larvae die subsequently from other causes, and females wind up with an average of about six pouch young. If food becomes scarce, undernourished females will save energy by ceasing milk production, and the young die in the pouch. Young release hold of the nipples after about two months, by which time they measure about 16 cm (6 in.) and weigh about 100 g (4 oz) each. They leave the pouch shortly thereafter, spending another week or two in a nest before ceasing to nurse entirely, at about three months of age.

Common opossums are thought to raise two litters a year in Costa Rica, principally in February and July. Females can be sexually mature by the age of six to seven months, males by eight months. Few individuals live more than two years in the wild.

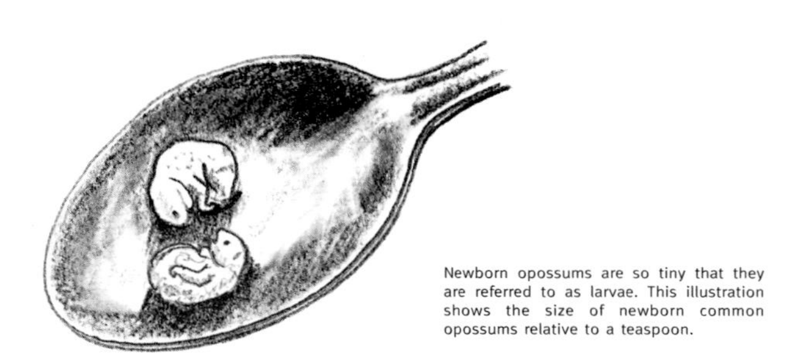

Newborn opossums are so tiny that they are referred to as larvae. This illustration shows the size of newborn common opossums relative to a teaspoon.

Sounds: Hisses, growls, screeches, and canine-tooth-clacking when threatened.

Mythology: For the indigenous Bribri of southern Costa Rica, the common opossum is a harbinger of death.

Conservation: The common opossum can do well in disturbed areas and is common through much of Costa Rica. Its meat and fur are not highly sought after, but it is sometimes shot as a pest, for it may feed on crops, poultry, eggs, or trash. Studies done in Mexico suggest that this species (and, to a lesser degree, *Philander opossum*) are important dispersers of cecropia seeds (*Cecropia* spp., illustrated on p. 62). Cecropias play a key role in forest regeneration, since they are among the first large trees to colonize open areas. The common opossum actively seeks out cecropia fruits, and, since it spends much of its time in open areas, tends to defecate the seeds in places apt for growth. Many of the other animals that eat cecropia drop the seeds in forested areas, where they are less likely to be successful.

References

Atramentowicz 1982, 1986; Bozzoli 1986; Fleming 1972, 1973, 1988; Gardner 1983a; Handley 1976; Janson et al. 1981; Medellín 1994; O'Connell 1979; Rasmussen 1990; Rasmussen & Rasmussen 1984; Reid 1997; Sunquist et al. 1987; Telford et al. 1979.

O: Didelphimorphia F: Didelphidae

VIRGINIA OPOSSUM

Didelphis virginiana

(plate 1)

Names: Sometimes referred to as the common opossum in the United States, where it is the only species. In Costa Rica, that name is generally used for *D. marsupialia. D. virginiana* used to be considered a subspecies of *D. marsupialia.* "*Virginiana*" because the type specimen was collected in the state of Virginia in the U.S. This species gave rise to the name opossum, which is derived from the North American Algonquin word *apasum*, meaning white animal. **Spanish:** *Zorro pelón norteño, zarigüeya norteña.*

Range: Southern Canada to northwestern Costa Rica (roughly down to the town of Cañas); from sea level to about 3,000 m (10,000 ft); in wet or dry forests and open habitats, especially around areas of human habitation. Introduced only relatively recently to New Zealand, and to the west coast of the U.S., from where it spread north to Canada.

Size: 43 cm, 2 kg (17 in., 4 ½ lbs); males larger than females.

Similar species: Upperparts vary from whitish to grayish to (rarely) blackish. This species is hard to distinguish from the common opossum (pl. 1, p. 32). It differs in having white cheeks that contrast with the darker tones of the rest of the head, and pale facial whiskers. Its tail is usually black for half its length or more (sometimes the tail is entirely black) and usually slightly shorter than the individual's head-and-body length. By contrast, the common opossum has cheeks that are a dirty pale color, black facial whiskers, and a tail that is typically black for less than half its length. Its tail is slightly longer than the head-and-body length. In Costa Rica, the two species occur together only in the northwest.

Natural history: This omnivore forages by night for insects, carrion, eggs, fruits, nuts, and small vertebrates including mice, birds, lizards, and snakes. It hunts on the ground or in trees.

The Virginia opossum is probably most famous for its occasional habit of feigning death, or "playing possum." When threatened, it may keel over on its side and lie motionless with its eyes closed, its mouth agape, and its tongue lolling on the ground. Sometimes, the opossum simultaneously emits foul-smelling secretions. For a period that may vary from a minute to six hours, it shows no response to being prodded or handled. Whether this state, termed catatonia, is a conscious performance or an involuntary response like, for example, fainting in humans, is not known. Whichever it is, catatonia is effective at deterring at least some predators—dogs tend to leave catatonic opossums alone. Catatonia is also used occasionally as a confrontation-avoiding submissive gesture to other opossums.

Virginia opossums are solitary and often react aggressively when they encounter others of their own kind. Nevertheless, like most other didelphids, they do not appear to defend a permanent territory. In the United States, they seldom use a den for more than one night before moving on, and they drift over ranges that typically encompass 12 to 39 ha (30 to 96 A), but which may be as large as 254 ha (630 A). Males tend to wander more than females. Virginia opossums den in cavities on or off the ground.

A gestation period of 12 ½ to 13 days typically produces around 20 (but up to 56) young, each about two-thirds the size of a honey bee. Only the young that attach themselves to one of the mother's 13 mammae have a chance of surviving. Successful young let go of the mammae after about 50 days, start to leave the pouch after 70 days, and stop nursing at three to four months. Both sexes reach sexual maturity at six to eight months. Females are not known to have young beyond the age of two. *D. virginiana* can live five years in captivity, but few survive three years in the wild.

Virginia opossum scat.
Each dropping is 1 to 3 cm long.

Sounds: Hisses, growls, and screeches when threatened; metallic lip-clicking sounds during courtship or by a mother when communicating with young; chirps by young when separated from mother.

Conservation: This species can thrive around areas inhabited by people, where it feeds on trash, fruit crops, poultry, and eggs. It is sometimes hunted for its pork-like meat and may be killed as a pest, but is common where it occurs in Costa Rica.

References

Allen et al. 1985; Collins 1973; Eisenberg & Golani 1977; Fitch & Shirer 1970; Francq 1969; Gardner 1982, 1983a; Gilmore 1977; Hilje & Monge 1988; Hunsaker 1977b; Hunsaker & Shupe 1977; Lowery 1974; McManus 1967, 1970, 1974; Reid 1997.

O: Didelphimorphia F: Didelphidae

CENTRAL AMERICAN WOOLLY OPOSSUM

Caluromys derbianus

(plate 2)

Names: *Central American* distinguishes this species from two others, the western woolly opossum (*C. lanatus*), and the bare-tailed woolly opossum (*C. philander*), both of South America. This species is sometimes referred to as Derby's woolly opossum, after a nineteenth-century biologist. Genus from the Greek words *kalos*, meaning beautiful and *mys*, meaning mouse. **Spanish:** *Zorro de balsa.*

Range: Veracruz, Mexico, to Ecuador; in South America it does not occur east of the Andes; from sea level to about 2,500 m (8,000 ft); in pristine or disturbed forests, but mostly in mature, evergreen, lowland forests.

Size: 25 cm, 300 g (10 in., 11 oz).

Similar species: Distinguished from the brown four-eyed opossum (pl. 2, p. 40) by more orangey-brown coloration, lack of pale patches above the eyes, and the tail, which is densely furred for a third to half its length rather than entirely naked, and which is mottled with brown spots on the naked portion. See also the silky anteater (pl. 4, p. 58). Near the Panamanian border, in the mountains above San Vito, *C. derbianus* is mostly or entirely pale gray with a brownish wash over the shoulders, and the dark facial markings are indistinct.

Natural history: Woolly opossums are generally solitary and nocturnal. In South America, at least, the activity patterns of woolly opossums are influenced by the moon cycle: the fuller the moon, the less time woolly opossums spend active, perhaps because moonlight makes them more conspicuous to predators.

The Central American woolly opossum feeds on fruit, nectar, insects, and small vertebrates. Favorite fruits include figs, *Piper* (illustrated on pl. 11), *Cecropia* (illustrated on p. 62), the eucalyptus relative *Eugenia salamensis*, and the basswood relative *Muntingia calabura*. Flowers visited for nectar include those of balsa (*Ochroma pyramidale*, illustrated on p. 38) and the poinsettia relative *Mabea occidentalis*. Captive woolly opossums take animal prey as large as mice, and wild individuals find the alarm cries of bats caught in mist nets irresistible—even to the point of using the head of one unsuspecting biologist to leapfrog over to an entangled Jamaican fruit bat!

Woolly opossums are almost exclusively arboreal, coming to the ground only to snatch prey they have spotted from above. Just as the kinkajou stands out among members of the raccoon family for its suite of monkeylike adaptations for life in the trees (see p. 230), so woolly opossums have evolved monkeylike specializations for their arboreal lifestyle. They are particularly similar to a family of primates found in Madagascar known as the Cheirogaleidae, or dwarf lemurs. Dexterous hands and a strong prehensile tail help woolly opossums move through thin treetop branches, where they find much of their food. Woolly opossums have larger and more forward-facing eyes than other didelphids, and may thus have better binocular vision and depth perception, valuable assets in their three-dimensional world. Unlike other didelphids, they climb and even leap with great agility. Woolly opossums studied at La Pacífica in Guanacaste were found to use all of these features to catch airborne moths. They would hang upside down from branches using their tail and hindlimbs, closely follow circling moths with their eyes, and swipe them with their hands. Like monkeys also, woolly opossums have larger brains than other didelphids (which are otherwise a relatively small-brained group). And they also differ from other didelphids in aspects of life cycle, producing relatively small litters that are slow to develop, and living longer (see below).

Central American woolly opossums often drink nectar from the flowers of balsa trees (*Ochroma pyramidale*).

In the study at La Pacífica, researchers Tab and Asenath Rasmussen documented how such arboreal adaptations can enable woolly opossums to exploit a niche that is not available to other large opossums. They found that while woolly opossums spent much of their time foraging at the tips of branches, the larger, clumsier common opossums foraged only on large branches and trunks, or on the ground.

Woolly opossums often seem little bothered by the beam of a flashlight, as long as the observer is quiet. Sometimes they nonchalantly lick their hands clean and then drag them from behind their ears down the sides of their faces, rather like mice do. While some other opossums remain passive when cornered, woolly opossums will defend themselves aggressively by lunging and biting.

In South America, woolly opossums have been found to wander 0.3 to 1 ha (¾ to 2 ½ A) each night, and many do not maintain a permanent home range. *C. derbianus* makes nests of dried leaves in hollow trees, in the crowns of palm trees, or in vine tangles. Pairs participate in lengthy chase rituals before mating. Gestation probably lasts about two weeks. Litters of one to six (but usually two to four) larvae—less than in most other opossums—are produced from at least early dry season through to early wet season, and possibly year-round. Young are carried in a well-developed pouch. In South America, woolly opossums (*C. philander*) carry the young for about two and a half months and then leave them in a nest, where they are nursed periodically. The young are not weaned until the age of about four months—almost twice the age at which the young of the only slightly larger *Philander opossum* are weaned. *C. derbianus* reaches sexual maturity at seven to nine months. Adult males can be distinguished by their conspicuous blue scrotums. Captive woolly opossums can live more than six years.

Sounds: Clicks and chirps during courtship and between mother and young; hisses, squeals, whines, and grumbling, short growls when threatened.

Conservation: This species probably disperses seeds and pollinates flowers. It used to be hunted for its beautiful pelt, but deforestation is undoubtedly its greatest foe. In areas where the forest has become fragmented, woolly opossums use electrical wires to cross roads and gardens, and frequently die of electrocution.

References

Atramentowicz 1982; Biggers 1967; Bucher 1975; Bucher & Fritz 1977; Bucher & Hoffmann 1980; Charles-Dominique et al. 1981; Collins 1973; Eisenberg et al. 1975; Emmons & Feer 1997; Emmons et al. 1997; Enders 1966; Fleming 1988; Gribel 1988; Hunsaker & Shupe 1977; Julien-Laferrière 1997; Miles et al. 1981; Phillips & Jones 1968; Rasmussen 1990; Rasmussen & Rasmussen 1984; Reid 1997; Steiner 1981; Timm et al. 1989.

O: Didelphimorphia F: Didelphidae

BROWN FOUR-EYED OPOSSUM

Metachirus nudicaudatus

(plate 2)

Names: Genus from the Greek words *meta*, meaning more recent, and *cheir*, meaning hand; species name from the Latin words *nudus*, meaning naked, and *cauda*, meaning tail. **Spanish:** *Zorro café de cuatro ojos, zorricí.*

Range: Chiapas, Mexico, and then from southeastern Nicaragua to Paraguay and northern Argentina; from sea level to at least 1,200 m (4,000 ft); in primary or secondary wet, lowland forest, often near streams. There are no records from the large area between Chiapas and southeastern Nicaragua.

Size: 26 cm, 400 g (10 in., 14 oz); males slightly larger than females.

Similar species: Distinguished from the much more abundant common gray four-eyed opossum (pl. 1, p. 41) by brown coloration (which can be difficult to distinguish at night); smaller, more yellowish, and more widely separated eye spots; proportionately longer hindlegs; and, most easily, by the tail, which is proportionately much longer, naked almost from the base, and brown or gray, getting slightly paler underneath and toward the tip. In the common gray four-eyed opossum, the tail is furred at the base for about a fifth of its length, and is dark gray, changing abruptly to white at the tip. Unlike female gray foureyes, female brown foureyes have no pouch. See also the Central American woolly opossum (pl. 2, p. 37).

Natural history: This elusive species forages by night, in trees or on the ground. It has a long tail for extra balance and support, a feature typical of arboreal animals. It also has relatively large hindlegs that help it run quickly, a feature typical of terrestrial animals. All 18 individuals captured in a study in Venezuela were taken on the ground.

The brown four-eyed opossum's diet includes fruit, snails, eggs, small vertebrates (including frogs, lizards, rodents, and birds), and insects (such as termites, beetles, and cicadas).

Unlike the common gray four-eyed opossum, to which, despite its similarity in name and appearance, it is not closely related, *M. nudicaudata* is shy and unaggressive. If startled, it runs away quietly and quickly along the ground. If cornered, it may employ mouth gaping and vocal displays, but tends to stand in a fixed spot, often trembling.

It dens beneath fallen palm fronds, in cavities under logs or rocks, or in balls of leaves and twigs lodged among tree limbs. Brown four-eyed opossums in Central America are thought to start breeding in November. They produce litters of one to nine larvae. As in mouse opossums, the females have no pouch. Young are able to stand alone by the time they are about 5 cm (2 in.) long, and are independent within about two months. The maximum lifespan for this species has been estimated at three to four years.

Sounds: Chatters teeth and hisses when threatened.

Conservation: Although this species is common in parts of South America, it is exceedingly rare in Costa Rica. Despite some fairly extensive mammal censuses, only a handful of specimens has ever been collected here—the first in 1876, the second in 1983! These and a few that have been collected since then were all taken in the Atlantic lowlands.

References

Collins 1973; Emmons & Feer 1997; Fleming 1973; Grand 1983; Handley 1976; Hunsaker 1977b; McPherson et al. 1985; Medellín et al. 1992; Miles et al. 1981; Reid 1997; Timm et al. 1989.

O: Didelphimorphia F: Didelphidae

COMMON GRAY FOUR-EYED OPOSSUM

Philander opossum

(plate 1)

Names: The long common name distinguishes this species from the brown four-eyed opossum (pl. 2, p. 39) and from two South American species, the black or Mcilhenny's four-eyed opossum (*P. mcilhennyi*) and Anderson's gray four-eyed opossum (*P. andersoni*). The genus and the identical English word—from the Greek *philos*, or loving, and *andros*, or man—mean to engage lightly in love affairs, presumably in reference to the species' wandering habits. The origin of the word opossum is explained on p. 35. Sometimes listed under the genus *Metachirops*. **Spanish:** *Zorro gris de cuatro ojos.*

Range: Southern Mexico to Paraguay and northern Argentina; from sea level to about 1,600 m (5,300 ft); mostly in evergreen forest, often near rivers or swamps.

Size: 30 cm, 450 g (12 in., 1 lb).

Similar species: The only other gray opossum is the yapok (pl. 2, p. 43). Although *P. opossum* is not as aquatic as the yapok, it is a strong swimmer and is often found in the vicinity of water feeding on aquatic animal life. It differs from the yapok in having white spots on the forehead and much larger ears, and in lacking both broad black bands across its back and webbed hindfeet. Unlike the yapok, *P. opossum* is a good climber; it is frequently found in trees, or takes refuge in them when disturbed. See also brown four-eyed opossum (pl. 2, p. 40).

Natural history: The common gray four-eyed opossum inhabits the lower levels of the forest. It forages by night, mostly on the ground. Its diet includes small vertebrates such as frogs, lizards, rodents, and birds; invertebrates such as insects, freshwater shrimp, snails, and worms; and a variety of other foods, including ripe fruits, nuts, nectar, and birds' eggs. Captives prefer animal over vegetable foods. *P. opossum* locates frogs by homing in on their calls. Once it has approached to within a meter or two, it stops to get a final bearing on the singer, turning its head from side to side and tilting its ears back and forth, sometimes while standing up on its back legs, and then dashes straight at the prey. In an experiment at a pond in Panama, four-eyed opossums would consistently pounce on speakers emitting recordings of the call of the mud puddle frog (*Physalaemus pustulosus*, illustrated on p. 101).

Studies done in Panama and French Guiana suggest that gray four-eyed opossums wander around, not maintaining permanent territories, and have home ranges that overlap with those of their neighbors. This loose territorial arrangement has been observed in a number of didelphid species.

P. opossum dens on or off the ground, in cavities among tree roots or rocks, under fallen palm fronds, in tree cavities, or on open branches. Branch nests consist of 30 cm (12 in.) wide balls of sticks and leaves lodged in a small tree or shrub, typically less than 10 meters (30 ft) from the ground. If surprised in its nest, or otherwise cornered, *P. opossum* usually reacts aggressively; although small, it has the reputation of being more feisty than other didelphids.

P. opossum produces litters of two to seven larvae. Larvae measure just 1 cm (⅓ in.) and weigh only 0.2 g (1/130 oz) at birth. They nurse inside a well-developed pouch, releasing the nipples after about two and a half months, by which time they measure about 10 cm (4 in.) and weigh about 25 g (1 oz) each. If food becomes scarce, undernourished females will save energy by ceasing milk production, and the young die in the pouch. Once out of the pouch, young remain in a nest and continue to nurse for another one to two weeks. Females can breed by the age of eight months, males by seven months. In Panama, Nicaragua, and French Guiana, common gray four-eyed opossums can be reproductively active almost year-round, and may produce two or three litters a year. Captives can live at least two years and four months; in the wild, few individuals live more than two years.

Sounds: Loud hisses and long, chattering chirps when threatened; clicking sounds by young.

Conservation: This species is hunted little, and it can prosper in disturbed habitats. In some areas it damages corn and fruit crops. Studies done in Mexico suggest that this opossum helps disperse the seeds of cecropia trees, which are important in forest regeneration (see p. 34).

References
Atramentowicz 1982, 1986; Charles-Dominique 1983; Collins 1973; Emmons & Feer 1997; Fleming 1972, 1973; Handley 1976; Hunsaker & Shupe 1977; Medellín 1994; Nowak 1991; Phillips & Jones 1969; Timm et al. 1989; Tuttle et al. 1981.

O: Didelphimorphia F: Didelphidae

YAPOK or WATER OPOSSUM

Chironectes minimus

(plate 2)

Names: The name yapok originated with native Guianans. Genus from the Greek words *nektos*, meaning swimming, and *cheir*, meaning hand, in reference to the yapok's webbed hindfeet. *Minimus* means smallest, a redundancy given that this is the only species of yapok; the species acquired this name in the eighteenth century, when it was considered to be a diminutive otter. **Spanish:** *Zorro de agua.*

Range: Southern Mexico to northwestern South America, and patchily through the rest of South America down to Paraguay and northern Argentina; from sea level to at least 1,800 m (6,000 ft); in or near streams or lakes, in forested or open areas.

Size: 28 cm, 650 g (11 in., 1 ½ lbs).

Similar species: See common gray four-eyed opossum (pl. 1, p. 41).

Natural history: The yapok hunts alone and mostly at night, in or around water. It feeds on crustaceans (such as crabs and shrimp), insects, fish, frogs, and other small vertebrates. It may also take aquatic vegetation and fruits.

The semiaquatic lifestyle of this species is unique among marsupials. Correspondingly, so are a number of anatomical features, including thick, water-repellent fur, webbed hindfeet, and unusually long, stout whiskers (like those of otters or seals) that are important tactile organs underwater. The yapok's forefeet are equipped with long toes, large pads on each fingertip, and palms with a rough, sandpaper-like surface, all features that help yapoks grip slippery prey. When swimming, yapoks propel themselves by dog-paddling with their hindfeet while holding their forelimbs straight forward to feel out or carry prey. All that is visible of a swimming yapok is the top of its head, but the eyes produce a distinctive bright yellow reflection when caught in the beam of a flashlight. Yapoks can climb, but probably do so seldom: their webbed hindthumbs are less opposable than the hindthumbs of other opossums, and their tail is only weakly prehensile.

Yapoks den in tunnels in river banks. The entrance, typically about 10 cm (4 in.) wide, is situated just above the high-water line. Yapoks may also make a nest of dried leaves or grass in shaded areas beside water. If disturbed in a ground nest, they may dive into the water and swim to the safety of the tunnel.

Litters normally contain two or three young, and rarely as many as five. The design of the female yapok's pouch ensures that the young stay dry while the mother swims. The pouch opening can be pulled shut by a sphincter muscle and is sealed by long hairs and an oily, water-resistant secretion. Since the pouch opens at the rear, it may be less susceptible to filling with water than it would be if the opening were at the front. (This feature is not unique to water opossums, however: common and Virginia opossums also have rear-opening pouches.) Experiments have shown that young opossums are remarkably tolerant of low levels of oxygen—a useful skill for a pouch-bound yapok during one of its mother's dives. Male yapoks also have pouches, which they use to house their scrotums while running or swimming. Yapoks can live about three years in captivity.

Sounds: Screechy barks when threatened.

Mythology: According to superstition of the Bribri tribe of southern Costa Rica, if you place the right rear foot of a water opossum in the hut of a pregnant woman, you bestow superior fishing and shrimping skills on her child.

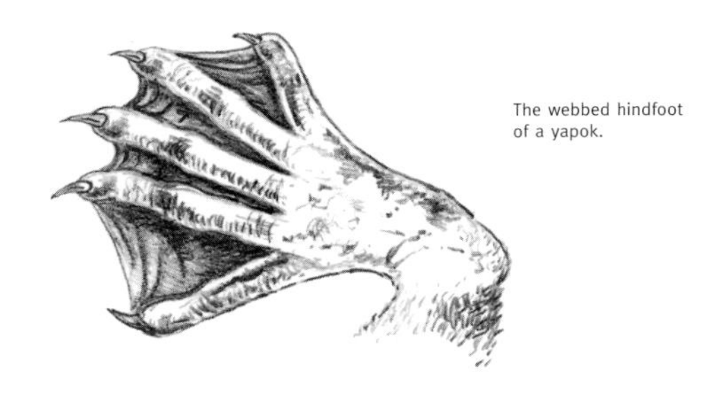

The webbed hindfoot of a yapok.

Conservation: Yapoks are apparently rare throughout their range. The clean ponds and streams that they inhabit are becoming increasingly scarce in Costa Rica.

References

Bozzoli 1986; Collins 1973; Handley 1976; Howell 1930; Hunsaker 1977b; Marshall 1978; Mondolfi & Medina 1957; Nowak 1991; Oliver 1976; Rink & Miller 1967; Rosenthal 1975; Zetek 1930.

O: Didelphimorphia F: Didelphidae

MOUSE OPOSSUMS

Micoureus alstoni and *Marmosa mexicana* (plate 3)

Names: Alston's mouse opossum (*M. alstoni*) and the Mexican mouse opossum (*M. mexicana*). *M. alstoni* was listed under the genus *Marmosa* until recently. Mouse opossums are sometimes referred to as murine opossums. **Spanish:** *Zorros ratones, zorricíes, zorrillos.*

Range: *M. alstoni*: Belize to northern Colombia; *M. mexicana*: Mexico to western Panama. Both species inhabit evergreen forests to about 1,600 m (5,300 ft), and *M. mexicana* may also be found in seasonally dry and disturbed areas. As a group, mouse opossums occur from Mexico to Argentina, with the vast majority of the more than 45 species occurring south of Panama.

Size: *M. alstoni*: 18 cm, 100 g (7 in., 4 oz); *M. mexicana*: 15 cm, 80 g (6 in., 3 oz); in *M. mexicana*, males tend to be larger than females, but the size of both is variable since they continue to grow throughout their lives.

Similar species: *M. mexicana* differs from *M. alstoni* in having more reddish upperparts, shorter fur, and a uniformly colored gray-brown tail. The tail of *M. alstoni* is brown on the basal half and white on the outer half. These are the only two mouse opossum species known to occur in Costa Rica. A third species, Robinson's mouse opossum (*M. robinsoni*), may occur here since it has been collected both north and south of Costa Rica, but there are no records to date. It might have been confused for *M. mexicana*, since the two are exceedingly similar—*M. robinsoni* tends to be larger, and has longer, more woolly fur, but can only be distinguished from *M. mexicana* with certainty by skull characteristics. Mouse opossums in general are easily confused with small rodents. They are best distinguished by their prehensile tails, black eye masks, and, when caught in the beam of a flashlight, by their ruby-red eyeshine.

Natural history: Few mouse opossum species have been studied in any depth, and relatively little is known about the natural history of either species in Costa Rica.

Mouse opossums are nocturnal and usually solitary. Both Costa Rican species are mostly arboreal. They have long, prehensile tails to help them move through trees, although they also forage on or near the ground. They feed predominantly on insects and fruit, but will also take small vertebrates such as lizards, birds, or even mice. They can be important predators on the eggs and hatchlings of birds such as house wrens in Monteverde, and parrots and woodpeckers elsewhere.

Mouse opossums spend the daytime resting in abandoned bird nests, or in leaf nests they construct themselves in small trees or ground burrows. Mexican mouse opossums have been recorded nesting in a 50 cm (20 in.) wide ball of moss hanging from a liana 20 m (65 ft) off the ground; in a rolled, dead cecropia leaf 3 m (10 ft) off the ground; and in a 15 cm (6 in.) wide chamber lined with dry leaves, about 40 cm (16 in.) down a 3 cm (1 in.) wide burrow in the ground.

Known reproductive patterns (based principally on studies of *M. robinsoni*) involve a brief courtship followed by a lengthy (up to five-hour) copulation. A gestation period of about two weeks results in litters of one to 14 (often seven to nine) larvae. *M. mexicana* has been found nursing litters of two to 13 young between March and June. Female mouse opossums have no pouch, but larvae remain firmly attached to their mother's nipples for about a month. They continue to nurse for a second month, but are left in a nest while the mother forages. They may start following the mother on some excursions when they are a month and a half old. Female Mexican mouse opossums will eat any young that die. Mouse opossums can become sexually active after five or six months. In females of *M. robinsoni*, sexual activity dies off about a year later, when they are 16 to 18 months old. If this species has only one year of reproductive activity and reproduces seasonally (as a study in Panama suggests), females may produce just one litter in their lifetimes. Other mouse opossum species have been found to breed year-round. Mouse opossums can live more than five years in captivity, but it is thought that few survive more than two years in the wild.

Sounds: Clicks during social encounters; churring-hissing sounds when threatened; chirps by young when separated from mother.

Conservation: Being small, arboreal, and nocturnal, mouse opossums are seldom seen, but both of Costa Rica's species may be locally common. They sometimes show up in houses, and at La Selva, where visual records of mouse opossums are scarce, bones of *M. mexicana* have been found frequently in the regurgitated pellets of spectacled owls. Mouse opossums can be minor pests on crops such as sugar cane, mangos, and bananas. They sometimes become inadvertent stowaways on banana boats, and as a result they have appeared in a number of ports and warehouses in the United States.

References

Alonso-Mejía & Medellín 1992; Aranda & March 1987; Barnes & Barthold 1969; Eisenberg 1988; Eisenberg & Maliniak 1967; Engstrom et al. 1993, 1994; Gardner & Creighton 1989; Hall & Dalquest 1963; Hunsaker 1977a; Kraatz 1930; Lowery 1974; Nowak 1991; O'Connell 1983; Reid 1997; Timm & Laval 2000a,b; Timm et al. 1989; Young 1996.

Anteaters, Sloths, & Armadillos (order Xenarthra)

Distribution and Classification

This bizarre order is found only in the New World, mostly in Central and South America. It contains four families and 29 species: the anteaters (Myrmecophagidae), with four species; the two-toed sloths (Megalonychidae), with two species; the three-toed sloths (Bradypodidae), with three species; and the armadillos (Dasypodidae), with 20 species. Seven species and all four families are represented in Costa Rica.

Until recently, this order was known as the Edentata, which means the toothless ones. This was a misleading name, since only the anteaters lack teeth. Indeed, one xenarthran, the giant armadillo (*Priodontes maximus*) of South America, has up to 100 teeth—more than any other mammal except for a few whales. Moreover, there is another order, the Pholidonta, or pangolins, of the Old World, that contains only toothless mammals.

The current name for sloths, armadillos, and anteaters refers to unusually shaped vertebrae in the lower back known as xenarthrales (from the Greek words *xenikos*, meaning strange and *arthron*, meaning joint). These bones are found in all members of this order but in no other mammal. Other features that are peculiar to the Xenarthra include a double post vena cava, the vein that returns blood from the hindquarters to the heart (other mammals have only one), and aspects of reproductive anatomy.

Evolution

The Xenarthra is thought to be one of the oldest mammal orders. Phylogenetic studies suggest that the order diverged from other placental mammals more than 100 million years ago. (For this reason, xenarthrans appear at or near the beginning of the placental mammals in most mammal lists and books.) Even the four xenarthran families appear to have separated some 80 million years ago, and may thus be as old as some mammal orders.

Several anatomical features also suggest that xenarthrans are an ancient group. The design of the reproductive tract, for example, has barely changed from that found in the most primitive mammals. In the earliest mammals, the reproductive and digestive tracts opened into a shared chamber known as the cloaca, while in most modern mammals the two tracts are separate. In xenarthrans, the two tracts are divided, but only by a thin layer of skin and tissue. This unique intermediate design is sometimes referred to as a pseudocloaca.

Xenarthrans evolved largely in South America, where they were isolated from the rest of the world for most of the last 60 million years. By about 60 million years ago, when mammals were just beginning to proliferate, there were already two distinct groups of xenarthrans. One, the suborder Palaenodonta, consisted of small, armorless animals that soon became extinct. The other, the suborder Xenarthra, was enormously successful. In eras past, this group included about 10 times the number of genera alive today, and some of the strangest mammals the world has ever seen. There were giant (up to 3 meter long) armadillos called glyptodonts whose carapaces (shells) are thought to have been used as shelters by South America's earliest human settlers. Some of them had massive spiked tails that resembled medieval weapons. And there were 6 meter (20 foot) tall giant ground sloths (*Megatherium*), which used their height to reach vegetation from the ground rather than hanging from trees like their close relatives, the two-toed sloths.

A giant ground sloth (left) and a glyptodont.

The disappearance of these and many other xenarthrans was probably due to several factors. Once the Panamanian land bridge was formed some seven to two million years ago, northern mammals invaded the previously isolated South America. These northerners included both carnivores that preyed on the xenarthrans, and insectivores and herbivores that outcompeted them. Some xenarthrans were able to withstand this invasion, and a few, including the ancestors of the species found in Costa Rica today, expanded their ranges north. Giant ground sloths, for example, which had crossed to North America by island-hopping before the land bridge was complete, spread as far north as Alaska and persisted in the United States until as recently as 9,000 years ago. The demise of giant ground sloths and other large xenarthrans may have been caused by climate change at that time—the end of the Ice Ages—or conceivably by primitive human hunters, who might have been arriving for the first time from the north. Xenarthrans may also have been prevented from spreading farther north because they depend on very specific foods that are not as abundant or diverse there, or because they are poorly adapted for cold climates. Whatever the reasons, only one xenarthran species, the nine-banded long-nosed armadillo, survives today north of Mexico.

Teeth and Feeding

Anteaters, sloths, and armadillos have quite different diets. Sloths are folivores (leaf-eaters). They lack incisors and canines, but have a set of cheek teeth for grinding up vegetation, and canine-like premolars that can be used in defense. Their teeth are not like those of most other mammals: they are rootless, peglike in shape, lack enamel, and grow constantly. Sloths have five teeth on each side of the upper jaw and four on each side of the lower jaw; their dental formula is usually written as simply 5/4 = 18.

Sloths face similar digestive problems as those faced by other folivorous mammals such as deer, howler monkeys, and manatees. Like these other folivores, sloths break down cellulose through a process of bacterial fermentation. They have a compartmentalized stomach that can take weeks to ferment a meal. A leaf diet poses another problem for sloths—such a low-energy, slow-to-digest food requires a very large stomach (almost a third of body weight), yet sloths must stay light enough to move along tree branches without breaking them. Sloths compensate for their large guts and low-energy diets by having relatively little muscle—much less than similar-sized terrestrial mammals.

Anteaters, of course, are mrymecophagous (they eat ants). They have no teeth. Their mouth is a tiny hole through which they can extrude a long tongue covered with microscopic, backward-pointing spines. An enormous gland lubricates the tongue with sticky saliva. The digestive tract of anteaters contains a muscular gizzard (similar to that found in chickens and some other birds) that crushes the hard exoskeletons of ants and termites.

Like sloths, anteaters use a food that is abundant. In Panama, average ant densities in forest-floor litter have been estimated at an astonishing 680 individuals per square meter, and 35 to 240 per square meter in the canopy. But, like leaves, ants (and termites) are not really such an easy food as they appear to be. Ants and termites have a wealth of chemical, structural, and behavioral defenses. The chemicals they use in defense include citrol, citronella, limonene, and a wide range of other noxious substances that anteaters are probably not immune to. Structural defenses include hard or spiny bodies, or tough, inaccessible nests. And behavioral defenses include biting, squirting chemicals, or simply retreating out of reach.

Apparently as a result of such defenses, anteaters usually feed for no more than a minute at any single nest. For example, it has been calculated that, if a silky anteater could sit around and suck up ants at its leisure, it could fulfill its nightly food needs in a single tree, yet it visits an average of 38 trees each night. Anteaters tend to travel in straight lines when foraging, and thus avoid winding up back at nests they have already pillaged, where the soldiers might still be irate and tastier castes hidden away. Extensive travelling may also enable anteaters to feed on a wider variety of species, and thus reduce the chances of receiving an excessive dosage of any one defensive chemical. (A similar theory has been suggested to explain the diversity of plants fed upon by some herbivores—see pp. 294-295.) Since anteaters raid each nest only briefly, they usually cause relatively little damage and can return to the same nests periodically for more.

Armadillos feed mostly on insects. Some, like naked-tailed armadillos, specialize on ants and termites; others, like long-nosed armadillos, are omnivorous, supplementing insects with other foods. Like sloths, armadillos have peglike, ever-growing teeth that lack roots and enamel, but their teeth are much smaller, more numerous, and more uniform in shape. Costa Rica's armadillo species have between seven and nine teeth in each quarter of the mouth.

Body Design

Xenarthrans have three to five toes on the hindfeet and two to five on the forefeet. All species have two or three enlarged claws on the forefeet that they use to dig, rip into ant nests, hang from trees, or defend themselves. Distinctive skeletal features include xenarthrales, which are vertebrae with an extra wing of bone on the front end that overlaps the adjacent vertebra, providing extra support. The xenarthrales, located in the lower back, reinforce the hindquarters. This reinforcement may contribute to armadillos' digging ability and help anteaters and sloths to stand (or hang) firmly with just the hindfeet when they need to free their forelimbs for feeding or defense.

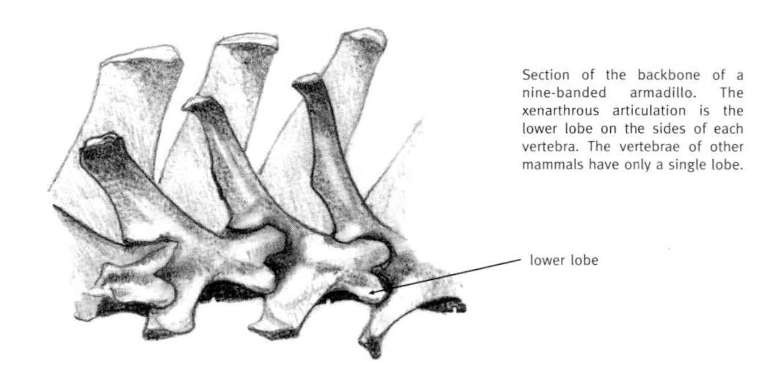

Section of the backbone of a nine-banded armadillo. The xenarthrous articulation is the lower lobe on the sides of each vertebra. The vertebrae of other mammals have only a single lobe.

Sloths have another peculiar skeletal feature. While almost all other mammals consistently have seven neck vertebrae, two-toed sloths have six to eight and three-toed sloths have eight or nine. Partly as a result, sloths are able to rotate their heads an impressive 180°, a skill the value of which any hammock user can understand.

Xenarthrans have unusually low body temperatures and metabolic rates. The two-toed sloth has the lowest and most variable body temperature of any mammal (24 to 33° C or 77 to 96° F). Low body temperatures and metabolic rates burn less energy, and thus compensate in part for the energy problems faced by all the species that feed on leaves and ants. On the other hand, these traits also make xenarthrans vulnerable to the cold. Sloths have to sunbathe each morning to warm up (and also to speed up digestion, since the warmer their body temperature, the faster the leaves in their stomachs ferment). Both sloths and anteaters have thick, shaggy fur, which seems inappropriate for the heat of the tropics, but which provides insulation. Resting posture helps, too. Sloths and silky anteaters conserve heat by curling into tight balls when resting, and giant anteaters fold their huge bushy tails over themselves as blankets. The shell of an armadillo—a sheath of bony scales known as scutes covered with a thin layer of horn—does not retain heat well, but armadillos' tight burrows provide considerable insulation. Despite all these features, xenarthrans do sometimes succumb to the cold. In North America, nine-banded armadillos often perish during cold spells; and in Costa Rica and Panama, dead sloths often appear during the wettest, cloudiest months, especially after temporales (prolonged periods of miserable weather).

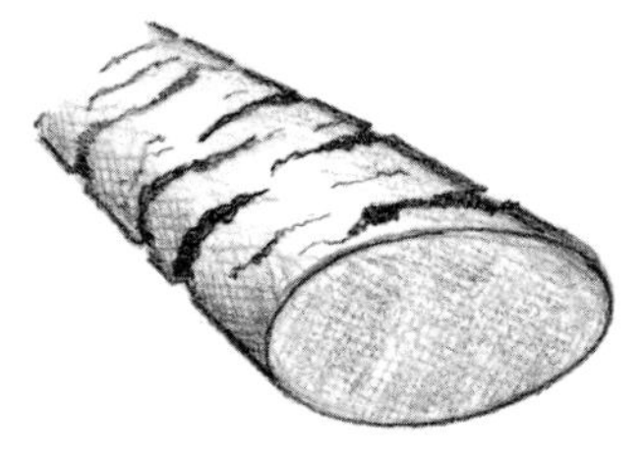

Cross section of the hair of two-toed (left) and three-toed sloths.

The structure of sloth hair is unique among mammals and appears to serve sloths another purpose. The hair of two-toed sloths has pronounced, evenly spaced, longitudinal corrugations, while the hair of three-toed sloths has less conspicuous, unevenly spaced, transverse cracks. These textures appear to encourage the growth of algae, which in turn may benefit the sloth. Many believe that algae and sloth participate in a symbiotic relationship in which the algae gains shelter and the sloth, camouflage. Mammals cannot normally produce green fur, but algae-covered sloths acquire a green tinge, especially during the wet season, that makes them even more difficult to see than they would be otherwise.

Researcher Annette Aiello has pointed out another possible benefit of algae to sloths. Algae could provide nutrients, such as nitrogen, that are otherwise deficient in the sloths' leaf diet. Sloths could acquire such nutrients either by licking their fur (although they seldom appear to do this) or by absorbing them through their hair and skin. Unlike the fur of other mammals, sloth hair soaks up water in a way that, if the tips get wet, water and perhaps nutrients spread up the hair to the base, where they could be absorbed. Aiello has pointed out that an analogous system can be found in orchid roots, which absorb water and nutrients through an outer layer of spongy, algae-covered cells known as velamen.

Senses

Xenarthrans rely on their excellent sense of smell for feeding and communicating. Anteaters and armadillos are thought to locate their food almost exclusively by smell. All xenarthrans have a well-developed pair of anal glands, and some also have glands on the ears, eyelids, and soles of the feet. Secretions from the anal glands are used to mark territories and may convey specific information about the marker. In tamanduas and armadillos, anal-gland secretions have a nauseating smell and may help deter predators.

Xenarthrans are generally thought to have poor senses of hearing and vision that are effective mostly at close range. Nonetheless, the whistles sloths and anteaters make may be long-distance signals. The same could be true of the black and yellow markings on the backs of male three-toed sloths. A male that was seen approaching and fighting with another male at La Selva reportedly seemed to find its competitor visually.

Conservation

A distinguished eighteenth-century biologist by the name of Compte de Buffon once purchased a pet sloth in Amsterdam and, having observed it move about its cage, wrote the following: "The inertia of this animal is not so much due to laziness as to wretchedness; it is the consequence of its faulty structure. Inactivity, stupidity, and even habitual suffering result from its strange and ill-constructed formation." He concluded that "one more defect and they could not have existed."

Compte de Buffon might have been surprised at the results of a census in Panama, from which mammalogists calculated that sloths represented two-thirds of the biomass and half the energy consumption of all terrestrial mammals in the study area. The study implies that, pound for pound, sloths are the most successful terrestrial mammals in the New World tropics. As such, they play a critical role in the nutrient cycle, converting leaves into forms accessible to numerous other organisms. Fortunately, these important mammals are generally left alone by Costa Rican hunters. They are inoffensive and inconspicuous, and their pelt and meat are not valuable (although sloths are eaten in some rural areas).

Close behind sloths on most Neotropical mammal-abundance lists comes the nine-banded long-nosed armadillo. While sloths have thrived thanks to specialization on an abundant but little-used food resource, nine-banded armadillos owe their success to flexibility. They inhabit forest and disturbed areas alike, eat a huge variety of foods, and even have a versatile reproductive strategy. They also have a faster reproductive turnover than other xenarthrans (see p. 72).

The rest of Costa Rica's xenarthrans—the anteaters and the naked-tailed armadillo—are not as fortunate; the giant anteater may already be extinct in Costa Rica. For reasons discussed above, anteaters have to move around much more than most mammals their size in order to cope with ant and termite defenses. In fact, they use home ranges about as large as those of similar-sized carnivores, and are thus especially vulnerable to habitat loss and hunting. The naked-tailed armadillo, also an ant and termite specialist, may be rare in part for the same reason, although its home range requirements have yet to be documented. The anteaters and the naked-tailed armadillo, like most xenarthrans, have relatively slow reproductive turnovers, so they have the additional problem of being slow to replenish depleted populations.

References

Aiello 1985; Bauchop 1978; De Jong et al. 1985; Dickman 1984c; Eisenberg & Thorington 1973; Engelmann 1985; Forsyth 1990; Gillette & Ray 1981; Glass 1985; Goffart 1971; Grand 1978; Greene 1989; William Haber, pers. comm.; Krantz 1970; Layne & Glover 1985; Long & Martin 1974; Lubin et al. 1977; McNab 1978, 1982, 1983, 1985; Meritt 1985a,b; Montgomery 1979, 1983a, 1985a,b,c; Montgomery & Lubin 1977; Montgomery & Sunquist 1975; Nowak 1991; Patterson & Pascual 1972; Redford 1985b, 1987; Reig 1981; Rich & Rich 1983; Sarich 1985; Webb & Marshall 1982; Wetzel 1985a,b.

SPECIES ACCOUNTS

O: Xenarthra F: Myrmecophagidae

GIANT ANTEATER

Myrmecophaga tridactyla

(plate 4)

Names: Genus from the Greek *myrmos*, meaning ant, and *phago*, to eat. *Tridactyla* means three-toed, although in reality this species has three large toes and one small one on each forefoot and five small toes on each hindfoot. **Spanish:** *Oso hormiguero gigante, oso caballo.*

Range: Guatemala and southern Belize to Ecuador and, east of the Andes, down to Uruguay and northern Argentina (although it is exceedingly rare or extinct in Central America); in lowland grasslands or forests.

Size: 130 cm, 30 kg (50 in., 66 lbs).

Similar species: The much more common northern tamandua (pl. 4, p. 55) is less than half the size of the giant anteater; has a broad black vest; possesses a naked, prehensile tail; and frequently forages in trees.

Natural history: The giant anteater is largely solitary and terrestrial. It forages by day or night. In some areas, it is almost exclusively nocturnal, perhaps due to human disturbance or to hot daytime temperatures. In a study in the llanos (grassy plains) of Venezuela, anteater experts Gene Montgomery, Yael Lubin, and colleagues found that most (96%) of the giant anteater's diet was made up of large soil- and wood-nesting ants, and the rest principally of termites. Occasionally the giant anteater ingests other insects, perhaps by accident, and in captivity it accepts fruit. The giant anteaters in Venezuela were found to feed on far fewer types of ants than do silky anteaters, and even somewhat less than tamanduas, for which ants constitute only about a third of the diet. More than 65% of the anteaters' food were ants in the genus *Camponotus*, and most of the rest were in the genus *Solenopsis*. Giant anteaters tend to avoid species such as army and leaf-cutting ants that have large jaws, strong chemical defenses, or spiny bodies.

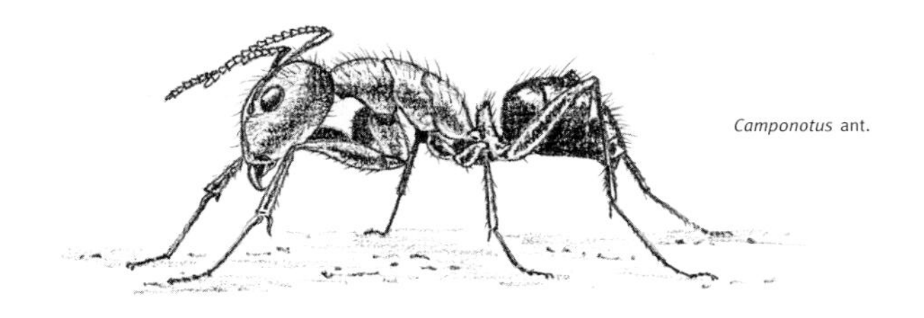

Camponotus ant.

Giant anteaters seldom feed on any given nest for more than about a minute (see p. 49), which is ample time for a giant anteater to down several thousand ants. The anteater's huge front claws quickly rip a hole into which its 50 cm (20 in.)-long snout can be introduced. Its tongue, which is anchored at the breast bone, can be extruded up to 60 cm (24 in.) at an astonishing rate of 150 times a minute. Over one pound of partly digested ants and termites has been found in the stomachs of some individuals, and it is estimated that adults consume as many as 30,000 ants each day.

In areas where food is relatively scarce, giant anteater home ranges can be as large as 2,500 ha (6,200 A), but in an extensive study in Brazil, home ranges were much smaller. Those of males averaged 270 ha (670 A), and those of females, 370 ha (910 A). Males were territorial, maintaining home ranges that overlapped very little, while neighboring females had home ranges that overlapped extensively. Giant anteaters settle their differences by circling one another, exchanging ritualized blows with their forefeet, and, on occasion, exchanging heavy blows that can cause injury.

Giant anteaters rest on the ground in shallow depressions that they sometimes scrape out themselves. They use their bushy tails as blankets, folding them back over their curled bodies (see p. 51).

Gestation lasts a little more than six months and results in a single young. The young is weaned at four to six weeks, but continues to ride on the mother's back for six to nine months. At the age of one year, a well-fed captive can be almost the same size as its mother, but wild yearlings in Brazil were only about half adult size. Giant anteaters reach sexual maturity at the age of two and a half to four years. Captives can live 26 years.

Sounds: Wheezes and growly roars; loud, rippled, descending, somewhat dolphinlike trills by young in distress.

Conservation: Giant anteaters may well be extinct in Costa Rica. At best, the population is so small that it is no longer viable. The most recent recorded sightings were in Corcovado National Park and in La Selva in the late 1970s, and in Santa Rosa National Park in 1984. Apparently this species was never common in Costa Rica, its abundance perhaps limited by its need for very specific foods (even by anteater standards) and large home ranges. Being both conspicuous and having a slow reproductive turnover, it was presumably unable to withstand the pressures of hunting and habitat loss.

References

Emmons et al. 1997; Jones 1982; Lubin 1983b; Meritt 1976; Merrett 1983; Montgomery 1985a,b; Montgomery & Lubin 1977; Nowak 1991; Redford 1985a; Redford & Eisenberg 1992; Rodríguez & Chinchilla 1996; Shaw et al. 1985, 1987; Timm & Laval 2000a,b; Timm et al. 1989.

O: Xenarthra F: Myrmecophagidae

NORTHERN TAMANDUA

Tamandua mexicana

(plate 4)

Names: Common name distinguishes this species from the southern tamandua, *T. tetradactyla*, which replaces it in northern and central South America. Also known as the collared anteater or lesser anteater. Genus from the Brazilian Tupi words *taa*, meaning ant, and *mandeu*, meaning to trap. *Mexicana* because the type specimen—that used to describe the species—was collected in Mexico. **Spanish:** *Oso hormiguero*; *oso mielero*; *oso colmenero*; *tamandua*.

Range: Southeastern Mexico to the northwestern tip of South America, and down to northern Peru; from sea level to about 1,500 m (5,000 ft); mostly in forest, occasionally in agricultural areas.

Size: 60 cm, 4.5 kg (24 in., 10 lbs).

Similar species: See giant anteater (p. 4, p. 53).

Natural history: Most of what is known about this species' behavior in the wild is derived from studies done on Barro Colorado Island, Panama, in the 1970s by Gene Montgomery and Yael Lubin. The northern tamandua can be active by day or night. Of 15 individuals radio-tracked in Panama, no two animals had the same activity pattern, but all were active about eight hours a day. Likewise, although tamanduas may forage in trees or on the ground, the Panamanian animals had consistent, individual preferences for one or the other. Unlike silky anteaters, tamanduas foraging off the ground do not use branches or lianas less than 5 cm (2 in.) in diameter.

Tamanduas feed on termites (such as *Nasutitermes* spp., the large, dark nests of which are abundant on tree trunks throughout the lowlands, and *Microcerotermes*), ants (such as *Azteca*, *Camponotus*, illustrated on p. 54, and *Montacis* spp.), and occasionally bees (including African killer bees). Lubin and Montgomery found that, on average, tamanduas ate about two-thirds termites and one-third ants, although individuals that spent more time in trees took more ants, and those that were more terrestrial ate more termites. Individual variation in diet might help reduce competition between neighboring tamanduas. In Panama, adults were estimated to consume an average of about 9,000 ants and termites, and to visit 50 to 80 colonies each day. In contrast to silky anteaters, they often attack large nests attached to limbs or located inside trunks; they rip open nests with their powerful forearms and foreclaws and probe the nest chambers with a long, sticky tongue that can be extruded 40 cm (16 in.). Differences between these two anteater species are discussed in more detail on pp. 58 and 59.

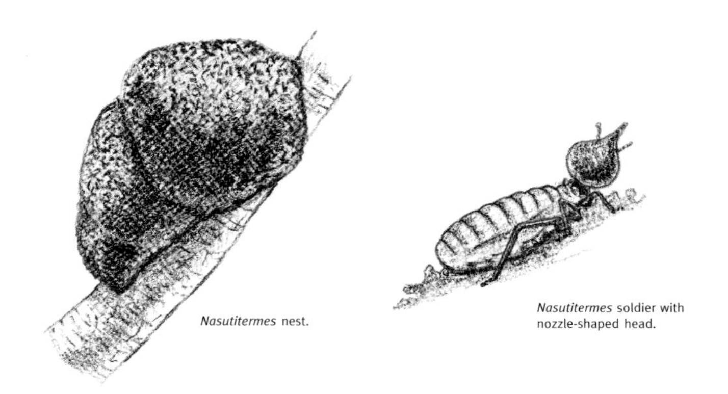

Nasutitermes nest.

Nasutitermes soldier with nozzle-shaped head.

Tamanduas tend to avoid ants and termites that have particularly spiny bodies, big jaws, or nasty chemicals, such as leaf-cutting or army ants, but none of their victims are entirely defenseless. Most nests are patrolled by large, aggressive, and armed soldier castes. *Nasutitermes* termite soldiers, for example, have nozzle-shaped heads through which they squirt a sticky, foul tasting secretion that smells strongly of turpentine. Consequently, tamanduas tend to raid nests quickly and move on before too many soldiers have rallied in defense (see also pp. 49-50), or they attack away from the nest—in a rotten log where termites are feeding, for example—where fewer soldiers are present. Avoidance of soldiers may also explain why tamanduas attack the nests of some species only seasonally. Tamanduas raid *Azteca* nests most frequently in the late wet season and termite nests mostly in late dry and early wet seasons. One Panamanian individual, for example, fed mainly on ants most of the year, but favored termites between March and May. At this time of year, termite nests are packed with almost fully grown reproductive broods that are about to make their annual emergence. These broods, made up of termites that grow wings and fly off en masse to mate, have been shown to be both more nutritious and heavier than the worker and soldier termites that predominate in nests during the rest of the year. Colonies may also produce fewer soldiers at this time of year, dedicating their resources instead to the reproductive generation. Thus, by choosing their moment, tamanduas may make their raids both more rewarding and less painful. As an additional benefit, such short, sporadic raids do not cause excessive damage to colonies, so tamanduas can come back periodically to the same self-replenishing nests for more.

Large gashes in ant and termite nests are generally the work of tamanduas, but smaller breaches may be caused by other animals. Circular holes about 8 to 12 cm (3 to 5 in.) across, or the beginnings thereof, are excavated in termitaries by various nesting birds (such as white-necked and pied puffbirds, and several species of trogons and parakeets), while 1 to 2 cm (⅓ to ¾ in.)-wide tunnels may be the work of euglossine bees.

The Panamanian tamanduas had home ranges of about 25 ha (60 A), although tamanduas (*T. tetradactyla*) in South America can have home ranges as large as 400 ha (1,000 A). The ranges of neighboring tamanduas in Panama overlapped little.

Females give birth to single young about once a year. Young can look quite different from the adults, varying from entirely off-white to black. During the nursing period, the young is left in a den (often in a hollow tree, abandoned burrow, or other cavity) while the mother forages. The mother may carry the young to a new den site at the beginning and end of her foraging. Young become independent when they are nearly half the size of the mother. Captives can live more than nine years.

Sounds: When threatened, tamanduas emit wheezes, prolonged soft whistles, and low-pitched, scratchy grunts and growls.

Mythology: The indigenous Bribri of southern Costa Rica admire the anteater for its ability to stand firm under attack. According to Bribri superstition, if anteater claws are placed in the house of a pregnant woman, then her child will be blessed with the same resilience.

Conservation: This is the most frequently observed of Costa Rica's three anteater species. It causes no harm to people, but may injure hunting dogs in self-defense. Although not an important game animal, it is occasionally hunted for sport and for its pelt.

References

Bozzoli 1986; Emmons et al. 1997; Jones 1982; Lubin 1983a,b; Lubin & Montgomery 1981; Lubin et al. 1977; Merrett 1983; Montgomery 1985a,b; Montgomery & Lubin 1977; Stiles & Skutch 1989; Timm et al. 1989; Wille 1987.

O: Xenarthra F: Myrmecophagidae

SILKY ANTEATER

Cyclopes didactylus

(plate 4)

Names: Also known as the pygmy anteater. The genus is derived from the Greek words *kyklos*, meaning circle, and *opsis*, meaning appearance, in reference to this species' habit of curling into a tight ball when resting. *Didactylus* means two-toed—the first, fourth, and fifth digits of its forepaws are not visible externally. **Spanish:** *Serafín del platanar*; *ceibita*; *tapacara*.

Range: Southeastern Mexico to Bolivia and Brazil, from sea level to about 1,500 m (5,000 ft), mostly in primary, wet, lowland forests.

Size: 17 cm, 200 g (7 in., 7 oz).

Similar species: Distinguished from the woolly opossum (pl. 2, p 37) by smaller size, more lethargic movements, inconspicuous ears, lack of facial markings, large foreclaws, and by the tail, which is furred all the way to the tip on the upper side.

Natural history: The silky anteater lives alongside tamanduas over much of its range. Radio-telemetry studies in Panama have shown that the two species can have overlapping home ranges and even use the same individual trees. The silky anteater differs from the tamandua, however, in both anatomy and behavior, and thus fills a different ecological niche.

Unlike the tamandua, it is exclusively nocturnal and arboreal. It is specialized for moving and feeding among thin lianas and branches. In Panama, about 85% of several hundred radio locations of silky anteaters were in trees with large loads of lianas, despite the fact that such trees represented only about 20% of the trees available. The smaller size of this anteater and a unique hand and foot design enable it to move along small stems that are inaccessible to tamanduas. In each hand, a wrist bone known as the pisiform is exceptionally long, extending into the base of the palm to form an opposable pad. This pad works against the two claws rather like a thumb, allowing the silky anteater to grip small stems tightly. The hindfeet also have a tight-gripping, padlike design, and are opposable, so they can be turned inwards and in opposite directions across a stem, and pull the stem taught for extra stability. The silky anteater can support itself with its hindfeet and prehensile tail alone, thus freeing its forelimbs for ripping into ant nests or defending itself.

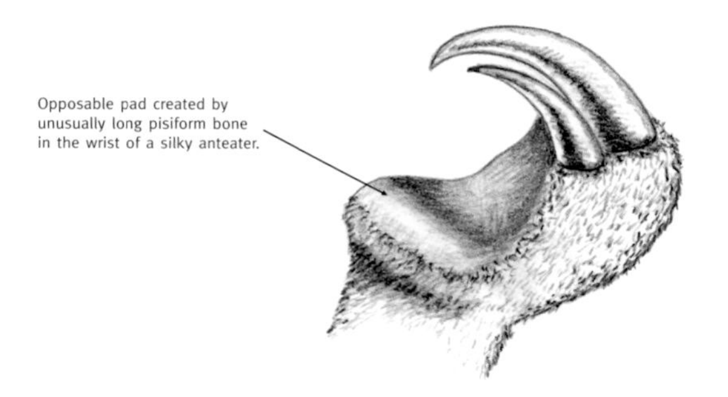

Opposable pad created by
unusually long pisiform bone
in the wrist of a silky anteater.

Since silky anteaters have weaker forearms and foreclaws than tamanduas, they do not try to breach the tough walls of termite nests. Tamanduas are able to feed heavily on termites, but silky anteaters feed almost exclusively on ants. Moreover, silky anteaters tend to eat different varieties of ants than do tamanduas, preferring smaller varieties and those living in stems. Of more than 100 varieties of ants eaten by the two anteaters in a study in Panama, only 11% were consumed by both.

Silky anteaters get at food by opening a small slit in an ant-infested stem with the large claws on their third fingers, prying the slit open and inserting their long, sticky tongue. Like other anteaters, they tend to feed only briefly at each colony (see p. 49). In Panama, adults were found to pass through an estimated 38 trees and to consume an average of about 3,000 ants each night. Both in Panama and in the Amazon, silky anteaters have been found to take a wide variety of ant species, but they feed predominantly on the genera *Crematogaster*, *Zacryptocerus*, *Pseudomyrmex*, *Solenopsis*, and *Camponotus* (illustrated on p. 54), all common varieties that build nests in hollow branches, leaves, or wood galls. Silky anteaters also ingest a few other types of insect, perhaps accidentally, and in captivity they accept fruit.

Silky anteaters spend the daytime curled in a ball on a shaded liana, usually within 20 m (65 ft) of the ground. A silky anteater barely meets its energy needs in a night of foraging, and when it rests during the day, it enters a torpid state: its metabolic rate and body temperature drop so much that it becomes almost comatose. Its thick fur coat and curled-up body position (for which it is sometimes referred to as the golden tennis ball) help it conserve heat while it rests.

Coiled, motionless silky anteaters are inconspicuous. If discovered, a silky anteater can defend itself by grasping a limb with its hindfeet and tail, raising its foreclaws to either side of its head, and tipping the top half of its body forward, claws-first, in a boxerlike lunge. If it manages to hook an aggressor, it then flexes its claws back and forth into the foe's flesh. Silky anteaters can be slow to react when torpid, however, and their boxing skills fail to impress a number of predators, including hawk-eagles, spectacled owls, and white-throated capuchin monkeys.

Opened kapok seed pod.

Legend has it that silky anteaters try to hide more effectively by sleeping in kapok trees (*Ceiba pentandra*), where they would be virtually indistinguishable from balls of silk-covered seeds that are revealed when the trees' seed pods open. Hence, one of the Spanish names for this species is *ceibita*, which means little kapok. If it exists at all, however, this strategy would seldom be of much use, since kapok trees fruit only briefly, principally in March and April, and not even every year.

Home ranges of silky anteaters can range from about 2 to 11 ha (5 to 27 A). Neither females nor males share home ranges with others of the same sex. Males have larger home ranges that incorporate at least part of the home ranges of several females.

The gestation period of this species is unknown. A single young starts feeding on ants when it is about a third the size of the mother and stops nursing altogether when it is about two thirds the size of the mother. During the nursing period, the mother leaves the young in a secluded spot such as a tree cavity when she goes out to feed. The mother, who changes her rest site almost nightly, returns to collect the young before dawn. Captive males have been seen carrying young and regurgitating ants for them, so it is possible they help raise young in the wild. When the young is about three-quarters adult size, it abruptly leaves the mother's home range, travelling in a straight line apparently until it finds a vacant territory. Little is known about silky anteaters' longevity, in part because captives are hard to maintain. One lived for two years and four months.

Sounds: Soft whistles; when threatened, a series of quickly repeated, dry, somewhat dolphinlike clicks that gradually turns into a prolonged, steady, scratchy retch; loud, scratchy distress calls by young when separated from their mother.

Mythology: Silky anteaters are sacred to the Bribri natives of southern Costa Rica, who believe that these anteaters guide the souls of the recently deceased to heaven.

Conservation: Silky anteaters are rarely seen, but since they are tiny, quiet, nocturnal, and arboreal, they could be more common than they appear. (See also p. 53.)

References

Baker 1983; Best & Harada 1985; Emmons et al. 1997; Federico Chinchilla, pers. comm.; Lubin 1983b; McNab 1982; Meritt 1976; Merrett 1983; Montgomery 1983b, 1985a,b,c; Montgomery & Lubin 1977; Nowak 1991; Sunquist & Montgomery 1973b; Taylor 1985; Timm et al. 1989.

O: Xenarthra F: Bradypodidae

BROWN-THROATED THREE-TOED SLOTH

Bradypus variegatus

(plate 5)

Names: The common name distinguishes this species from the two other varieties of three-toed (*Bradypus*) sloths—the pale-throated three-toed sloth (*B. tridactylus*), which replaces the brown-throated in northeastern South America, and the maned three-toed sloth (*B. torquatus*) of the coastal forests of southern Brazil. Genus from the Greek *brados*, meaning slowness, and *podus*, meaning foot. *Variegatus* means varied, since this species' coloration varies from dark to blond. **Spanish:** *Perezoso* (or *perico ligero*) *de tres dedos.*

Range: Honduras through most of northern South America as far as northern Argentina; from sea level to at least 2,400 m (7,900 ft) in some areas; in primary and secondary forest, and some town parks.

Size: 60 cm, 4 kg (24 in., 9 lbs).

Similar species: See Hoffmann's two-toed sloth (pl. 5, p. 65).

Natural history: Most of what is known of this species' behavior in the wild comes from an extensive study conducted in Panama in the 1970s by researchers Gene Montgomery and Mel Sunquist. Three-toed sloths are active by day or night. They feed on leaves, twigs, and buds. An individual might feed on as many as 40 trees, but spends most of its time in just a few. Three-toed sloths radio-tracked by Montgomery and Sunquist were found to have eight or fewer favorite trees in which they spent more than half their time, and most individuals had one preferred, or modal, tree in which they spent as much as 20% of their time.

The diet of three-toed sloths is less diverse than that of twotoeds. The extra reach provided by three-toed sloths' proportionately longer arms may compensate for this in part by improving three-toed sloths' ability to pull down terminal branches and feed on the choice young leaves that grow there. (Young leaves generally contain more protein, less fiber, and fewer toxins than do old leaves.)

A commonly heard myth is that three-toed sloths feed almost exclusively on the leaves of cecropia trees (*Cecropia* spp.). Some do feed heavily on cecropia, but others eat little or none, and no sloth can eat cecropia exclusively—one pet sloth fed only cecropia died within a few days. Another reason that three-toed sloths are seen more often in cecropias than in any other tree is that they are more conspicuous in the open crowns of cecropias than they are in the denser foliage of most of the other trees they frequent.

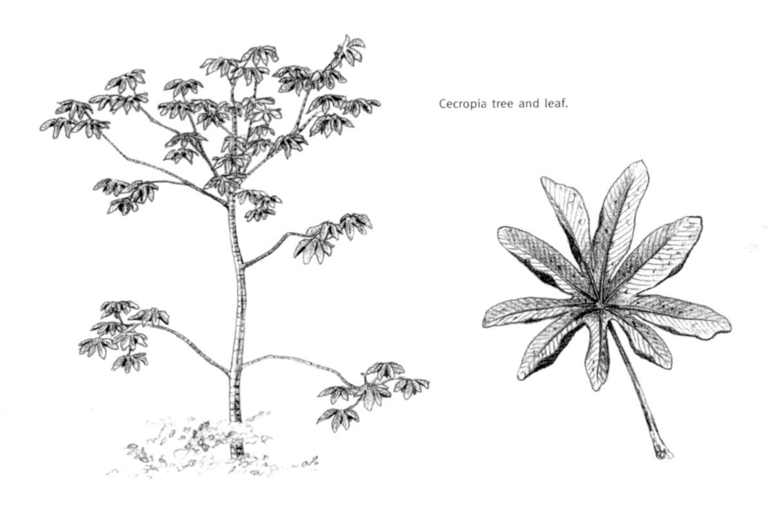

Cecropia tree and leaf.

Sloths have one of the slowest food passage times of any mammal. While food passes through most herbivores in a few hours or days—about five days in cows, for example—it can take four weeks or more to pass through a sloth. Since the type and age of a leaf greatly effects both its digestibility and nutritional value, sloths that do not select their food carefully can die of starvation even though their stomachs are full. Given all this, it is not surprising that a disproportionate number of the Panamanian sloths were found to die in the late wet season, when trees produce few new leaves and when limited sunshine and abundant rain may make it hard for sloths to maintain body temperatures adequate for digestion (see pp. 50-51).

Three- (and two-) toed sloths have the peculiar habit of descending to the ground about once a week to urinate and defecate. They do not evacuate at all between visits, and can shed as much as a third of their body weight in fecal pellets and urine in a single session! A visit to the forest floor is a huge undertaking for animals with the energetic limitations of sloths;

it is also dangerous, for sloths become more vulnerable to predators when they move out of the canopy vegetation. What benefit, if any, could outweigh such large costs and thus cause tree-descending sloths to prevail is a mystery, and has been the cause of much speculation.

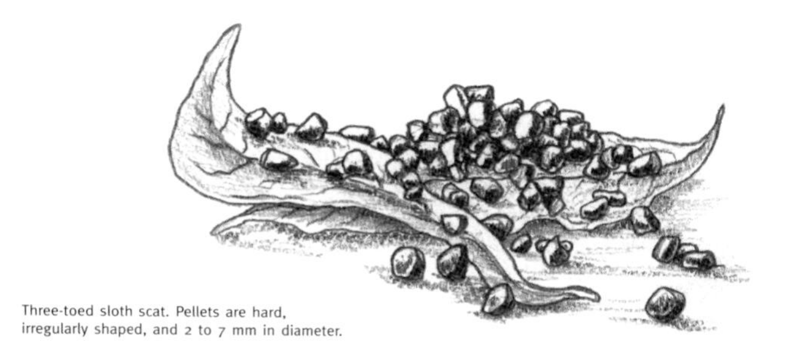

Three-toed sloth scat. Pellets are hard, irregularly shaped, and 2 to 7 mm in diameter.

Montgomery and Sunquist have suggested that one benefit might be the recycling of nutrients to food trees. Modal trees can lose almost 10% of their annual leaf production to a single sloth, and some trees are used by more than one individual. If the careful placement of feces ensured that nutrients were absorbed by food trees, then sloths could be returning as much as half the nutrients they consume to the trees. One problem with this theory is that it is hard to imagine how coming to the ground would greatly increase the probability of fertilizing specific trees, since the active roots of a tree are often far from the trunk, and since the roots of neighboring trees and other plants overlap extensively.

Another theory is that, by descending to defecate, sloths make themselves less vulnerable to predation. If sloths simply defecated from where they were feeding or resting, predators might be able to use the distribution of sloth droppings on the ground to ascertain the locations of sloths up in the trees. Since sloths move slowly and spend a lot of time in certain trees, they could be vulnerable to this type of predation. The three-toed sloth does conceal its droppings somewhat by depositing them in a shallow hole it digs with its tail, and then sweeping debris over the top with its hindlimbs as it departs. (Curiously, however, the two-toed sloth lacks a tail and does not go to such trouble, simply defecating on the open ground.)

A third possible explanation of sloths' mysterious pooping habits lies with a host of arthropods—including beetles, moths, and mites—that live in sloth fur and pass the egg and larval stages of their life cycles in sloth dung. As adults, some of them feed on the algae that grows on sloth fur (see pp. 51-52). A single sloth can host more than 100 moths and 1,000 beetles. It is conceivable that sloths depend in some way on some such arthropod that, in turn, depends on sloths' defecation trips to complete its life cycle.

Three-toed sloths spend more time active than do two-toed sloths—about 10 hours a day—but they are less mobile. Sunquist and Montgomery found that only 11% of the three-toed sloths moved more than 38 meters (125 feet) each day, and that the home ranges of threetoeds were smaller than those of twotoeds, averaging 1.5 ha (4 A).

Three-toed sloths give birth to a single 200 to 250 g (½ lb) baby after a gestation period of about six months. The young starts to eat leaves, in the form of fragments licked off the mother's lips, at the age of just two weeks. It spends the first four months of its life clinging to its mother, feeding on whatever she feeds on. After about six months, the mother abruptly abandons it in the area in which it was raised and moves to another part of her territory. The young usually leaves the mother's territory by the time it is about a year old, but keeps the inheritance of her taste in leaves. Thus, a sloth's food preferences are greatly influenced by its mother, and different sloth lineages often have quite different tastes. (Hence, for example, the species' variable dependence on cecropia.) Such variation in diet reduces competition between neighboring, unrelated sloths, and may help explain why sloths can reach densities as high as eight individuals per hectare. Potential longevity for this species has been estimated at 20 to 30 years.

A sloth moth (*Cryptoses choloepi*) amidst the fur of its host. This moth is only about 12 mm long.

Sounds: High-pitched, monotone whistles emitted through the nostrils; hisses and wheezes when threatened; mournful bleats when distressed.

Conservation: This is one of the most common mammal species in Costa Rica's lowland forests (see pp. 52 and 67).

References

Bauchop 1978; Eisenberg & Thorington 1973; Greene 1989; William Haber, pers. comm.; Leigh & Windsor 1996; Meritt 1985b; Merrett 1983; Molina et al. 1986; Montgomery 1983a; Montgomery et al. 1973; Montgomery & Sunquist 1974, 1975, 1978; Parra 1978; Sunquist & Montgomery 1973a; Waage & Best 1985; Waage & Montgomery 1976.

O: Xenarthra F: Megalonychidae

HOFFMANN'S TWO-TOED SLOTH

Choloepus hoffmanni

(plate 5)

Names: Common name distinguishes this species from the only other two-toed (*Choloepus*) sloth—the southern two-toed sloth (*C. didactylus*) of northeastern South America. Genus from the Greek words *cholos*, meaning lame, and *podus*, meaning foot. Species name after Carl Hoffmann, a German medical doctor who did substantial work on the taxonomy of Costa Rica's flora and fauna in the mid-1800s. He served Costa Rican forces as an army surgeon in the tide-turning battle against North American mercenary and slave trader William Walker at Santa Rosa in 1856. Around that time he became fascinated with the huge diversity of bats in Guanacaste and started the first scientific research ever done on bats in Costa Rica. About a dozen Costa Rican plant and animal species bear Hoffmann's name. **Spanish:** *Perezoso* (or *perico ligero*) *de dos dedos.*

Range: Honduras to northwestern South America down to Bolivia; from sea level to at least 3,300 m (11,000 ft); in primary and secondary forests. This species is most common at mid- to high-elevations. In wet, lowland forests, it is generally seen much less often than the three-toed sloth; in the seasonally dry forests of northwestern Costa Rica, it is rare. In a seasonally dry, lowland forest in Panama, it was found to be outnumbered by its three-toed cousin by four to one.

Size: 60 cm, 6 kg (24 in., 13 lbs).

Similar species: The brown-throated three-toed sloth (pl. 5, p. 61) has dark eye bands and a "smiling" mouth, relatively longer forelimbs, a short, stubby tail, and three toes on each forefoot. (Note that both species have three toes on the hindfeet.) Male three-toed sloths have a yellow and black patch in the center of the back.

Natural history: Since this sloth is mostly nocturnal, usually spending the daytime immobile amidst densely vegetated tree crowns, it is even less conspicuous than the three-toed sloth. Biologists have found that even individuals that have been precisely radio-located can be impossible to see. This species likes to rest in the middle of liana tangles; in such places it is hard for predators to approach without moving the lianas and alerting the sloth to their presence. Although slow, the two-toed sloth can defend itself with its claws and canine-like premolar teeth. When active, the two-toed sloth generally stays in the upper levels of the forest, descending to the ground only to defecate, about once a week (see pp. 62-63).

The two-toed sloth feeds mostly on leaves, although it also eats buds, flowers, twigs, and fruit. In captivity, it accepts meat, and in the wild might occasionally eat insects, bird eggs, or small vertebrates. The two-toed sloth feeds on a wider variety of plants and other foods than does the three-toed sloth, a characteristic that makes it easier to maintain in captivity.

In Panama, two-toed sloths were found to spend less time active than three-toed sloths—an average of about seven and a half hours per day—but they moved around more. They tended to have larger home ranges, often of 2 to 3 ha (5 to 7 A), and they changed trees some four times more often than threetoeds, seldom resting in the same tree on consecutive days.

Two-toed sloths have an exceedingly long gestation period of about 11 ½ months. Single young nurse for about a month. They stop clinging to the mother after about five months, but may stay near her for up to two years. Females are sexually mature at about two years of age, males probably later (not until four and a half years in the South American *C. didactylus*). Females appear to greatly outnumber males in the wild. This bias might compensate in part for the long gestation period and increase reproductive efficiency: if two-toed sloths mate only infrequently, few males are needed to get the job done. In three-toed sloths, which have a much shorter gestation period, females and males appear to be equally abundant. Captive two-toed sloths can live 32 years.

Sounds: Hisses, wheezes, and tooth clacks when threatened; loud, mournful bleats when distressed, or by young separated from their mother.

Harpy eagles are major sloth predators, but are now almost extinct in Costa Rica.

Conservation: Sloths cause no direct harm to people or crops, although in Panama the two-toed sloth was found to be the major vertebrate host of *Leishmaniasis*, a skin disease that is transmitted to people by sand flies.

Like all forest-dwelling creatures, sloths have suffered vast losses of habitat over recent decades. Where forest remains, however, sloths are among the most abundant mammals (see p. 52), even though most individuals go unnoticed. Indeed, sloths in forest remnants may have benefited from the elimination of large predators, one of the main causes of sloth mortality in more pristine areas. Jaguars, which have been found to feed heavily on sloths at La Selva and in Corcovado, are now scarce over most of Costa Rica. Harpy eagles, one pair of which was recorded feeding 26 sloths to their young in just 10 months, have all but disappeared in Costa Rica. Sloths were especially vulnerable to the harpy eagle because at sunrise, the harpy eagle's main hunting time, they climb out to exposed limbs to sunbathe.

References

Braker & Greene 1994; Chinchilla 1997; Eisenberg & Maliniak 1985; Evans 1999; Izor 1985; Jones 1982; Meritt 1985b; Meritt & Meritt 1976; Molina et al. 1986; Montgomery & Sunquist 1978; Nowak 1991; Reid 1997; Rettig 1978; Shaw 1985; Sunquist & Montgomery 1973a.

O: Xenarthra F: Dasypodidae

NORTHERN NAKED-TAILED ARMADILLO

Cabassous centralis

(plate 6)

Names: Common name distinguishes this species from three other naked-tailed (*Cabassous*) armadillo species that inhabit South America. Members of this genus are also known as eleven-banded armadillos (although the number of movable bands around the midbody varies from 10 to 13), or five-toed armadillos. *Cabassous* is a name used by native South Americans. *Armadillo*, which comes from Spanish, means little armored one. **Spanish**: *Armadillo zopilote*; *armadillo* (or *cusuco*, or *armado*) *de once bandas*.

Range: Chiapas, Mexico to the northwestern tip of South America; from sea level to about 1,800 m (6,000 ft); in forest or open habitats.

Size: 40 cm, 3 kg (16 in., 7 lbs).

Similar species: Distinguished from the nine-banded long-nosed armadillo (pl. 6, p. 69) by smaller size; a broader, more flattened body shape; a naked rather than armored tail; greatly enlarged claws on the forefeet; a broader and shorter snout; and by the ears, which are positioned on the sides of the head and point sideways. In nine-banded armadillos, the ears are positioned close together and point upward.

Natural history: The northern naked-tailed armadillo has never been studied in the wild, and barely even in captivity, so little is known of its natural history. Like the nine-banded armadillo, this species is solitary and mostly nocturnal. Occasionally, the naked-tailed armadillo is active after sunrise, but only in the immediate vicinity of its burrow. It is more fossorial (spends more time underground) than most other armadillos and is an ant and termite specialist. Studies of South American naked-tailed armadillos suggest that these insects form 90 to 98% of the diet. Having located an underground nest, apparently by smell alone, *C. centralis* uses its massive foreclaws to dig and cut through roots, sometimes completely burying itself during an excavation. Like anteaters, it has a long, extrudible tongue with which to extract ants and termites from their elaborate nests.

With its large claws, the naked-tailed armadillo is a clumsy walker, treading only with the soles of its hindfeet and the tips of its foreclaws. When threatened, it may run a short distance, but usually tries to escape by burrowing or jumping into water. One harassed by dogs in Monteverde dug its way into a bank in under two minutes. When its hole started to fill with water, it flipped on its back and started to dig upward with equal speed. If a naked-tailed armadillo is caught before it manages to bury itself, it may urinate and defecate on its predator, spattering the excrement about by twirling its tail. It also releases nauseating secretions from its anal glands.

Naked-tailed armadillos den in roughly 16 cm (6 in.)-wide burrows in old termite mounds or in banks. The burrow entrance is almost exactly the same shape and size as the armadillo. The soil around the entrance, which is usually hard-packed rather than mounded, is free of litter, vegetation, and scat. *Cabassous* sometimes dens under houses. Active nests give off a pungent odor. Unlike nine-banded armadillos, naked-tailed armadillos apparently do not construct nests within their burrows. They change burrows almost nightly.

Little is known of this species' reproductive biology. Unlike the nine-banded armadillo, it gives birth to single young. Newborns are pink, soft-skinned, and tiny, weighing only about 100 g, or 3% of the mother's body weight.

Sounds: Wheezes and low buzzing sounds when digging; loud growls and piglike snorts and grunts when threatened.

Mythology: For the indigenous Bribris of southern Costa Rica, the naked-tailed armadillo (but not the long-nosed armadillo) is a harbinger of death. Curiously, the modern Spanish name for this species in Costa Rica is *armadillo zopilote*, which translates as vulture armadillo. Given that there is nothing conspicuously vulturelike in this species' appearance or natural history, and that the vulture is also a harbinger of death in Bribri and other cultures, it seems likely that the modern Costa Rican name is rooted in the Bribri superstition.

Conservation: Studies on anteaters suggest that mammals specialized for feeding on ants and termites need unusually large home ranges to meet their nutritional requirements (see pp. 49-50 and 53). This may be a factor that limits the abundance of the naked-tailed armadillo, which is apparently rare even in pristine habitats. The strong musk produced by this species' anal gland is said to permeate the meat and make it unappetizing; in contrast to the nine-banded armadillo, it is generally left alone by hunters. All the same, the naked-tailed armadillo is legally protected as an endangered species within Costa Rica.

References

Aranda & March 1987; Bozzoli 1986; Carrillo & Wong 1992; Marc Egger, pers. comm.; Meritt 1985a; Nowak 1991; Redford 1985b; Reid 1997; Wetzel 1980, 1982, 1985a.

O: Xenarthra F: Dasypodidae

NINE-BANDED LONG-NOSED ARMADILLO

Dasypus novemcinctus

(plate 6)

Names: Genus from the Greek *dasys*, meaning hairy, and *podus*, meaning foot. *Novemcinctus* means nine-banded, although the number of movable bands in the middle of the carapace varies from seven to ten. In Costa Rica, nine-banded armadillos most often have eight bands. **Spanish:** *Armadillo* (or *cusuco*, or *armado*) *de nueve bandas.*

Range: South-central and southeastern United States to Peru, Argentina, and Uruguay—the broadest range of any of the world's 20 or so armadillo species, or any other member of this order; mostly from sea level to about 1,500 m (5,000 ft), rarely up to 2,600 m (8,500 ft); in forest or open habitats. This species, the only xenarthran found north of Mexico, has greatly expanded its range through the United States during the last century. Although it was introduced to some areas to eat crop pests, it has extended its range largely on its own merit, thanks to its great versatility and its ability to cross large rivers. It manages the latter either by inflating its stomach with air and swimming, or by holding its breath and walking along the bottom.

Size: 45 cm, 4 kg (18 in., 9 lbs); males larger than females.

Similar species: See northern naked-tailed armadillo (pl. 6, p. 67).

Natural history: The nine-banded armadillo has been studied extensively in North America, but relatively little in the tropics. It is terrestrial and mostly solitary. It uses its keen sense of smell to locate prey in the leaf litter or underground, and roots out choice items with its nose and large foreclaws. Its ability to hold its breath for as long as six minutes helps it keep soil out of its nostrils while it searches.

Nine-banded armadillos dig for food in the forest leaf litter and on lawns (shown here). They leave a distinctive trail of shallow, snout-shaped holes.

In North America, the nine-banded armadillo has a diverse diet that includes beetles, ants, termites, caterpillars and other larvae, earthworms, millipedes, centipedes, snails, small vertebrates (such as frogs, lizards, and snakes), bird eggs, fungi, tubers, fallen fruits, and carrion. In one analysis of 160 armadillos, animal matter was found to comprise more than 90% of the stomach contents. At some times of year, apparently when normal foods are scarce, the armadillo may feed heavily on fruit.

In the tropics, the diet of these armadillos may be somewhat different. Various studies suggest that tropical nine-banded armadillos feed much more heavily on ants, termites, and other social insects than their North American counterparts. Several authors have pointed out that nine-banded armadillos may have even evolved for eating ants, just like their cousins, the anteaters, as well as the naked-tailed and a few other armadillos. The mouth design, for example, approaches that of anteaters in that nine-banded armadillos have a proportionately smaller jaw, and fewer and smaller teeth, than any other armadillo. Perhaps such features evolved for a largely ant and termite diet in the tropics, and then, when the Panamanian land bridge formed and allowed this armadillo to spread north, the specialized mouth and jaw turned out to be adequate for a variety of other soft foods that were more abundant in North America.

Nine-banded armadillo scat.
Each dropping is 2 to 4 cm long.

The nine-banded armadillo has poor eyesight, and it may be so engrossed in its rummagings that it fails to notice approachers. When the armadillo is surprised at close range, a common "startle" tactic is to leap straight up in the air before dashing to safety. If caught just inside its burrow, the armadillo arches its back, wedging its ridges of armor into the dirt roof and thus making it very difficult for any pursuer to pull it out by its legs or tail. The armadillo's armor, which constitutes almost one fifth of its body weight, also helps protect it against abrasive vegetation and small predators, such as hawks. The armor does not, however, provide significant protection against large predators, such as pumas or jaguars, and armadillos are an important component in the diets of these cats.

Biologists have studied the territorial arrangement of the nine-banded armadillo in the United States. The home ranges of neighboring females do not overlap, while males' home ranges overlap those of neighbors of either sex and are, on average, about 50% larger than females' home ranges. Home range sizes vary from 1 to 14 ha (2 ½ to 35 A). Armadillos den in burrows, each individual usually having four to 12 different burrows that are used alternately. Most burrows have at least two entrances, with well-worn trails radiating out from each. Burrows are typically semicircular in shape, about 18 cm (7 in.) wide, and up to 8 m (26 ft) long. They end in a chamber that is often lined with dry grass or leaves. This armadillo's habit of lining nests with insulating material is thought to have helped it survive in the seasonally colder climates of North America.

A nine-banded armadillo hole. This one was about 27 cm wide.

Reproduction in nine-banded armadillos is peculiar. The females are able to prolong their pregnancies by delaying implantation of the fertilized egg in the uterus wall. In North America, nine-banded armadillos typically mate at the end of the summer, but delay implantation three or four months. After implantation, true pregnancy lasts a little over two months, and the young are thus born at the optimum moment—the following spring.

Leprosy researchers in England inadvertently discovered that, under stress, nine-banded armadillos can prolong their pregnancies even longer. As expected, some of the female armadillos in their custody—which had been isolated from males since they had been captured in the wild in Florida—produced young the first spring after their capture. But the researchers were astonished when others gave birth in their second spring, and one even in her third, some 32 months after her last opportunity to mate. While various other mammals, such as some bats and members of the weasel family, can delay their pregnancies, no other mammal is known to be able to delay this long. This reproductive flexibility may be another factor that has helped nine-banded armadillos survive in temperate climates of North America, much further north than any other xenarthran.

Reproduction in nine-banded armadillos is curious for another reason, too. Once fertilized, the egg of a female nine-banded armadillo divides, and then divides again, to produce four embryos. Thus, nine-banded armadillos produce litters of four identical, same-sex armadillolets.

Young reach sexual maturity after six months to a year, about twice as quickly as other types of armadillo. Potential longevity is up to 15 years, although few nine-banded armadillos more than four years old have been recorded in the wild.

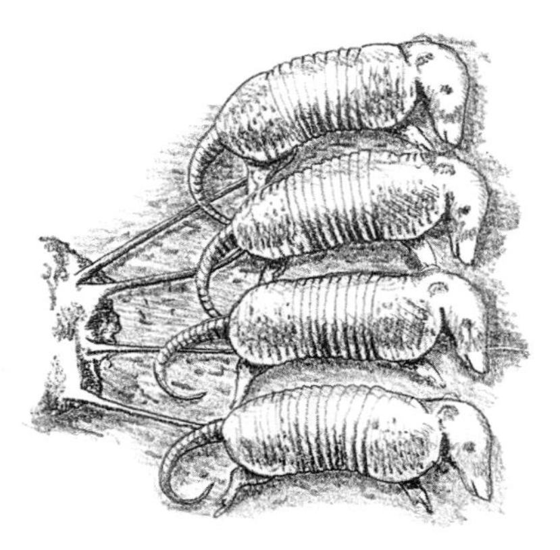

Nine-banded armadillo quadruplets: four embryos attached to the same placenta.

Sounds: Wheezes and grumpy mumbles when digging; grunts and squeals when startled; rarely, when walking, bizarre noises resembling those made by large drops of water dripping slowly into a pond.

Conservation: The nine-banded armadillo sometimes eats eggs and can cause erosion, but its impact is seldom serious. It helps people by eating insect pests. It has proved valuable in many areas of medical research, including multiple births, birth defects, organ transplants, and various diseases, in large part because the genetically identical quadruplets permit unusually precise comparative studies. It has been particularly valuable in the study of leprosy, for it is the only animal other than *Homo sapiens* that suffers naturally from the disease. Abandoned armadillo burrows serve as shelter for numerous animals. Nine-banded armadillos have good-tasting, porklike meat and are popular game throughout Costa Rica. Nonetheless, thanks to their flexible diet, tolerance of disturbed habitats, and high reproductive turnover, they are common in many areas.

References

Emmons et al. 1997; Horwich & Lyon 1993; Kalmbach 1943; Layne & Glover 1977, 1985; McBee & Baker 1982; McDonough & Loughry 1997; McNab 1980; Nowak 1991; Redford 1985b; Storrs & Burchfield 1985; Storrs et al. 1989; Wetzel 1982, 1983; Wetzel & Mondolfi 1979; Yager & Frank 1972.

Shrews (order Insectivora)

Distribution and Classification

During the last century, mammal taxonomists used the Insectivora as a sort of miscellaneous category where they would place a variety of smallish creatures that did not necessarily have very much in common, but did not fit conveniently into any other order. Since then, some such creatures, such as elephant shrews and tree shrews, have been given orders of their own (Macroscelidea and Dermoptera respectively), but experts still debate over the status of the roughly 360 species and seven families that are currently considered Insectivora.

Members of the Insectivora are distributed almost worldwide. The best known members of the order are hedgehogs, moles, and shrews. More obscure members include gymnures, moonrats, desmans, golden moles, tenrecs, otter shrews, and solenodons. Voles, which people often associate with this group, are rodents. Only about five species of insectivores inhabit Costa Rica, all belonging to the Soricidae, the shrew family. The Soricidae is by far the largest and most widespread Insectivora family, with about 290 species dispersed through Eurasia and Africa, and in the New World from Alaska to northern South America.

Evolution

Members of the Insectivora have many features of external and internal anatomy in common with the earliest known mammals, and are thus considered the most primitive living placental mammals. The oldest fossils of insectivore-like mammals (in the form of teeth and bone fragments) date back 80 million years to the Cretaceous period. The oldest fossils of shrewlike creatures are from the north, dating back about 54 million years in Europe and 38 million years in North America.

Teeth and Feeding

The dental formula for Costa Rica's shrews is I 3/1, C 1/1, P 2/1, and M 3/3 = 30. The lower incisors and first upper incisors are large and curved, and work together like a pair of pincers. This feature is well-suited for pulling worms and larvae out of the earth or for prying the soft interior out of the hard-shelled bodies of large insects. The canines are inconspicuous. Costa Rica's shrews belong to a subfamily (the Soricinae) that is commonly referred to as the red-toothed shrews because one of the subfamily's characteristics is reddish brown stains on the teeth.

Shrews are voracious eaters. Small mammals require proportionately more energy to stay warm than do large mammals, because their bodies have more surface area relative to volume; bodies with more outside and less inside simply disipate heat faster. Shrews are among the smallest mammals in the world, and must eat a lot to fuel the high metabolic rate they need to stay warm. In fact, some experiments suggest that shrews have even higher metabolic rates than other mammals their size. Thus, they consume at least their own body weight each day, and they perish if they spend more than about three hours without eating.

Like most members of the Insectivora, shrews are not exclusively insectivorous. Most species eat other invertebrates, and some can handle small vertebrates. Like rabbits and some rodents, shrews sometimes eat their own feces, from which they may acquire nutrients such as vitamins K and B that they are unable to absorb during their food's first passage.

A few species (the solenodons and some shrews) produce venemous saliva from a gland in the front of the lower jaw. These species are the only mammals known to have a venemous bite (although the male of another mammal species, the duck-billed platypus, *Ornithorhynchus anatinus*, has a venemous spur on each hindfoot which it employs in defense and prey capture). The saliva contains neurotoxins not unlike those found in the venom of coral snakes and some vipers. As the saliva enters the veins of a victim, it causes lowered blood pressure, a slowed heart beat, and inhibited respiration. The neurotoxin is quite powerful; that housed in the salivary glands of one North American species would be sufficient to kill 200 mice if it were injected intravenously, and a small bite can cause severe swelling in humans. The poison is used to immobilize insects for later consumption, to subdue large prey, and in defense.

Body Design

All insectivores are small. No member of the order is larger than a house cat, and no shrew is larger than a rat. Indeed, the shrew family contains one of the world's two smallest mammals (the pygmy shrew, *Suncus etruscus*, from the Mediterranean and Asia, which weighs a mere 2 grams). Other external characteristics of the Insectivora include long, thin, flexible snouts, short limbs, and plantigrade feet, with five clawed digits on each foot.

With their frenzied foraging and high metabolic rate, shrews are restless, nervous animals. It has been calculated, for example, that the heart of the pygmy shrew beats about the same number of times during the shrew's one-year lifespan as does the heart of an elephant over the course of 70 years. When shrews are alarmed, their heart may beat 1,200 times a minute, and captives can die of fright from loud noises such as thunder.

Senses and Scent Glands

Like most other members of the Insectivora, shrews have tiny eyes and their vision is poor. Most shrews have small ears, but their hearing is excellent, and vocalizations are thought to be important in communication. Some species produce ultrasound, which they may use in a crude form of echolocation. European shrews, at least, can echolocate their burrow entrances from a distance of about 20 cm (8 in.). Shrews have good senses of touch and smell. They probably use odors produced by glands on their flanks and belly in communication since the odors are reported to be strongest during breeding season. When threatened, shrews also produce foul-tasting skin secretions that appear to be successful in deterring some predators. In many parts of the world, canids and felids kill and then discard shrews without eating them, and some members of the weasel family have been found to avoid shrews unless there is insufficient alternative prey.

Reproduction

The reproductive anatomy of the Insectivora is one of the features that is considered primitive. Like reptiles, birds, and the earliest known mammals, insectivores have a reproductive tract that opens out into a cloaca (the Latin word for sewer), a chamber that is also used as an emptying space for the digestive tract. In all other living placental mammals, the reproductive and digestive tracts are independent.

Shrews seldom live more than two years, but they have a fast reproductive turnover. Gestation periods lasting only a few weeks result in litters of up to 10. The young reach sexual maturity when they are just a few months old, and many species produce two or three litters a year.

Studies on European shrews have brought to light some intriguing details of shrew breeding biology. Since shrews disperse very short distances once they leave their mothers and start breeding immediately, the chances that they will inbreed are high. Inbreeding is generally detrimental to the health of offspring; inbred shrews have been shown to weigh less at birth and to die younger than outbred shrews. Although it is disadvantageous, shrews do breed with relatives, perhaps because they are unable to distinguish them. In one population, one third of all matings were found to occur between close relatives. Female shrews apparently compensate for this problem by mating with numerous males, to the extent that young within the same litter may have as many as six different fathers. Thus females increase their chances that at least some of their offspring will be fathered by an unrelated male.

Conservation: See pp. 78 and 79.

References

Barnard 1984; Choate 1970; Gould 1969; Macdonald 1995; Martin 1981; Pearson 1942, 1946; Rodríguez & Chinchilla 1996; Tomasi 1979; Wroot 1984; Yates 1984.

SPECIES ACCOUNT

O: Insectivora F: Soricidae

LEAST SHREWS

Cryptotis spp.

(plate 23)

Names: Also known as small-eared shrews. Genus from the Greek words *kryptos*, meaning hidden, and *otus*, meaning ear. **Spanish:** *Musurañas, ratas arañeras, ratas sordas.*

Range: North, Central, and northern South America; from sea level to at least 3,000 m (9,900 ft); mostly in forested and/or wet habitats.

Size: The five species vary from 6 to 9 cm, and 4 to 7 g (2 ½ to 3 ½ in., ¼ oz).

Similar species: Five species of least shrews are currently listed for Costa Rica: *Cryptotis parva*, *C. nigriscens*, *C. gracilis*, *C. jacksoni*, and *C. merriami*. *C. parva* is brown, while the other four are all blackish and very hard to tell apart by external anatomy alone. Shrews differ from small rodents in having long snouts and five rather than four toes on each forefoot.

Natural history: Most least shrews are forest dwellers, although *Cryptotis parva* prefers open fields and marshes. Shrews forage by day or night, pushing through the leaf litter or the upper levels of the soil with their forepaws and long snouts. Their diet includes insects, larvae (*C. parva* has been known to raid beehives to get at the brood), earthworms, spiders, and other invertebrates; small vertebrates (including lizards, frogs, and other shrews); carrion; and occasionally fungi and seeds. Food is sometimes stashed in tunnels before being eaten.

Least shrews are generally solitary, although *C. parva* shows some social tendencies. In captivity, adults have been observed helping each other dig tunnels, and males stay with females while young are being raised.

Least shrews den under logs or rocks, in tunnels they dig themselves, or in those abandoned by other animals. Nests consist of 5 to 12 cm (2 to 5 in.)-wide balls of dry grass and leaves, typically situated beneath logs or rocks. The only Costa Rican species whose reproductive habits have been studied in any detail is *C. parva*. A gestation period lasting 21 to 22 days results in litters of one to nine young. The young triple their weight in their first week of life; they reach adult size and stop nursing by the time they are about three weeks old. To facilitate this speedy development, they grow and shed their milk teeth before they are even born. They reach sexual maturity at the age of one to two months. In captivity, least shrews can live two and a half years, although the average lifespan is only about eight months.

Sounds: Soft clicking noises during friendly social interactions; ultrasonic clicking noises possibly used in echolocation (see p. 76); chirps.

Conservation: In eras past, shrews were thought to be responsible for poisoning livestock and laming horses. Consequently, they gave rise to the term *shrewd*, an adjective which, in its original sense, was no compliment: it was applied almost exclusively to criminals. As far as is known, however, shrews do humankind no harm. They do not damage crops, and there is certainly no evidence that they affect livestock. Indeed, they help us by eating agricultural pests.

In part because of their secretive habits, little is known of the distribution or abundance of shrews in Costa Rica and Central America.

They are seldom caught in biologists' mouse traps, either because they are not attracted by mouse bait or because they are more cautious than mice. Many of the records of Costa Rican shrews are of individuals found dead on trails. Predators such as coyotes, gray foxes, and small cats frequently kill and then abandon shrews without eating them, perhaps because shrews produce distasteful skin secretions when threatened. The remains of shrews are found frequently in the stomach or scat of some predators such as owls, however, so shrews may not be uncommon. *C. parva* is thought to be abundant in some areas, such as at Barva Volcano.

References

Barbour & Davis 1974; Broadbooks 1952; Choate 1970; Gould 1969; Macdonald 1995; Mock & Conaway 1975; Nowak 1991; Reid 1997; Rodríguez & Chinchilla 1996; Timm & Laval 2000b; Timm et al. 1989; Tomasi 1979; Woodman 1992; Woodman & Timm 1993.

Bats (order Chiroptera)

Distribution and Classification

About 925 species of bats comprise the order Chiroptera. They are split into two suborders. One, the Megachiroptera, contains a single family, the Pteropodidae or flying foxes, all 166 species of which inhabit the Old World tropics. All Costa Rica's bats belong to the other, much more diverse and widespread suborder, the Microchiroptera, which contains 16 families and about 759 species, distributed almost worldwide. As their names suggest, the Megachiroptera contains the world's largest bats (some flying foxes have wingspans of almost 2 meters) and the Microchiroptera contains the world's smallest (some Asian bats weigh only 2 grams), but there is considerable size overlap between members of the two suborders.

Worldwide, bats are the second most diverse order, being outnumbered by rodents by more than two to one. In Costa Rica, however, the reverse is true. The 109 bat species (as opposed to just under 50 rodent species) that have been recorded in Costa Rica to date represent just over half the country's 216 or so mammal species, and about 12% of the world's bat species.

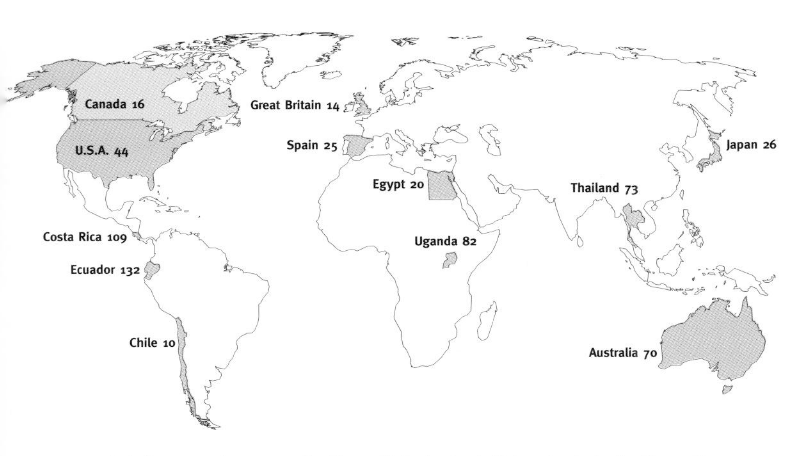

Costa Rica represents a mere 0.03% of the world's land surface but is home to about 12% of the world's bat species. This map shows the number of bat species in selected countries around the globe.

There are probably bat species in Costa Rica that have not yet been noticed. Unlike, say, birds, most bats cannot be identified visually unless they are captured. Bats can be captured either when discovered at a roost or with the use of mist nets. Mist nets are stretches of fine mesh, generally attached to poles, that are strung out along trails or other potential bat flyways to entangle passing bats. (In the daytime, mist nets are also used to catch birds.) Entangled bats can then be carefully extracted from the mist net, examined, and released. Unfortunately, some species are not prone to be captured. Some species may be able to detect and avoid mist nets, while others may forage in areas where mist nets are seldom placed, such as above the canopy. Any of these may also roost in inaccessible or inconspicuous spots.

A recent and still-evolving technique for censusing bats is the use of bat detectors. These devices, which enable researchers to record and, potentially, identify the ultrasonic calls of bats, are now being used in Costa Rica for the first time. In the United States, bat detectors have been shown to record about 30% more species than standard capture techniques, and, likewise, a bat-detector census at Barro Colorado Island in Panama, a site that has been mist-netted extensively for decades, turned up five new species. In some parts of the world, innovative researchers have even attached bat detectors to helium-filled balloons, and have thus been able to record species foraging hundreds of meters off the ground.

The 30 species depicted and described in the following pages provide a cross-section that includes distinctive, common, or representative members of each of the nine families found in Costa Rica. For those interested in identifying bats at the species level, two excellent references are now available: an updated technical key (Timm & LaVal 1998, or its revised version in Spanish, Timm et al. 1999) and a color-illustrated field guide (Reid 1997).

Evolution

The history of the Chiroptera is poorly understood. Bat bones are so small and delicate that they only rarely survive as fossils. Nevertheless, the scant fossil record shows that, compared to many other mammal orders, bats have been around a long time. The oldest bat fossils, found in Wyoming in the United States, in Germany, and in Pakistan, date back 50 to 60 million years to the Eocene period. Since these species were already well-developed, resembling modern day bats physically, and even, apparently, having evolved the use of echolocation, it is presumed that the earliest batlike mammals appeared considerably earlier, perhaps 70 to 100 million years ago. The earliest forms may have evolved from arboreal, insect-eating, shrewlike creatures that learned to glide and, eventually, to fly.

Fundamental differences between the Megachiroptera and the Microchiroptera have led some to believe that the two suborders evolved independently—in other words that bats, or flight in mammals, evolved twice. One hypothesis, based on the presence of unusual nerve channels between the eyes and the brains of megachiropterans, has it that members of that suborder are more closely related to us primates than they are to microchiropterans. Other studies, however, including molecular comparisons, suggest that the two groups are indeed each other's closest relatives. For the time being, theories about the early evolution of bats can only be speculative.

Teeth, Skull, and Feeding

In temperate zones, most bats feed on insects, which they catch in the air. In the tropics, bat diets and foraging styles are much more varied. About half of Costa Rica's bats feed primarily on something other than insects. At least 25% feed principally on fruit, almost 10% eat mostly nectar and pollen, at least 7% feed mostly on vertebrates or blood, and at least 5% are highly omnivorous. The insectivores themselves can be split into numerous feeding categories. They may prefer to hunt inside the forest, either in the understory or in the canopy; in open areas; or over water. They may be best adapted for catching prey that is either flying or perched, soft- or hard-shelled, or large or small. It is scarcely surprising, then, that the design of bats' teeth and skulls is highly variable. Peculiarities of teeth, skulls, and feeding habits are presented in the family accounts.

Body Design

Bats are the only mammals capable of true flight (as opposed to gliding). The name Chiroptera is derived from the Greek words *cheir*, meaning hand, and *pteron*, meaning wing because, as illustrated below, the wing is a modified hand. The hand bones serve as a supporting frame for membranes made up of a thin layer of nerves, muscle fibers, and blood vessels enclosed between two layers of skin. The membranes are extremely elastic, contracting rather than folding when the wing is closed.

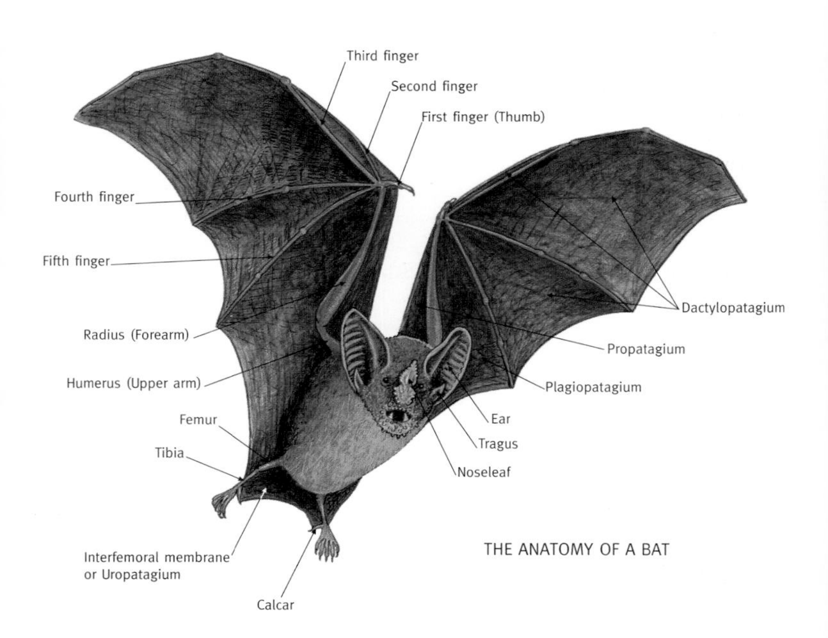

THE ANATOMY OF A BAT

As in birds, the shape of the wing varies with the flight and feeding habits of the bat. Relatively short, broad wings, such as those of tailless fruit bats (pl. 11-12, p. 109), allow for a slow but maneuverable flight that can be useful for catching slow-moving or stationary objects in a cluttered forest environment. Long, thin wings, such as those of free-tailed bats (pl. 16, p. 121), allow for less maneuverable but fast flight, appropriate for bats that hunt in open areas.

The leg design of bats is also unique among mammals. Bats' femurs are rotated 180°, such that the knees and feet face in the opposite direction to those of other mammals. This design is appropriate for several reasons. It helps bats hang upside down on vertical surfaces; it is important in flight, for the legs anchor and control a tail membrane, or uropatagium, that is used much like a rudder; likewise, the legs can be used to arch the tail membrane beneath the body, forming a scoop that is used by some bats to bag airborne food, or even their own babies—some females turn the right way up in their perch just before delivery and use the membrane to catch their newborn pup.

A potential problem with backwards-facing feet is that they leave a bat that is hanging from a vertical surface face-to-face with the wall. Microchiropterans have solved this problem, too, however, by evolving highly modified neck vertebrae that enable them to arch their heads straight backward.

Like their "hands", bats' feet have five digits. Bats' toes are specialized for a hanging lifestyle. The toes of all bats except thumbless and disk-winged bats (pl. 14, p. 117) and the white-winged vampire (pl. 13, p. 113) are equipped with locking tendons that let the bats hang without expending energy, and prevent them from falling when they are asleep.

Echolocation and Hearing

Bats have remarkable hearing. By bouncing loud, high-frequency sounds (which are mostly inaudible to us) off their surroundings, many bats are able to form an aural image of what is around them in the dark. This process, known as echolocation, is used by all microchiropterans to navigate and feed, although some, such as insectivorous bats, rely on it much more heavily than others. Microchiropterans emit sounds through the mouth or, in the case of most leaf-nosed bats and a few others, through the nose, and pick up the echoes with their large, sensitive ears. The timing and pattern of the echoes enables the bats to determine the location and shape of surrounding objects. The most advanced echolocators can detect strands as thin as spider webs. Among megachiropterans, on the other hand, only a handful of species are able to echolocate. Those few differ from microchiropterans in that they produce the sounds using the tongue rather than the larynx, and they use them only for navigation.

Echolocating bats face an interesting dilemma. Sound diminishes very quickly once it is emitted, so a call has to be incredibly intense in order to go out, bounce off sometimes tiny objects, and return with enough clarity to convey precise information to a bat. Even though calls are generally inaudible to us, according to one calculation some bats produce sounds that, at close range, have about the same energetic intensity as the noise of a jet plane. The problem for bats is that the incredibly sensitive hearing apparatus they need to pick up incoming echoes would be ravaged by such deafening outgoing calls. As a solution, bats have evolved specialized ear muscles. The muscles contract as calls are emitted, forming a kind of muffler around the ear bones, and then reopen to receive the echoes, all within a fraction of a second.

Threatened by such sophisticated weaponry, bats' insect prey have had no choice but to evolve similarly sophisticated defenses. Many moths, lacewings, crickets, katydids, and some mantids have evolved "bat radar," in the form of sound-sensing organs. Most moths, lacewings, and mantids are unable to produce sound, and may therefore use their "ears" solely in defense, while female crickets and katydids also use their ears to find their noisy males. Moth ears are located on either side of the abdomen or thorax, or, in the case of some hawkmoths (Sphingidae), on either side of the mouth. Lacewings hear through their wing veins, crickets and katydids hear through their forelegs, and mantids have single hearing organs on their bellies. All but the mantids hear in stereo, and can thus gauge not only an incoming bat's proximity, but also its direction. Moths and lacewings fly in the opposite direction as soon as they detect a bat; if the bat closes in, they fold their wings and try to nose-dive out of the way.

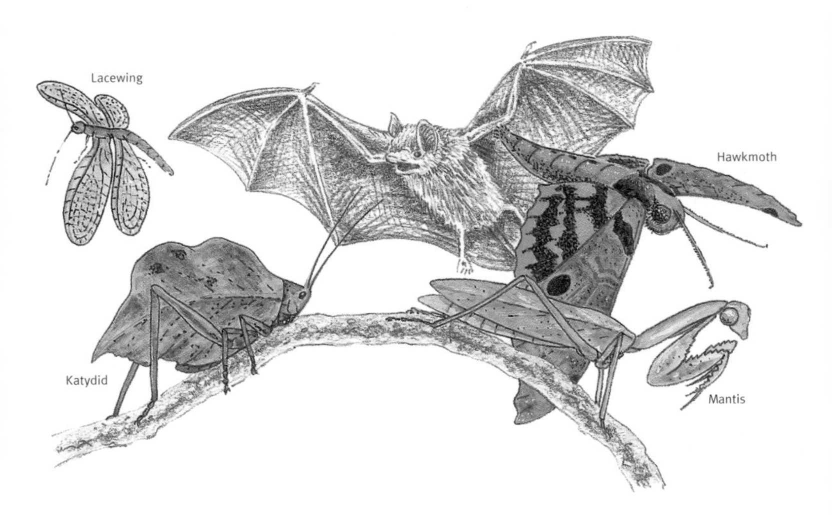

Insectivorous bats have influenced the evolution of many groups of insects. Some insects are able to hear bat wingbeats or echolocation calls and take evasive action. The lacewing hears through its wing veins, the katydid through its knees, the mantis through its belly, and the hawkmoth through organs on either side of its mouth.

A few moths take this arms race to a still higher level. Tiger moths (members of the Arctiidae family) are unpalatable because they acquire noxious chemicals from plants. During the daytime, they are able to warn or remind predators that they are distasteful with bright colors—wasplike patterns, for example—that predators, hopefully, have learned to associate with bad experiences and tend to avoid. Since colors don't work at night, however, these moths have evolved sound-producing organs—a feature that is highly unusual among moths—that emit warning clicks. The moths click whenever they sense a bat, or when they are grabbed by any other animal. A number of studio experiments have shown that bats will veer away from calling tiger moths. If the moths' sound-producing organs are tampered with, the bats catch them, but then spit them out. There is some evidence that moth calls may repel bats in other ways, such as by interfering with echoes and disorienting the bats, or simply by startling the bats. However, the first theory is weakened by the fact that bats avoid clicking moths regardless of whether the bats are homing in using echolocation or vision, and the second by the fact that bats in studios quickly learn to ignore moth clicking sounds. Moreover, only bad-tasting moths are known to produce sounds. All the same, the three ideas are not mutually exclusive.

Tiger moths have evolved not only hearing organs with which to detect bats but also sound-producing organs with which to remind bats that they are distasteful. (When molested, many tiger moths also secrete droplets of a noxious liquid, visible in this illustration, through a pair of glands just behind the head.) Defense against bats may be the sole function of tiger moths' hearing and sound-producing organs.

Bats may use ultrasound not only in echolocation but also in communication. The pup of a Seba's short-tailed fruit bat (pl. 11, p. 106) that was collected by a biologist from its temporary nocturnal roost, for example, was relocated by its mother, some distance away, while still in a cloth collecting bag. Lower frequency sounds that are within our own hearing range are important to bats, too. The Jamaican fruit bat (*Artibeus jamaicensis*), for example, emits raucous yipping sounds when captured. The call attracts other bats, whose swirlings might distract a predator and enable the captive to escape.

Jamaican fruit bat (*Artibeus jamaicensis*) in the talons of a black and white owl. Jamaican fruit bats emit audible (not ultrasonic) calls when captured. The calls attract other bats, whose swirlings might distract a predator and enable the captive to escape.

Some species also use lower frequency sounds to find prey. The frog-eating bat (pl. 9, p. 100) locates and even identifies frogs by listening to their calls, while part of the brain (the colliculus) of the common vampire bat (pl. 13, p. 113) is thought to be specialized for detecting the steady breathing of sleeping mammals.

Other Senses

Many microchiropterans have such sophisticated hearing that vision plays only a secondary role. Many species have tiny eyes. No microchiropterans are known to be able to see color. And, unlike most nocturnal mammals (and birds), none are known to possess a tapetum lucidum—a reflective layer behind the retina that improves night vision (although a few stenodermatine bats appear to have this feature, since their eyes shine when illuminated). The phrase *blind as a bat* is without foundation, however, and, in fact, some bats—such as members of the large leaf-nosed bat family—have excellent vision.

Some species can recognize patterns, and perhaps thus locate stationary targets such as fruits, flowers, or nonflying insects. Vision is probably also important for long-distance navigation, since echolocation is only effective at close range. Megachiropterans, most of which lack the ability to echolocate, have excellent vision. They have large eyes, can see color, and possess a tapetum lucidum.

Smell is also important to bats. Some nectivorous bats seem to discern flowers containing nectar, and some frugivorous bats locate ripe fruits by odor. In one experiment, researchers found Seba's short-tailed fruit bats (pl. 11, p. 106) would rip into sealed bags containing ripe fruit while consistently ignoring bags of unripe fruit hung alongside. Other bats, such as the false vampire (pl. 9, p. 100) and common vampire (pl. 13, p. 113), are thought to use smell to find prey, too. Scent glands around bats' faces and elsewhere can play an important role in courtship, breeding, and other social interactions. The white-winged vampire (pl. 13, p. 113), for example, has a pair of cup-shaped glands in its mouth that it directs at antagonists when disturbed. The glands produce a foul odor that my deter rival vampires or predators.

Roosting

Bats roost in places that afford them protection from the elements and from predators such as snakes, carnivorous mammals, and birds of prey. Bats of the same or different species often share roost sites. In such cases, different species of bats are typically segregated, and families or other social groups within species may also separate into distinct groups. Some species form tight clusters, a behavior that helps them maintain higher and more constant body temperatures. (Since bats are so small, they lose heat relatively quickly, so this device can be important to them.)

One of many bat predators, the lyre snake (*Trimorphodon biscutatus*) is common in northwestern Costa Rica.

Caves and buildings are perhaps the best known bat roosting sites, but since both tend to be scarce in pristine Costa Rican forests, many species roost in trees. Some bats roost inside hollow trunks, while others cling to trees' exteriors, between buttress roots, on the undersides of large limbs, or amidst canopy foliage. One group of bats even constructs shelters by modifying the shape of large leaves (see pp. 110-112).

Reproduction

Compared to other mammals their size, such as rodents, bats have a slow reproductive turnover. Long pregnancies of three to seven months result in small litters; the vast majority of species produce a single pup per litter. While the young of most mammals are weaned by the time they are about 40% adult size, bats nurse their young until they are about 90% adult size.

Temperate-zone bats typically produce just one litter a year in the summer, for with the onset of winter they must either undertake long migrations or go into hibernation. Many species mate in autumn before entering hibernation and use various strategies to delay birth until the beginning of the following summer. Costa Rican bats may enter a state of torpor resembling hibernation for brief periods as a response to miserable weather, and some species appear to make seasonal altitudinal migrations. In Monteverde, for example, populations of several species of fruit- and nectar-eating bats—like those of many birds and butterflies—fluctuate seasonally, and the fluctuations seem to correspond to the availability of fruits and flowers. But Costa Rican bats are essentially free of the pressures of lengthy hibernation or latitudinal migration that temperate zone bats face, and, partly as a result, they have been able to evolve more varied reproductive strategies.

Many insectivorous bats in Costa Rica produce a single litter at the beginning of the rainy season, although a few produce two or, rarely, as in the case of the black myotis (pl. 15, p. 119), three litters each year. Many fruit-eating species also reproduce more than once a year, often producing two back-to-back litters in the late dry and early wet seasons, when fruit tends to be most abundant, and holding off during the rest of the year. As in temperate zones, some species prolong the duration of their pregnancies so that young will be born at a favorable time of year. The Jamaican fruit bat (illustrated on p. 87), for example, becomes pregnant immediately after giving birth to a litter in July or August, but delays embryo development until the beginning of the following dry season, eventually giving birth in March or April. It then has time for a normal pregnancy that results in the July-August litter before starting the cycle again.

Regional weather patterns, and their effect on food abundance, strongly influence the breeding cycles of most species. Bat experts Richard LaVal and Henry Fitch compared bat communities in three very different Costa Rican habitats. They found that, in general, bats in Guanacaste, where there is a marked and severe dry season, had the shortest breeding period; bats in Monteverde, where the dry season is less pronounced, had a longer breeding period; and those in the Atlantic lowlands, where it is wet year round, had the most extended breeding period.

Bats give birth while hanging in their roosts. Young are born tail first. Once they are partially emerged, they grab their mother's fur with their hindfeet and help pull themselves out of the womb. Since newborn bats frequently weigh 25% of the mother's bodyweight, and occasionally as much as 40%, the mother needs all the help she can get. As an additional aid, the ligaments that hold together the pelvic bones of female bats are particularly elastic, and they stretch to accommodate the passage of the large baby. Although bats have an unusually slow reproductive turnover, they also have unusually long lifespans compared to other mammals their size. They frequently live more than 10 years, and have been known to surpass 30.

Conservation

Bats are often depicted in the vicinity of graveyards, and in many cultures they have a mythologically embedded reputation as harbingers of evil spirits, or, worse, as blood-sucking killers. Although imaginative early taxonomists gave a plethora of bats scientific names beginning with the letters *Vamp*, after the demon of Slavic legend, only three of the world's 925 or so bat species feed on blood. Only one of these hematophagous species, the common vampire, takes blood from humans, and it does so only rarely. Since this bat feeds predominantly on livestock, sometimes decimating herds by spreading rabies, it is undoubtedly a pest in agricultural areas. The same could not be said, however, of most of the rest of the world's bats.

Insect-eaters play a critical role in natural ecosystems by controlling insect populations, and they do us a service by including in their diets disease-bearing mosquitoes and species of weevils and other beetles that are prevalent agricultural pests. Other bats pollinate or disperse hundreds of species of Costa Rican plants, including many that are of economic importance. Bats help seeds not only by dispersing them: the passage of a seed through a bat's gut can increase the seed's chances of germinating. In a laboratory study at La Selva, for example, researcher Jorge López found that wild black pepper (*Piper*) seeds defecated by *Carollia* bats and fig (*Ficus*) seeds defecated by *Artibeus* bats germinated much more frequently than seeds that had not passed through bats.

Fruit-eating bats play an especially important role in forest regeneration. While birds tend to drop seeds while perched in areas that already have trees, bats often defecate over open areas. Consequently, they play a crucial role in dispersing pioneer plants (the plants that first colonize open areas), such as black peppers (*Piper* spp.), wild tomatoes (various members of the Solanaceae), or *Cecropia*, into natural treefall gaps and abandoned pastures. Those plants in turn provide shelter and food for other animals, and those animals deposit a whole new variety of seeds. Thus, the regrowing area can start to regain diversity. Fruit-eating bats participate in latter stages of reforestation, too. López found that almost a quarter of the seeds defecated by fruit-eating bats he caught in regrowing areas came from primary forest plants.

All around the world, bat populations are in decline. The main cause is deforestation, which has destroyed many of the richest bat habitats and fragmented much of the remaining forest. Forest fragmentation may hinder bat movements, and it increases bats' exposure to other problems, such as direct persecution by people.

More subtle threats are also at play. Pesticides used on the insects that, ironically, bats help control, become more concentrated as they work their way up the food chain. Such poisons have been implicated in many bat declines, including some of the vast losses seen at the largest bat roosts in the world (see pp. 122-123). Pollution may also affect bats (and other wildlife) by contributing to global warming. At Monteverde, several species of bats that, during decades of previous censusing, had only been recorded from lower (and warmer) elevations are now appearing higher up the mountain. Some researchers believe that these changes, which have also been noticed in birds and other animals, are the result of global warming. Threats such as these are particularly worrying, for they can affect even apparently pristine, protected environments and the bats that inhabit them.

In Costa Rica, these problems have taken an unmeasured toll on one of the most fascinating and diverse bat assemblages in the world. Of course, a few species always benefit from human disturbances. It is ironic that the replacement of tropical forests with cattle pastures has probably benefited the common vampire, the one bat we have any (very small) reason to fear, more than any other species.

References

Bennett et al. 1988; Blest 1964; Charles-Dominique 1986; Crerar & Fenton 1984; Dinerstein 2000; Dunning 1968; Fenton 1992; Fenton & Griffin 1997; Fleming 1971a; Fleming 1988; Fullard et al. 1979; Goodwin & Greenhall 1961; Greenhall 1988; Handley et al. 1991; Janzen & Wilson 1983; Kalko et al. 1996; Kunz 1987; Kunz et al. 1994; Kunz & Pierson 1994; LaVal 1977; LaVal & Fitch 1977; López 1996; Macdonald 1995; McDonnell & Stiles 1983; Miller 1975, 1991; Moiseff & Hoy 1983; Morrison 1983; Norberg & Rayner 1987; Novacek 1985; Nowak 1994; O'Farrell & Gannon 1999; Pettigrew 1986; Pettigrew et al. 1989; Pounds et al. 1999; Quinn & Baumel 1993; Reid 1997; Rodríguez (B.) & Wilson 1999; Rodríguez (J.) & Chinchilla 1996; Roeder & Treat 1961; Schutt 1993; Simmons 1994; Suriykke & Miller 1985; Timm 1984, 1987; Timm & LaVal 1998, 2000b; Timm et al. 1999; Turner 1975; Wilson 1997; Wolff 1981; Yager & Hoy 1986.

SPECIES ACCOUNTS

O: Chiroptera F: Emballonuridae

SAC-WINGED or SHEATH-TAILED BATS

(plate 7)

Names: The long-nosed or proboscis bat (*Rhynchonycteris naso*), the greater white-lined bat (*Saccopteryx bilineata*), the gray sac-winged bat (*Balantiopteryx plicata*), and the northern ghost bat (*Diclidurus albus*). Members of this family are known as sac-winged bats because some species—such as *S. bilineata* and *B. plicata*—have pouch-shaped scent glands located just in front of the forearm (hence the genus *Saccopteryx*, from the Greek for sack and wing). The glands secrete a smelly red substance used in territorial and breeding displays. The northern ghost bat has a two-chambered gland on its tail that probably serves a similar function (hence the genus *Diclidurus*, from the Greek for two-valved and tail). Emballonurids are also known as sheath-tailed bats because the end of their tail emerges from a hole in the top of the tail membrane. The family's scientific name, which refers to the same feature, is derived from the Greek words *oura*, meaning tail, and *emballo*, meaning slipped through.

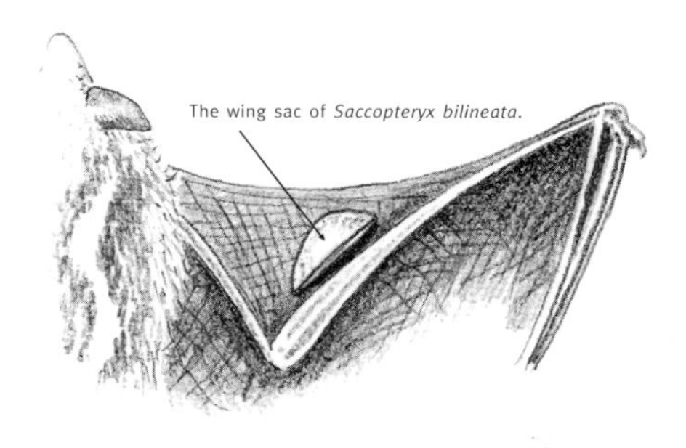
The wing sac of *Saccopteryx bilineata*.

Range: There are about 48 species of emballonurids, distributed through both the New and Old World tropics. Ten species inhabit Costa Rica. The long-nosed, greater white-lined, and ghost bats are found from Mexico to Brazil, the long-nosed bat to about 300 m (1,000 ft), the greater white-lined bat to about 500 m (1650 ft), and the ghost bat to about 1,500 m (5000 ft). The ghost bat has never been recorded in northwestern Costa Rica, although it may occur there. The gray sac-winged bat is found from Mexico to northwestern Costa Rica, from sea level to about 1,500 m (5,000 ft).

Size: *R. naso*: 4 cm, 4 g (1 ½ in., 1/10 oz); *S. bilineata*: 5 cm, 7 g (2 in., ¼ oz); *B. plicata*: 5 cm, 6 g (2 in., 1/5 oz); *D. albus*: 7 cm, 20 g (3 in., ¾ oz); females usually larger than males in all four species.

Similar species: The greater white-lined bat is easily confused with the lesser white-lined bat (*S. leptura*), a slightly smaller, brown rather than blackish species that roosts in small colonies (typically of less than nine bats) in more exposed sites. The long-nosed bat is the only other bat in Costa Rica with white lines on its back. The gray sac-winged bat is the only gray bat with sacs and without a noseleaf. The ghost bat is larger than the only other white bat in Costa Rica, the white tent bat (pl. 12, p. 110), and lacks a noseleaf.

Natural history: The long-nosed bat is one of the most frequently observed bats in Costa Rica because it roosts in relatively conspicuous colonies alongside slow-moving waterways, often on large tree trunks, but also amidst over-hanging roots, on rocks or cliffs, under bridges, or under the curled dead leaves of banana and heliconia plants. On some substrates, the bats look like patches of lichen. Colonies usually consist of five to 15 bats, but occasionally as many as 45. Each group has three to six roosting sites. Within each colony, there are about as many males as females, but one male is dominant. At night, *R. naso* catches insects over water, typically foraging within 3 m (10 ft) of the surface. Breeding females and young use the richest, most central part of the foraging area, other bats find what they can elsewhere, and the dominant male generally stays around the edge, chasing away any intruders from neighboring colonies.

The greater white-lined bat roosts in tree hollows or well-lit caves, on large buttress roots, or on the walls of buildings in harem groups typically consisting of a single male and one to eight females. Occasionally it forms colonies containing several harems and as many as 50 bats. The species emerges just before dusk to forage for insects in the forest understory, mostly in relatively open spots about 3 to 8 m (10 to 26 ft) above the ground. Each harem has a feeding territory, and each female a territory within that area. Insect abundance can be patchy and seasonal, so the bats must move between different feeding patches over the course of a year. Each colony uses an area of about 6 to 18 ha (15 to 45 A), depending on habitat and season. When the females return to their roost at dawn, and periodically throughout the day, the male displays to them by singing elaborately and hovering before them, fanning them with scents from his wing sacs. Both the song and the scents are clearly detectable by humans. Neighboring males within a colony participate in similar displays, called "scent fights", where their territories meet.

The gray sac-winged bat catches tiny insects during slow, steady sallies 15 to 20 m (50 to 65 ft) above the canopy or ground. It roosts in caves, hollow trees, and buildings, often in large colonies; 1,500 to 2,000 of them (arranged in distinct groups of 50 to 200 each) were reported from a single cave in Guanacaste.

The northern ghost bat forages for insects high above the ground in open areas, such as above rivers or around tall light posts, sometimes emitting a musical twittering as it flies. It roosts on the undersides of palm leaves, especially those of coconuts, and occasionally on tree trunks. In Mexico, it roosts alone most of the year, but in small groups consisting of a male and a few females during its January to February breeding season.

Conservation: The long-nosed, greater white-lined, and gray sac-winged bats are common; the northern ghost bat appears to be much rarer, but since it forages high off the ground, well above most mist nets, its abundance is hard to ascertain.

References:

Arroyo-Cabrales & Jones 1988; Bradbury 1983a; Bradbury & Emmons 1974; Bradbury & Vehrencamp 1976a,b, 1977a,b; Ceballos & Medellín 1988; Ceballos & Miranda 1987; Jones 1966; McCarthy 1987b; Plumpton & Jones 1992; Rodríguez (B.) & Wilson 1999; Rodríguez (J.) & Chinchilla 1996; Sánchez & Chávez 1985; Starrett & Casebeer 1968; Timm & LaVal 1998; Yancey et al. 1998.

O: Chiroptera F: Noctilionidae

FISHING or **BULLDOG BATS**

(plate 8)

Names: The greater fishing bat (*Noctilio leporinus*) and the lesser fishing bat (*N. albiventris*). They are also referred to as bulldog bats because they have puffy, folded faces and large canine teeth. These are the only two members of the family Noctilionidae (from the Latin meaning nocturnal ones).

Range: *N. leporinus* ranges from Mexico to Brazil and northern Argentina; from sea level to about 200 m (700 ft); mostly around fresh water and along the coast. *N. albiventris* occurs from Mexico to northern Argentina; from sea level to about 1,100 m (3,600 ft); mostly near fresh water.

Size: *N. leporinus*: 12 cm, 70 g (5 in., 2 ½ oz), with a wingspan of about 55 cm (22 in.); *N. albiventris*: 7 cm, 30 g (3 in., 1 oz), with a wingspan of about 45 cm (18 in.).

Similar species: Many bats swoop to water to drink, and a few skim insects off the surface, but no other Costa Rican bats are known to fish. In fact, only one other New World bat species—*Myotis vivesi*, from the Pacific coast of Mexico—and a small handful of Old World species—*Megaderma spasma*, *Myotis adversus*, and possibly a couple of other *Myotis* species—include fish in their diets. Both *Noctilio* fishing bats tend to forage in small groups. *N. leporinus* flies in a zigzag pattern just above the water surface. When it locates a fish, it dips its feet into the water several times in quick succession. If it is successful, there is a small splash and the bat then rises away from the water. The fur color of *Noctilio* bats varies with location, with the extent of bleaching ammonia fumes (from decomposing urine and guano) at roosts, with the age of the bat, and, in some areas, with the gender of the bat. Both species can be browner or grayer than those depicted. Both species have a pale stripe on the back, although in some individuals it is barely visible.

Natural history: The greater fishing bat hunts mostly over calm freshwater and along the coast. Although it doesn't normally start foraging until dusk, it has been seen accompanying pelicans in the late afternoon, perhaps to catch fish they disturb. It has a number of specializations for preying on fish. It uses echolocation to detect ripples made by fish swimming near the surface (rather than trawling randomly, as was once thought). Having tracked the course and speed of a fish, it rakes through the fish's predicted path with its large, clawed hindfeet. The long, hooked claws, flattened laterally to reduce resistance as they slide through the water, impale prey. Short, water-repellent fur prevents the bat from becoming water-logged. The bat passes the unfortunate fish to its mouth. Fish may be eaten immediately, while the bat is in flight, or stored in the bat's modified cheek muscles, which serve as pouches. The pouches help the bat carry prey to temporary feeding perches and perhaps also to bring solid food to its young. (If a fish were left in the mouth, it would block echolocation pulses and hinder navigation.) An individual may enjoy some 30 to 40 fish of about 2 to 8 cm in length during a night, but the species is by no means exclusively piscivorous. It also eats insects, shrimp, crabs, and frogs, sometimes foraging away from water. Flying insects, which it scoops out of the air with its wings and tail membrane, can constitute more than half the diet at certain times of year.

The greater fishing bat roosts in colonies of up to several hundred in dark caves, rock fissures, hollow trees, and, occasionally, buildings. Colonies have been found to consist either exclusively of males—groups of bachelors who see little reproductive action—or harems consisting of a single adult male and several females and their young. A captive greater fishing bat lived 11 ½ years.

The lesser fishing bat also echolocates and then skims prey off the surface of water, but it feeds mainly on insects, rather than fish, and often catches them in the air. Like its larger cousin, it has short, dense, water-repellent fur and cheek muscles that are modified into pouches, but it lacks the huge, rakelike feet. Water beetles, especially predaceous diving beetles (Dytiscidae), make up much of the diet; they are supplemented with other beetles, moths, flies, crickets, and, occasionally, fish and fruit.

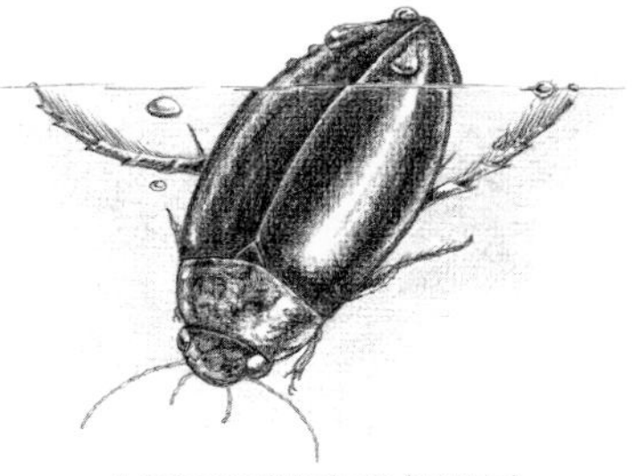
A predaceous diving beetle (Dysticidae).

The behavior of the lesser fishing bat may illustrate how fishing evolved in bats. The most primitive bats were insectivorous. Many of them, like numerous insectivorous bats today, probably foraged over water, where there are few obstructions to maneuver around and where insects, many of the larvae of which are aquatic, can be abundant. It is easy to imagine how such bats could learn to scoop insects from the surface of water, as does *N. albiventris*, and how some of these could then use their skills to catch surface-dwelling fish and evolve further specializations for this purpose, as has *N. leporinus*.

Roosts of lesser fishing bats have been found in hollow trees, foliage, and buildings. Roosts of both species (as well as the bats themselves) give off a strikingly pungent odor.

Sounds: The greater fishing bat often emits audible chirps while foraging; on still nights, a distinctive hissing sound made by the claws skimming through water can also be heard.

Conservation: Either species can be common in the right habitat, such as on the canals of Tortuguero and on the Puerto Viejo River at La Selva. They are among the few bats that can be readily identified and observed while they forage, especially at water illuminated by dock lights.

References

Altenbach 1989; Bloedel 1955a; Brandon 1983; Brooke 1994; Brown et al. 1983; Fleming et al. 1972; Hood & Jones 1984; Hood & Pitocelli 1983; Hooper & Brown 1968; Howell & Burch 1974; Jones 1982; Jones et al. 1988; Kalko et al. 1998; LaVal & Fitch 1977; Murray & Strickler 1975; Nowak 1994; Reid 1997; Rodríguez (B.) & Wilson 1999; Rodríguez (J.) & Chinchilla 1996; Schnitzler et al. 1994; Suthers & Fattu 1973; Timm & LaVal 1998; Timm et al. 1989; Wenstrup & Suthers 1984; Whitaker & Findley 1980.

O: Chiroptera F: Mormoopidae

LEAF-CHINNED or MUSTACHED BATS

(plate 8)

Names: Parnell's mustached bat (*Pteronotus parnellii*) and Davy's naked-backed bat (*P. davyi*). Mormoopids are known collectively either as leaf-chinned bats because they have folds of skin below the bottom lip, or as mustached bats because they have a mat of thick bristles above the upper lip. Two species are known as naked-backed bats because their wings attach along the middle of the back, covering the fur beneath. Mormoopids were formerly considered members of the leaf-nosed bat family (pl. 9-13, pp. 99-116) and referred to as the Chilonyterinae. The current family name is no compliment; it stems from the Greek words *ops*, meaning face, and *mormo*, meaning a profoundly frightening object.

Range: There are eight species of Mormoopids, distributed from the southern United States to Peru and Brazil. Four species have been recorded in Costa Rica. *P. parnellii* and *P. davyi* both range from Mexico to Brazil. *P. parnelli* has been found as high as 3,000 m (10,000 ft), and *P. davyi* as high as 2,300 m (7,500 ft), but both species are most common beneath about 500 m (1,650 ft). *P. parnelli* occurs almost countrywide; *P. davyi* seems to prefer the seasonally dry forests of northwestern Costa Rica, but it has also been captured at La Selva and in the Osa Peninsula.

Size: *P. parnellii*: 7 cm, 18 g (3 in., ⅔ oz); *P. davyi*: 5 cm, 8 g (2 in., ¼ oz).

Similar species: Mormoopids can be distinguished by their flared lips; tiny eyes; long, thin bodies and wings; and broad tail membranes. *P. parnelli* is distinguished from the only other species with a similar face shape, the lesser mustached bat (*P. personatus*), by its larger size. Davy's naked-backed bat is distinguished from the only other naked-back, the big naked-backed bat (*P. gymnonotus*), by its slightly smaller size.

Natural history: In contrast to the phyllostomids (see p. 99), mormoopids emit their echolocation sounds through the mouth rather than through the nose. They have greatly enlarged lips which form a megaphone shape and help amplify echolocation sounds. This feature takes the place of the leafnose found in phyllostomid bats (literally also—mormoopid nostrils are situated in the upper lip!). The mouth may also act as an insect scoop, perhaps aided by the stiff mustache hairs, whose function might be to direct airflow toward the mouth.

The head of *Mormoops megalophylla*—a species that has never been recorded in Costa Rica, although it is found both north and south of the country. The lips of mormoopids are flared to form a megaphone shape that aids in echolocation. Their nostrils are located in the upper lip.

Mormoopids have long, narrow wings that allow them to fly fast and for prolonged periods. Parnell's mustached bat, which has been clocked flying at about 18 km/h (11 mph), may be away from its roost for up to seven hours in a night of foraging. Leaf-chinned bats studied in Mexico were found to travel 3.5 km (2 miles) or more each night just to reach their feeding grounds.

Mormoopids are almost exclusively insectivorous. Parnell's mustached bat forages amidst thick vegetation beneath the canopy, often using trails as flyways. It has a particularly sophisticated echolocation system to help it navigate and find prey in this cluttered environment; it is possibly the only New World bat capable of distinguishing fluttering targets. It is fond of scarab beetles but, in captivity at least, avoids bad-tasting insects such as tiger moths and stink bugs. Davy's naked-backed bat often forages over water. Its diet includes moths, earwigs, and flies.

Both species form large colonies that roost in large, humid caves and mines. A colony comprised of four species of mormoopid bats in a cave system in Mexico was estimated to contain 400,000 to 800,000 bats. Typically, mormoopids roost in the darkest portion of caves, well away from the entrance.

Conservation: Both species can be common where appropriate roosting sites are available. Mormoopids can have a big impact on insect populations in some areas. The aforementioned mormoopid colony in Mexico was estimated to consume some 1,900 to 3,800 kg (4,200 to 8,400 lbs) of insects each night.

References

Adams 1989; Bateman & Vaughan 1974; Eisenberg 1989; Fenton 1992; Handley 1976; Herd 1983; LaVal & Fitch 1977; Reid 1997; Rodríguez (B.) & Wilson 1999; Rodríguez (J.) & Chinchilla 1996; Smith 1972; Starrett & Casebeer 1968; Timm & LaVal 1998; Timm et al. 1989; Vaughan & Bateman 1970; Whitaker & Findley 1980.

O: Chiroptera F: Phyllostomidae

AMERICAN LEAF-NOSED BATS

With about 140 species, the Phyllostomidae is the largest bat family in the New World. American leaf-nosed bats are found from the southwestern United States to northern Argentina. At least 62 species occur in Costa Rica, more than half the country's bat fauna.

The family name comes from the Greek words *phyllon*, meaning leaf, and *stoma*, meaning opening, since most phyllostomids possess noseleaves. Noseleaves consist of a U-shaped flap of skin beneath the nostrils and a leaf-shaped flap of skin above the nostrils. While other bats emit echolocation calls through the mouth, phyllostomids emit them through the nostrils, and are thought to use their noseleaves to help direct the sounds. The only members of the family that lack noseleaves are the three true vampire species (pl. 13, p. 113) and the wrinkle-faced bat (pl 12, p. 109). In the New World, the only non-phyllostomids that also have noseleaves are a handful of Vespertilionids (the so-called plain-nosed bats, pl. 15, p. 119).

The phyllostomidae is divided into five subfamilies. Due to the great size and diversity of the phyllostomidae, each subfamily is described individually in the following pages. (Note that, if the group name ends in *-inae*, it is a subfamily, rather than a family, which is denoted by the ending *-idae*.)

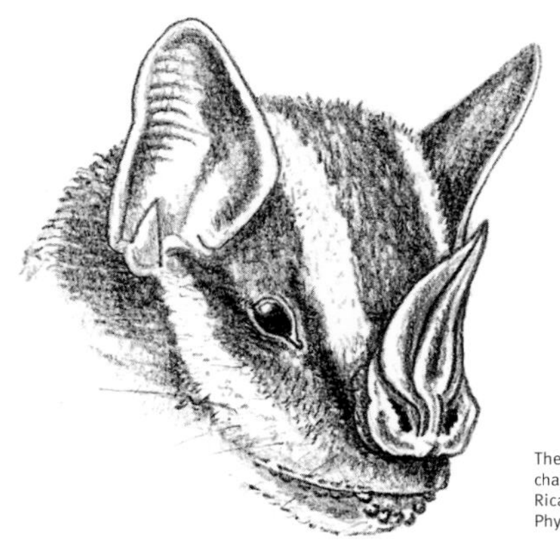

The presence of a noseleaf characterizes most members of Costa Rica's largest bat family, the Phyllostomidae.

O: Chiroptera F: Phyllostomidae (S: Phyllostominae)

SPEAR-NOSED BATS

(plate 9)

Names: Tome's long-eared bat (*Lonchorhina aurita*), the frog-eating or fringe-lipped bat (*Trachops cirrhosus*), and the false vampire or spectral vampire bat (*Vampyrum spectrum*). The family name comes from the Greek words *phyllon*, meaning leaf, and *stoma*, meaning mouth; all phyllostomines, like most other members of the Phyllostomidae family, possess noseleaves.

Range: There are about 30 species of phyllostomines, distributed through the New World tropics. Twenty-one species have been recorded in Costa Rica. All three species illustrated here range from Mexico to Brazil. *L. aurita* is found from sea level to about 1,500 m (5,000 ft); *T. cirrhosus* to about 1,400 m (4,600 ft); and *V. spectrum* to about 1,650 m (5,400 ft).

Size: *L. aurita*: 6 cm, 12 g (2 ½ in., ½ oz); *T. cirrhosus*: 9 cm, 30 g (3 ½ in., 1 oz); *V. spectrum*: 15 cm, 200 g (6 in., 7 oz)—the New World's largest bat.

Similar species: No other phyllostomid has such a proportionately long noseleaf as *Lonchorhina*. The wartlike protrusions around the mouth of *Trachops* are unique. There are two other large carnivorous bats—the woolly false vampire (*Chrotopterus auritus*) and the greater spear-nosed bat (*Phyllostomus hastatus*)—that are of a similar shape to *Vampyrum*, but they are less than half the size.

Natural history: Phyllostomines have proportionately short, broad wings that allow them to fly slowly but with great maneuverability through forest interiors and pluck food off vegetation with their mouth.

Most of the smaller species, such as Tome's long-eared bat, feed on insects, and some also take small vertebrates and fruit. The long-eared bat's superb noseleaf may reflect superior echolocating abilities. The species has been observed slowing down in front of mist nests and either landing on them or turning around and flying away, while other bats fly straight into the same nets. It roosts in caves and tunnels, sometimes singly and inconspicuously amidst large colonies of other bats, sometimes in groups. The largest recorded group, found in the back of a tunnel in a dense Panamanian forest, contained at least 500 individuals.

Larger bats in this subfamily tend to be more carnivorous. The frog-eating bat feeds on insects, lizards, and, especially, anurans. It is the only frog specialist bat in the New World, although there are several others in the Old World.

Mammalogist Merlin Tuttle and herpetologist Michael Ryan, who teamed up to study the frog-eating bat on Barro Colorado Island in Panama, made a series of fascinating discoveries about how the bat finds and distinguishes frogs and how its hunting technique affects the behavior of the frogs themselves. They found that *Trachops* is much more sensitive to low-pitched sounds than other bats, and the low frequency it hears best is that at which most frogs call the loudest. *Trachops* uses this sensitivity to home in on calling male frogs (female frogs are generally silent). Its dependence on calls is so great that individuals in Panama would often pass within inches of silent frogs without noticing them, yet the bats would land on speakers emitting frog-call playbacks and even try to pry away the speaker covering!

Tuttle and Ryan also found that *Trachops* can distinguish calls, and thus avoid frogs that are either poisonous or too large. Bats would consistently ignore speakers emitting the calls of inedible frogs. Curiously, the bats also correctly distinguished between recordings of edible frogs and inedible toads from other areas—species that the bats could not possibly have heard before. This suggests that the calls of inedible species have a certain subtle characteristic in common that the bats learn or are genetically programmed to avoid.

Trachops creates a Darwinian quandary for edible male frogs. The more conspicuously males sing, the more likely they are to attract females and thus maximize reproductive success; but they are also more likely to attract a *Trachops*, which clearly would be detrimental to their reproductive success. Frogs deal with this problem in different ways.

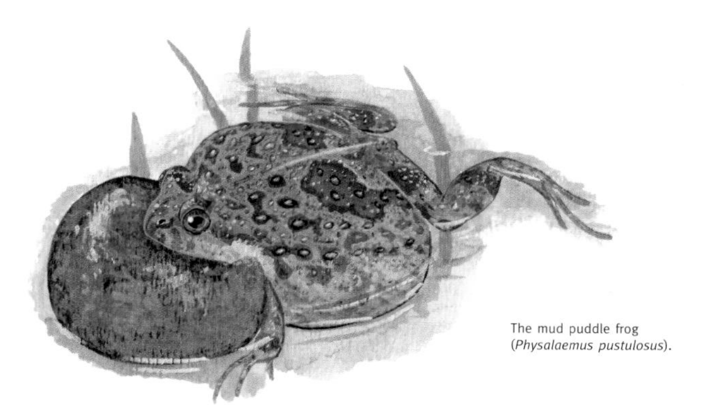

The mud puddle frog (*Physalaemus pustulosus*).

Years before Tuttle and Ryan set up their study, herpetologist Stanley Rand had noticed that mud puddle frogs (*Physalaemus pustulosus*) use their full call (a whine followed by a few chucks) only when surrounded by lots of other males. When the frogs are alone, they switch to a simpler sound (just the whine) and call more quietly and less frequently. Since less complex, less frequent, and quieter calls are less attractive to females,

Rand had reasoned that something other than female preference was influencing the frogs singing behavior, and proposed that a call-oriented predator might be that factor. In fact, Tuttle and Ryan found two such predators, for their speakers were besieged not only by *Trachops*, but also by common gray four-eyed opossums (see p. 41). Presumably, mud puddle frogs use their full calls when in the company of other males, either because they are obliged to compete for females, or because they gain safety in numbers, or both.

Another edible frog species, the pug-nosed smilisca (*Smilisca sila*), uses different strategies. It does not gather in large aggregations like the mud puddle frog, but when two or more males are in the same vicinity, they call in perfect synchrony, such that it is hard for a predator to pinpoint any one of them.

The pug-nosed smilisca (*Smilisca sila*).

Tuttle and Ryan also noticed that edible frogs would often fall quiet as soon as a *Trachops* flew by, while inedible species would just keep on singing, unconcerned. They determined that the frogs could see the bats in the air on all but the darkest nights and react accordingly. Just as the bats were able to distinguish frog calls, so the frogs had some ability to distinguish bats, for they did not hush up when smaller insectivorous bats passed overhead. The pug-nosed smilisca, in addition to calling in synchrony, takes maximum advantage of its ability to spot bats. On moonlit nights, it calls with gusto from exposed perches, diving to cover whenever a frog-eater comes close; on dimly lit nights, it uses simpler and less frequent calls that are harder to locate; and on the darkest nights, when it is unable to detect bats, it remains quiet.

Little is known of the social and reproductive biology of the frog-eating bat. It roosts in small groups in hollow trees, caves, tunnels, and buildings.

The false vampire was once thought to be a true vampire that used its long canines to stab its victims and its leafnose to suck out the blood. Given that it has a wingspan of some 80 cm (almost 3 ft), thank goodness it isn't. Rather, it is a carnivore that feeds on birds, rodents, other bats (it is sometimes attracted by the distress calls of bats caught in mist nets), lizards, and, rarely, insects and fruit. Prey can weigh up to 150 g (5 oz), which is almost the bat's own size. A group of false vampires in Guanacaste was found to prey on at least 18 bird species, including parakeets, doves, trogons, motmots, and cuckoos. *Vampyrum* also takes nestlings. In the Peñas Blancas Valley east of Monteverde, a false vampire was observed alighting on the nest of a chestnut-headed oropendula, breaching the nest wall, and killing and eating most of the breast meat of the well-grown chick inside.

Since several of the prey species, such as anis and motmots, are notably odiferous, *Vampyrum* may use scent cues when hunting. Its skull and teeth resemble those of larger carnivorous mammals; large canines kill and secure prey, knifelike cheek teeth tear flesh, and a pronounced sagittal crest along the top of the skull anchors powerful jaw muscles.

The false vampire roosts in the tops of hollow trees in small groups presumed to be adult pairs and their young. Roosts may be revealed by a characteristic chittering sound made by disturbed false vampires and by piles of bones and feathers on the ground. There appear to be strong social bonds within groups. When the older bats leave to forage, the youngest bat in the group is left with a babysitter; and when the bats reunite after foraging, they greet each other by interlocking mouths, rather like some canids. Prey is sometimes brought back to the tree and shared between roost members.

Sounds: The false vampire periodically emits a loud screech while foraging.

Conservation: None of these three phyllostomines is common. The false vampire, like all top carnivores, is rare. As such, it is considered an indicator species—one that is likely to be among the first to disappear when a habitat is disturbed—and indeed its numbers have declined at La Selva, and probably in numerous other areas, over recent decades. Its roosts should be observed quietly and not disturbed.

References

Bloedel 1955b; Bradbury 1983b; Debbie Derosier & Koki Porras, pers. comm.; Fleming et al. 1972; Goodwin & Greenhall 1961; Greenhall 1968; Howell & Burch 1974; Lassieur & Wilson 1989; LaVal & Fitch 1977; McCarthy 1987a; Navarro & Wilson 1982; Nelson 1965; Peterson & Kirmse 1969; Reid 1997; Rodríguez (B.) & Wilson 1999; Rodríguez (J.) & Chinchilla 1996; Ryan & Tuttle 1984; Ryan et al. 1983; Timm 1994b; Timm & LaVal 1998; Tuttle 1982; Tuttle & Ryan 1981; Vehrencamp et al. 1977.

O: Chiroptera F: Phyllostomidae (S: Glossophaginae)

NECTAR-FEEDING or LONG-TONGUED BATS
(plate 10)

Names: Pallas' long-tongued bat (*Glossophaga soricina*), Geoffroy's tailless bat (*Anoura geoffroyi*), and the orange nectar bat (*Lonchophylla robusta*). *Glossophaginae* is the Greek for "those that feed with their tongues." Some authors place *Lonchophylla* and two other genera of nectar-feeders (*Lionycteris* and *Platalina*) in their own, separate subfamily, the Lonchophyllinae.

Range: Thirty-three or so species of nectar-feeders are found from the southwestern United States to northern Argentina. Ten species have been recorded in Costa Rica. *G. soricina* occurs from Mexico to Argentina, from sea level to about 2,600 m (8,500 ft), mostly in seasonally dry lowland forests. *A. geoffroyi* occurs from Mexico to Brazil, from sea level to at least 2,550 m (8,400 ft). *L. robusta* frequents wet forests from Nicaragua to Ecuador, from sea level to about 1,600 m (5,300 ft).

Size: *G. soricina*: 5 cm, 10 g (2 in., ⅓ oz); *A. geoffroyi*: 7 cm, 15 g (3 in., ½ oz); *L. robusta*: 7 cm, 18 g (3 in., ⅔ oz).

Similar species: *L. robusta* is the only nectar-feeding bat with orange fur. Bats in the genus *Anoura* (with two species in Costa Rica) stand out among nectar-feeders in having very hairy legs and almost no tail membrane. At the species level, however, most nectar-feeders are distinguished from one another only by small differences in tooth and head anatomy, and in the design of the tail, tail membrane, and calcars.

Natural history: Nectar-feeding bats are characterized by their narrow muzzles, small noseleaves and ears, and long, extensible tongues with brushlike tips. A notch in the lower lip and absent or reduced lower incisors provide a channel for the tongue. Short, broad wings facilitate slow, precise flying and hovering. All these features are, of course, particularly appropriate for extracting nectar from flowers, but nectar forms only part of the diet of these bats.

Since nectar lacks protein, an essential component in the diet of any mammal, the bats must also seek other foods. One source of protein, also found in the flowers these bats are so well-adapted for visiting, is pollen. As the bats probe flowers for nectar, pollen sticks to their fur. Some biologists have argued that nectar-feeding bats have evolved a microscopic hair design that makes the fur especially prone to pick up pollen. Subsequent studies, however, have demonstrated that this is unlikely, since many other bats have similar fur structure. Moreover, since pollen from many flowers is sticky, it may adhere to fur regardless of hair structure. In any case, when nectar-feeding bats return to roost, they groom themselves and one another, eating the pollen. Pollen is tough and hard to digest, which may explain why some of these bats drink their own urine; the urine increases acidity in the stomach, helping pollen digestion.

Flowers tend to be seasonal, so many nectar-feeding bats migrate or switch to foods other than nectar or pollen when flowers become scarce. Studies in Guanacaste and on Barro Colorado Island in Panama (both areas with marked seasonal changes in climate) have found that Pallas' long-tongued bats feed on nectar and pollen during the dry season, when flowers are abundant, but switch to a diet that consists almost exclusively of insects or fruit during the wet season. Geoffroy's tailless bat and the orange nectar bat also feed heavily on insects in some areas.

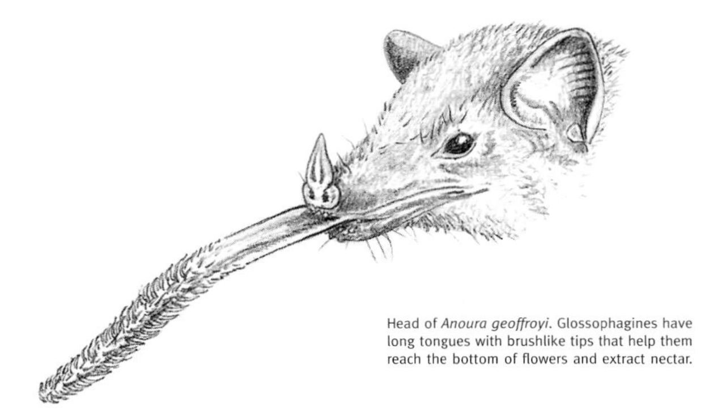

Head of *Anoura geoffroyi*. Glossophagines have long tongues with brushlike tips that help them reach the bottom of flowers and extract nectar.

The diet of each species can vary geographically. Bat expert Richard Laval has noted that, in Monteverde, the feces of Geoffroy's tailless bats never contain insect remains. Diets may also be influenced by competition between species. In Brazil, Pallas' long-tongued bats were found to reduce competition with Goldman's nectar bats (*Lonchophylla mordax*, a species also found in southwestern Costa Rica) by feeding almost exclusively on fruit.

Partly as a result of flower-feeding specializations such as small noseleaves and ears, the echolocating ability of nectar-feeding bats is far inferior to that of insect specialists, and mist-net data suggests they forage little on foggy nights. All three species roost in caves and tunnels; Pallas' long-tongued bat also uses hollow trees and buildings. Nectar-feeding bats can live at least 10 years in captivity.

Sounds: Nectar-feeding bats sometimes emit audible high-pitched chattering calls as they chase rivals away from flowers.

Conservation: Nectar-feeding bats play an important role in the ecology of tropical forests, and propagate many plants of commercial value. Nectar-feeding bats do not eat all the pollen stuck to their fur, and they thus pollinate numerous plants, including kapok trees (*Ceiba pentandra*, illustrated on pl. 10), which are used to make wooden furniture, life preservers, upholstery stuffing, and soap; chicle (*Manilkara zapota*), a major source of latex and timber; balsa (*Ochroma lagopus*), used in rafts, airplanes, and insulation products; gourd trees (*Crescentia alata*, illustrated on p. 293), the seed pods of which are used to make water containers and crafts; and bull's eye vines (*Mucuna* spp., illustrated on pl. 10), whose seeds are used to make jewelry and whose leaves are food for caterpillars of the spectacular morpho butterfly. Some nectar-feeding bat species probably also disperse seeds.

References

Alvarez et al. 1991; Arita & Santos del Prado 1999; Bonaccorso 1979; Fenton 1992; Gardner 1977; Handley 1976; Howell 1974, 1983; Howell & Burch 1974; Howell & Hodgkin 1976; Jiménez 1973; Jones 1982; Richard Laval, pers. comm.; LaVal & Fitch 1977; Lemke 1984; López 1996; Reid 1997; Rodríguez (B.) & Wilson 1999; Rodríguez (J.) & Chinchilla 1996; Thomas et al. 1984; Timm & LaVal 1998; Timm & LaVal 2000a; Timm et al. 1989; Tuttle 1970; Webster 1993; Willig 1986.

O: Chiroptera F: Phyllostomidae (S: Carollinae)

SHORT-TAILED FRUIT BATS

(plate 11)

Names: Seba's short-tailed fruit bat (*Carollia perspicillata*). Genus and subfamily names from the Greek words *kara*, meaning head, and *ollos*, meaning different, although neither the head nor any other anatomical feature of these bats is astonishingly unique.

Range: The subfamily contains seven species, distributed from Mexico to northern Argentina. Four species inhabit Costa Rica. *C. perspicillata* occurs from Mexico to Paraguay; from sea level to about 1,000 m (3,300 ft); mostly in disturbed habitats.

Size: 6 cm, 20 g (2 ½ in., ¾ oz).

Similar species: *Carollia* bats look similar to *Glossophaga* nectar bats (pl.10, p.104), but they have shorter snouts, lack a groove in the center of the lower lip, and have a distinctive U-shaped row of warts around a larger central wart on the chin. When mist-netted, they often reveal themselves to be fruit-eaters by pooping seeds on their handlers. *Carollia* bats have short tails, while the other fruit specialists—the Stenodermatinae—lack tails. *C. perspicillata* is usually gray-brown, but bats that roost in certain poorly ventilated caves are bleached a rusty color due to ammonia fumes emanating from bat guano accumulated on the floor.

Even in the hand, *Carollia* bats are exceedingly difficult to distinguish from one another. They are identified by subtle differences in size, color, and dentition. One such feature is patterns of banding on the hairs, which can be appreciated in hand-held individuals by blowing apart the fur. Crouched students blowing on these bats and gazing at scientific keys in a defeated fashion is a common scene in the vicinity of mist nets. Such scenes are sometimes punctuated by occasional cries of pain, as *C. perspicillata* tends to be rather vicious.

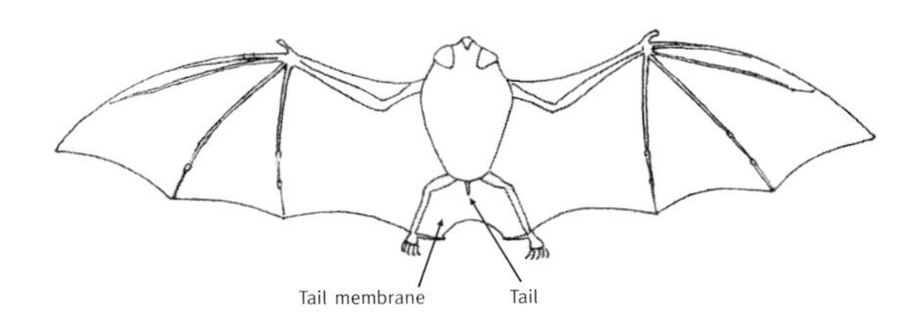

Members of the Carollinae have short tail membranes and very short tails. Relatively short, broad wings facilitate precise flying at low speeds and are apt for maneuvering around vegetation to pluck fruits.

Natural history: Thanks in large part to a 10-year-long study in Guanacaste led by Theodore Fleming, Seba's short-tailed fruit bat is one of the best understood bats in the Neotropics.

The species is known to feed on the fruits of over 50 plant species. Most fruits are plucked from low-growing shrubs, especially various black pepper relatives in the genus *Piper*. The bat identifies ripe fruit stalks by smell, pinpoints them using echolocation, and snaps them off in flight. Fruit stalks are carried to feeding roosts, where they are eaten much like we would eat corn on the cob. A predilection for *Piper* fruits may help *Carollia* bats coexist alongside other fruit-eating bats. In a study at La Selva, researcher Jorge López found that fruit-eating bats of different genera partition resources, with *Carollia* bats (*C. perspicillata*, *C. castanea*, and *C. brevicauda*) feeding predominantly on *Piper* fruits, *Artibeus* bats (*A. jamaicensis*) taking mostly fig and cecropia fruits, and *Glossophaga* (*G. comissari*) preferring *Vismia* fruits. *C. perspicillata* also takes nectar or pollen from about a dozen plant species and gleans insects from foliage, especially during the dry season when fruit is less abundant.

In radio-tracking studies, many bats were found to spend whole nights away from their day roosts, but they did not travel very far and spent relatively little time on the wing. They foraged mostly within 1 km of their day roosts in the wet season, and within 2 and 3 km in the dry season. They spent some 86% of their time away resting in the temporary roosts rather than foraging and feeding.

C. perspicillata roosts in hollow trees, caves, crevices, and buildings in colonies of up to several hundred. In large roosts, the bats form distinct clusters, each cluster containing either a harem of a male and up to 18 females, or a bachelor group of unlucky adult males and adolescents of both sexes. Females move between harems frequently, and mate with various harem males. The harem males, which tend to be among the older bats in the colony, actively defend their harem areas against other males and monopolize reproductive activity. One male in a cave near the Santa Rosa casona (the site of several human territorial disputes, including the Costa Rican defeat of North American slave-trader William Walker in 1856) maintained control of a harem territory for over four years. As a result of this kind of arrangement, all mating may be done by only 12 to 17% of the males.

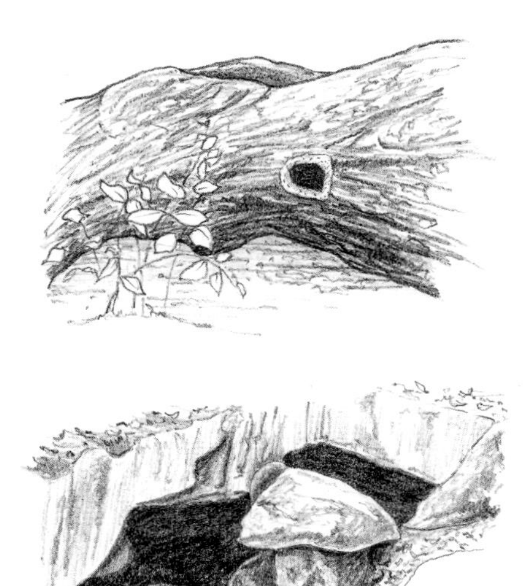

Two roosts used by *Carollia perspicillata* near the Santa Rosa casona: the hollow, horizontal, one-meter-tall base of a *Bombacopsis* treetrunk (top), and a tunnel created by stream erosion (bottom). *C. perspicillata* shares this tunnel with several hundred nectar bats (*Glossophaga soricina*), over 100 vampire bats (*Desmodus rotundus*), and a few gray short-tailed bats (*Carollia subrufa*) and Parnell's mustached bats (*Pteronotus parnellii*).

Most adult females give birth to single pups twice a year, after four-month pregnancies. Once weaned at the age of a month to a month and a half, many young emigrate to new roosts, and then generally stay in that roost for the rest of their lives. Wild *C. perspicillata* in Guanacaste have an average lifespan of two and a half years, although some individuals can survive more than a decade.

Conservation: Short-tailed fruit bats help us by pollinating flowers and dispersing numerous seeds. They play an especially important role in forest regeneration (see p. 90). *C. perspicillata* thrives in disturbed habitats and is among the most common bats in Costa Rica. Nonetheless, it depends on many primary forest trees for fruits, flowers, and roosting sites.

References

Charles-Dominique 1986; Fleming 1983a, 1988; Fleming et al. 1977; Fleming et al. 1972; Gardner 1977; Handley 1976; Hayes & Laval 1989; Heithaus & Fleming 1978; Heithaus et al. 1975; Richard Laval, pers. comm.; Laval & Fitch 1977; López 1996; Nowak 1991; Pine 1972; Porter 1978; Reid 1997; Rodríguez (B.) & Wilson 1999; Rodríguez (J.) & Chinchilla 1996; Thies et al. 1998; Timm & LaVal 1998; Timm et al. 1989.

O: Chiroptera F: Phyllostomidae (S: Stenodermatinae)

TAILLESS or NEOTROPICAL FRUIT BATS
(plates 11 & 12)

Names: The highland epauleted bat (*Sturnira ludovici*), the common tent-making bat (*Uroderma bilobatum*), the white tent bat (*Ectophylla alba*), and the wrinkle-faced or lattice-winged bat (*Centurio senex*). Bats in the genus *Sturnira* are called epauleted bats because males have a gland on each shoulder, the secretions from which sometimes stain the fur, creating yellow or orange shoulder patches. The subfamily name comes from the Greek *stenos*, meaning narrow, and *dermatos*, or skin, in reference to the bats' short tail membranes. Other characteristics include broad muzzles, large eyes, and the absence of a tail.

Members of the Stenodermatinae (as well as the three vampire bat species and some nectar-feeding bats) have very short tail membranes and lack tails.

Range: The Stenodermatinae is the largest leaf-nosed bat subfamily, with about 64 species distributed from Mexico to northern Argentina. At least 24 species inhabit Costa Rica. *S. ludovici* is found from Mexico to northern South America up to about 2,000 m (6,600 ft), but mostly above 800 m (2,600 ft). *U. bilobatum* occurs up to about 1,500 m (5,000 ft), and *C. senex* to about 1,400 m (4,600 ft). Both species range from Mexico to South America. *E. alba* is restricted to the Caribbean slope from Honduras to Panama, up to about 800 m (2,600 ft).

Size: *S. ludovici*: 7 cm, 20 g (3 in., ¾ oz); *U. bilobatum*: 6 cm, 16 g (2 ½ in., ½ oz); *E. alba*: 4 cm, 6 g (1 ½ in., 0.2 oz); *C. senex*: 6 cm, 20 g (2 ½ in., ¾ oz).

Similar species: There are four very similar bats in the genus *Sturnira* in Costa Rica. They are distinguished only by subtle differences in size, color, hairiness, and dentition. *U. bilobatum*'s stripy pattern is similar to that of several other Costa Rican stenodermatines. *E. alba* is one of only two bat species with all white fur (the other is *Diclidurus albus*, illustrated on pl. 7). *C. senex*, with its bizarre face and barred wings, is unmistakable.

Natural history: Tailless fruit bats feed mainly on fruit, supplemented occasionally by nectar, pollen, flower parts, and insects. Plucked fruits and leaves are carried to feeding roosts where juice and pulp can be extracted and fibers and larger seeds spit out. Repeated use of such roosts (which typically consist of a branch protected beneath dense foliage or a large leaf) can leave conspicuous piles of fruit goop and leaf pellets on the forest floor. A few members of this subfamily, including *Artibeus jamaicensis* (illustrated on p. 87), also feed on leaves. The leaves these bats choose tend to be high in protein and may compensate for the low quantities of protein found in most fruits.

Centurio (so named because it has a wrinkled face like that of a hundred-year-old person) may have especially peculiar feeding habits. The species has barely been studied in the wild, but its face is thought to act as a juicer when plunged into soft or rotting fruits, the strange grooves channeling goop toward the mouth. As though in shame, it stretches folds of skin on its chin over its face when roosting. A swollen ridge along the bat's forehead prevents the mask from slipping. Males have two semitransparent window panes where the fold covers the eyes. Males also differ from females in having a bolder pattern of stripes on their wings, and in emitting a strong, musky odor from their chins.

Wrinkle-faced bats roost amidst dense foliage, singly or in small, often single-sex groups. The roosting habits of *Sturnira* bats are a mystery. Although these bats are caught frequently in mist nets, almost no roosts have been found, which has led some to believe that they may roost high in the canopy.

At least 15 species from this subfamily, including *Uroderma bilobatum* and *Ectophylla alba*, construct their own roost sites by modifying the shape of large leaves. By severing leaf veins with their teeth, these bats cause portions of leaves to fold downward, such that the leaf acquires a tentlike shape. When roosting, the bats cling to the underside of the tent roof. They may use the same tent for days or weeks at a time, or alternate between several tents. Tents provide the bats shelter from the elements and concealment from predators, and even come with a built-in alarm system. Since most large leaves are situated at the end of long stems, it is difficult for nonflying predators to approach without shaking the leaf and scaring the bats away (but see the squirrel-monkey account on p. 132).

Here are a few tent designs to look out for. The illustrations are based mainly on drawings and descriptions in Timm 1984 and 1987, articles that contain extensive information on bat tents for those interested. Red markings indicate where the bat has chewed the leaves. Needless to say, bat tents should be admired silently and not disturbed—mostly out of respect for the bats, but also because abandoned bat tents often house wasp nests!

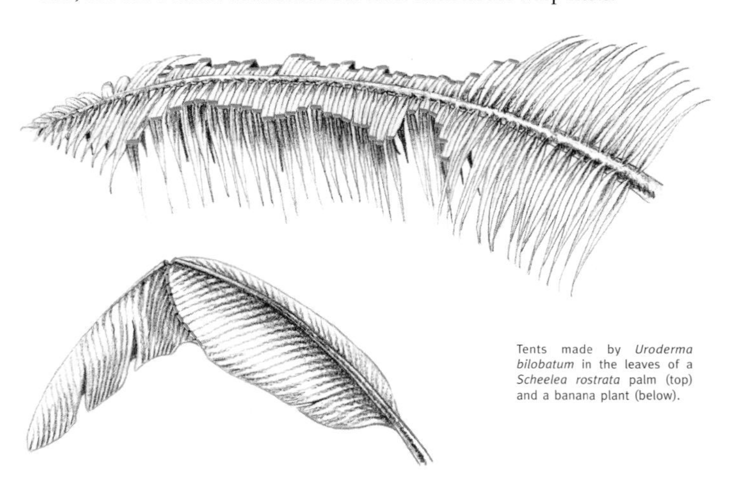

Tents made by *Uroderma bilobatum* in the leaves of a *Scheelea rostrata* palm (top) and a banana plant (below).

Uroderma bilobatum cuts through the midrib of banana (*Musa* sp.) leaves or severs the leaflets of large palms such as coconuts (*Cocos nucifera*) or royal palms (*Scheelea rostrata*). It also makes tents out of clusters of simple leaves. It usually roosts singly or in small groups, but, sometimes, in groups of up to 59 under large tents.

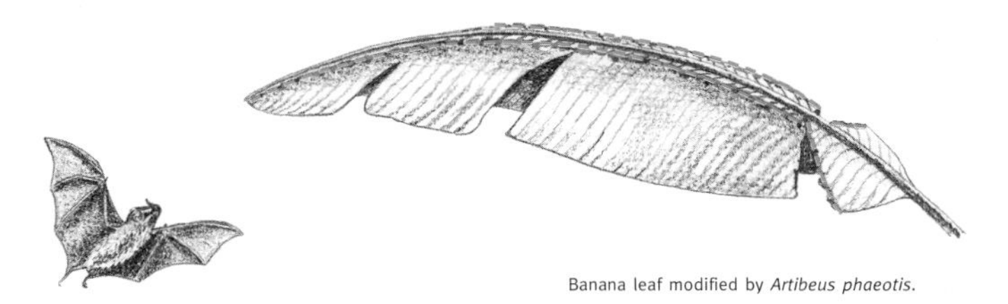

Banana leaf modified by *Artibeus phaeotis*.

Ectophylla alba constructs tents close to the ground in banana leaves or the similar leaves of plants such as *Heliconia* and *Calathea*, where it roosts in groups of four to eight. It tends to pick more or less horizontal leaves, and folds them by partially severing the tissue and veins about 5 mm either side of the leaves' midribs. In the center of the folded section, where it roosts, it also chews small holes in the roof. At La Selva, males and females usually roost together, although the males leave temporarily when young are born in the second half of April. Colonies may have access to several tents at any given time, and are quick to fly off if the tent is disturbed. Tents similar to those made by *Ectophylla* are also made by the pygmy fruit-eating bat (*Artibeus phaeotis*) and Thomas's fruit-eating bat (*Artibeus watsoni*). Neither of these species is illustrated in the plates, but they both closely resemble the common tent bat in coloration.

Tents made by *Artibeus watsoni* in the *Carludovica palmata* cyclanth (left), and the *Asterogyne mariana* palm (below).

Thomas's fruit-eating bat also makes tents out of rattlesnake plants (*Calathea* spp.), philodendrons, and, as illustrated above, cyclanths and palms. Its tents can be common at mid-elevations and, especially, at wet low elevations such as La Selva and Corcovado. Bat-tent expert Robert Timm once found 90 tents belonging to this species in a single day along a trail in Corcovado.

A final design to look out for is that illustrated to the right, made from an *Anthurium* leaf by the toltec fruit-eating bat (*Artibeus toltecus*). These tents can be found in mid-elevation forests, from about 600 to 1,500 m (2,000 to 5,000 ft).

Anthurium caperatum leaf modified by *Artibeus toltecus*.

The coloration of these tent-making bat species may be cryptic: the white stripes of *Uroderma* and *Artibeus* bats break up their outline and make them blend in with the leaf veins of their tents, while the pale fur of *Ectophylla* reflects a green, leaflike hue.

Ectophylla has another interesting coloration feature, a band of melanin in the skin of its scalp, which is unique to the species. This dark pigmentation may protect it from solar radiation filtering through the thin roof of its roost.

Conservation: Many tailless bats, including *S. ludovici* and *U. bilobatum*, are common, and are important seed dispersers in both pristine and disturbed habitats. In the Monteverde area alone, for example, *S. ludovici* was found to take fruit from at least 27 plant species, including both primary forest plants and pioneer species such as wild tomatoes (*Solanum* spp.). *Ectophylla* was known from only two records, both from the nineteenth century, prior to its discovery at La Selva in 1961. It is now listed as common at La Selva and is known to occur at a number of other localities in the Atlantic lowlands of Costa Rica. *Centurio* is rare.

References

Baker & Clark 1987; Brooke 1990; Dinerstein 1986; Fleming et al. 1972; Foster & Timm 1976; Gardner & Wilson 1970; Goodwin & Greenhall 1961; Handley et al. 1991; Kunz & Diaz 1995; Kunz et al. 1994; López 1996; Lowry 1989; Morrison 1983; Reid 1997; Rodríguez (B.) & Wilson 1999; Rodríguez (J.) & Chinchilla 1996; Snow et al. 1980; Timm 1982, 1987, 1994b; Timm & LaVal 1998, 2000a; Timm & Mortimer 1976; Wilson 1997.

O: Chiroptera F: Phyllostomidae (S: Desmodontinae)

VAMPIRE BATS

(plate 13)

Names: The common vampire bat (*Desmodus rotundus*), the white-winged vampire bat (*Diaemus youngi*), and the hairy-legged vampire bat (*Diphylla ecaudata*). These are the sole members of the leaf-nosed bat subfamily, the Desmodontinae. This name is derived from the Greek for reduced teeth. While most other bats have more than 30 teeth, *Diphylla* possesses just 26, *Diaemus*, 22, and *Desmodus*, only 20. With their liquid diet, vampires don't need a lot of teeth and in fact only use four of them—the upper incisors and canines.

Range: *D. rotundus* ranges from Mexico to central Chile, from sea level to at least 2,700 m (8,900 ft); *D. youngi* from Mexico to northern Argentina, to at least 1,150 m (3,800 ft); and *D. ecaudata* from Texas to Peru, to at least 1,900 m (6,200 ft).

Size: *D. rotundus* and *D. youngi*: 8 cm, 35 g (3 in., 1 ¼ oz); *D. ecaudata*: 8 cm, 25 g (3 in., 1 oz).

Similar species: The flattened faces and M-shaped noseleaves of true vampires do not resemble those of any other bats. Within the three species, *Diphylla* can be recognized by its short, broad ears, large eyes, and heavily-haired legs, and *Diaemus* by its white-tipped wings (*Desmodus* sometimes has pale-tipped, but not white-tipped wings). Other features used to distinguish the three include tail membrane and thumb designs, and dentition.

Natural history: These three species are the only bats in the world that drink blood. Common vampire bats occasionally take blood from poultry, wild mammals, people, or other vertebrates, but their principal targets are livestock. They emerge mostly during periods of little moonlight, flying silently about a meter off the ground and usually following the course of a stream to the vicinity of their prey. The farther cattle herds are from such streams, the lower the incidence of vampire bites.

Once a common vampire has landed on or near its victim, it hops or crawls in a spiderlike fashion (using feet, wing thumbs, and elbows) to a feeding spot. A heat-sensitive area on its noseleaf helps it detect veins beneath the prey's skin. It often selects areas with little or no hair, such as the feet, neck, ears, nipples, anus, or vagina. Although the vampire has large canines, a prominent feature in any good Count Dracula, its incisors are its most important feeding teeth. Having licked the bite site, the bat uses its lobed lower incisors to secure a firm hold on the skin, and its razor-sharp upper incisors to shave away any fur and cut out a shallow, roughly 5 by 8 mm strip. While vampires can inflict a deep and painful wound when biting in defense, intrepid biologists have reported that the feeding bite is virtually painless. The bat laps rather than sucks at a feeding wound; grooves on the underside of the tongue draw in blood through capillary action. Vampires have anticoagulants in their saliva that keep blood flowing for up to eight hours, but feeding seldom lasts more than 30 minutes. By this time the vampire may have ingested more than half its body weight and be so bloated it can barely fly. A feeding wound may be revisited on successive nights, sometimes by more than one bat.

In Guanacaste, vampires have been found to prefer certain types of cattle. In mixed-species herds, for example, Brown Swiss cattle are preferred over Zebu, perhaps because the Swiss react less when bitten and/or because they tend to bed down around the edges of clusters of Zebu, where they are more vulnerable. On the rare occasions that they feed on people, vampires also seem to prefer certain individuals, often taking blood from the same person on successive nights while ignoring others nearby.

Common vampires roost in caves and other cavities in colonies that can contain 2,000 individuals, but usually less than 100. The roost sites, which may be shared with other bat species, can be recognized by a strong ammonia odor and smatterings of gooey blood and orange excrement. This species tends to scuttle into crevices rather than fly when disturbed. In large roosts, vampires separate into clusters containing either bachelor males or a harem of eight to 20 females and young guarded by a single male. Males fight viciously for access to harems. Harem members stay together for extended periods and form close social bonds, sharing feeding wounds on cattle and regurgitating blood for one another in the roost. In captivity, a young that loses its mother is adopted by another adult female.

Unlike most other bats, common vampires breed at any time of year, presumably because the abundance of their food does not fluctuate seasonally. Nonetheless, a female can produce only one pup each year because young are very slow to become independent. Although young can forage for blood from the age of four months, they are not completely weaned until they are nine or 10 months old, some six months later than most other microchiropterans. The common vampire can live at least nine years in the wild and almost 20 in captivity.

The white-winged vampire bat prefers bird to mammal blood, but it sometimes dines on livestock. It approaches roosting birds by crawling along the underside of a branch and licks a bite area for several moments, usually on the victim's feet or anus. The bat's saliva may contain anesthetic or digestive chemicals, or it may soften the scaly skin on birds' feet. Then the bat makes a tiny incision, most often on the backward-pointing toe, where the bat may be more concealed from its donor. If the victim becomes agitated during feeding, *Diaemus* ducks back under the branch or clings to the birds vent to avoid being pecked. Feeding sessions typically last about 15 minutes. In one experiment, captives fed on a wide variety of birds weighing from 15 to 200 g (½ to 7 oz), and invariably drained their victims to death in a single bout. They were not interested in smaller prey, such as hummingbirds.

Not much is known of the social or reproductive biology of white-winged vampires. They roost in caves and hollow trees in groups of up to 30, often in clusters. Observations on captives suggest that male white-wings establish dominance hierarchies, and they do so in a bizarre fashion. When previously unacquainted males are placed together, they perform a variety of threat displays, including rising up on their feet and thumbs, boxing with their closed wings, hissing, and spraying one another with a musky secretion emitted from two large glands in the mouth. Eventually, one signals its submission by crouching and, sometimes, by holding its mouth agape to expose the oral glands. If so, the dominant male inserts its tongue into the other male's mouth for several seconds before the two separate. The musk may help individuals recognize, and perhaps learn about, one another. Since musk is often sprayed in response to other threats, researchers believe that it may also serve as a deterrent to predators—like a sort of built-in Mace.

The hairy-legged vampire bat is thought to be the most specialized of the three vampire species since captives accept only birds' blood. It feeds at the feet or around the anus of birds, the latter by landing on a bird's rump and then latching on to the bird's tail, head-downward. *Diphylla* roosts in caves and hollow trees, usually singly, although several individuals may inhabit different parts of the same roost. There is one record of several hundred roosting in a cave within a few inches of one another, although not in contact.

Mythology: According to legend of the Bribri tribe of southern Costa Rica, the world's surface was originally barren and covered only with rock. The first soil was created when a vampire bat flew to the center of the earth to feed on a jaguar cub (or, in other versions, a baby girl), and returned to the surface to drop guano over the rock. Sibú, the Bribri God-Creator, was then able to plant seeds in the guano.

Conservation: The two bird specialists are rare. *Diphylla* has been known to take blood from poultry, which it thus weakens and exposes to secondary infections. In pristine areas, the common vampire is uncommon, too. It is used as a resource in some areas. Nambiquara tribes in western Brazil smoke common vampires out of tree roosts and eat them. In the vicinity of cattle ranches, however, this bat can be abundant and is a prevalent pest. Unnaturally high vampire populations may monopolize roosts, displacing other, beneficial bat species. Above all, the common vampire leaves wounds on livestock that are vulnerable to infection or parasitization, and is a major vector of paralytic rabies and other diseases. The species causes an estimated $100 million of damage to livestock each year in Central and South America. Sadly, indiscriminate eradication programs such as dynamiting caves have resulted in the destruction of numerous other bat species. A more precise method is to capture vampires with mist nets placed around herds, and smear them with fatal anticoagulants. The bats are then released, and before they die, they spread the fatal chemicals to other vampires at their roosts through mutual grooming.

References

Acha & Alba 1988; Aguilar 1986; Arellano-Sota 1988; Bozzoli 1986; Fenton 1992; Greenhall 1988; Greenhall et al. 1983; Greenhall & Schmidt 1988; Greenhall et al. 1984; Greenhall et al. 1971; Greenhall & Schutt 1996; Hilje & Monge 1988; Hoyt & Altenbach 1981; López-Forment 1980; Lord 1988; Nowak 1994; Reid 1997; Rodríguez (B.) & Wilson 1999; Rodríguez (J.) & Chinchilla 1996; Schutt 1995; Schutt et al. 1999; Setz & Sazima 1987; Timm & LaVal 1998; Turner 1975, 1983, 1984; Uieda 1994; Uieda et al. 1992; Wilkinson 1985a,b, 1988.

O: Chiroptera F: Natalidae, Furipteridae, Thyropteridae

FUNNEL-EARED, THUMBLESS, and DISK-WINGED BATS
(plate 14)

Names: The Mexican funnel-eared bat (*Natalus stramineus*); the thumbless bat (*Furipterus horrens*), also frequently and confusingly referred to as the smoky bat, confusingly, since that name is also used for the emballonurid, *Cyttarops alecto*; and Spix's disk-winged bat (*Thyroptera tricolor*). These bats represent three small and poorly known bat families that contain tiny, somewhat similar-looking bats with large ears, small eyes, and no noseleaf. Male funnel-eared bats have an additional characteristic known as the natalid organ, a structure on the face that is unique to the family. Its function is unknown. Furipterids are known as thumbless bats because their thumbs are largely hidden within the wing membrane. The thyropterids, or disk-winged bats are sometimes referred to as New World sucker-footed bats. (There is one Old World sucker-footed bat, on the island of Madagascar, which is thought to have evolved independently and is in its own family, the Myzopodidae. Several Old World species in the family Vespertilionidae also have wrist and/or ankle suckers.)

Range: *N. stramineus* is the only funnel-eared bat in Costa Rica and mainland Central America. It is found from Mexico to Brazil, mostly in relatively dry, lowland areas. In Costa Rica, it has only been recorded in Guanacaste and Monteverde. There are four other natalids that occur on islands or in South America. *F. horrens* occurs from Costa Rica to Brazil, up to 2,500 m (8,200 ft), but mostly in wet, lowland areas. In Costa Rica it is known only from La Selva. The only other thumbless bat species is restricted to South America. *T. tricolor* ranges from Mexico to Brazil, to at least 1,600 m (5,300 ft). One other rare species, *T. discifera*, was recently confirmed to occur in Costa Rica. The only other member of the family, *T. lavali*, was described for the first time in 1993 and is known only from Peru.

Size: *N. stramineus*: 5 cm, 5 g (2 in., 0.2 oz); *F. horrens*: 3.5 cm, 3 g (1 ⅓ in., 0.1 oz); *T. tricolor*: 4 cm, 4 g (1 ½ in., ⅛ oz).

Natural history: Members of these three families are thought to feed principally on insects caught on the wing. They forage near the ground, with an agile, fluttery flight pattern that, along with their small size, gives them the appearance of large moths.

The Mexican funnel-eared bat is most active during the first hours after dark, later retiring to temporary, nocturnal roosts. During the daytime, it roosts in deep caves. It has been found both in small groups and in colonies containing several hundred individuals.

Stomach analyses of the thumbless bat have shown it to be particularly fond of moths. It roosts in caves, hollow logs, and rock crevices. A colony of at least 60 males was found in a hollow log at La Selva, and colonies containing 250 thumbless bats have been found in South America.

Spix's disk-winged bat has the peculiar habit of roosting on the insides of the young, rolled, tube-shaped leaves of heliconia and banana plants. Unlike most bats, it roosts head upward, using the suckerlike disks on its wrists and ankles to cling to the smooth leaf surface. Captives can hang from a window pane with one sucker alone, periodically licking the disks to sure up their grip! In the wild, these bats must move roost constantly, for the leaves are only appropriate for use in the last day or two before they unfurl, when the opening at the top is about 5 to 10 cm (2 to 4 in.) wide. A leaf may shelter anywhere from one to nine bats.

Close-up of suction cups on the undersides of the left wrist (left) and ankle (right) of *Thyroptera tricolor.*

In a mark and recapture study on the Osa Peninsula, bat experts James Findley and Don Wilson found that leafmates stay together as they move around, possibly using high-frequency calls to locate one another in new homes. When an entire group of nine was captured and then released one at a time from a bag, the first individual to be let go circled and even landed several times on the bag containing its buddies, before disappearing into a rolled leaf nearby; of the remaining bats, four went straight to the same leaf, and four went to another, apparently because the first leaf was too crowded.

Conservation: Since these species are adept at avoiding mist nests, their abundance is hard to gauge. Spix's disk-winged bat can be common around heliconia stands at Tortuguero and on the Osa Peninsula.

References

Eisenberg 1989; Fenton 1992; Findley & Wilson 1974; Laval 1977; Pine 1993; Reid 1997; Rodríguez 1993; Rodríguez (B.) & Wilson 1999; Rodríguez (J.) & Chinchilla 1996; Timm & LaVal 1998, 2000a; Timm et al. 1989; Tschapka et al. 2000; Uieda et al. 1980; Watkins et al. 1972; Wilson 1978, 1997; Wilson & Findley 1977.

O: Chiroptera F: Vespertilionidae

PLAIN-NOSED BATS
(plate 15)

Names: The black myotis (*Myotis nigricans*), the southern yellow bat (*Lasiurus ega*), and the western red bat (*Lasiurus blossevillii*). The most distinctive feature of the Vespertilionidae, or plain-nosed bat family, is a long tail that is encased to the tip in a broad, V-shaped membrane. The family name comes from the Latin *vespertilio*, meaning animal of the evening.

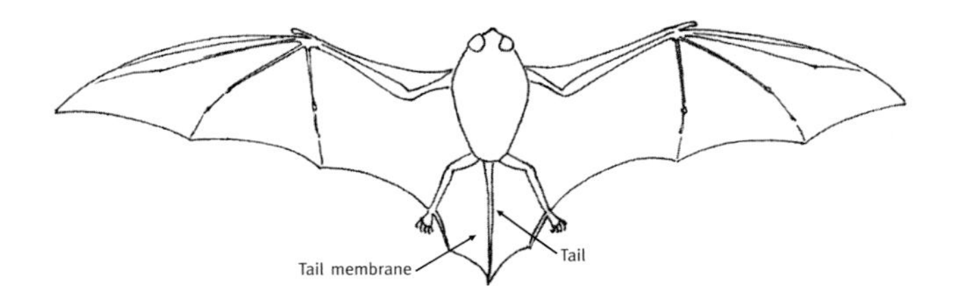

Vespertilionids (as well as the funnel-eared bat, the disk-winged bat, and some leaf-nosed bats) have long tails that are encased to the tip in a large, V-shaped membrane.

Range: The Vespertilionidae has been successful in temperate areas as well as in the tropics. It is represented on every continent except Antarctica, and is the world's largest bat family, with almost 300 species. The roughly 100 species in the genus *Myotis*, in particular, may have a broader distribution than any other genus of wild, terrestrial mammal. While vespertilionids are the predominant bats in North America, they comprise only 15 of Costa Rica's 109 bat species. *M. nigricans* ranges from Mexico to northern Argentina, from sea level to at least 3,150 m (10,300 ft); *L. ega* is found from southern U.S. to Uruguay, to at least 1,650 m (5,500 ft); and *L. blossevillii* is found from Canada to Argentina, to at least 2,500 m (8,200 ft).

Size: *M. nigricans*: 4 cm, 5 g (1 ½ in., 0.2 oz); *L. ega*: 6 cm, 12 g (2 ½ in., ½ oz); *L. blossevillii*: 5 cm, 8 g (2 in., ⅓ oz).

Natural history: Vespertilionids are strong fliers that can spend long hours on the wing. They use this ability to catch airborne insects, their primary food, and, in temperate areas at least, to undertake seasonal migrations of up to several hundred kilometers. Red bats in North America, for example, sometimes appear far out to sea, taking rest stops on boats and islands, while the black myotis in Panama can return to its roost just two days after being released 50 km (30 miles) away.

The black myotis often forages over streams, light gaps, and trails. Where insects are abundant, it can catch several hundred mosquito-sized bugs in an hour. It roosts in buildings, caves, and hollow trees, sometimes in colonies of over 1,000 bats. Females with young cluster together in separate maternity roosts. During the cool, early morning hours these bats tend to cluster together in the higher (and therefore warmer) parts of the roost; as the roost starts to heat up around midday, they separate and move downward. Curiously, while this bat forages on arthropods outside the roost, the tables can be turned inside the roost. A large orb spider produces silk strong enough to entangle these small bats, and has been observed feeding on them at a roost in Panama.

Droppings of insectivorous bats and discarded, inedible insect parts.

Most information on bats in the genus *Lasiurus*, such as the southern yellow and western red bats, comes from studies done in North America; they have barely been studied in the tropics. Hairy-tailed bats (as *Lasiurus* bats are known collectively) tend to forage with a fast, steady flight some 6 to 15 meters (20 to 50 feet) off the ground, catching most of their prey on the wing. Occasionally they land on vegetation to pluck a perched insect. Moths are important in their diets.

Although hairy-tailed bats may form flocks of several hundred when migrating in temperate areas, they are generally solitary. They roost alone in trees, protected only by foliage, often just a few meters from the ground. They are thought to mate while on the wing. While most female bats have two nipples, hairy-tailed bats have four, and are the only bats in the world that commonly produce more than two young per litter. Red bats have even been found with litters of five.

Conservation: Vespertilionids can eat their own weight in insects each night, and play an important role controlling populations of mosquitoes and other pests. For example, individuals of one species (*Eptesicus nilssoni*) studied in Sweden were found to consume about 1,000 mosquitoes each hour, while in North America a colony of 150 big brown bats (*Eptesicus fuscus*), a species also found in Costa Rica, ate so many cucumber beetles that they reduced the local population of rootworms (the larvae of cucumber beetles and a major agricultural pest) by an estimated 18 million in a single summer.

The black myotis is common in Costa Rica. The southern yellow bat and the western red bat are seldom recorded here, possibly in part because they tend to forage above the height of most mist nets.

References

Baker et al. 1988; Barbour & Davis 1969; Dinerstein 1985; Eisenberg 1989; Fenton 1992; Hamilton and Stalling 1972; LaVal 1973a; Nowak 1994; Reid 1997; Rodríguez (B.) & Wilson 1999; Rodríguez (J.) & Chinchilla 1996; Rydell 1989; Shump & Shump 1982; Timm & LaVal 1998; Timm et al. 1989; Wilson 1983, 1997; Wilson & Findley 1972; Wilson & LaVal 1974.

O: Chiroptera F: Molossidae

FREE-TAILED or MASTIFF BATS
(plate 16)

Names: The Brazilian free-tailed bat (*Tadarida brasiliensis*) and the Sinaloan mastiff bat (*Molossus sinaloae*). Molossids are known as free-tailed bats because a short tail membrane encases only about the first half of the tail, or as mastiff bats, after their doglike faces. The latter feature also gives rise to the scientific family name, which comes from the Greek word for mastiff hound, *molossus*, (after a Greek region called Molossis that was famous for its dogs). Other distinctive features include unusually shaped ears, narrow wings, and long, curved bristles on the toes.

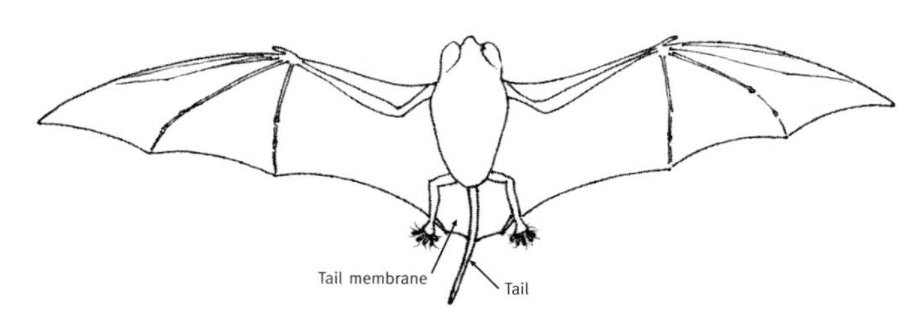

Molossids are known as free-tailed bats because the latter half of the tail protrudes beyond the tail membrane. Relatively long, thin wings facilitate swift, sustained flight and are apt for the pursuit of insects in open areas, such as above the canopy.

Range: The roughly 90 species in this family inhabit tropical and temperate regions of both the New and Old Worlds. At least 12 species occur in Costa Rica. *T. brasiliensis* ranges from Oregon in the United States to central Chile, to at least 3,000 m (10,000 ft); *M. sinoloae* ranges from Sinaloa, Mexico, to Surinam, up to 2,400 m (7,900 ft).

Size: *T. brasiliensis*: 6 cm, 12 g (2 ½ in., 1/2 oz); *M. sinoloae*: 8 cm, 20 g (3 in., ¾ oz).

Natural history: Long, thin wings afford molossids the ability to fly fast for long periods. *T. brasiliensis* has been clocked at an average speed of 40 km/h (25 mph), with a maximum speed of 105 km/h (65 mph), and may fly 65 km (40 miles) just to reach a foraging area each night. This wing design does not permit great maneuverability, and molossids catch their insect food on the wing, in open areas or above the canopy, sometimes at great heights. Members of a *T. brasiliensis* colony that roosts near an airforce base in Texas have appeared on radar at altitudes of up to 4,500 m (15,000 ft). Likewise, long, thin wings do not work well at low speeds, and molossids must drop from a perch in order to take off. In this sense, molossids are much like birds that have long, thin wings such as swifts, which must also drop from perches, or albatrosses, which must run along the surface of the sea into an oncoming wind to get airborne.

T. brasiliensis is most famous, however, for the vast colonies it forms in caves in the southern U.S., which are the largest, most concentrated aggregations of a single mammal species in the world. In the 1960s, 13 caves in Texas housed about 100 million of these bats, while a single cave in Eagle Creek, Arizona, was estimated to contain between 30 and 50 million. Ammonia fumes emanating from the vast quantities of decomposing guano and urine at poorly ventilated roosts can reach levels that are sufficient to incapacitate or even kill most other mammals, including humans. Free-tailed bats survive such conditions in part by producing copious quantities of mucous.

These huge colonies are maternal roosts, consisting almost entirely of females and young. When females go out to forage, the young form writhing clumps on the cave walls known as crèches. Whether or how a female locates her own pup among millions upon her return is a question that has intrigued a number of biologists. Studies have shown that mothers do usually locate their young: they go straight to the area where the pup was left, and then, since pups often wander while mothers are gone, use vocal and olfactory cues to track them down. The pups, meanwhile, try to get milk out of any female that lands in their vicinity, so the females must fend them off with wing blows if they are to reach their young. Only if a female loses her pup will she nurse the young of others.

In Costa Rica, free-tailed bats have been found in much smaller colonies, often in buildings. Their flattened bodies enable them to squeeze into narrow spaces behind walls or between sheets of corrugated iron in roofs, where they often rest in a horizontal position. Colonies of up to 200 *T. brasiliensis* and 79 *M. sinoloae* have been found in Costa Rica. The latter species sometimes shares roosts with other *Molossus* bats and with lesser fishing bats (*Noctilio albiventris*). Such roosts often have a musky odor, perhaps emanating from large scent glands that are situated on the throats of male molossids.

Sounds: Some species emit echolocation clicks that are clearly audible to humans.

Conservation: Molossids benefit people by eating insect pests, and guano produced by *T. brasiliensis* colonies has been used for fertilizer. Humans have twice used *T. brasiliensis* for bellicose purposes. During the Civil War in the United States, its guano provided sodium nitrate for gunpowder; and during World War II, United States forces used the species to deliver fire bombs. Since the fire bombs seldom reached their intended targets, once even setting fire to an army barracks and a general's car, that project was eventually scrapped.

Some *T. brasiliensis* populations have been decimated over the course of this century. The population at one famous site, Carlsbad Caverns in New Mexico, dropped from an estimated 8.7 million bats in 1936 to 200,000 in 1973, and the Eagle Creek cave contained only about 600,000 bats by 1970. The causes of such declines are unknown. Vandalism was suspected at Eagle Creek, but in many areas DDT, a pesticide that contaminates bats' insect food, is considered the most likely culprit. DDT was banned in 1972 in the United States, and subsequently in Costa Rica, but it is still used in Costa Rica due to poorly enforced laws. Numerous other harmful pesticides are still used in both countries.

Since molossids tend to forage high off the ground, they are not prone to fly into mist nets, and their status in Costa Rica is therefore hard to gauge. Indeed, two species (the big-crested mastiff bat, *Promops centralis*, and the broad-eared bat, *Nyctinomops laticaudatus*) that are found both north and south of Costa Rica and almost certainly occur here but have yet to be recorded.

References

Barbour & Davis 1969; Eisenberg 1989; Fenton 1992; LaVal 1973b; LaVal & Fitch 1977; McCracken & Gustin 1987; Nowak 1994; Reid 1997; Rodríguez (B.) & Wilson 1999; Rodríguez (J.) & Chinchilla 1996; Timm & LaVal 1998; Timm et al. 1989; Tuttle 1995; Wilkins 1989; Williams et al. 1973.

Monkeys (order Primates)

Distribution and Classification

Primates are found mainly in tropical regions, in all parts of the world except Australia. About 50 of the world's 150 or so species are native to the New World. All but *Homo sapiens* comprise the infraorder Platyrrhini, or broad noses, and are characterized by having widely-spaced, sideways-pointing nostrils. We belong to the Old World infraorder, the Catarrhini, or narrow noses, which have closely-spaced, downward-pointing nostrils.

The New World monkeys are split into three families. The largest and most diverse family, with more than 30 species, is the Cebidae, which includes the four monkeys found in Costa Rica. All but one of the remaining New World monkeys belong to the Callitrichidae—the marmoset and tamarin family. Marmosets and tamarins differ from cebids in being tiny (their body weight ranges from a mere 100 to 600 g, or 3 ½ to 20 oz), in having tufts of colorful fur around their heads, in having claws rather than nails on their fingers, in lacking prehensile tails, in producing more than one young per litter, and in various other aspects of natural history and internal anatomy. A tamarin is illustrated on p. 126. The third family, the Callimiconidae, contains just one rare species, Goeldi's monkey (*Callimico goeldii*), which resembles a tamarin but differs in several aspects of its natural history and internal anatomy. It is found just east of the Andes in South America.

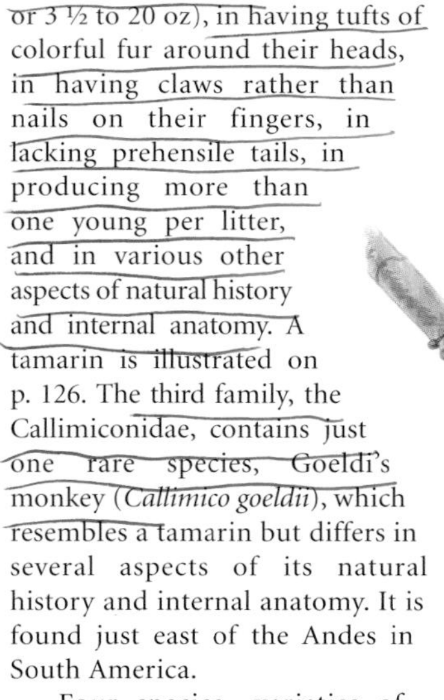

Goeldi's monkey (*Callimico goeldii*).

Four species—varieties of squirrel, capuchin, howler, and spider monkeys—are known to inhabit Costa Rica. Two other species, the western night monkey (*Aotus lemurinus*) and Geoffroy's tamarin (*Saguinus geoffroyi*), appear sporadically on Costa Rican mammal lists, but it is doubtful they occur here.

The night monkey has recently been listed as possible in the Atlantic lowlands, based on a specimen that may or may not have been collected in Costa Rica around 1870 and on various reported sightings over recent years in the vicinity of La Selva and Bribri. This species does occur close to the Costa Rican border in the Atlantic lowlands of Panama and could conceivably have been overlooked in Costa Rica for several reasons. It is nocturnal (the only such species in the New World); it forages in the canopy; it lives in small family groups, consisting of monogamous pairs and their young, that subsist in very small home ranges; it is generally quiet, although its owl-like, often two-syllable hoots give its presence away in areas where it is common; it tends to shy away from lights quickly; and if not seen well is easily mistaken for the only slightly larger olingo (pl. 31).

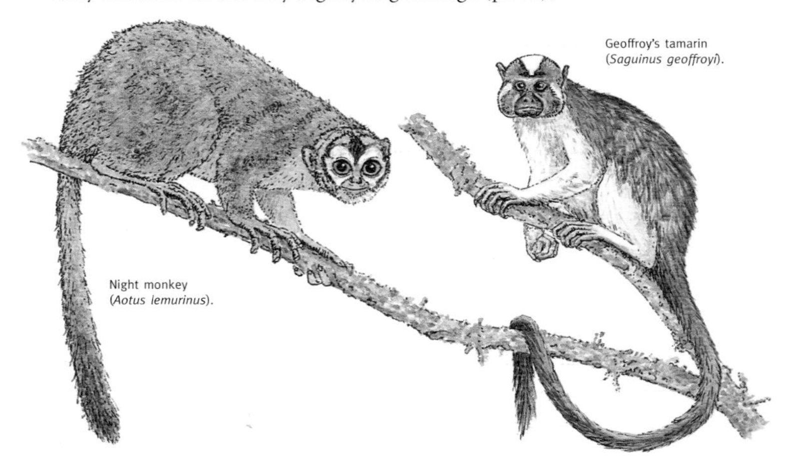

Geoffroy's tamarin (*Saguinus geoffroyi*).

Night monkey (*Aotus lemurinus*).

The second species of doubtful occurrence here, Geoffroy's tamarin, was reported as a rare inhabitant of the Coto region of southwestern Costa Rica in 1935 by a highly respected biologist. Even though there has been no evidence of its presence ever since, it continues to appear on some Costa Rica lists. Since it is conspicuous, often frequenting forest edge and secondary habitats and foraging close to the ground, since it is hard to confuse with any other mammal, and since the nearest it is known to occur today is more than 200 km (125 miles) away in eastern Panama, it is highly unlikely to inhabit Costa Rica. Confusion about its occurrence here may have been started and perpetuated by the local use of the name *mono tití* to refer to both the tamarin and the squirrel monkey, the latter of which does occur in the Coto region.

Evolution

Fossils indicate that the earliest clear ancestors of today's New World monkeys arrived in South America about 35 million years ago. Since South America was already isolated from other land masses at that time, there has been some debate as to where these ancestors came from.

Some scholars have argued that they were derived from primitive, now extinct primates that inhabited North America, and that they drifted down on chunks of floating vegetation. Others, perhaps the majority, believe that the New World monkeys are derived from primitive Old World monkeys that drifted across the Atlantic from Africa. They point out that, during that era, the Atlantic was still quite narrow and littered with islands, while a journey from North America would have been longer and probably not favored by sea currents.

Dizzyingly extensive comparisons of the anatomy of New and Old World monkeys have not resolved this issue. Those peddling the North American theory claim that anatomical differences between the two monkey infraorders reflect independent origins and that similarities are caused simply by parallel evolution, while those beating the African drum assert that similarities are the result of common ancestry and that differences reflect selective divergence since the groups became separated. (There is a similar debate concerning the earliest origins of caviomorph rodents; they also arrived in South America about 35 million years ago, could have come from North America or Africa, and have many features in common with certain Old World species—see pp. 151-152.)

Regardless of how the New World monkeys reached South America, they were subsequently isolated until about seven to two million years ago, when the uplifting of Panama connected South America with Central America. Thus, they were unable to disperse into Central America until relatively recently, which might explain why only five of the 50 or so species of New World monkeys occur north of Panama.

Probably by at least 12,500 years ago, Costa Rica's monkeys were joined by an unusually intelligent, prolific, and destructive Old World cousin—*Homo sapiens*. When, how, and from where the first humans arrived in the New World is still not clear, but most experts believe they arrived between about 14,000 and 30,000 years ago from Asia, either using a land bridge that existed where the Bering Straight lies today or traveling by boat. These early Americans were hunter-gatherer nomads: they did not know how to cultivate food and had no permanent settlements. They traveled in small family groups.

Between about 10,000 and 6,000 years ago, however, humans figured out how to cultivate plants. Settlements started to become larger and more permanent, and with more reliable sources of food and shelter, the population started to grow. From their earliest times in Central America, people may have caused or contributed to the disappearance of certain large mammals, such as giant ground sloths (see pp. 48-49). But it was not until this century that humans became sufficiently numerous and sophisticated to have truly devastating effects on the natural environment.

Teeth and Feeding

The dental formula of the cebids is I 2/2, C 1/1, P 3/3, and M 3/3 = 36. (That of *Homo sapiens* is the same except that we have one less molar in each quarter of the mouth.) Monkeys' teeth are not very specialized, which helps them eat a wide variety of foods. There are nonetheless clear dietary differences between Costa Rica's four monkey species. Although all eat fruit, the spider monkey relies on it much more heavily than the others. Only the howler feeds extensively on leaves, and only the capuchin and squirrel monkeys rely on insects as a staple. Of the latter two, whose ranges overlap only in a small area in southwestern Costa Rica, the squirrel monkey takes smaller foods, tends to forage closer to the ground, and prefers more disturbed habitats. These differences help Costa Rica's monkeys coexist without out-competing one another. Indeed, it is not unusual to see different species peacefully foraging together.

Senses

Monkeys have good eyesight. Primates' eyes are positioned at the front of the head and face forward, so the field of vision of each eye overlaps considerably with that of the other, more so than in any other mammals (see p. 24). This stereo, or binocular, vision aids greatly in depth perception, which is essential for jumping about in trees. The disadvantage of forward-facing eyes is that they have a narrower total field of vision than do laterally-positioned eyes such as those of rabbits, for example, and this makes monkeys more vulnerable to predators. Since monkeys live in troops, they are able to compensate by watching each others' backs.

Some of the adaptations associated with binocular vision, such as a shortened muzzle, have reduced monkeys' ability to smell, and monkeys' sense of smell is inferior to that of most terrestrial mammals. Monkeys do rely partly on smell to select certain foods, such as ripe fruits, however, and the sense is also important in communication. Scents from glands located in areas such as the head, chest, forearms, and anus are rubbed onto branches and mixed with urine and feces. They serve to mark territories and aerial runways, and help individuals assess one another. A number of studies, mostly on tamarins and marmosets, have demonstrated that these scents contain extraordinarily complex combinations of chemicals. One study identified 249 chemicals in the scent marks of a South American marmoset! Different chemical combinations have been shown to identify species, subspecies, individuals, gender, degree of familiarity, social status, and reproductive status. As an indication of how important these chemical messages can be, males that have their sense of smell temporarily blocked during experiments show no response to females in heat until the sense is restored.

Monkeys' hearing is also important in communication, as evidenced by the huge variety of sounds made by many species. Monkeys' ears are not very mobile and, like their eyes, are directed forward, so monkeys must turn their heads toward sounds to hear them well.

Body Design

If one considers that the ancestors of all mammals resembled shrews (see p. 14), then one can interpret the diversity of modern mammalian forms according to how much they have diverged from the primitive, shrewlike design. From this perspective, body designs of New World monkeys represent a fascinating progression of adaptations to life in the trees.

All New World monkeys are arboreal. The smallest monkeys, the callitrichids, resemble primitive mammals the most. They have relatively long back legs for springing along branches, strong claws that can support the entire weight of their body and enable them to cling to vertical trunks, and long tails for balance.

Cebids' bodies have evolved further from the design of those earliest shrewlike ancestors. They tend to move more deliberately, gripping and climbing through branches rather than scampering along them. Since they rely more heavily on their forelimbs to pull themselves around, they have evolved proportionately longer and stronger arms than the tamarins, as well as longer, more flexible fingers. Squirrel, capuchin, and howler monkeys have evolved partially opposable thumbs with which to grasp small branches. Claws would be a hindrance for this style of locomotion, so cebids simply have nails. Being heavier and clawless, they cannot readily cling to vertical trunks like the tamarins.

To varying degrees, cebids have also evolved prehensile (gripping) tails. Newborn squirrel monkeys have limited prehensility in their tails but lose it after a few weeks. Capuchins have prehensile tails but seldom use them for anything more than partial support. Howler and spider monkeys, which need to reach the tips of canopy branches to harvest the leaves and fruit they eat, have strong prehensile tails equipped with a naked pad on underside of the tip for extra grip. They often hang by their tails alone.

Of Costa Rica's species, the spider monkey has evolved the furthest from the primitive mammalian design. In order to facilitate its peculiar swinging style of locomotion, it has evolved exceptionally long fingers and arms, has all but lost its thumbs, and possesses the strongest and most dexterous prehensile tail (see p. 147).

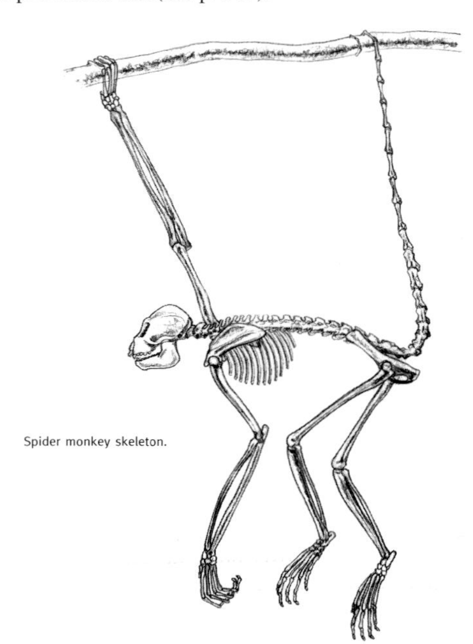

Spider monkey skeleton.

Conservation

At present, more than a third of the world's primate species are listed as threatened by the International Union for the Conservation of Nature, and about one in seven are in imminent danger of disappearing. Lest anyone think such lists are alarmist, it should be pointed out that a number of species (including, for example, six entire genera of lemurs on Madagascar) are already extinct.

Costa Rica's monkeys have not been hunted as heavily as many primates in other parts of the world, nor as heavily as many other Costa Rican mammals, but they have not been entirely ignored. Spider monkeys are killed for their meat, and capuchin and squirrel monkeys are sometimes persecuted as agricultural pests. The same three species were popular as pets in Costa Rica until recently. International trade in monkeys for use as pets or for research also took its toll in decades past. An estimated 200,000 primates were exported to the United States each year during the 1950s, for example. By far the greatest threat to Costa Rica's monkeys today is habitat loss and fragmentation, which has exterminated or isolated numerous populations.

Forest fragmentation doesn't just hinder the movement of troops and the complex migrations of individuals between troops that are described in the species accounts; it also increases the monkeys' exposure to other problems. A human yellow fever epidemic that swept through Central America in the 1950s, for example, decimated monkey populations in areas such as La Selva and Monteverde. The vector of yellow fever—the mosquito *Aedes aegypti* (also famous for transmitting dengue fever)—arrived and thrived in the New World thanks to humans. It came on boats from Africa and proliferates in urban areas.

In a twisted way, people may also have caused monkey populations to increase in some areas by exterminating or greatly reducing many of the principal monkey predators, such as harpy eagles, pumas, or boa constrictors. But monkeys can still be slow to replenish reduced populations. At La Selva, researchers have noted that there are still fewer monkeys than there were prior to the yellow fever epidemic or than occur today in similar habitats elsewhere in the Neotropics. Such observations are cause for rather glum speculation when one considers the decidedly predictable trends of population growth for Costa Rica's Old World primate species. Between the early 1600s and the 1920s, the world's human population increased from about 600 million to two billion, and Costa Rica's from about 10,000 to 400,000. Today, there are some six billion people in the world and about four million in Costa Rica. And at current rates, these figures will double over the next 30 years.

References
Baldwin & Baldwin 1972a; Belcher et al. 1989; Ciochon & Chiarelli 1980; Coimbra-Filho & Mittermeier 1981; Estrada & Coates-Estrada 1988; Fishkind & Sussman 1987; Hershkovitz 1977; Izor 1985; Mittermeier 1978, 1988; Mittermeier et al. 1988; Molina & Palmer 1998; Nowak 1991; Oppenheimer 1977; Parfit 2000; Rasmussen 1990; Reid 1997; Rodríguez & Chinchilla 1996; Schaik 1983; Smith et al. 1994; Stoner 1993; Timm 1989; Timm et al. 1989; Timm & Laval 2000b; Washabaugh & Snowdon 1998; Wille 1987.

SPECIES ACCOUNTS

O: Primates F: Cebidae

CENTRAL AMERICAN SQUIRREL MONKEY
Saimiri oerstedii
(plate 17)

Names: Genus from the Guyanan Tupí word *saimarí*, which means small monkey. Species name after Danish collector Aders Sandoe Ørsted, who worked in Costa Rica from 1846 to 1848 and subsequently published one of the first major works on the flora and fauna of Central America. The well known orchid genus *Oerstedella* was also named after him. There is only one other species of squirrel monkey, the common squirrel monkey (*S. sciureus*) of South America. Since *S. sciureus* varies geographically in color, some taxonomists think it should be split into as many as seven species, but recent studies suggest that all forms intergrade. **Spanish:** *Mono tití*; *mono ardilla.*

Range: Forests with abundant low- and mid-level vegetation (mostly secondary or partially logged primary forests) along a small stretch of Pacific lowlands, up to at least 300 m (1,000 ft), in southern Costa Rica and, perhaps, western Panama. None of its range in Panama is protected, and it may recently have become extinct there. There is a strange several-hundred-kilometer gap between the ranges of *S. oerstedii* and its close South American relatives, which has led some biologists to believe it may have been introduced by humans in eras past. There are two subspecies, which can be distinguished both by location and by coloration: *S. o. citrinellus*, which is found north of the river Térraba, is paler and grayer than *S. o. oerstedii*, which is found to the south.

Size: 30 cm, 0.9 kg (12 in., 2 lbs).

Natural history: Much of what is known about *S. saimiri* derives from a suite of studies conducted by Sue Boinski and colleagues in Corcovado and Manuel Antonio national parks, and a long-term survey by Grace Wong in Manuel Antonio and the unprotected, fragmented habitat that surrounds it.

Squirrel monkeys form larger troops than do any other New World monkey. South American varieties can form troops of over 100 individuals, while Costa Rica's squirrel monkeys travel in troops typically containing between 20 and 70 individuals.

Squirrel monkeys forage mainly in the middle and lower levels of the forest for insects and their larvae, spiders, small fruits, leaves, flowers, nectar, bark, and small vertebrates (including lizards, birds, frogs, and bats). They capture tent-making bats in an ingenious fashion: they peer cautiously under folded leaves, and, when they find roosting bats, they climb and jump on the leaf from above. Dislodged bats that do not fly away quickly enough are caught on the ground and eaten. Favorite fruits include those of cecropias (*Cecropia* spp., illustrated on p. 62), legumes (*Inga* spp.), black pepper relatives (*Piper* spp., illustrated on pl. 11), coffee relatives (*Palicourea* sp.), palms (*Scheelea* sp.), cerillo (*Symphonia globulifera*), quiubra (*Pseudolmedia spuria*), yayo flaco (*Xylopia sericophylla*), and the wild cashew espavél (*Anacardium excelsum*).

In Corcovado, three bird species—the double-toothed kite, the gray-headed tanager, and the tawny-winged woodcreeper—habitually follow troops of squirrel monkeys to pick off flushed insects and small

In Corcovado, the double-toothed kite, the gray-headed tanager, and the tawny-winged woodcreeper (pictured from left to right) habitually follow squirrel monkey troops and try to catch insects and other small animals that the monkeys flush out.

vertebrates. Various other birds, including other woodcreepers and birds of prey, motmots, and trogons also join in on occasion. Many of these birds also follow army ants, for the same reason. The birds follow squirrel monkeys most often during the wettest months, when arthropods in Corcovado are at their scarcest.

Home ranges used by 10 troops that Wong studied varied from 35 to 63 ha (85 to 155 A). Wong found that the home ranges of neighboring troops living within Manuel Antonio park tend to overlap extensively, with several troops sharing almost half their home ranges. The home ranges of neighboring troops living in more fragmented habitats outside the park, however, tend to overlap less, or not at all.

Boinski's studies have revealed that this species has an unusual social arrangement. Females leave their natal troops before their first breeding season, while males stay put for part or all of their adult lives. In most other primates, including South American squirrel monkeys, females remain in their natal troops and males migrate. Likewise, there is far less antagonism between troop members of this species than there is in most other monkeys. Males of the same age hang out with one another in distinct groups, known as cohorts, within each troop. Members of the oldest cohort look out for the rest of the troop: they spend much of their time gazing at the sky, scanning for hawks, and attempt to drive off any predators found in trees.

Male cohorts periodically leave their troops for a few hours to go and investigate neighboring troops. Such visits are frequent during the breeding season, especially right before and after females in the cohorts' own troops are sexually receptive. Females are receptive to such visitors, and will mate with visitors that get past resident males. Not surprisingly, there can be much antagonism between visiting and resident cohorts. If the second oldest cohort of one troop is in prime condition—often about five or six years old—it may permanently oust the oldest cohort of another troop it visits and stay with that troop. Usurped males wander alone until they die.

S. saimiri has a highly synchronized breeding season. Most breeding takes place around September. Gestation lasts a little less than six months, and most births occur around March. The single young are independent after about a year. Females first mate at two and a half years, males not until at least five. Females produce one baby a year during their first reproductive years, and subsequently one every other year. Captives can live at least 15 years.

Sounds: Squirrel monkeys have an extensive vocal repertoire; foraging troops reveal themselves with constant, high-pitched, birdlike peeps and twitters.

Conservation: Squirrel monkeys probably disperse seeds, and may pollinate some flowers, such as the passion flower *Passiflora adenopoda*. They are not significant agricultural pests, although they occasionally eat corn, coffee, bananas, mangos, and guanábanas.

The squirrel monkey is the most endangered monkey species in Central America. In the 1960s and early 1970s, thousands of squirrel monkeys were exported alive (mostly from South America, but also from Central America) for the pet trade and for biomedical research. Over 173,000 were sent to the United States alone between 1968 and 1972. Small numbers of squirrel monkeys are still being captured illegally in Costa Rica for the pet trade.

An additional and far greater problem for the Costa Rican species has been habitat loss. Squirrel monkeys can tolerate, and may even benefit from, moderate forest disturbance, for they spend more time in secondary and partially logged forests than in primary forests in areas where they have the choice. But they cannot persist in agricultural monocultures. In 1938, the United Fruit Company started massive deforestation in southwestern Costa Rica after Sigatoka disease, soil depletion, and labor disputes caused it to abandon banana plantations in the Atlantic lowlands. In just a few decades, the Central American squirrel monkey's habitat, which extended only a few hundred square kilometers to begin with, has been largely replaced by United's banana and African oil palm plantations, as well as by countless smaller beef cattle and other farms (see maps on pl. 17).

Although Corcovado National Park provides refuge for the southern subspecies, the northern subspecies has little habitat outside the small Manuel Antonio National Park. According to recent estimates, Central American squirrel monkeys now number a mere 3,000 individuals, with less than 250 of these in just six troops surviving inside Manuel Antonio Park. Detailed surveys have shown that squirrel monkeys are no longer found in a continuous band through southwestern Costa Rica, but as a series of disjunct populations separated from one another by as much as 40 km. As forest remnants become ever smaller and farther apart and troops become isolated, the complex migrations of individuals between troops described above, and the genetic diversity that this ensures, become impossible. Habitat fragmentation also increases squirrel monkeys' exposure to other dangers. Chemicals sprayed on crops that neighbor forest fragments kill or contaminate the squirrel monkeys' food, and a number of squirrel monkeys have been found dead from such poisoning. Some squirrel monkeys have also been found dead from electrocution after trying to move between forest patches along electric cables.

The status of the Central American squirrel monkey is particularly dire because its range was so small to begin with. Sadly, beyond that, its plight is really no more than a microcosm of the plight of most of the world's tropical mammals.

References

Arauz 1993; Baldwin 1970; Baldwin & Baldwin 1971, 1972b, 1976, 1981; Boinski 1985, 1987a,b,c,d; Boinski & Mitchell 1994; Boinski & Newman 1988; Boinski & Scott 1988; Boinski & Timm 1985; Costello et al. 1993; Emmons & Feer 1997; Happel 1983, 1986; Mitchell et al. 1991; Vaughan 1983a; Vaughan & McCoy 1984; Wong 1990; Wong et al. 1994.

O: Primates F: Cebidae

WHITE-THROATED CAPUCHIN

Cebus capucinus

(plate 17)

Names: Frequently called the white-faced capuchin, despite its pink face. Genus from the Greek *kebos*, meaning long-tailed monkey. Species and English name after the similarly colored outfit of Capuchin monks (hence, also, cappuccino coffee). There are three other species of capuchins (*Cebus* spp.), all found in South America. **Spanish**: *Mono carablanca; mono cariblanco.*

Range: Belize to northwestern Colombia and northern Ecuador; from sea level to about 3,000 m (10,000 ft); in evergreen or seasonally dry forests, second growth, and mangroves.

Size: 40 cm, 3 kg (16 in., 6 ½ lbs); males larger than females.

Natural history: Capuchins travel through all levels of the forest in troops of up to 30 individuals. They are omnivorous, but fruit and insects are their staples. A wide variety of fruit (95 different species in a study in Panama) can comprise from half to more than two thirds of the diet, depending on the troop. Capuchins check fruits by smelling, prodding, and tasting them, and often eat only the ripe ones. Having extracted the juice and fruit pulp in their mouth, they spit out large seeds and fibers.

Capuchins also eat other plant parts, such as flowers and new leaf shoots, especially when fruit is scarce. Bromeliads are favorite targets. Capuchins search in them for animal snacks and sometimes drink from

White-throated capuchin sign: a torn up bromeliad with chewed leaf bases.

the water reservoirs that are cupped between the leaves. They also pluck bromeliad leaves and chew on the fleshy bases, much as we would eat an artichoke. Bromeliads are strongly built and do not typically come apart when they fall to the ground, so scattered, chewed bromeliad leaves on the forest floor are a good sign that capuchins are about.

Favorite insect food includes beetle larvae, butterfly and moth caterpillars, and ants and wasps and their larvae. Sometimes capuchins take on larger animal prey, such as birds (as well as nestlings and eggs), small mammals, lizards, frogs, and, along the coast, crabs and oysters. In Guanacaste, capuchins pluck brooding white-crowned parrots from their tree-hole nests, and snatch white-throated magpie jays that don't keep their distance while mobbing. They also team up to chase squirrels, often knocking the squirrels to the ground and then pouncing on them, and have become so proficient at raiding coati nests and eating the young that they almost prevent coatis from raising any young successfully in parts of Guanacaste. Capuchins have even been seen taking bites out of a torpid silky anteater and the tail of a large, live iguana. The amount of vertebrates eaten by capuchins varies between troops and also between sexes. In Guanacaste, some troops eat relatively few vertebrates, while others are among the most carnivorous primates in the world. Males, which spend more time foraging on or near the ground than females, tend to be more carnivorous.

Foraging capuchins can attract other animals. Collared peccaries, agoutis, and coatis may follow on the ground, and pick up dropped food, while double-toothed kites, gray-headed tanagers, and slaty-tailed trogons may perch nearby in trees to pick off flushed insects and other small animals. Capuchins attract fewer birds than do squirrel monkeys, however, perhaps because they forage in smaller, less hyperactive troops and thus stir up less prey, or because they sometimes grab and eat the birds themselves.

Like spider monkeys and coatis (see pp. 147 and 226), capuchins appear to use some plants for pharmaceutical purposes. In Curú Wildlife Refuge on the Nicoya Peninsula, researcher Mary Baker observed capuchins rubbing themselves with various citrus fruits (including lemons, limes, sweet oranges, sour oranges, and mandarins), various vines (including old man's beard, *Clematis dioica*, and the black pepper relative *Piper marginatum*), and the hairy seed pods of *Sloanea terniflora*, better known as monkey comb. With the citrus fruits, capuchins abrade the peel by biting the fruits and scraping and banging them on branches, and then either rub themselves with the fruit intact, or break the fruit open and douse themselves with juice and pulp. With the vines, capuchins tear apart the leaves of *Piper* or the stems of *Clematis* with their hands and teeth, and mix the pieces with saliva to form an ointment. They use the monkey combs without any prior treatment. Capuchins rub each comb on their fur until all the comb's fuzzy hairs are rubbed off, then discard it and start with another hairy comb. (Curiously, the monkey comb acquired its name—both in English and in Spanish—long before its use by capuchins was documented by scientists.) In other areas, white-faced capuchins have also been seen rubbing themselves with dumb cane (*Dieffenbachia* sp., illustrated on p. 311), custard apple (*Annona* sp.), and even with ants and millipedes.

How, exactly, this behavior benefits capuchins is not known. Many of the items selected contain chemicals that could serve as insect repellent and/or to treat various skin conditions. Both *Citrus* and *Piper* plants contain insecticides, antiseptics, fungicides, and anesthetics, among other substances, while *Clematis* contains bactericides, analgesics, and anti-inflammatory chemicals. *Dieffenbachia*, ants, and millipedes all contain toxic chemicals that could also deter insects. Baker noticed that the Curú capuchins fur-rubbed most frequently in the wet season, which is when many insects and skin ailments would be most common. Alternatively, fur-rubbing could be a form of scent-marking, since many of the items selected have pungent odors. Collective fur-rubbing by troop members could create a "group odor," as urine-washing is thought to do for various other primates. And there is always the alternative that capuchins fur-rub simply because it feels good, especially since Baker noted that her capuchins seemed "almost frenzied" as they fur-rubbed. Monkey comb hairs and dumb cane sap, for example, are irritating to the human skin, but perhaps they cause a pleasurable sensation to monkeys.

Until recently it was thought that only Old World primates such as chimpanzees, orangutans, and ourselves had evolved the ability to use tools. In a sense, however, the plants and insects with which capuchins (and spider monkeys, with one plant anyway) rub themselves are medicinal tools. Various other types of tool use have also been recorded in capuchins. In captivity, capuchins use and even make tools to get at food and to attack or defend themselves. One innovative individual was recorded using a squirrel monkey as a projectile, hurling it at a human observer! A wild capuchin in Santa Rosa was seen dropping a stick repeatedly on a boa in an unsuccessful attempt to get the boa to release its stranglehold on a young troop member, and another in Manuel Antonio National Park was observed beating an almost two-meter-long fer-de-lance to death with a club. Capuchins, like other tool-using primates, have proportionately larger brains than other members of their family. The smaller-brained squirrel monkeys do not learn to use tools, even when placed together with tool-making capuchins.

Fer-de-lance (*Bothrops asper*).

The social organization of capuchins in Costa Rica has been studied by Linda Fedigan and Lisa Rose in Santa Rosa, and by Susan Perry in Lomas Barbudal. Female white-throated capuchins usually stay in their natal troops, while males disperse. This pattern is common among Old World primates, but in the New World it has been observed only in capuchins and the South American common squirrel monkey. Species that exhibit this pattern are sometimes termed female-bonded because females of such species tend to hang out together and form close ties. This may be because at least some of the females are related. There is usually a clear hierarchy among females of female-bonded species, defined and reinforced by the extensive social interactions. The research in Costa Rica revealed these traits in female white-throated capuchins.

Although male white-faced capuchins do not bond as closely as the females, neither do they get along as poorly as the males of some other female-bonded species, including their close relatives *Cebus olivaceus* and *Cebus apella*. In these South American capuchins, one male (typically the oldest and largest) tends to be dominant to the point where he alone fathers a troop's young. Adult males of these capuchins generally have antagonistic relationships, unless they are brothers, and resist the immigration of other males to their troops. The research in Costa Rica revealed that, in white-throated capuchins, adult males in the same troop have less-competitive relationships than the South American capuchins; they even associate with one another and cooperate to drive away predators. Dominant males are not necessarily the oldest individuals, and they do not monopolize reproduction. Ranking behind the dominant male, in variable order, are the dominant female and the beta male and female.

Recorded troop ranges for white-throated capuchins have been between about 30 and 160 ha (75 and 400 A). Single young or, rarely, twins are born mainly in the dry season after a gestation period of about five and a half months. Young capuchins have gray rather than white hair on their fronts until they are a month old. They move independently at five to six months. Females reach sexual maturity at about 4 years, males not until they are eight. Females produce young about every other year on average, although this is variable. A captive lived 47 years.

Sounds: Hoarse barks and coughs when alarmed; squeals, mostly when playing or feeding.

Conservation: Intelligent and adaptable, many capuchins have fallen victim to the pet trade. Although the same qualities help them survive outside pristine areas and they are conspicuous in some areas of Costa Rica, deforestation has decimated or isolated many populations. Such populations sometimes feed on corn and other crops and are occasionally persecuted by farmers. At the same time, deforestation has caused the demise in Costa Rica of one of the capuchins main predators, the harpy eagle. Various studies in South America have found capuchins (along with sloths) to be the favorite prey of harpies.

Capuchins influence the balance of ecosystems in numerous ways. Since they spit out larger seeds whole and pass smaller seeds through their guts unharmed, they are probably important dispersers. Studies have shown that some seeds germinate more quickly if they have been through a capuchin. By pruning the new growth of trees such as the brazil nut relative *Gustavia superba* (illustrated on p. 159) and the peeling tourist tree *Bursera simaruba*, capuchins cause the trees to branch more heavily and perhaps thus to produce more fruit. They may help some trees by eating insect pests, and limit the success of others by destroying flowers. They sometimes kill bull thorn acacias (*Acacia collinsii*) by ripping apart the branches and consuming larvae of the shrubs' ant colonies.

References

Baker 1996; Baldwin & Baldwin 1972b; Boinski 1988, 1989; Boinski & Scott 1988; Chapman 1986; Cheney & Seyfarth 1990; Fedigan 1990, 1993; Fedigan & Rose 1995; Fontaine 1980; Fowler & Cope 1964; Freese 1976b, 1977, 1978, 1983; Freese & Oppenheimer 1981; Greenlaw 1967; Happel 1986; Jiménez 1970; Jones 1982; Longino 1984; Moscow & Vaughan 1987; Oppenheimer 1968, 1996; Oppenheimer & Lang 1969; Perry 1996; Perry & Rose 1994; Rettig 1978; Rodríguez (M.) 1985; Rose 1994a,b; Smuts 1987; Stott & Selsor 1961; Timm et al. 1989; Timm & Laval 2000b; Westergaard & Fragaszy 1987; Wille 1987; Wrangham 1980, 1987.

O: Primates F: Cebidae

MANTLED HOWLER MONKEY

Alouatta palliata

(plate 18)

Names: The common name refers to the mantle of blond fur on the back that distinguishes this species from five other howler species (*Alouatta* spp.) found in Central and South America. Genus is the Greek for "another that moves using the tip of its tail"; *palliata* is the Latin for mantled. This and the Yucatan black howler (*A. pigra*) of northern Central America were formerly considered the same species, *A. villosa*. **Spanish**: *Mono congo*; *mono aullador*.

Range: Southeastern Mexico to northern Peru and Colombia west of the Andes; in primary and mature secondary forests; from sea level to at least 2,500 m (8,200 ft).

Size: 50 cm, 5 kg; (20 in., 11 lbs); males larger than females.

Natural history: The mantled howler monkey has been studied quite extensively, particularly by Kenneth Glander, Margaret Clarke, and Clara Jones at Finca La Pacífica in Guanacaste; by Katherine Milton in Panama; and by Alejandro Estrada and Rosamond Coates-Estrada in Mexico. Glander's study, which started in 1970 and continues to this day, is one of the longest-running studies of any tropical mammal.

Howlers forage by day, although they sometimes start moving before dawn. At night they sleep on canopy branches, singly or in small clusters, with troop members often dispersed through more than one tree. They do not use any sort of nest, nor do they sleep in the same trees every night. When active, they stay mostly in the upper levels of the forest, although they do come to the ground to cross open areas between forest fragments or to swim across rivers that cannot be traversed via canopy vegetation.

Howlers differ from all other New World monkeys—and, indeed, all New World arboreal mammals except sloths—in feeding heavily on leaves. It seems strange that such an abundant resource should be so rare in the diets of arboreal mammals. In reality, however, leaves are a very challenging food, for they are full of cellulose and many contain toxins as well. Since no mammalian stomach enzyme can break down cellulose, mammals that consume large quantities of it must break it down through fermentation, which is a slow and space-consuming process. Howlers ferment leaves in the hindgut—in the cecum and in the colon—and do not digest leaves as efficiently as ruminants such as deer or cows (see pp. 304-305).

Howlers eat the leaves of dozens of different tree species, but they select the leaves carefully. About two thirds of the foliage eaten by Glander's howlers were young leaves, which tend to contain less cellulose and toxins and more protein than mature leaves. Howlers eat the leaves of many trees—such as carne asadas (*Andira inermis*), guapinols (*Hymenaea courbaril*), or figs (*Ficus* spp.)—only at certain times of year, when new leaves are available. When new leaves are scarce, and howlers have to rely on mature leaves, they select those that contain the most protein and the least toxins and fiber. Some trees respond to being eaten by producing more toxins. An individual tree that is palatable to a howler at one time may not be shortly thereafter, so howlers cannot necessarily feed on all individuals of a given tree species. Glander noticed that this was the case, for example, with the peeling tourist tree (*Bursera simaruba*) and the mata ratón (*Gliricidia sepium*).

Mata ratón trees (*Gliricidia sepium*).

Howlers are not exclusively folivorous (leaf-eating). Flowers and fruit comprised about a third of the diet of Glander's howlers, and have been found to comprise half the diet or more in other areas. At La Pacífica, howlers enjoy the flowers of many of the plants they get leaves from, including the peeling tourist tree and the mata ratón, as well as raintrees (*Pithecellobium saman* and *P. longifolium*), the wild cashew espavél (*Anacardium excelsum*), and alcornoque (*Licania arborea*). Both there and in other areas, howlers tend to feed on flowers most heavily during the dry season.

Favorite fruits at La Pacífica include those of the basswood relative capulín (*Muntingia calabura*), sapotes (*Manilkara sapota*), hogplums (*Spondias mombin*), and cecropias (*Cecropia peltata*). Fig fruits (and leaves) are important to howlers in many areas because they do not appear synchronously and are thus available through much of the year. In a study near San Ramón, Costa Rica, some troops were found to dedicate more than 70% of their feeding time to figs alone. Researchers working with the black howler monkey (*A. caraya*) in Argentina have noticed that howlers eat many fruits—especially figs and members of the avocado family (Lauraceae)—that are infested with insect larvae. This habit may provide howlers important proteins that their other foods lack.

Since flowers, fruits, and palatable leaves are not equally abundant throughout the year, howlers have to vary their diet seasonally. At La Pacífica, howlers feed predominantly on new leaves and flowers during the late dry and early wet seasons, but depend mostly on fruits and mature leaves during the late wet season. During the early dry season, food there is scarcest, and howlers have to eat leaves that are less nutritious and more toxic. Weak individuals sometimes die at this time of year as a result.

Howlers get most of the water they need from their food, but they do drink from cisterns in trees during some portions of the year. Strangely, Glander found that howlers at La Pacífica drink only during the wet season, apparently because their main source of water is succulent new vegetation, which is abundant only during the dry season.

With their low-energy diet, howlers are more lethargic than other monkeys. They spend as much as three quarters of the day and all night resting or asleep. Spider monkeys, by comparison, spend only about one fifth of the day resting. Howlers like to rest sprawled across exposed branches where heat from the sun saves them energy and speeds up the fermentation process. They sometimes contest good early-morning sunning spots with turkey vultures. Troops seldom move more than 2 km in a day, and usually cover less than 1 km. They occupy home ranges of only about 3 to 76 ha (7 ½ to 190 A). Being more sedentary, howlers do not usually flush enough prey to attract the birds that follow white-faced and squirrel monkeys, but they do drop food morsels that attract Jesus Christ lizards (*Basiliscus basiliscus*) on the ground.

Jesus Christ lizard (*Basiliscus basiliscus*).

Howler troops typically do not defend exclusive territories; they simply tend to stay clear of one another. If two troops happen to meet, however, they will often contend the area they are in. Although frequently indifferent, howlers are sometimes irked or scared by people, and they like to express such emotions by urinating or defecating on their observers, often with startling accuracy. It is thus unwise to view howlers from directly beneath.

Mantled howler troops contain anywhere from 2 to 45 individuals. All males are dominant over all females. Within the sexes, there is an unusual hierarchy. In contrast to other primates and most other social mammals, the youngest adult of each sex is dominant. The hierarchy continues roughly according to age, with the oldest individuals at the bottom of the ladder. Each alpha female enjoys, on average, about a year at the top of the hierarchy before she is replaced. There are usually far fewer adult males in a troop than adult females, and the alpha male changes only about once every four years. High ranking individuals get first choice of food and resting sites, and the alpha male generally mates with all his troop's females. Only if the alpha male is distracted does the beta male get lucky.

Males solicit sex with a Mick Jagger-like tongue-flicking display. If the female is impressed, mating quickly ensues and lasts on average about 30 seconds. Gestation lasts about six months. Single young or, occasionally, twins, are born at any time of year, in the treetops. The mother gives birth without assistance or even interest from other troop members, cradling and lightly pulling the baby's head herself as it emerges from her birth canal. The baby is able to cling to its mother without help immediately after birth. Babies sometimes fall to the ground, where they are retrieved by their mothers, usually without serious injury. Despite their disinterest during the birth process, fellow troop members soon flock around to investigate the newborn.

The fur of baby howlers is silver at birth, but turns pale or golden within a few days. The fur gradually darkens from then on, reaching adult coloration by the time the infant is about three months old. The mother carries the baby almost constantly until it is about two months old, at which time the mother starts to push it off, especially when she is feeding. The baby continues to get rides until it is four or five months old. Other troop members, mostly adult males and juveniles, take spells carrying or caring for the baby.

Not all troop members are friendly toward young. High-ranking, nonrelated adult females will sometimes kidnap babies, removing them forcibly from their mothers and then mistreating them, sometimes even causing them fatal injuries. By harassing or killing other babies, the kidnappers may reduce competition for their own offspring. The arrival of a new alpha male means almost certain death for a troop's young, for the new male generally kills all infants less than eight months to a year old—most or all of which would have been fathered by his rival predecessor. The mothers of the murdered babies come into heat again within just two weeks, mate with the new male, and become pregnant with the new male's own young. With all these threats to babies within the troop, let alone the constant danger of other predators such as cats, weasels, snakes, and eagles, only about 30% of young at La Pacífica survive more than a year.

Young lucky enough to survive stop nursing all together around the time they reach adult size, at a little over a year. Young males leave their natal troop at just over a year to three years of age, and then wander alone for up to three years before joining a new troop. Males reach sexual maturity at about three years, but usually do not mate until they are five or six, by which time they are strong enough to dominate a troop. Males prepare to take over a troop by hanging around nearby and associating with subordinate females, and then try to oust the resident alpha male with a brief fight. Skirmishes for the alpha-male position are silent and brief, lasting only about 90 seconds, but the loser usually winds up bleeding and wounded. Adult males are easily distinguished by their larger size and conspicuous white scrotums.

Females depart their natal troop between about two and three years of age, and integrate with a new troop within a year, by which time they are sexually mature. Like males, females prepare to work their way into a troop by hanging out nearby. They woo and then mate with a dominant male, thus securing his protection and other troop members' respect. They give birth on average every one and a half to two years from then on.

Howlers can live at least 25 years in the wild; several of Glander's howler's died just before or after their twenty-fifth birthday.

Sounds: Grunts and intimidatingly loud, pulsating roars lasting four to five seconds, often repeated several times, given mainly by adult males. The calls of females are softer and higher pitched. Howlers call at dawn and dusk, and periodically throughout the day in response to the calls of neighboring troops or other, presumably similar noises, such as those made by engines, thunder, or Arenal volcano. The calls can be heard several kilometers away. In male howlers, the hyoid—a hollow bone next to the vocal chords that serves as an amplifying chamber rather like, for example, the body of a drum—is 25 times larger than that of other similar-sized monkeys. Given their unusual diet, howlers may have evolved such astonishing vocal capacity to save energy; the roars enable troops to locate one another without moving around, and to minimize physical confrontations.

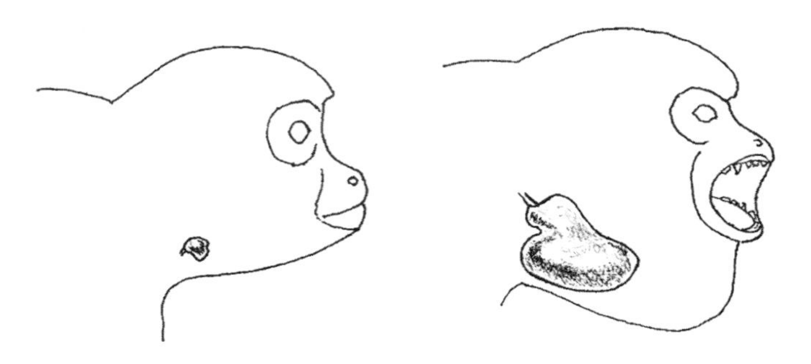

The hyoid bone is situated next to the vocal chords and serves as an amplification chamber. The hyoid bone of howlers (right) is about 25 times larger than that of other similar-sized monkeys (left).

Conservation: The mantled howler is the most common monkey in Costa Rica. Although it is hunted and eaten in a few areas, and its broth used as a supposed remedy for chest congestion, it is left alone in most parts of the country. Since howlers feed heavily on leaves, they are situated relatively low in the food pyramid. Thus they are more common and need less space than most mammals their size. In parts of Costa Rica where howlers coexist with white-faced and spider monkeys, howlers can represent 70% or more of the biomass of all the monkeys put together. These traits have enabled howlers to survive in forest fragments where many other mammals have disappeared.

It is clear, however, that howlers have complex needs. Forest fragmentation has reduced and isolated howler populations, and it can affect the survivors in unseen ways, such as by hindering the flow of individuals between troops and the genetic diversity that this ensures. Analysis of blood samples taken from dozens of howler monkeys in Guanacaste over the last few years has revealed that the genetic diversity of howlers in that region is about half what it should be. Forest fragmentation can also impede access to seasonal foods. Some howler researchers have wondered whether increased exposure to sun and wind in forest fragments could affect howlers or the plants howlers depend on. Others worry that, in some fragments, dust from dirt roads and eroded fields may accumulate on leaves and cause excessive wear on howlers' teeth. In a study of a large number of fragments averaging about 10 ha (25 A) each in Veracruz, Mexico, less than half the fragments were found to contain howler monkeys. The surviving troops were found to be smaller (averaging only about four individuals) and to have a significantly lower proportion of young than troops from more pristine habitats.

The disappearance of howler monkeys can affect an ecosystem in countless ways. One group of animals that is known to depend heavily on howlers is dung beetles. In Mexico, researchers have discovered that, in fragments where howlers have disappeared, they can find only 10% of the quantity of dung beetles that they capture in areas that still have howlers. Dung beetles are important secondary dispersers for some of the seeds contained in dung, and, at densities of up to 2,000 beetles per hectare in pristine conditions, are also major nutrient recyclers. Howlers themselves are important seed dispersers for some plants. Being large, they are among the few animals that can transport the large seeds of certain primary forest trees, such as avocados (*Persea americana*) or sapotes (*Pouteria sapota*). And some smaller seeds, such as those of figs, are known to have a better chance of germinating if they have passed through a howler. The success of figs and other plants in turn is critical for numerous other animals.

A dung beetle.

References

Baldwin & Baldwin 1972a, 1976; Bravo et al. 1995; Bravo & Zunino 1998; Cantero 2000; Clarke (M.) 1981, 1982, 1983; Eisenberg 1989; Estrada 1984; Estrada, Anzures et al. 1999; Estrada & Coates-Estrada 1984, 1985, 1986, 1991, 1996; Estrada, Juan-Solano et al. 1999; Fedigan & Rose 1995; Ferrari 1991; Freese 1976a; Froehlich et al. 1981; Glander 1974, 1975, 1978a,b, 1979, 1980, 1983, 1996; González 1992; Haber et al. 1996; Hladik 1978; Hladik & Hladik 1969; Horwich & Lyon 1993; Jones 1978, 1980a,b, 1985; Keleman & Sade 1960; Kinzey 1997a; Milton 1978, 1980, 1984, 1996; Milton et al. 1979; Neville et al. 1988; Peck & Forsyth 1982; Rodríguez (M.) 1985; Sánchez 1991; Sekulic 1982; Silva et al. 1993; Smith 1977; Stoner 1993; Young 1982.

O: Primates F: Cebidae

CENTRAL AMERICAN SPIDER MONKEY

Ateles geoffroyi
(plate 18)

Names: Also known as the black-handed spider monkey. Common name distinguishes it from three other spider monkey species (*Ateles* spp.), all found in South America. Some believe that all four spider monkeys are the same species, *A. panicus. Ateles* means incomplete, in reference to spider monkeys' thumbless hands. Species name after Isidore Geoffroy, a nineteenth-century French zoologist who was the first to collect and describe several species of Neotropical mammals. **Spanish**: *Mono colorado*; *mono araña.*

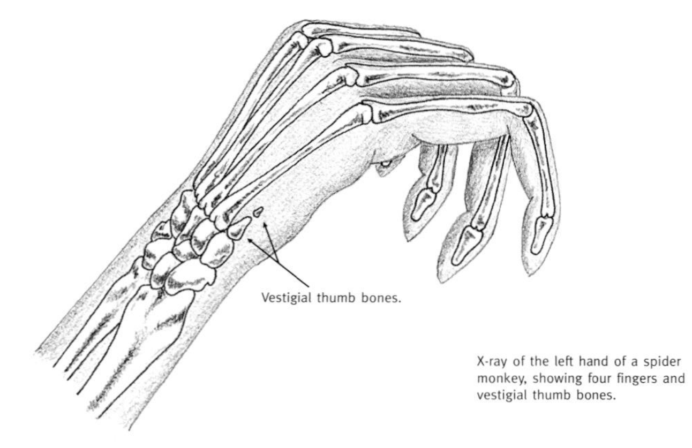

X-ray of the left hand of a spider monkey, showing four fingers and vestigial thumb bones.

Range: Northeastern Mexico to northwestern Colombia; from sea level to at least 2,800 m (9,200 ft) on Cerro de la Muerte; mostly in large expanses of primary forest.

Size: 50 cm, 7 kg (20 in., 15 lbs); males larger than females.

Natural history: Spider monkeys forage by day in the upper levels of forest. They are more exclusively frugivorous (fruit-eating) than Costa Rica's other monkeys; various studies have found that they spend between 70 and 80% of their feeding time eating fruit. They feed on dozens of varieties, but prefer fleshy, ripe fruits when these are available. Leaves can make up most of the rest of the diet. Young leaves—of trees such as figs, breadnuts (*Brosimum alicastrum,* illustrated on p. 158, and

Pseudolmedia spp.), and cecropias (*Cecropia* spp., illustrated on p. 62)—are preferred, and may be an important protein supplement in their otherwise protein-poor diet. Spider monkeys don't just eat leaves; they have been seen rubbing a mixture of masticated leaves of lime trees (*Citrus aurantifolia*) and saliva over their fur, possibly as a form of homemade insect repellent. On occasion, spider monkeys also eat flowers, bark (especially wet bark from the underside of limbs), aerial roots, insects, and honey. South American spider monkeys can spend almost 10% of their feeding time eating decayed wood and, curiously, will visit the same decayed trees or parts thereof repeatedly, sometimes for years on end.

Fruit is a challenging staple, for it is seasonal and patchily distributed, and spider monkeys have evolved a suite of adaptations to help them deal with this problem. Various anatomical features enable them to move easily and quickly between scattered fruiting trees and reach the branch tips where fruit grows. Strong arms that are about 25% longer than the legs—the opposite of most mammals—allow spider monkeys to brachiate, which means to swing hand to hand beneath branches like a gibbon. Unusually long, strong fingers serve as grapple-like tools for hooking branches. The thumbs on spider monkeys' hands are reduced and barely visible, for they would hinder brachiation and are not necessary for manipulating fruit. And the spider monkey has the strongest and most dexterous prehensile tail of any mammal in the world. A palmlike naked pad on the underside of the tip increases the tail's sensitivity and helps with grip. Truly a fifth limb, the tail alone can propel its owner forward, and it is sometimes used to pick fruits or scoop water out of tree holes.

Spider monkeys have curious social habits that might also be adaptations for frugivory, at least in part. Spider monkeys live in groups of about 20 to 40 individuals that are often referred to as communities rather than troops, for they tend to separate into small subgroups during the day and often even to sleep at night. Subgroups most often contain two to four individuals, and seldom more than six. By splitting up, communities are probably able to search for scattered fruit trees more efficiently. In a study at Santa Rosa Park, researcher Colin Chapman found that, when fruit is most abundant, spider monkeys tend to split up much less, presumably because at those times of year they don't need to bother searching for smaller, more distant food resources.

Food availability is not the only factor that influences the composition of spider monkey subgroups. Chapman and other primatologists working elsewhere have found that most subgroups contain individuals of the same sex. Male subgroups travel farther and faster than female ones, spending more time around the edge of the communal range, perhaps scouting for potential mates or defending the range, or both. In Santa Rosa, such male groups stay united even when food is at its scarcest and most other community members have split into smaller groups, perhaps because they cannot successfully defend their territory otherwise. Female subgroups, on the other hand, especially those with young, tend to stay nearer the center of the communal range. Females with young also tend to hang out in small subgroups, or alone. Young in small, centrally located subgroups may be safer, for they are less likely to be attacked by adults, a common event among spider monkeys. For all these reasons, individuals tend to use only portions of the communal ranges. The widest-ranging Santa Rosa male, for example, used only about 60% of his community's 170 ha (420 A) range. The few home-range estimates for communities of this and other spider monkey species vary from 100 to 390 ha (250 to 960 A).

Spider monkeys mate with the male and female seated on a branch, facing in the same direction. The male wraps his arms around the female's chest and his legs around her waist. During mating, the female purses her lips and partly closes her eyes. She shakes her head constantly, periodically turning to gaze into the male's face. Sometimes she reaches back to caress the male with one of her long arms. Their passionate embrace lasts anywhere from 8 to 25 minutes, and females in heat will mate three or four times a day. Apparently females are able to choose their mates. One female of a South American species was observed breaking off copulation with one male in order to rush off and mate with another that was calling nearby.

Single young are born after a gestation period of about seven and a half months. Births occur at any time of year, but more often in the dry season in Guanacaste. Young ride on the mother's chest for the first one and a half to two months and then on her back. Young are dark or black all over until the age of about five months, at which time the fur on their backs starts to turn reddish. After about three months, young can move independently and start to eat solids, but they continue to nurse until they are almost a year old. When an immature spider monkey reaches a gap in the canopy vegetation that is too wide for it to cross, an adult stretches itself across the gap and allows the young to traverse across its back. (Howler monkeys also use this tactic, but less often.)

Females reach sexual maturity at about four years, males at about five. As in squirrel monkeys, females leave their natal troops, while males stay put. Thus the males in a given troop are likely to be related, which could help explain why they team up. Once settled in a new community, females have babies only once every two to four years. Females are easily mistaken for males; they have a conspicuous, penislike clitoris, while males' genitals are not usually visible. Captives can live at least 33 years.

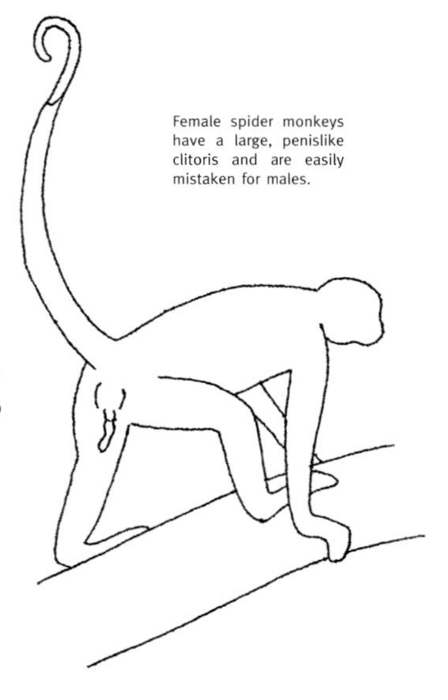

Female spider monkeys have a large, penislike clitoris and are easily mistaken for males.

Sounds: Barks and prolonged raspy notes in alarm; squeaks, squeals, and horselike whinnies during social interactions; whinnies and prolonged screams as long-distance calls given at dawn and dusk, in distress, and periodically during foraging. By recording hundreds of calls and comparing the sonograms, biologists in Santa Rosa ascertained that each monkey makes a consistent and distinct whinny sound. Spider monkeys may therefore be able to recognize each other by call alone, a useful skill given their habit of splitting into small, scattered groups.

Conservation: Spider monkeys are probably excellent seed dispersers. They swallow many seeds rather than discarding them, and pass them through their digestive tracts in just a few hours, so that most are defecated intact. Studies have shown that fruits of fig trees and the mahogany relative *Trichilia*, for example, germinate after passing through spider monkeys. Spider monkeys also tend to feed quickly at each tree and then move on. One South American spider monkey was recorded eating about 100 *Pouteria* fruits (measuring about 2 to 2.5 cm each) in just seven minutes! Seeds scattered widely in small piles may have a better chance of germinating than those that fall or are dropped to the ground in large concentrations because they are less conspicuous to seed predators. Spider monkeys scatter seeds effectively both because they move around a lot (covering as much as 5 km each day), and, on a smaller scale, because their feces, consisting mostly of loose seeds, tend to break apart when falling from the canopy. Spider monkeys may also disperse larger seeds, such as cannonball avocados (*Persea americana*), simply by carrying them away from the parent tree rather than by swallowing them.

Spider monkeys do not survive outside large tracts of primary forest and have been greatly affected by deforestation. Since they are good-tasting, large, and conspicuous, they have been hunted heavily in many regions. In some regions they used to be captured for the pet trade. The protection of this species by international law—it is now listed as an endangered species in CITES Appendix I—has reduced hunting pressure. Spider monkeys' reproductive turnover is so low, however, that depleted populations cannot easily replenish themselves. They have thus disappeared from many regions where they used to be common, even in areas that are now protected, such as the Cabo Blanco Reserve, Finca La Pacífica, and parts of Monteverde.

References

Chapman 1988a,b, 1990a,b; Chapman & Chapman 1990; Chapman & Lefebvre 1990; Chapman & Weary 1990; Dare 1974; Eisenberg 1973, 1976, 1983a; Estrada & Coates-Estrada 1988; Fedigan & Rose 1995; Freese 1976a; Gompper & Hoylman 1993; Haber et al. 1996; Hladik & Hladik 1969; Klein & Klein 1976, 1977; Lippold 1988; Mittermeier 1978; Mittermeier et al. 1988; Nowak 1991; Oppenheimer 1977; Richard 1970; Rodríguez (J.) & Chinchilla 1996; Rodríguez (M.) 1985; Silva et al. 1993; Symington 1988a,b; Timm & Laval, 2000b; Van Roosmalen 1980; Van Roosmalen & Klein 1988.

Rodents (order Rodentia)

Distribution and Classification

With about 2,000 species worldwide, rodents represent almost half the planet's mammals, and easily make up the largest mammal order, with about twice as many species as the next largest order, the bats. In Costa Rica, the relative abundance of rodent and bat species is reversed, for there are just under 50 types of rodents and about 116 of bats. Rodents occur naturally on every continent except Antarctica, and they have been introduced to many islands, such as New Zealand. Their most distinctive feature is their incisor teeth; the order name comes from the Latin verb *rodere*, which means to gnaw.

Rodents traditionally have been split into three suborders, which are distinguished principally by the arrangement of the jaw muscles. These suborders are somewhat artificial, since jaw musculature alone is not necessarily an indicator of relatedness. The suborder Sciuromorpha, or squirrellike rodents, is represented in Costa Rica by five species of squirrel, four pocket gophers, and three spiny pocket mice. The suborder Myomorpha, or mouselike rodents, is represented here by about 30 species in the family Muridae. And the suborder Caviomorpha (known also as the Hystricomorpha), or cavylike rodents, includes Costa Rica's single species of porcupine, agouti, and paca and two species of spiny rats.

Evolution

The rodent species that inhabit Costa Rica today are distinctly Neotropical. The ranges of more than three quarters of Costa Rica's rodents are restricted to the area between Mexico and northern South America, and about a third of those are endemic to Costa Rica and northwestern Panama. The ancestral origins of these rodents, however, are diverse.

Primitive rodents apparently spread to different parts of the world before the continents drifted too far apart, and then evolved separately for millions of years. Ancestors of Costa Rica's sciuromorph and myomorph rodents evolved largely in the north. North American fossils of sciuromorphs dating back about 60 million years to the Paleocene period are among the earliest-known rodents, while myomorph fossils date back about 25 million years in North America and about 35 million years in Europe.

The caviomorphs, meanwhile, came to Costa Rica from South America, where fossils date back about 35 million years to the Oligocene. Since extensive fossil collections suggest that caviomorphs did not inhabit South America prior to the Oligocene, and since South America by that time was isolated from other land masses, how and whence the caviomorphs first arrived there has been much debated. Curiously, monkeys arrived in South America at exactly the same time, so monkeys and caviomorph rodents are often lumped together in this debate. The main theories are summarized in the section on monkey evolution on pp. 126-127. Regardless of where the caviomorphs originated, the Panamanian land bridge that rose out of the sea seven to two million years ago enabled the caviomorphs and their northern cousins to mingle once again.

Teeth

Rodents are characterized above all by their unique dental arrangement. Their most important teeth are their four incisors, two upper and two lower. The incisors grow constantly, as quickly as several millimeters per week, and are simultaneously worn down since the upper and lower incisors grind against each other and against the hard foods that prevail in most rodents' diets. Only the incisors' front edge is covered with enamel, and it wears down more slowly than the softer, back part of the tooth. This design, much like that of a chisel, ensures that the incisors stay sharp. The incisors are rootless, and emerge from a deep, curved canal, arching inward as they grow. If rodents do not wear them down—if, for example, the upper and lower incisors are not aligned—the incisors will become so long that they impede feeding; eventually they can curl around so far that they penetrate the skull.

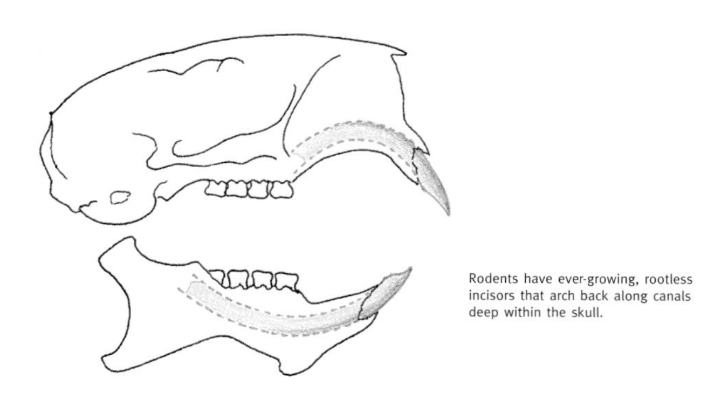

Rodents have ever-growing, rootless incisors that arch back along canals deep within the skull.

Behind the incisors, there is a large gap or "diastema," for rodents lack outer incisors, canines, and anterior premolars. Rodents can pull their lips into this gap, thus shutting out the incisors and preventing gnawed, inedible debris from falling into the mouth. Behind the diastema, rodents have three to five cheek teeth in each quarter of the mouth—premolars and molars in sciuromorphs and caviomorphs, and molars only in myomorphs. On these teeth too, enamel ridges are worn down slower than areas with softer dentine, creating patterns on the grinding surfaces that are sometimes used in classification (see p. 171, for example).

Body Design

Rodents have five digits on each forefoot (although many species' thumbs are almost invisible) and three to five digits on each hindfoot. Different species may have big claws and short limbs for digging, sharp claws for climbing, hooflike claws for running and scraping for food, long hind limbs for hopping like a kangaroo, or webbed feet for swimming.

Most rodents have conspicuous tails. Bushy tails, such as those of squirrels, may be used for balance in trees, as a cloak for insulation, or as a flag in territorial or breeding displays. Naked tails, too, can help rodents regulate their body temperature. Hispid cotton rats (p. 171) that have lost their tails have markedly different body temperatures and metabolic and water loss rates compared to hispid cotton rats that have their tails intact. Naked tails may play an important sensory function in some rodents, such as pocket gophers. Sometimes rodent tails are fragile, like those of certain lizards, and break off in the mouths or claws of predators, giving the rodent a chance to escape. In other species, the skin covering the tail slips off the flesh and bone, with the same effect. The tail owner then bites off the exposed portion, allowing the stump to heal. For us, rodent tails often provide the best means of distinguishing species. The length and color is an obvious distinguishing feature, but even naked tails, when viewed with a magnifying lens, reveal unique patterns of scales and hair growth.

Senses

Some rodents have good eyesight. Squirrels, for example, need good vision to leap around in treetops, and also rely on the sense to find food and detect predators. Their conspicuous tail-shaking visual displays are an important part of courtship. All rodents have an acute sense of smell, which is particularly important in communication and reproduction. Odors in males' urine can bring sexually quiescent females into heat, and odors in females' urine in turn alert males when females are receptive. Rodents also have excellent hearing. Many of the sounds they use in communication are ultrasonic and inaudible to humans.

Reproduction

Rodents, the smaller ones anyway, are notoriously fecund. Except for the larger species, such as Costa Rica's caviomorphs, rodent gestation periods last weeks rather than months, and their large litters require equally little time to reach sexual maturity. Lemmings (which are found only in the Northern Hemisphere) are a famous, if extreme, example. Females have been found pregnant at the age of 14 days, and one pair was recorded to produce eight litters in less than six months (at which point the male died). House mice can produce up to 14 litters a year with up to 12 young per litter, and Norway rats up to 12 litters a year with as many as 22 young per litter!

As in many other mammals, males tend to be more promiscuous than females. In gregarious species such as the house mouse, dominant males have been observed mating with as many as 20 females in just six hours. This pattern may exist because mating is cheaper for males than it is for females. To sire young, all males have to do is mate, and they perpetuate their genes most effectively by mating with as many females as possible. Females, on the other hand, have to gestate and then nurse young, and are thus more limited in how many young they can produce. They maximize success at perpetuating their genes not by being promiscuous, but by selecting a male who will give them healthy and successful young, and this may be best achieved by mating repeatedly with the top stud.

Most rodents don't live very long. Even in captivity, few rodents are likely to live more than five years. In the wild, most rodents fall victim to seasonal food shortages or to one of their many predators before they are two years old. Only around areas of human habitation, where food may be constant and abundant, and predators limited, does this system break down.

Conservation

From one perspective, this section might more appropriately be titled *eradication* than *conservation*. It has been estimated that there are approximately the same number of rodents in the world as people. They eat or spoil more than 40 million tons of food each year, and, by spreading diseases, have caused more deaths over the past 5,000 years than have all the world's wars during the same period. The vast majority of this damage is caused by a small handful of species that tend to associate with people (and whose abundance and geographic range, thanks to this association, has ballooned in a very short period). Where introduced, these species can be detrimental to the natural world as well, destroying native species that are evolutionarily unprepared for their presence.

But what about all the other species? A native rodent in its natural habitat is as important as any other creature. Rodents greatly influence the abundance of thousands of plant species by destroying or dispersing seeds. A few species may pollinate flowers or disperse fungus spores. And rodents form a critical link in food chains, for they are the staple food of numerous carnivorous animals.

Rodents can help us directly, too. In a perverse way, laboratory rats have compensated for the widespread health problems caused by some of their cousins by contributing enormously to our understanding of issues that range from psychology and genetics to stomach ulcers and cancer. (Rats and humans have many biological similarities, which is why we are susceptible to so many of their diseases). It is somewhat surprising how little we have exploited rodents for food. Many species, such as Costa Rica's paca, agouti, spiny rats, and some squirrels, have excellent meat and could probably be farmed profitably in an environmentally friendly manner.

References

Corbet 1984a,b; Dubock 1984; Eisenberg & Kleiman 1977; Gray 1967; Hilje & Monge 1988; Kirksey et al. 1975; McPherson 1986; Nowak 1991; Rodríguez & Chinchilla 1996; Rowlands & Weir 1974; Smith 1974; Smythe 1987; Stoddart 1984.

SPECIES ACCOUNTS

O: Rodentia F: Sciuridae

ALFARO'S PYGMY SQUIRREL
Microsciurus alfari
(plate 19)

Names: Also known as Alfaro's or the Central American dwarf squirrel. Some authors reserve the term *pygmy squirrel* for the genus *Sciurillus*, which contains a single species, the Neotropical pygmy squirrel (*S. pusillus*) of South America. The family name, genus, and the word *squirrel* all come from the Greek word for squirrel, *skiouros*, which in turn is derived from the words *skia*, meaning phantom or shadow, and *oura*, meaning tail. *Microsciurus* means small squirrel. The species name is a dedication to Anastasio Alfaro, one of Costa Rica's pioneer naturalists who was instrumental in founding Costa Rica's National Museum in 1887, and who, at the age of only 22, became its first director. The museum currently houses one of Costa Rica's most important mammal collections, among many other things. **Spanish**: *Ardilla enana.*

Range: The Sciuridae contains about 270 species, distributed almost worldwide; the squirrel family is absent only from Australia and Antarctica. There are four species in the genus *Microsciurus*, distributed from Nicaragua to Peru and the Amazon. They tend to be most common at mid- to high-elevation forests. *M. alfari* is found from southern Nicaragua to northwestern Colombia near Panama; from sea level to at least 2,600 m (8,500 ft); in wet, relatively undisturbed forests.

Size: 14 cm, 100 g (5 ½ in., 3 ½ oz).

Similar species: The diagnostic features of Alfaro's pygmy squirrel are its small size; narrow, tapering tail; inconspicuous ears that do not protrude above the crown of the head; and underparts that vary from gray to buffy. Montane squirrels (pl. 19, p. 164) also have inconspicuous, low-set ears, but are somewhat larger, have a bushy, nontapering tail lightly frosted with orange, and more orangey underparts. Deppe's squirrels (pl. 19, p. 157) resemble Alfaro's pygmy squirrels in coloration but are larger, have a bushy, nontapering tail, and have ears that protrude above the crown. Red-tailed (pl. 19, p. 159) and variegated (pl. 20, p. 161) squirrels are much bigger and have large, bushy tails and conspicuous ears.

Natural history: Alfaro's pygmy squirrels have not been studied in any depth, and details of their natural history are largely unknown. They forage by day at all levels of the forest, often on or near the ground. Their diet includes fruits and nuts, bark, tree sap, and insects. They are known to be fond of the fruits and nuts of various palms (including *Scheelea*), and of the legume almendro (*Dipteryx panamensis*). They sometimes feed heavily on the sap of other leguminous trees (*Inga* spp.) and of the kapok relative *Quararibea costaricensis*. In Monteverde, pygmy squirrels cling to trunks of *Quararibea* and peel back bark to get at the sap. They may come back to the same feeding spot repeatedly, leaving bare patches as large as 50 cm^2 (8 $in.^2$).

Pygmy squirrels tend to move quickly and quietly, and they make impressively long leaps between branches. Their speed and agility are due in part to the design of their limbs, which are proportionately longer than those of *Sciurus* squirrels. Longer limbs also improve the pygmy squirrel's ability to move up and down tree trunks. When on vertical trunks, squirrels spread their forelimbs sideways around the trunk in order to increase leverage and grip. The long forelimbs of pygmy squirrels provide greater mechanical advantage in this position and help the squirrels cling to certain large trunks (such as *Quararibea*) or smooth trunks (such as *Scheelia*) that might otherwise be inaccessible to them.

Alfaro's pygmy squirrels often travel in pairs and have been seen on occasion in larger groups. Otherwise, almost nothing is known of the social or territorial arrangement, or of the reproductive biology, of this species.

Sounds: In alarm, bouncy, birdlike notes that somewhat resemble the call of the white-throated spadebill, but which are a bit lower pitched and sound more like a *chuck* and less like a whistle, repeated in an erratic series.

Conservation: Pygmy squirrels are inconspicuous but can be common in mid- to high-elevation wet forests, such as at Monteverde. They are less common in the wet lowlands of the Atlantic and south Pacific slopes, and absent from the seasonally dry forests of northwestern Costa Rica.

References

Eisenberg 1989; Emmons et al. 1997; Enders 1935; R. Guindon, unpublished data; Hayes & Laval 1989; McPherson 1985; Reid 1997; Rodríguez & Chinchilla 1996; Thorington & Thorington 1989; Timm 1994a; Timm et al. 1989; Timm & Laval 2000a,b.

O: Rodentia F: Sciuridae

DEPPE'S SQUIRREL
Sciurus deppei
(plate 19)

Names: Has also been referred to as the Miravalles squirrel (the first specimen from Costa Rica was captured at Miravalles Volcano, and the Costa Rican subspecies, originally thought to be a separate species, is *S. d. miravallensis*) or the Guanacaste squirrel. Species name after F. Deppe, a nineteenth-century biologist. **Spanish**: *Ardilla* (or *chisa*) *de Deppe*; *ardilla del Miravalles*; *ardilla de Guanacaste.*

Range: Veracruz, Mexico, to the Guanacaste mountain range and surrounding lowlands in northwestern Costa Rica; from sea level to about 3,000 m (10,000 ft); in or near humid, dense, dark, and undisturbed forest. Over much of its range, it is most common in the lowlands. Although Deppe's squirrel appears on some mammal checklists for La Selva, Monteverde, and the Osa Peninsula, there are no specimens from those areas. It now appears that all records may have been mistakenly identified red-tailed squirrels, specimens of which have been collected at all three sites.

Size: 20 cm, 200 g (8 in., 7 oz).

Similar species: The upperparts of Deppe's squirrel vary from dark reddish to olive brown, and the underparts and tail frosting from whitish to ochraceous gray. Red-tailed squirrels (pl. 19, p. 159) are slightly larger, have more orangey underparts and tail frosting, and bushier tails. Redtails and Deppe's squirrels may be allopatric (have nonoverlapping ranges). Variegated squirrels (pl. 20, p. 161) are larger, and tend to inhabit more open habitats; the only subspecies of the variegated squirrel found within Deppe's squirrel's known range in Costa Rica is colored black and white. Alfaro's pygmy squirrels (pl. 19, p. 155) are smaller and have inconspicuous ears and a narrow, tapering tail. Deppe's and montane squirrels (pl. 19, p. 164) are allopatric. Deppe's squirrels have only three pairs of functional mammae, while all other *Sciurus* squirrels have four pairs.

Natural history: Deppe's squirrel has not been studied in any depth, and details of its natural history are largely unknown. It is most active in the early morning and late afternoon. It uses all levels of the forest, but spends a significant proportion of its foraging time (30 to 60% in a population studied in Mexico) on the ground. It feeds on fruits, seeds, leaves, fungi, and insects. Known foods include figs (*Ficus* spp.), acorns (*Quercus* spp.), palm nuts (Arecaceae), sapotes (*Manilkara zapota*), breadnuts (*Brosimum alicastrum*), and mastate (*Poulsenia armata*).

Ojoche (*Brosimum alicastrum*). Each fruit is about 2 cm in diameter.

Home range size has been estimated at 1.5 ha (4 A). Deppe's squirrels den in hollow trees or in roughly 23 to 30 cm (9 to 12 in.)-wide balls of leaves constructed on large branches or in dense tangles anywhere from 2 meters (7 feet) off the ground to up in the canopy. Litters contain two to eight, but usually four, young. Over their range as a whole, Deppe's squirrels may breed at any time of year. Nothing has been published on the timing or frequency of breeding in Costa Rica.

Sounds: Steadily repeated, high-pitched, *chuck* sounds in alarm; birdlike trills.

Conservation: In some areas, Deppe's squirrel damages corn planted next to forests. It is too small to be hunted extensively for its meat, and is still common in parts of its fairly extensive range. It is more sensitive to deforestation than many other squirrels, however, disappearing from areas that are heavily disturbed. Within Costa Rica, it inhabits only a small area that has been further reduced by deforestation, and it is now listed as an endangered species here.

References

Best 1995b; Coates-Estrada and Estrada 1986; Emmons & Feer 1997; Emmons et al. 1997; Hall & Dalquest 1963; Hall & Kelson 1959; McPherson 1985; Moore 1961; Reid 1997; Timm 1994a,b; Timm et al. 1989.

O: Rodentia F: Sciuridae

RED-TAILED SQUIRREL

Sciurus granatensis

(plate 19)

Names: Also known as the tropical red squirrel. *Granatensis* means from Granada, because the first specimen to be described was collected in New Granada, the colonial name for the region that now encompasses Colombia and Venezuela. **Spanish**: *Ardilla* (or *chisa*) *coliroja*.

Range: Northeastern Costa Rica (and probably southeastern Nicaragua) to the northern tip of South America, south to Ecuador and east to Venezuela; to at least 3,200 m (10,500 ft); in primary forest or disturbed habitats that have trees, but not in areas with a severe dry season.

Size: 24 cm, 400 g (9 in., 14 oz); females slightly larger than males.

Similar species: See other squirrel species descriptions.

Natural history: Red-tailed squirrels forage mostly in the morning and late afternoon at all levels of the forest, including the ground. They feed principally on seeds and fruits. Studies done on Barro Colorado Island in Panama found redtails there to feed on 58 plant species, but 73% of the diet was comprised of just four species: two palms (*Attalea butyracea* and *Astrocaryum standleyanum*), the legume known as almendro (*Dipteryx panamensis*), and a brazil-nut relative (*Gustavia superba*). Redtails eat the nuts of the palms and the almendro, and the fruits of *Gustavia*. The nut casings of the palms and almendro are so strong, and the fruits of *Gustavia* so large and hard, that few other vertebrates go after these foods.

A *Gustavia superba* fruit (which is about 7 to 10 cm in diameter) hanging from its tree (right) and the same fruit cut in half to reveal the seeds (below).

Studies in Panama have also revealed that the feeding behavior of this species differs between the sexes, possibly as a result of different territorial arrangements. Females do not share territories with one another, except in periods of exceptional food abundance. Restricted to foraging within relatively small, 0.3 to 1 ha (¾ to 2 ½ A) territories, they do not have access to favorite foods year-round. They hoard seeds for retrieval in times of food scarcity, when they also feed on alternative foods such as soft fruits, fungi, bark, flowers, young leaves, sap, and insects. There is even a record of a redtail eating the eggs of a rain frog (*Eleutherodactylus* sp.). Young also include these alternative foods in their diet. Since some such foods are high in protein, they may be selected because they are an important nutritional supplement for young and for reproductive females.

Red-tailed squirrel scat. Each pellet is about 1 to 2 cm long.

Males, meanwhile, share territories with squirrels of either sex and tend to have larger, 1 to 4 ha (2 ½ to 10 A) territories. With more room to roam, they have a better chance of finding favorite foods throughout the year. They seldom hoard seeds themselves, but help themselves freely to females' caches in times of food scarcity.

Redtails are usually solitary, but groups form during the breeding season. In Panama, the breeding season lasts from December to August, which is also when the staple foods are available. Clusters of up to 15 males chase any female in heat, often for hours at a time, until she selects one of them for a 10-second-long copulation. The ensuing gestation period lasts about a month and a half and most often results in two young. Young are usually born in a tree hollow. They open their eyes when they are about one month old, and leave the nest about two weeks later. Females produce one to three litters a year. The Panamanian studies suggest that the availability of favorite foods determines the timing and frequency of reproduction. In the wild, few redtails live more than two years, although one wild female lived at least seven years.

Sounds: Short, hoarse, steadily repeated *chuck* sounds in alarm; grunts, squeals, and hiccuping sounds by males during mating chases.

Conservation: In some areas, redtails are hunted for food and sport, but they are common over much of their range. Since redtails gnaw into even the hardest nuts, and since they eventually retrieve and destroy most of the nuts they hoard, they are important seed predators. They can damage crops including cacao, mango, avocado, banana, corn, rice, cypress, palms, and eucalyptus. They may also disperse some seeds, however. Ironically, one plant they are thought to help disperse is naturally growing cacao, the origin of commercial cacao, so while we can begrudge redtails for raising the price of chocolate bars by damaging crops, we can also thank them for allowing chocolate to exist in the first place.

References

Bonaccorso et al. 1980; Emmons et al. 1997; Garber & Sussman 1984; Glanz 1984; Glanz et al. 1996; Heaney 1983; Heaney & Thorington 1978; Hershkovitz 1977; Hilje & Monge 1988; Lawrence 1991; Monge 1989; Nitikman 1985; Reid 1997; Skutch 1980; Timm 1994a,b; Timm et al. 1989; Toxopeus 1985.

O: Rodentia F: Sciuridae

VARIEGATED SQUIRREL

Sciurus variegatoides

(plate 20)

Names: The origin of the genus is given on p. 155. Species name from the Latin *varius*, meaning variable. **Spanish**: *Ardilla* (or *chisa*) *variable.*

Range: Chiapas, Mexico, to the Panama Canal; mostly below about 1,500 m (5,000 ft), but up to as high as 2,500 m (8,200 ft) on Poás Volcano and 3,000 m (10,000 ft) on Cerro de la Muerte; mostly in wooded but relatively open habitats such as dry and secondary forests.

Size: 28 cm, 700 g (11 in., 1 ½ lbs).

Similar species: One of the most curious aspects of the variegated squirrel is the great variety of color forms that occur within its relatively small range, and especially within Costa Rica, where seven of the 14 described subspecies occur. Regardless of the color form, *S. variegatoides* is larger than other squirrels and has a proportionately longer tail. The closest in size is the red-tailed squirrel (pl. 19, p. 159), which differs in having a brown tail frosted with orange; all similar forms of the variegated squirrel have blackish tails frosted with white. Deppe's squirrels (pl. 19, p. 157) are about a third smaller and may have pale ochre-colored, but never deep orangey underparts. The montane squirrel (pl. 19, p. 164) is a little smaller still, has less orangey underparts, and inconspicuous, low-set ears. All these other squirrels prefer dense, evergreen forests where variegated squirrels tend to be rare or absent.

Natural history: The variegated squirrel forages by day at all levels of the forest, but it seldom comes to the ground. It feeds on fruits, seeds, and flowers, and occasionally fungi, bark, and insects. It usually favors a few staple food plants that tend to be among the most common species and vary with location. About half the diet of a population studied in Panama consisted of mango (*Mangifera indica*, illustrated on pl. 12), hogplum (*Spondias mombin*, illustrated below), and guácimo (*Guazuma ulmifolia*, illustrated on p. 310), while at Hacienda Curú on the Nicoya Peninsula, about three quarters of the diet of *S. v. atrirufus* was found to consist of coconut (*Cocos nucifera*), beach almond (*Terminalia catappa*), and the fruits of the flamboyant tree (*Delonix regia*). Other foods include guavas (*Psidium guajava*), guitite (*Acnistus arborescens*), and figs (*Ficus* spp.).

Hogplums (*Spondias mombin*). The fruits are yellow or orange and 2 to 4 cm long.

Many of these foods are relatively soft, and this may be a key difference between the diets of the variegated squirrel and Costa Rica's other large squirrel, *S. granatensis*. When two apparently similar animals occur together, there is often a subtle ecological difference that reduces competition between the two; if not, one, in theory, would eventually replace the other. Although variegated squirrels prefer drier or more open habitats than do red-tailed squirrels, there are many areas where the two occur together. In such areas in Panama, variegated squirrels were found to take even more soft food than normal, eating about 60% fruit, while redtails in those areas were eating more than 70% seeds.

Variegated squirrels produce one or, perhaps, two litters a year and are thought to give birth to an average of four to six young in each litter. In Curú, mating chases involving single females and four or more males take place early in the morning from February to April, and most young are born at the beginning of wet season in May or June. Young are born in balls of leaves and twigs that are 30 cm (12 in.) or more in diameter and lodged between branches in tree crowns 15 meters (50 feet) or more off the ground.

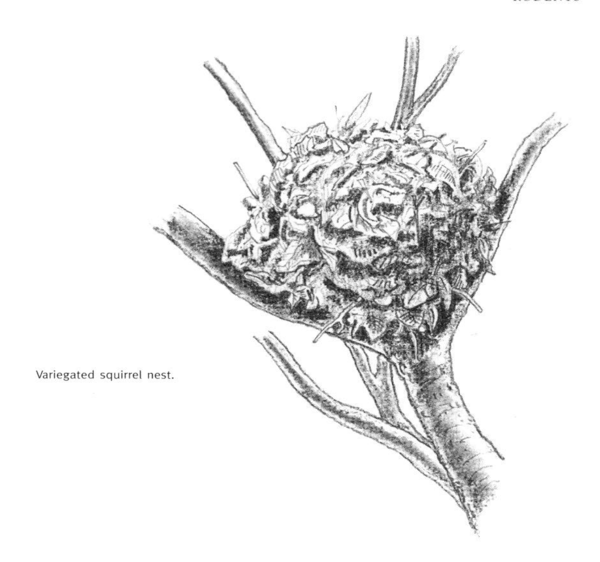

Variegated squirrel nest.

Sounds: Quickly repeated, bouncy, high-pitched *chucking* notes in alarm.

Conservation: Variegated squirrels have good-tasting, white meat, and are popular game animals. Nonetheless, they do well in disturbed areas; they are common and conspicuous over much of their range. They are active seed predators and can damage commercial plants such as coconuts, mangoes, bananas, papayas, macadamias, avocados, chayotes, and carrots. They also chew the bark of pochote trees (*Bombacopsis quinatum*), one of the trees most frequently used for living fence posts in Guanacaste. Careful observations of their behavior with acorns (*Quercus oleoides*) and the seeds of gourd trees (*Crescentia alata*, illustrated on p. 293) in Costa Rica suggest that, unlike red-tailed squirrels, variegated squirrels do not carry away and cache food. Thus they may play little or no role in seed dispersal.

References

Best 1995a; Boucher 1981; Emmons et al. 1997; Giacalone et al. 1987; Glanz 1984; Gómez 1983; Hilje & Monge 1988; Janzen 1982b; Monge 1989; Reid 1997; Rodríguez & Chinchilla 1996; Timm et al. 1989.

O: Rodentia F: Sciuridae

MONTANE or POAS SQUIRREL
Syntheosciurus brochus
(plate 19)

Names: Also known as the groove-toothed squirrel. Previously known as *S. poasensis*. Genus translates as "placed together with squirrels"; *brochus* is the Latin for with projecting teeth. **Spanish**: *Ardilla del Poás.*

Range: Known only from Poás Volcano and Tapantí, Costa Rica, and an area in the Chiriquí mountains of western Panama; from 1,250 m (4,100 ft) at Tapantí to at least 2,600 m (8,500 ft); in forest, thick brush, and the edges of pastures. This squirrel was first collected at the beginning of this century in Chiriquí, Panama. The specimen had unusually protrusive upper incisors, and each incisor was marked with a well-defined groove, a feature that is absent or occasionally only very faint in all other squirrels, so it was given a genus all of its own: *Syntheosciurus.* In 1930, an almost identical squirrel was collected in Costa Rica at Poás Volcano, but it lacked the grooves, so it was named a different species, *S. poasensis*, or the Poás squirrel. More recent studies in Panama, however, showed the incisor grooves to be a variable characteristic, so the disjunct populations of *Syntheosciurus* squirrels are now considered to be the same species. In 1985, a specimen was collected, and several more were observed, in the vicinity of the Tapantí reserve just southeast of San José, and part of the range gap was thus filled in.

Size: 17 cm, 150 g (7 in., 5 oz).

Similar species: The most similar species is Alfaro's pygmy squirrel (see pl. 19, p. 155). Deppe's squirrel (pl. 19, p. 157) is not known to occur in the same area; the red-tailed squirrel (pl. 19, p. 159) and the variegated squirrel (pl. 20, p. 161) are much larger and have longer ears.

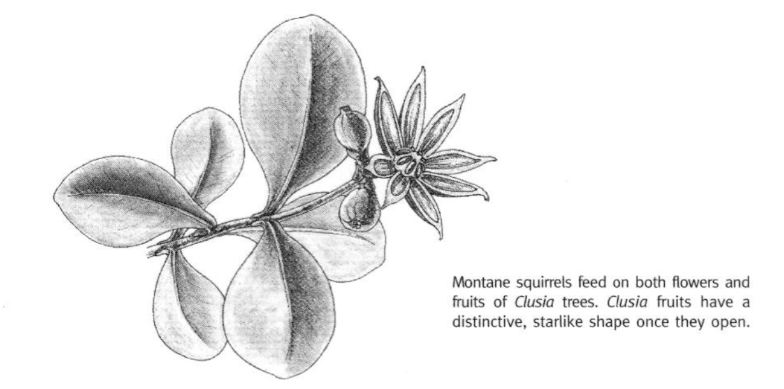

Montane squirrels feed on both flowers and fruits of *Clusia* trees. *Clusia* fruits have a distinctive, starlike shape once they open.

Natural history: Most of what is known of the natural history of this species is derived from a brief (three-month) study conducted at Poás Volcano in 1984 by Jacalyn Giacalone, Nancy Wells, and Gregory Willis. Montane squirrels forage by day at all levels of the forest, including the ground. They are usually alone, but sometimes travel in pairs or small family groups. They have been recorded feeding on nectar, pollen, sap, fruit, insect larvae, and bark. Favorite foods at Poás Volcano during the dry season include nectar and pollen from *Clusia* flowers, sap from holes left by yellow-bellied sapsuckers in cipresillo trees (*Escallonia poasana*), gleanings from bromeliads, and bromeliad flower buds. Other foods include *Clusia* fruits, oak tree flowers (*Quercus* spp.), sweet gale fruits (*Myrica phanerodonta*), gleanings from bamboo (*Chusquea* sp.), and bark from cow's tongue trees (*Miconia coriacea*).

Montane squirrels have unusually narrow, pointed snouts and protruding incisors. These features could be specializations for probing flowers or sapsucker holes, or for searching under bark.

To get at sap, the montane squirrel gnaws at holes drilled into *Escallonia poasana* trees by yellow-bellied sapsuckers. The pagodalike, flaky-barked *Escallonia* trees are abundant at Poás Volcano; the sapsucker, a type of woodpecker, visits Costa Rica only seasonally, to escape the North American winter.

Montane squirrels den in tree hollows such as abandoned woodpecker holes 6 meters (20 feet) or more off the ground. Since they often den in pairs, it has been suggested that montane squirrels may form prolonged pair bonds, which is unusual or perhaps even unique among tree squirrels. Observations, both at Poás and in Panama, of mating chases (involving single females and six to eight males), of lactating females, and of dens with young suggest that montane squirrels mate in the dry season and produce two to five young in the early wet season.

Conservation: In broad terms, this squirrel, known only from three small and geographically isolated areas, is clearly rare. At Poás Volcano, however, it is the predominant species of squirrel; in the 1984 study, more than half of over 300 squirrel sightings were of this species. It is also possible that inconspicuous, undetected populations exist in other parts of the 250 km (155 mile) band that is or was its range.

References

Enders 1980; Giacalone et al. 1987; Heaney & Hoffmann 1978; Morúa 1986; Rodriguez & Chinchilla 1996; Wells & Giacalone 1985.

O: Rodentia F: Geomyidae

POCKET GOPHERS

(plate 21)

Names: Four pocket gopher species are known from Costa Rica: the Chiriquí (*Orthogeomys cavator*), the variable (*O. heterodus*), Cherrie's (*O. cherriei*), and Underwood's (*O. underwoodi*). They are among some 37 pocket-gopher species that constitute the family Geomyidae. The family name comes from the Greek words *geios*, meaning of the earth, and *mys*, meaning mouse. The genus derives from the same, as well as the Greek *orthos*, meaning straight, perhaps in reference to the gophers' tunnels. Pocket refers to their cheek pouches, and the term *gopher* is derived from the French word *gaufre*, or honeycomb, in reference to their elaborate tunnel systems. **Spanish**: *Taltuzas*.

Range: Pocket gophers are found from Canada to the northern tip of South America, in forested or disturbed habitats, especially in agricultural areas. In Costa Rica, they are found from sea level to about 2,400 m (7,900 ft). The Chiriquí pocket gopher is endemic to southern Costa Rica and western Panama (which means it is native to that region and found nowhere else). The other three species are endemic to Costa Rica.

Size: *O. cavator* and *O. heterodus*: 25 cm, 650 g (10 in., 1 ½ lbs); *O. cherriei* and *O. underwoodi*: 20 cm, 350 g (8 in., 12 oz); males larger than females.

Natural history: Gophers are seldom seen, for they spend most of their lives underground. A series of tunnels just beneath the surface gives them access to the plant roots and tubers they feed on, while deeper tunnels lead to food storage, sleeping, and even bathroom chambers. They are most abundant in agricultural areas, where they feed on a variety of crops. *O. cherriei* also occurs in pristine habitats at La Selva, where its diet includes *Welfia* palms. Gophers locate food plants by smell and either nip off the bases or pull the whole plant down into the tunnel.

Each tunnel system is the permanent home of a single gopher, and is modified continually. In a study of *O. heterodus* near Irazú Volcano,

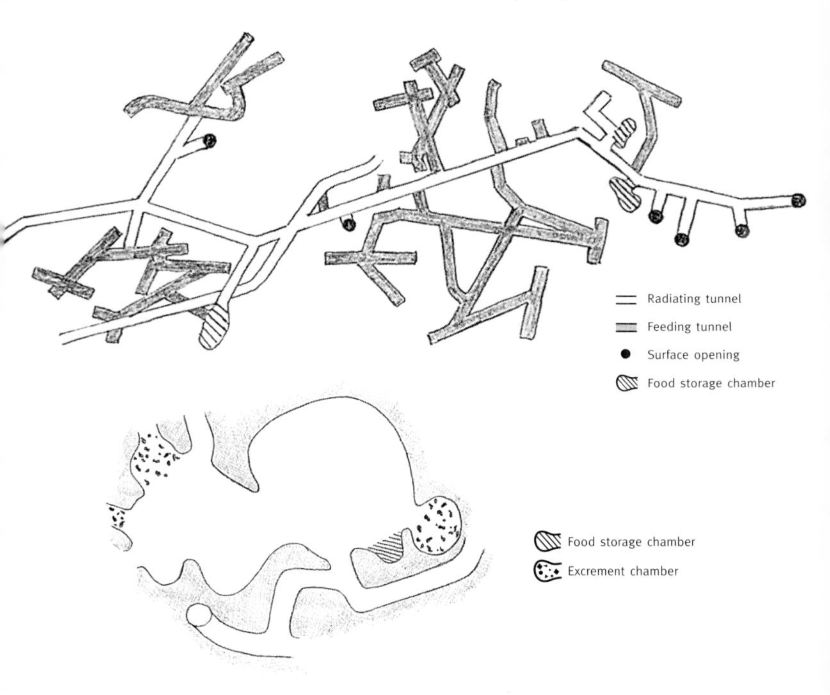

Researchers Thomas Sisk and Christopher Vaughan documented the underground world of Costa Rica's endemic variable or giant pocket gopher (*Ortheogeomys heterodus*) for the first time by carefully excavating and drawing tunnel systems they found in Concepción de San Rafael de Heredia. This redrawing of their work shows a roughly seven-meter-long section of one tunnel system (top) and a nest chamber (bottom), the latter about a meter wide at its broadest point.

researcher Never Bonino found the tunnel systems of males to cover an average of 325 m^2 (3500 ft^2), and those of females, 233 m^2 (2500 ft^2). Gophers excavate tunnels with their large front claws and incisors, pushing the dirt behind them as they go. Once loose soil has accumulated, they flick themselves around by pushing off their tail, put their forepaws together, and bulldoze it to the surface, where it is deposited in distinctive, fan-shaped mounds. The small entrance hole is always plugged before the gopher returns to digging.

Two types of gopher mound: a large, fan-shaped mound with a visible "plug" over the tunnel entrance, or simply a small pile of dirt right over the entrance.

A number of unusual features help gophers in their peculiar lifestyle. Their lips close naturally behind their incisors, preventing soil and gnawed debris from entering their mouths. Unusually large tear glands secrete a thick fluid that washes dirt out of their eyes. External cheek pouches enable gophers to transport vegetable segments to storage chambers without tying up their hands. The pouches can be turned inside out for emptying or cleaning. Finally, gophers may be the world's only mammals that can run backward almost as fast as they can run forward, a skill that presumably is useful should they to come face to face with one of their predators, such as a snake, weasel, or coyote.

One feature that is harder to explain is the bold dark and white coloration of *O. cherriei*, *O. underwoodii*, and some individuals of *O. heterodus*. Such coloration is rare among gophers, and even among mammals in general. Some believe that, since gophers spend most of their lives underground, their coloration is non-adaptive; in other words, their coloration is irrelevant to their success and is simply the result of random genetics. Others have pointed out that young gophers dispersing from their area of birth and, occasionally, adult gophers, do travel above ground; even if only for brief periods, bold coloration could have helped these species to prevail. Individuals are unlikely to meet above ground, so it is improbable that markings would play a role in territorial or courtship displays. The markings could, however, help to deter predators in some way, for example by serving as a warning, or giving the gophers a vague resemblence to skunks (or both).

Gophers are particularly unsociable animals. If they happen to burrow into a neighbor's tunnel system, they quickly retreat and fill in the hole. Chance meetings usually result in vicious fighting, sometimes to the death. Of course, they do need to reproduce, so males are allowed to enter females' burrows during breeding season. Young gophers start their solitary existence at the age of about two months. In North America, males first breed when about one year old, while females of some species can start when they are barely two months old. In the wild, gophers can live about five years, but the average lifespan is just over a year.

Sounds: Clicking noises made by the incisors; hisses or squeals in defense.

Conservation: Gophers can be serious agricultural pests. The Chiriquí gopher damages banana, yuca, and rice crops in southern Costa Rica. The variable gopher can wreak havoc on corn, onions, and potatoes in the Central mountain range, and also enjoys carrots, oats, and coffee, among other crops. *O. cherriei* attacks bananas, coffee, cacao, yuca, tiquisque, corn, beans, and rubber trees in the Santa Clara plains and the Matina and Sarapiquí river valleys. In some years, the species has caused the loss of as much as 80% of these crops.

Gophers are controlled most effectively through the use of traps. They have a pleasant-tasting, dark, fine-grained meat, but are seldom eaten in Costa Rica.

References

Barker 1956; Bonino 1990, 1993, 1994; Bonino & Hilje 1992; Delgado 1990, 1992; Greene & Rojas 1988; Hafner 1991; Hafner & Hafner 1987; Hilje & Monge 1988; Nowak 1991; Patton 1984; Sisk & Vaughan 1984; Timm et al. 1989.

O: Rodentia F: Heteromyidae

SPINY POCKET MICE

(plate 22)

Names: Salvin's spiny pocket mouse (*Liomys salvini*), the forest spiny pocket mouse (*Heteromys demarestianus*), and the mountain spiny pocket mouse (*Heteromys oresterus*). The genus and family name come from the Greek words *heteros*, meaning different and *mys*, meaning mouse. What makes heteromyids different from other rodents is the combination of a long tail, external, fur-lined cheek pouches (the pockets—a feature unique to heteromyids and pocket gophers), and their forefeet which, unlike those of pocket gophers, are not adapted for heavy digging. When present, spines are largely hidden by the fur (as in *L. salvini* and *H. demarestianus*); many species (such as *H. oresterus*) lack spines. It has been suggested that *H. oresterus* may have evolved finer, denser, spineless fur to improve insulation, given that it inhabits cold, highland forests. **Spanish**: *Ratón de bolsas.*

Range: The 60 or so species in this family are distributed from Canada to northern South America. About a third of the family are the kangaroo rats (so named because of their long back legs and hopping gait) that inhabit the deserts of the southwestern United States and northern Mexico. Only three species are found in Costa Rica. *H. oresterus* is a Costa Rican endemic, known only from highland oak forest in a small area near El Copey de Dota in the Talamancas, from 1,800 to 2,650 m (6,000 to 8,700 ft). *H. demarestianus*, abundant in wet forests throughout the country, ranges from Mexico to northwestern Colombia, from sea level to about 2,400 m (7,900 ft). Its dry forest counterpart, *L. salvini*, is found from northwestern Costa Rica to Mexico, from sea level to about 1,500 m (5000 ft), in deciduous forest and disturbed habitats.

Size: *L. salvini*: 13 cm, 50 g (5 in., 1 ¾ oz); *H. demarestianus*: 14 cm, 60 g (5 ½ in., 2 oz); *H. oresterus*: 15 cm, 80 g (6 in., 3 oz).

Natural history: Spiny pocket mice forage at night on the ground. They eat seeds, fruits, fungi, and insects.

L. salvini is particularly fond of the seeds of the buttercup tree (*Cochlospermum vitifolium*) and is the main predator on fallen seeds of Costa Rica's national tree, the Guanacaste (*Enterolobium cyclocarpum*, illustrated on pl. 22). The latter seeds contain toxins that are deadly to the other predominant rodent in northwestern Costa Rica, the hispid cotton rat (pl. 23, p. 171). Experiments have shown that *L. salvini* can thrive on a diet of Guanacaste seeds alone, as long as the seeds have germinated; if it feeds too heavily on ungerminated seeds, it too can die. *L. salvini* has evolved a response to this problem—it makes small cuts in hoarded seeds, a treatment that softens the seeds and makes them more likely to germinate in the soil of its storage chambers. *L. salvini* is so efficient at harvesting Guanacaste seeds that, where the rodent is common, Guanacaste trees may be unable to reproduce. *Liomys* can remove more than 90% of a crop of seeds, even sniffing out seeds that have passed intact through horses' guts and lie buried in dung.

L. salvini lives in elaborate, multichambered tunnels. It produces one or two litters between January and June, with an average of about four young per litter. When very small, the young may be carried around in the mother's cheek pouches. Males first breed when about six months old, and females can do so even earlier. Few live more than a year.

H. demarestianus looks much like *L. salvini* but, living in wetter habitats, differs in many aspects of its natural history. It enjoys most types of palm nuts, diamond seeds (*Meliosma* spp.), wild nutmeg (*Virola sebifera*), and figs (*Ficus* spp.). Occasionally it snacks on the poisonous seeds of the gavilán legume (*Pentaclethra macroloba*). It dens in tunnels or in hollow logs. It is reproductively inactive only during extended dry periods, and can produce up to five litters a year, with an average of three young per litter. Males first breed at about nine months, females at about eight. In the wild, most die within a year but a few can survive two or three years.

The natural history of *Heteromys oresterus* has not been studied.

Conservation: *L. salvini* and *H. demarestianus* are among the most abundant mammals in their habitats, while *H. oresterus* is apparently uncommon. *L. salvini* and *H. demarestianus* are thought to be important both as seed predators and as seed dispersers. They stash seeds in or near their dens, retrieving and destroying most later on, but probably forgetting about others or dying before they retrieve the seeds. In a recent study of rodents in Monteverde, researcher Federico Chinchilla has found *H. demarestianus* to be much more common in forest fragments than in continuous forest, and has noted that it would be interesting to investigate how the abundance of such an important seed predator and disperser might affect the success of seeds in forest fragments.

References

Federico Chinchilla, pers. comm.; Flemming 1974a,b, 1977, 1983b,c; Fleming & Brown 1975; Janzen 1982a,c,d, 1983f; Reid 1997; Rogers 1990; Rogers & Rogers 1992; Timm 1994a,b.

O: Rodentia F: Muridae (S: Sigmodontinae)

NATIVE MURID RODENTS
(plates 23 & 24)

Names: The vesper rat (*Nyctomys sumichrasti*), Alston's singing mouse (*Scotinomys teguina*), the hispid cotton rat (*Sigmodon hispidus*), the dusky rice rat (*Melanomys caliginosus*, formerly *Oryzomys caliginosus*), Goldman's water mouse (*Rheomys raptor*), and Watson's climbing rat (*Tylomys watsoni*). With over 1,300 species—two-thirds of all rodents—the Muridae is by far the largest mammal family in the world. It includes hamsters, gerbils, voles, lemmings, and, above all, true mice and rats. (There is no technical distinction between the latter two; large mice simply tend to be termed rats.) The family name comes from the Latin word for mouse, *muris*. All murid rodents native to the New World, a little over 400 species, belong to the subfamily Sigmodontinae. About 30 species inhabit Costa Rica. Sigmodontines are so named because the pattern of enamel on the grinding surface of their molars resembles the Greek letter Sigma (Σ). **Spanish**: *Ratas*; *ratones*.

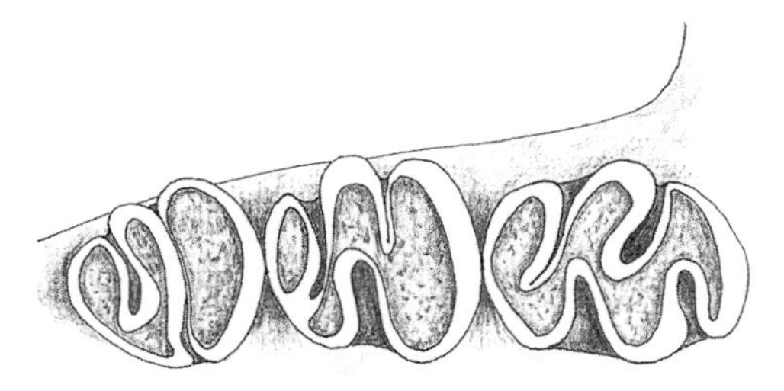

The molars of a sigmodontine rodent (*Sigmodon hispidus*) showing the sigma (Σ)-shaped patterns of enamel on the crowns.

Range: *N. sumichrasti*: Mexico to Panama; from sea level to about 1,800 m (6,000 ft); in primary or secondary forest. *S. teguina*: Mexico to Panama; in or near mid- and high-elevation forests to about 2,900 m (9,500 ft). *S. hispidus*: central United States to the northern tip of South America; from sea level to 2,700 m (8,900 ft), but mostly at low and mid-elevations; in disturbed, wet or dry habitats. *M. caliginosus*: Honduras to Ecuador; from sea level to about 700 m (2,300 ft); mostly in disturbed, wet habitats. *R. raptor*: restricted to Costa Rica and the western edge of Panama; from about 800 to 1,800 m (2,600 to 6,000 ft); in and around fast-moving forest streams. *T. watsoni*: Costa Rica and Panama only; to about 2,700 m (8,900 ft); in wet forests.

Size: *N. sumichrasti*: 12 cm, 50 g (5 in., 1 ¾ oz); *S. teguina*: 8 cm, 12 g (3 in., ½ oz); *S. hispidus*: 15 cm, 60 g (6 in., 2 oz); *M. caliginosus*: 12 cm, 50 g (5 in., 1 ¾ oz); *R. raptor*: 12 cm, 20 g (5 in., ¾ oz); *T. watsoni*: 24 cm, 230 g (10 in., 8 oz).

Natural history: The vesper rat is nocturnal and arboreal. Since it often freezes when caught in the beam of a flashlight, it is one of the few small rodents that can be identified and admired without the aid of a trap. It feeds on fruits, seeds, and insects. It dens in tree hollows or in exposed nests of twigs and shredded bark. New nests are sometimes stacked on top of old ones.

Alston's singing mouse is diurnal and mainly terrestrial. Insects, especially beetles, comprise most of its diet, but it also feeds on seeds, fruit, and nectar. Males and females work together to build a tightly-woven nest and raise young.

The hispid cotton rat is active by day or night and is terrestrial. Worn down runways beneath vegetation in overgrown fields may belong to this species. It feeds principally on fruit, fungi, and leaves, and it cuts grasses and weeds at their bases and feeds on the new shoots as they emerge. The hispid cotton rat dens on the ground or in tunnels in cup- or ball-shaped nests that it constructs with woven grass.

The dusky rice rat is most active in the late afternoon. It feeds mainly on insects, supplemented with fruit, seeds, and plant shoots.

Goldman's water mouse, which is semiaquatic, has partially webbed hindfeet. Its diet includes caddisfly larvae, beetles, and spiders. It dens in long, narrow tunnels in exposed stream banks.

Watson's climbing rat frequently resides in forest cabins. Some people find it unnervingly large, but it is docile and harmless. It can, however, make quick work of soap and chocolate supplies. Little is known of its natural history in the wild.

Sounds: Native murid rodents emit ultrasonic calls that we cannot hear, but also distinctive, audible calls that can even be recognized to species. Alston's singing mouse rises up on its back legs, holds its forefeet out in front of its body, and cocks its head backward—like a soprano reaching the climax of her aria—to give a 6- to 10-second-long trill, which slows, loudens, and drops in pitch before ending abruptly. This performance is given mainly at dawn and dusk. The call resembles an incredibly high-pitched version of that of the silvery-fronted tapaculo, a vociferous bird found throughout Costa Rica's highlands. Goldman's water mouse emits loud, low-pitched tongue clicks. The vesper rat emits short, stuttered series of loud, birdlike, chirpy squeaks that resemble the sounds made by the small, screw-in-wood bird-calling gismos. One function of these long-distance calls may be to attract mates. A pair of vesper rats in Monteverde were seen homing in on each other using chirps, and when they met, one tried to mate with the other.

Conservation: The hispid cotton rat and the dusky rice rat are among the most common rodents in Costa Rica. They are both pests on crops such as rice, sugar cane, and corn and important food for carnivorous animals (cotton rat fur is prevalent in coyote dung at Santa Rosa and Palo Verde, for example). Alston's singing mouse is also common. The vesper rat is seldom seen, perhaps because it is arboreal. The water mouse is rare. Watson's climbing rat is uncommon but conspicuous.

Some native murid rodents may help disperse seeds. In Monteverde, several species have been found to take nectar from the flowers of the hemi-epiphytic melastome, *Blakea clorantha*, although it is not clear whether or not they pollinate the flowers. And these rodents may even disperse the spores of mycorrhizal fungi, which grow in association with the roots of plants and without which many plants would be unable to grow. Studies in North America, Peru, and Monteverde, Costa Rica, have shown that rodents that eat fungi can pass viable spores in their feces.

References

Bakarr 1990; Baker 1983; Cameron & McClure 1988; Cameron & Spencer 1981; Emmons et al. 1997; Fleming 1970; Gardner 1983b; Hall & Dalquest 1963; Hayes & Laval 1989; Hooper 1968; Hooper & Carlton 1976; Janos et al. 1995; Langtimm 2000; Langtimm & Unnasch 2000; Lumer 1983; Lumer and Schoer 1986; Maser et al. 1978; Monge 1992; Reid 1997; Reid and Langtimm 1993; Timm 1994a,b; Timm & Laval 2000a; Vaughan & Rodríguez 1986.

O: Rodentia F: Muridae (S: Murinae)

NON-NATIVE MURID RODENTS
(plate 25)

Names: The roof or black rat (*Rattus rattus*), the Norway or brown rat (*Rattus norvegicus*), and the house mouse (*Mus musculus*). All belong to the Old World subfamily, the Murinae (from the Latin *murinus*, meaning mouselike), which includes more than 500 species. **Spanish**: *Ratas*; *ratones*.

Range: These are the only three nondomesticated mammal species known to have been introduced to Costa Rica. Thanks to their association with humans, all three species are now found worldwide. The black rat is originally from India and southeastern Asia, and probably reached the New World during early colonial times. The Norway rat is no more Norwegian than it is Costa Rican: it hails from China and Siberia. The species spread west much later than the roof rat, reaching Europe around 1727 and

America around 1775. Being considerably larger, it has replaced the roof rat through much of Europe and North America (to the point that the roof rat now appears on some European endangered-species lists!), although the roof rat is still the prevalent species in Latin America. The house mouse is thought to be native to most of Europe and Asia, and probably reached the New World on colonial ships.

Size: *R. rattus*: 17 cm, 110 g (7 in., 4 oz); *R. norvegicus*: 22 cm, 300 g (9 in., 11 oz); *M. musculus*: 8 cm, 12 g (3 in., ½ oz).

Unnatural history: The roof rat inhabits towns and agricultural areas. A good climber, it has also penetrated the forest in some areas. Its diet includes all human foods as well as an astonishing variety of other substances, such as soap and paper. It is most famous for the role it has played in spreading the bubonic plague, or black death. The rat can carry this disease without succumbing to it. People contract the plague principally when they are bitten by fleas that have fed on infected rats. The most famous outbreak of the bubonic plague ravaged western Europe between 1345 and 1350, claiming an estimated 43 million lives and leaving about 200,000 towns completely uninhabited. Bubonic plague is not a thing of the past, as many people think, and still has the potential to cause catastrophes in developing countries. An outbreak in India at the beginning of the twentieth century claimed over three million lives in just 10 years.

The Norway rat is an even greater problem. Unlike the black rat, it is essentially a burrowing species that has been particularly successful in the underground worlds of large towns and cities. Like the roof rat, the Norway rat eats almost anything, but it prefers animal matter, including eggs, chickens, and even young lambs and pigs. It also thrives in areas of food storage. It has been calculated that a single rat eats about 12 kg (27 lbs) of food and leaves about 25,000 droppings each year. Droppings and urine are the principle causes of numerous food-borne diseases including trichinosis (pork threadworms), leptospirosis, and the 600 or so varieties of salmonella poisoning. These rats have also spread schistosomiasis to more than 200 million people worldwide. The Norway rat even occasionally attacks humans—according to one estimate, there are some 14,000 attacks each year within the United States, a few of which are fatal.

Economic damage caused directly by the roof and Norway rats is vast. It was estimated at $500 million to $1 billion per year for the United States alone in the late 1970s. As nonnatives, both rat species also have a huge impact on the nonhuman world, for they can ravage native ecosystems evolutionarily unprepared for their presence. They have, for example, extinguished huge colonies of puffins and other birds on European islands, and caused the extinction of an Irish frog species.

The house mouse lives in small cavities between rocks, down tunnels, or, mainly, in hidden nooks and crannies of houses. Although not nearly as detrimental as the two rats, it also spreads disease, damages houses and crops, contaminates food, and displaces native rodent species. As some form of compensation, the famous albino house mice used in laboratories have greatly furthered our understanding of medicine and genetics.

References

Barker 1956; Bronson 1979; Dubock 1984; Gray 1967; Hayes & Laval 1989; Nowak 1991; Pratt et al. 1977; Reid 1997; Timm et al. 1989; Wace 1986.

O: Rodentia F: Erethizontidae

MEXICAN HAIRY PORCUPINE

Coendou mexicanus

(plate 26)

Names: Also known as the prehensile-tailed porcupine. The genus is derived from the name *coendu*, used by Mexican indians; *mexicana* because the type specimen—the specimen used to describe the species—was collected in Mexico. This porcupine is sometimes listed under the genus *Sphiggurus*. The family name Erethizontidae translates roughly as irritating girdle, in reference to porcupines' spines. The family contains all 10 species of New World porcupines. (Another 11 species of porcupines are found in the Old World and belong to the family Hystricidae.) **Spanish**: *Puercoespín* (both this name and the word porcupine are derived from the Latin for spined pig); *cuerpoespín*.

Range: Veracruz, Mexico, to western Panama; from sea level to about 3,200 m (10,500 ft); in primary or dense secondary forest.

Size: 45 cm, 2 kg (18 in., 4 ½ lbs).

Natural history: The Mexican hairy porcupine has barely been studied. It is usually solitary. It emerges after dark to feed on seeds, fruits, leaves, flowers, and flower buds. It feeds mostly in the middle layers of the forest, rather than in the canopy. In Mexico it was found to spend about 80% of its feeding time consuming fruits and seeds and about 20% enjoying leaves. It cuts and crushes seeds and unripe fruits such as those of ingas (*Inga* spp.) with its sharp incisors and strong molars and thus predates rather than disperses plants. Like the howler monkey (p. 140), it prefers young leaves, presumably because these contain more protein and less fibers and toxins than older ones. Favorite leaves include those of figs (*Ficus* spp.), cecropias (*Cecropia* spp., illustrated on p. 62) and breadnuts (*Brosimum* spp., illustrated on p. 158).

A prehensile tail and a movable pad on the hindfoot in place of the thumb (see track illustration on pl. 26) help the porcupine grip flimsy branches and move slowly but securely through trees. The prehensile tail of this and several other New World porcupines differs from that of other mammals in that the gripping pad is on the top side, so the tail is coiled upward rather than downward around branches. The tail is spineless and, unlike that of the North American porcupine, is not lashed out in defense.

The entire upperparts of the porcupine's body are covered in sharp yellow quills. When relaxed, the porcupine holds the quills flat against its body, where they are largely hidden by its long, dark fur. When threatened, however, it erects the quills such that they stick out perpendicular to the body. This behavior doesn't just serve as a physical deterrent; it also serves as a visual one, by making the brightly colored spines more conspicuous and by making the porcupine appear much larger than it is. This porcupine could thus be said to have aposematic (warning) coloration, which, although common in animals such as insects or snakes, is rare among mammals. Armored rats (*Hoplomys gymnurus*) often have pale-tipped or white spines, too (see pl. 27, p. 184). In the New World, the only other mammals with coloration that is clearly aposematic are the skunks (see pp. 236-237).

Contrary to popular belief, no porcupine can throw its quills, but the quills do become detached at the slightest touch. The quilltips of the Mexican hairy porcupine are covered with minute, backward-pointing scales (visible only through a microscope) that act like the barbs of a harpoon. The scales make embedded spines hard to withdraw, but allow quills to work their way into the flesh of potential predators at a rate that can exceed 1 mm per hour. Despite this, some predators, such as the larger cats, are known to prey on porcupines, presumably by flipping them onto their backs and attacking the quill-free underparts.

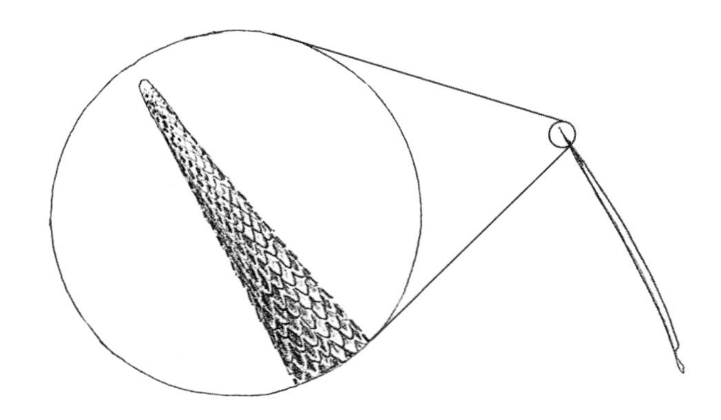

Mexican hairy porcupine quills have smooth yellow shafts and scaly black tips. The keel-shaped, backward-pointing scales act like the barbs of a harpoon, making embedded spines hard to withdraw.

C. mexicanus spends the daytime concealed in a tree crown or hollow trunk. Roosting sites, which can be used for months at a time, may be revealed by their pungent odor and scatterings of 2 cm (¾ in.) long, roughly oval-shaped fecal pellets on the ground. Mexican hairy porcupines studied in Mexico were found to have home ranges of about 10 ha (25 A). Other porcupines in this genus have been found to use home ranges of between 8 and 38 ha (20 to 94 A). Male Mexican hairy porcupines fight in captivity and may have exclusive territories in the wild.

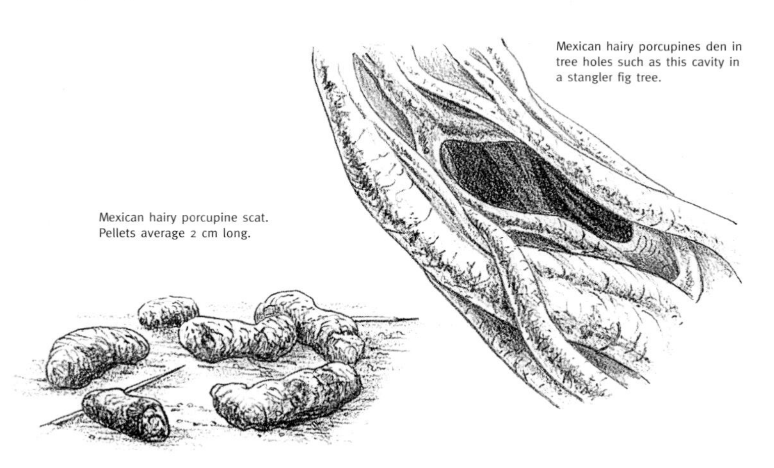

Mexican hairy porcupines den in tree holes such as this cavity in a stangler fig tree.

Mexican hairy porcupine scat. Pellets average 2 cm long.

Gestation lasts about six and a half months, producing single young which, for obvious reasons, are born with soft spines. The spines harden within about a week. Although there is little information for *C. mexicanus*, young of the closely related *Coendou prehensilis* of South America start to eat some solid food at about one month, are weaned at about three and a half months, and reach adult size at about a year and four months. Females reach sexual maturity at about a year and a half, and, in captivity, can reproduce continuously until they are at least 11 years old; one captive produced 10 litters in eight and a half years. Captive *Coendou* porcupines can live at least 17 years.

Sounds: Hisses and clicking sounds when threatened; plaintive, low-pitched whines by young.

Conservation: Porcupines are sometimes hunted for their porklike meat, or rarely as a nuisance to farmers. They are scarce throughout the Atlantic lowlands and absent from southwestern Costa Rica, but can be relatively common elsewhere.

References

Chinchilla 1997; Coates-Estrada & Estrada 1986; Emmons et al. 1997; Estrada 1984; Estrada & Coates-Estrada 1985; Hilje & Monge 1988; Janzen 1983a; Nowak 1991; Reid 1997; Roberts et al. 1985; Timm 1994a; Timm & Laval 2000a; Timm et al. 1989; Vaughan 1980.

O: Rodentia F: Dasyproctidae

CENTRAL AMERICAN AGOUTI

Dasyprocta punctata
(plate 26)

Names: *Dasyprocta* is the Greek for bushy-rumped, in reference to long rump hairs that the agouti erects when startled. *Punctata* means spotted, presumably in reference to the finely grizzled hair color. *Agouti* comes from the Guaraní language of South America. **Spanish**: *Guatusa*; *cherenga*.

When an agouti is startled, long hairs on its rump stand on end.

Range: The Dasyproctidae contains about 14 other species: 12 species of agouti (four in Central America and Mexico, and nine in South America), and two species of acouchi (similar-looking animals found in the Amazon basin). The Central American agouti ranges from southern Mexico to Ecuador; from sea level to at least 2,400 m (7,900 ft); in all types of forest or brush.

Size: 50 cm, 3 kg (20 in., 7 lbs).

Similar species: Rabbits (pl. 28) have large ears and are smaller. See paca (pl. 26, p. 181).

Natural history: Central American agoutis are exclusively ground-dwelling and active mostly by day. They feed on fruits and seeds, and, occasionally, fungi, insects, and crabs. Sometimes agoutis follow beneath troops of monkeys to pick up fallen food. Favorite foods include the seeds and/or fruits of palms such as *Astrocaryum standleyanum*, pejibaye

(*Bactris gasipaes*, illustrated on p. 184), or welfia (*Welfia georgii*, illustrated on pl. 27), and of trees such as almendro (*Dipteryx panamensis*), sapote (*Pouteria fossicola*), guapinol (*Hymenaea courabil*, illustrated on p. 298), cedro macho (*Carapa guianensis*, illustrated on pl. 27), or the balsa relative *Matisia ochrocalyx*.

An agouti dropping; the main clump is about 3 cm long.

The availability of fruit fluctuates seasonally. In an extensive study on Barro Colorado Island, Nicholas Smythe and others found that the amount of fruit falling at the end of wet season was insufficient to meet agoutis' dietary needs. Consequently, agoutis bury seeds throughout the year and dig up their caches when fruit is scarce. This strategy differs from that used by the agouti's similar-looking cousin, the paca (see p. 181).

Agoutis live in apparently monogamous pairs that share a territory but generally sleep and forage alone. They den in hollow trees, tunnels, or under logs or dense piles of brush. In Panama, each pair had a territory of about 2 ha (5 A), usually centered on a stream. (Home-range size may vary according to habitat, however: a lone, radio-collared female in Palo Verde had a home range of at least 4 ha, or 10 A.) When food is plentiful pairs tend to stay near the center of their territories; when food is scarce they need to roam further, and defend their territories more vigorously. If marking, raising rump hair, and foot stamping fail to work, territories can be defended violently, especially by males, whose bites and leaping kicks can inflict serious injury to rivals.

Females bear litters of one, two, or rarely three or four young two or three times a year. Males' courtship displays include showering the females with urine. Receptive females respond by going into a frenzied dance that culminates in mating. Gestation lasts about four months. The mother gives the well-developed young a tour the first morning after birth, during which they select their own den site and line it with leaves and twigs. The den is much smaller than that of the mother and thus inaccessible to large predators. The mother, unable to enter the den herself, calls the young out for nursing, at which time she also stimulates them to urinate and defecate by licking them and eats the excretions. Otherwise, the den would quickly become filthy; accumulated excrement might also attract predators, such as ctenosaur lizards.

Young start to accompany the mother at the age of about three weeks and abandon their nests at about two months. When food is plentiful, they continue to forage with the mother until they are four or five months old; when food is scarce, the mother quickly sends them packing. Forced into independence in a season of food shortage, and without a territory of their own, abandoned subadults have little chance of surviving. Even when young are not let go prematurely, only about 30% survive their first season of food shortage. Weakened by starvation, the rest are easy prey for carnivores such as male coatis. In the wild, adults have an average lifespan of two or three years. In captivity, they can live 17 years.

Sounds: Barks and grunts as threat or in alarm; loud squeals if captured.

Conservation: The agouti is an important seed disperser. It hoards seeds in small, widely scattered clumps that are less likely to be located by other seed-eating animals, such as peccaries or beetles, than large or concentrated seed stashes. An agouti is unlikely to retrieve all buried seeds, however, either because it never relocates them or because it dies. Young agoutis, which start burying seeds as soon as they are independent, are particularly likely to disperse seeds in this way because they have a high mortality rate. Neglected agouti seed stashes are set up perfectly to germinate and grow.

The agouti is common in many areas. Its relatively fast reproductive turnover and small range requirement make it somewhat resilient to both hunting and deforestation. It is a popular game animal. Its habit of running in circles within its territory when pursued can make it easy to hunt, and some (but not all) Costa Rican hunters claim that its meat resembles veal and tastes good. It is possible the agouti could be farmed profitably, and in a less destructive, more sustainable fashion than beef cattle.

References

Eisenberg 1974; Emmons et al. 1997; Hallwachs 1986; Jones 1982; Kleiman 1974; Kleiman et al. 1979; Larson & Howe 1987; Reid 1997; Rodríguez (J.) 1985; Smythe 1970a, 1978, 1983; Smythe et al. 1996; Timm et al. 1989.

O: Rodentia F: Agoutidae

PACA

Agouti paca

(plate 26)

Names: Genus from the Guaraní language of South America. Species and common names originated with Peruvian natives. Note that, confusingly, the family Agoutidae (sometimes considered a subfamily of the Dasyproctidae), does not contain the agoutis. The family contains only one other species, the mountain paca (*A. taczanowskii*), which inhabits the Andes above the tree line. Previously listed under the genus *Cuniculus*. **Spanish**: *Tepezcuintle.*

Range: Southeastern Mexico to northern Paraguay; from sea level to about 3,000 m (10,000 ft); in relatively undisturbed forest, usually near streams or other fresh water.

Size: 70 cm, 9 kg (28 in., 20 lbs); males larger than females.

Similar species: Central American agoutis (pl. 26, p. 178) are smaller, lack white spots, and are diurnal. Young tapirs (pl. 36, p. 297) have a prehensile snout, smaller eyes, and white-tipped ears.

Natural history: Pacas are exclusively nocturnal; captives reach peak activity at around midnight. Their diet overlaps to a large degree with that of their similar-looking cousins, the agoutis: both eat fallen fruit when it is available. Among the pacas' favorite foods are cedro macho seeds (*Carapa guianensis*, illustrated on pl. 27), guavas (*Psidium guajava*), avocados (*Persea americana*), and mangoes (*Mangifera indica*, illustrated on pl. 12).

A paca dropping; the main clump is about 4 cm long.

When fruit is scarce, however, pacas and agoutis cope in different ways. Agoutis bury seeds that are hard enough not to rot and retrieve them when fruit is scarce (see p. 179). They are able to eat hard seeds because, like squirrels, they squat on their hindlegs while feeding, freeing their hands to hold the seeds still while they gnaw at them. Pacas, however, stay on all fours while feeding and do not take food that needs to be manipulated. When fruit is scarce, they rely on seedlings, leaves, and roots.

Foliage is a relatively low-energy food that is slow and space-consuming to digest. A few key adaptations, not found in agoutis, help pacas deal with this partly herbivorous diet. Pacas are larger than agoutis and have longer large intestines. Like rabbits, they are coprophagous: they eat their fecal pellets in order to extract important nutrients not available during the food's first passage. They are more sedentary than agoutis, seldom running more than a few meters when startled. Being more sedentary, they are able to put on more fat than agoutis when food is plentiful, and they also rely on these fat reserves during the season when fruit is scarce. They have relatively low metabolic rates. Finally, like howler monkeys, they are able to produce astonishingly loud sounds that may help them fend off rivals or aggressors without expending too much energy.

Pacas live in monogamous pairs that share a roughly 3 ha (7 ½ A) territory but, like agoutis, they sleep and often forage apart. They den in roughly 3 to 9 meter (10 to 30 ft)-long, 20 cm (8 in.)-wide burrows that are frequently located in steep banks. They may dig their own or modify those of other animals, such as armadillos. Each burrow has a main entrance and one or more secret exits, the latter stuffed full of leaves for concealment. The hidden exits are so legendary to Costa Rican hunters that they have been given a specific name: *uzús*. If a predator intrudes down the main entrance, the paca explodes out of an *uzú*. Pacas are good swimmers; they tend to den near water, where they often take refuge if flushed. As a last line of defense, pacas have a fragile layer of outer skin and blubber that may make it difficult for predators to get a firm hold on them.

Pacas breed at any time of year. Courtship involves a twisting, hopping dance, similar to that of agoutis, during which the male tries to spray the female with urine. The tibia of the paca is almost as long as the femur, and provides tremendous leverage for jumping high off the ground (a skill that is also useful for eluding predators). At first the female avoids the male and his urine, or even attacks him, but once sprayed a few times she crouches and allows him to approach.

Gestation lasts about three and a half months, usually resulting in a single offspring. Young are weaned by the time they are three months old, but may stay with the mother until almost her size. Females become sexually active at about nine months, males after about a year. Captives can produce two or three litters in a year, but one litter is more common. Pacas can live 13 years in the wild and at least 16 in captivity.

Sounds: Pacas have an unusually swollen zygomatic arch (the cheek bone) that acts as a resonating chamber, a feature unique among mammals. The amplified grunts, growls, barks, and tooth-grinding sounds that pacas emit may help intimidate rivals or predators. Young pacas make a meowing sound until they are about a month old.

Conservation: The paca has become rare or extinct over much of the country due to habitat loss and hunting. Widely considered the best tasting mammal in Costa Rica, it is the number one target of most hunters. Researcher Teresa Zúñiga interviewed dozens of hunters from Barro Colorado in northeastern Costa Rica and found that 95% of them identified the paca as their preferred game species. She estimated that those hunters killed an average of almost 900 pacas each year in a 146 km^2 (56 mi^2) area. Pacas are relatively easy to hunt since, in pristine habitats, they show little fear toward people; their well-used trails can make dens obvious; and their strategy of running a short distance and then sitting absolutely still when attacked away from the den does not work well against hunters and dogs. Moreover, with their low reproductive turnover, pacas are slow to replenish reduced populations and are easily wiped out.

Several dozen small-scale paca farms exist in Costa Rica. A few are simply fronts for illegal paca hunters, and many are not proving to be economically viable, but paca farming could potentially be profitable and environmentally friendly.

References

Chacón 1996; Collett 1981; Eisenberg 1974; Emmons et al. 1997; Freiheit 1966; Hershkovitz 1955; Kleiman 1974; Kleiman et al. 1979; Matamoros 1980, 1981, 1982; McHargue & Hartshorn 1983b; McNab 1982; Pérez 1992; Quesada & Hanan 1988; Smythe 1983, 1987; Smythe et al. 1996; Zúñiga 1994.

O: Rodentia F: Echimyidae

SPINY RATS

Proechimys semispinosus and *Hoplomys gymnurus*
(plate 27)

Names: Tomes' spiny rat (*P. semispinosus*) and the armored rat (*H. gymnurus*). *Proechimys semispinosus* translates roughly as "partially spined, hedgehoglike mouse"; *Hoplomys* stems from the Greek words *hoplon*, meaning weapon, and *mys*, meaning mouse, while *gymnurus* comes from the Greek *gymnos*, meaning naked, and *oura*, or tail. The family name comes from the Greek words *echinos*, meaning hedgehog, and *mys*, or mouse. Despite their name and appearance, echimyids belong to the caviomorph rodent suborder with porcupines, agoutis, and pacas, rather than to the myomorph suborder with true mice and rats, such as spiny pocket mice (see p. 169). **Spanish**: *Ratas espinosas.*

Range: Most of the 70 species in this, the largest caviomorph family, are found in northern South America; these are the only two species found as far north as Costa Rica. Both species range from eastern Honduras to Ecuador west of the Andes (with Tomes' spiny rat also inhabiting the Amazon Basin); from sea level to about 800 m (2,600 ft); mostly in forested areas near water.

Size: Both species: 25 cm, 450 g (10 in., 1 lb).

Similar species: Spiny rats are distinguished by their large size, ear shape (the rear edge is indented, forming a horizontal M shape), and, in both Costa Rica's species, the presence of spines. Spiny pocket mice (pl. 22, p. 169) are smaller and have round ears. Spiny rat spines are not barbed like those of porcupines. *Proechimys* has thin spines that lie flat, hidden amidst the fur, while *Hoplomys* has thick spines that protrude conspicuously. Another defensive characteristic found in most echymyids is a brittle tail, which helps them escape, albeit tailless, from predators. The tail does not grow back, and spiny rats with reduced or missing tails are common.

Natural history: Both species are nocturnal and terrestrial. They feed on fruits, seeds, fungi, insects, and leaves. Both species eat the fruits and seeds of the welfia palm (*Welfia georgii*, illustrated on pl. 27), and the armored rat is fond of the fruit of the pejibaye palm (*Bactris gasipaes*, illustrated below) and the germinating seeds of cedro macho trees (*Carapa guianensis*, illustrated on pl. 27).

Fallen fruits and trunk of the pejibaye palm.

Hoplomys has more specific microhabitat requirements than *Proechimys*. *Hoplomys* prefers very wet habitats, often in steep and rocky areas along streams in relatively undisturbed areas, less often in young forests or away from streams. *Proechimys* uses a broader range of habitats, having more tolerance for drier or disturbed forests. In an area in Panama where the two species occur together, as they do over most of their range, *Hoplomys* was found only along undisturbed streams, while

Proechimys was found almost everywhere, including along ridgetops and disturbed streams where *Hoplomys* was absent. *Proechimys* was absent in that area only along streams where *Hoplomys* was common.

Mammalogist Louise Emmons discovered that habitat use by *Proechimys* can also be influenced by the moon cycle. She found that the amount of *Proechimys* spotted while walking along trails dropped by 60% or more during the five days before and after each full moon. Such "lunar phobia," which has also been observed in other small rodents and bats, may be due to increased risk of predation on moonlit nights, perhaps especially by owls. Owls, which are among the major predators of small rodents and bats, are descended from a largely diurnal order—birds—and still rely heavily on vision to locate prey. Studies have demonstrated that owls are much more successful at finding prey visually, and indeed are more active, on moonlit nights. By following radio-collared spiny rats, Emmons determined that *Proechimys* is no less active than normal on moonlit nights. Thus, *Proechimys* is less conspicuous on moonlit nights either because it is just better at spotting approaching biologists and taking evasive action, or because it restricts its activities to areas where it is better concealed. Studies on other species of rodents in captivity suggest the latter.

Recorded home-range sizes for *Proechimys* have varied from 0.1 to 1.5 ha (¼ to 4 A). Where the species is abundant, home ranges overlap extensively; where *Proechimys* is less common, female home ranges overlap little. *Proechimys* dens in tunnels in the ground, in hollow logs, or under fallen brush. *Hoplomys* dens in burrows that are often located in stream banks. The tunnels are typically horizontal, widening at the end into a chamber lined with dry vegetation. A second chamber is used as a bathroom. Both species hoard food in their burrows.

Like other caviomorph rodents, and unlike the true myomorph rats, spiny rats have relatively long gestation periods, resulting in small litters of well-developed young that are slow to become independent. *Proechimys* has a gestation period of just over two months. It can produce up to four litters a year, with one to five young per litter. Young become independent by the time they are about two and a half months old, when they are almost adult size, and females can start reproducing shortly thereafter. *Hoplomys* produces litters of one to three young. Its young do not grow spines until they are about a month old. *P. semispinosus* can live about five years in captivity.

Sounds: Tomes' spiny rat growls and chatters its teeth when threatened, and twitters and whimpers when alarmed or during courtship. Alarmed armored rats squeal, whine, and emit quickly repeated sounds that resemble muffled dog barks.

Conservation: In South America, spiny rats are valued for their meat, which is sometimes sold in markets. Wild carnivorous animals also value them for their meat. Scat and radio-telemetry studies have shown that

Proechimys is the staple food of bushmasters (*Lachesis muta*) at La Selva. These snakes habitually wait beneath the *Welfia* palms where *Proechimys* comes to feed. Likewise, a year-long study in Corcovado National Park found that more than a quarter of all puma scats and half of all ocelot scats contained *Proechimys* remains. These studies and other observations suggest that, in the lowlands of Costa Rica, *Proechimys* is common, while *Hoplomys* is uncommon.

Bushmasters (*Lachesis stenophrys*) wait under *Welfia* palms to ambush spiny rats that come to eat Welfia seeds.

References

Adler et al. 1997; Chinchilla 1997; Clarke (J.) 1983; Dice 1945; Eisenberg 1974; Eisenberg 1989; Emmons 1982; Emmons & Feer 1997; Emmons et al. 1989; Emmons et al. 1997; Fleming 1971b; Gliwicz 1984; González & Alberico 1994; Greene 1988; Greene & Santana 1983; Jones 1982; Kleiman et al. 1979; Maliniak & Eisenberg 1971; McHargue & Hartshorn 1983a,b; Morrison 1978; Reid 1997; Seamon & Adler 1999; Tesh 1970; Timm 1994a,b; Tomblin & Adler 1998; Vandermeer 1983a,b; Vandermeer et al. 1979; Weir 1974.

Rabbits (order Lagomorpha)

Distribution and Classification

Lagomorphs are found on all the continents except Antarctica. They have been introduced to many areas, including Australia, New Zealand, many small islands, parts of Europe, and southern South America. The order Lagomorpha (the Latin for hare-shaped ones) includes about 70 species, which are split into just two families: the Ochotonidae, or pikas, and the Leporidae, or rabbits and hares.

A pika, a rabbit, and a hare (from left to right).

There are 22 species of pikas, of which only two inhabit the New World, and these are restricted to rocky, high-altitude, or tundra regions of the United States and Canada. There are 47 species of rabbits and hares, 24 of which inhabit the New World. Of the New World lagomorphs, only three species of rabbit occur south of Mexico; all three are found in Costa Rica.

Technically, rabbits differ from hares in being smaller, and in having shorter ears without black tips and a somewhat differently shaped skull. While hares give birth to precocial young that are born fully furred, open-eyed, and able to run around within a few minutes of birth, rabbits produce altricial young that are born naked or only somewhat furred, blind, and helpless. All true hares belong to the genus *Lepus*. Nonetheless, popular names do not always follow these rules, and a number of species familiar to Europeans or North Americans, including the Belgian hare, jack rabbit, and snowshoe rabbit, are misnamed.

Lagomorphs are distinguished from all other mammal orders by their dental arrangement, which is described below. They resemble rodents in a number of ways (such as body shape, dentition, and high reproductive turnover), and were originally considered a suborder of the Rodentia. Many still believe that the two orders are closely related, although other authorities have collected evidence that suggests the lagomorphs may be more closely related to deer and peccaries (order Artiodactyla), or even to elephant shrews (order Macroscelidea).

Evolution

Most lagomorph evolution and diversification took place in Eurasia and North America, where the greatest number of species are still found today. Leporids not greatly different from those around now are known to have existed in Asia about 55 million years ago and at least 45 million years ago in North America, from where they eventually spread south to Costa Rica. Although one European species introduced since colonial times has established wild populations in South America, Costa Rica's wild rabbits are native.

Teeth and Feeding

The dental formula of the lagomorphs, with the exception of one Asian species, is I 2/1, C 0/0, P 3/2, M 3/3 = 28. Like rodents, lagomorphs have two pairs (one upper and one lower) of ever-growing incisors, which are kept worn down through use, and a large gap between these and the cheek teeth. Lagomorphs differ from rodents, however, in having their incisors fully covered with enamel, and in possessing a second pair of upper incisors situated directly behind the larger front pair. The latter feature is diagnostic of the order. Lagomorphs' cheek teeth are good for grinding up vegetation, which is essentially all they eat (although some hares eat voles, young lagomorphs, and other meat on occasion).

Rather like perissodactyls such as tapirs, lagomorphs ferment leafage in their hindguts, and this hinders their ability to make the best of their food's nutritional value (see p. 294). To compensate for this limitation, lagomorphs, like some rodents and shrews, eat their own feces. They produce two types of fecal pellets. Fecal pellets emerging after food's first passage are larger, wetter, and lighter-colored than the pellets one finds deposited on the ground, and are eaten immediately. They are swallowed unchewed, bypass the front of the stomach where fresh food is stored, and settle again in the rear portion of the stomach for further fermentation. Microbes in the digestive system then permit the extraction of important nutrients, including vitamins K and B, that are not assimilated during the material's first passage. This process, known as coprophagy (from the Greek *kopros*, meaning dung, and *phago*, meaning to eat), also helps lagomorphs obtain the maximum amount of water from their food, which helps some species survive in very dry habitats.

Body Design

Lagomorphs are digitigrade, and have five toes on the forefeet and four on the hindfeet. Individual digits are seldom visible in tracks. Small but strong claws on all digits help lagomorphs excavate burrows and forms. Otherwise, since lagomorphs do not climb trees or use their feet to feed, their limbs are designed almost exclusively for terrestrial locomotion. Practically defenseless, lagomorphs' only hope of escaping predators is to run for it, using their strong, long hindlegs to kick them up to speeds that can reach 80 km/h (50 mph) in the largest hares and diving into the nearest cover. If they do not find shelter, they may abruptly end their sprint and sit motionless. Many species have tails with white undersides that flash conspicuously while the animal is in flight, and which may help warn neighboring leporids of danger. If the leporid suddenly sits still, the tail (and the leporid) vanish. This disappearing act may temporarily confuse predators.

Senses

Leporids have excellent senses of smell and hearing that they use to detect predators and to communicate. Males have glands on their chins with which they rub branches and other objects. Studies have shown that dominant males tend to have greater quantities of sex hormones circulating in their blood, and that these hormones in turn increase scent and sperm production. By following scent signposts, females may be able to home in on dominant, virile males, and thus increase the chances that their offspring will inherit the same desirable characteristics. Leporids have a limited vocal repertoire, but use sound to communicate in an unusual way. When they sense danger, they sometimes drum the ground with their hindlegs, creating sounds that are audible to humans only at close range, but which can alert others of their own kind from some distance. Leporids have poor visual acuity (the ability to see objects at a distance), but their bulging, laterally positioned eyes afford them a broad field of view for detecting predators (see illustration on p. 24).

Reproduction

Leporids breed like rabbits: in a single year, a typical leporid female (or doe) has between two and seven litters, each with two to 15 young. Few leporids defend exclusive territories, but males (or bucks) must compete if they want to mate. Despite their cuddly image, lagomorphs can be quite aggressive, raising up to box with their forefeet, kicking with their hindfeet, and even biting chunks out of each other's rumps. In many species, a buck that has won a female's attention will join her in an elegant, leaping courtship dance. He may also urinate on her. If love prevails, the buck then has the chance to use his somewhat unusual reproductive apparatus: Lagomorphs' penises are situated behind the scrotum, rather than in front, as in all other mammals except marsupials. The doe may allow her partner to mate with her repeatedly for a few hours or days, after which she drives him away with bites and kicks.

Like many leporids, Costa Rica's rabbits nest in forms, which are depressions roughly 7 cm (3 in.) deep and 14 cm (6 in.) in diameter that they scrape out of the ground. Young are often covered with a blanket made of fur that the mother plucks from her belly, mixed with grass. These materials, also used for nest lining, provide both camouflage and insulation. Rabbit young (or fawns) are left unattended in the form for long periods, although the mother often sits nearby in vigil. Although normally timid, cottontails defending their young have been known to attack and drive away dogs, cats, snakes, and people. In all leporids that have been studied, mothers visit and nurse their fawns just once a day, typically for less than five minutes. Rabbit milk is extremely nutritious, however (it contains more fat and protein than cow or goat milk), and most rabbits are weaned after just three to five weeks.

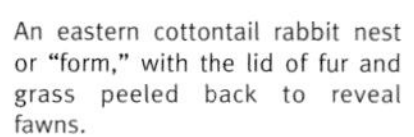

An eastern cottontail rabbit nest or "form," with the lid of fur and grass peeled back to reveal fawns.

Conservation

Since they are so prolific, leporids are often a primary source of food for meat-eaters, and native species at least represent an important link in food chains. Humans value them for a number of other reasons. Rabbits' left hindfeet are used as goodluck charms: legend has it that rabbits are born with their eyes open and are thus mystically alert and capable of shooing away evil spirits. (The biological part of this superstition is incorrect, since, as mentioned above, only hares are born with their eyes open.) More importantly, they are among the world's top game animals, for they are easy to hunt, and they have good-tasting meat and fur that is used to make garment linings and felt.

On the other hand, leporids can be major pests in agricultural areas; in parts of the world where they have been introduced, they can cause damage to pristine ecosystems as well. The most famous example is *Oryctolagus cuniculus*, the European rabbit. Native only to southwestern Europe, the species was introduced as a game animal all over the rest of continental Europe by the Romans, to the British Isles by the Normans, and to Chile in the 1700s and Australia and New Zealand in the 1800s by colonial Europeans. In Australia and New Zealand, it has caused massive losses to the important sheep farming industry, for it feeds in sheep pastures. It has also thrived in nonagricultural areas there and, without its natural predators to keep populations in check, has displaced several species of native marsupials. On a number of oceanic islands, where the European rabbit was introduced to provide food for passing sailors, it has been detrimental to nesting marine birds by eating nest cover and causing cliffs to erode. Perhaps the misconceived superstition about rabbits' feet has been a bad omen. Few creatures have better demonstrated how humans can alter the course of thousands of years of evolution in a geological instant.

References

Chapman & Ceballos 1990; Diersing 1984; Flux & Fullagar 1983; Grzimek 1975; Hartenberger 1985; Howard & Amaya 1975; Macdonald 1995; Marsden & Holler 1964; McKenna 1975; Rue 1968.

SPECIES ACCOUNTS

O: Lagomorpha F: Leporidae

TAPITI and **DICE'S COTTONTAIL**

Sylvilagus brasiliensis and *S. dicei*

(plate 28)

Names: The tapiti (*S. brasiliensis*) is also known as the forest rabbit or Brazilian rabbit. Dice's cottontail (*S. dicei*) is also known as the greater forest rabbit. Dice's cottontail was formerly considered a subspecies of the tapiti, but was separated because it is consistently larger; even where the two species occur close to one another, there are no intermediate forms. **Spanish**: *Conejo de monte, conejo de bosque, tapetí.*

Range: The tapiti occurs from eastern Mexico to northern Argentina; from sea level to 2,500 m (8,200 ft)—but in Costa Rica only to about 1,500 m (5,000 ft) in the Tilarán mountains and about 1,100 m (3,600 ft) in the Talamancas; mostly in open areas along the edge of wet forest or second growth. Dice's cottontail is restricted to the Talamanca range in southern Costa Rica and western Panama; from 1,100 m (3,600 ft) to at least 3,500 m (11,500 ft) on Chirripó mountain; in habitats like those used by the tapiti, as well as in páramo (the scrubby habitat found above the tree line). Rabbits seen at 2,600 m (8,500 ft) in Braulio Carrillo National Park, north of Dice's cottontail's known range, could also be this species, but no specimens have been collected.

Size: *S. brasiliensis*: 38 cm, 850 g (15 in., 2 lbs); *S. dicei*: 42 cm, 1 kg (17 in., 2 lbs and 3 oz); females slightly larger than males in both species.

Similar species: Dice's cottontail is slightly larger and tends to be darker than the tapiti, but there is little visible difference between the two. This may be irrelevant to identification, since the two species appear to be allopatric (i.e., their ranges do not overlap). The only other rabbit in Costa Rica is the eastern cottontail (pl. 28, p. 193), which occurs only in northwestern Costa Rica, outside the range of either species. The tapiti and Dice's cottontail differ from the eastern cottontail in having shorter ears; a shorter, less-conspicuous tail that is buff rather than white on the underside; usually darker fur; and a slightly smaller patch of orange on the nape. All three species give off a reddish eye glow when caught in the beam of a flashlight. Note that only one eye is visible when a rabbit is in profile.

Natural history: These two rabbit species have barely been studied. Both are nocturnal and crepuscular (active at twilight). They feed predominantly on grass and other green vegetation. Dice's cottontail is fond of the young shoots of high-elevation bamboo (*Swallenochloa subtessellata*) that emerge after fires. The tapiti also feeds on fungus, including *Leccinum chromapes* at Irazú Volcano and *Tylopilus alboater* in oak forests at San Cristóbal, and is reported to be attracted by salt and urine.

Although often referred to by the name forest rabbit, the tapiti favors the forest edge. It is seen most often in disturbed, grassy habitats, such as around the buildings at La Selva, or in scrubby second growth around the Poco Sol field station of the Children's International Rainforest. In pristine habitats, it forages around the edge of swamps or along the flood lines of large rivers. Tapitis hollow out nest chambers in piles of dried grass. A central chamber may be connected to smaller chambers via runways concealed under thick ground vegetation. In Chiapas, Mexico, tapitis breed year-round, producing litters of two to eight fawns after gestation periods of about a month. Above the treeline in the Venezuelan

Andes, tapitis have longer gestation periods of about a month and a half. Fawns of Dice's cottontails have been observed only between September and April, although the species may breed year-round.

Tapitis are the natural vectors of myxomatosis, a viral disease that has no effect on them or other rabbits in the genus *Sylvilagus*, but which is deadly to the European rabbit. The disease has been deliberately introduced to Australia and other areas where the European rabbit is a non-native pest (see p. 191). Such introductions have been very effective in reducing populations, although the rabbits have now become resilient to the disease in many areas.

Conservation: Both species are undoubtedly important prey for a variety of carnivorous animals. The tapiti is known to be preyed upon by the fer-de-lance (illustrated on p. 137), and Dice's cottontail was found be the principal food of coyotes in a study done recently in the Costa Rican highlands. In parts of the Caribbean lowlands, the tapiti is also hunted by people, either for its meat or because it can damage some commercial plants by eating the leaves of seedlings or gnawing bark. Since the tapiti likes disturbed habitats, it may have benefited from deforestation, but it is seldom abundant in Costa Rica. Most of the range of Dice's rabbit lies within Chirripó and Amistad National Parks, where the species is common and relatively unaffected by human activities.

References

Chapman & Ceballos 1990; Diersing 1981; Durant 1983, 1984; Emmons & Feer 1997; Glanz 1996; Gómez 1983; Hilje & Monge 1988; Hoogmoed 1983; Janzen 1983f; Reid 1997; Timm & Laval 2000a; Timm 1994a; Timm et al. 1989; Vaughan & Rodríguez 1986.

O: Lagomorpha F: Leporidae

EASTERN COTTONTAIL

Sylvilagus floridanus

(plate 28)

Names: Genus from the Latin *silva*, meaning forest, and *lagus*, meaning hare. *Floridanus* because the type specimen (the specimen after which the species was originally described) is from Florida. **Spanish**: *Conejo cola de algodón*.

Range: Southern Manitoba and Quebec, Canada, to northwestern Costa Rica, and in the plains of Venezuela and Colombia (strangely, it is absent from Panama); from sea level to about 3,300 m (10,800 ft), but mostly below about 1,000 m (3,300 ft) in Costa Rica; in deciduous forest, forest edge, and open areas with low cover. This species has a larger natural range than any other lagomorph.

Size: 38 cm, 1 kg (15 in., 2 lbs and 3 oz); smaller here than in North America.

Similar species: In Costa Rica, the eastern cottontail is known only from the northwestern lowlands, where it is the only species of rabbit. See p. 192 for physical differences between the eastern cottontail, the tapiti (found only at higher elevations in the northwest and on the Atlantic slope), and Dice's cottontail (found only in the highlands of the Talamancan and, possibly, Central mountain ranges).

Natural history: Eastern cottontails are mainly crepuscular and nocturnal, spending the daytime hidden amidst ground vegetation or brush in rest sites known as forms. They are basically solitary, although they may reach high densities in good feeding areas. Their diet includes leaves, buds, twigs, bark, and fruits. In North America, they are able to stomach toxic plants such as poison ivy (*Rhus radicans*).

Eastern cottontail droppings. Each pellet is 7 to 10 mm in diameter.

Eastern cottontails often come out to open areas to feed on grass, but usually stay within a few bounds of cover, where they disappear at the slightest disturbance. They must be alert constantly, for they have numerous predators. If caught in the open, they can run fast for short distances, reaching speeds of up to 40 km/h (25 mph). If they do not find cover, they may suddenly freeze, hiding their conspicuous white tails and sitting absolutely motionless for up to 15 minutes before sneaking off.

Typically, *S. floridanus* has a home range size of less than 5 ha (12 A), although males may wander further during the breeding season. This species does not defend exclusive territories, but males fight with each other during breeding season to establish a mating hierarchy.

Studies have shown that the life cycle of cottontails in the tropics is different from that of cottontails in temperate areas. In temperate areas, gestation lasts a little less than a month. Temperate-zone cottontails breed only seasonally, but can give birth to as many as seven litters, with an average of five fawns per litter, in a single season—a turnover that is high even among rabbits. Fawns are weaned and independent after about a month, but most do not breed until the following season. Temperate cottontails have an average lifespan in the wild of about 15 months.

In South America, on the other hand, cottontails breed year-round but have a longer gestation period (a little over a month) and smaller litters (averaging two or three fawns). Most females first become pregnant when they are only about two and a half months old. And they are shorter-lived than their temperate cousins, seldom surviving more than a year.

Sounds: Squeaks during courtship, throaty growls by females disturbed on their nests, and screams when captured. Imitations of the latter are used by hunters to attract larger game.

Conservation: This species is hunted both for its meat and as an agricultural pest. It can damage crops such as rice, beans, cabbage, and lettuce, and occasionally causes damage to fruit trees by gnawing bark from the bases of trunks. Its high reproductive turnover makes it resilient to hunting pressures. It thrives in highly disturbed habitats and has probably benefited from deforestation. Population densities may fluctuate substantially according to food availability and outbreaks of disease.

References

Chapman 1983, 1984; Chapman et al. 1977; Chapman et al. 1980; Marsden & Holler 1964; Nowak 1991; Ojeda & Keith 1982; Rue 1968.

Carnivores (order Carnivora)

Note: The word *carnivore* can mean two different things: members of the order Carnivora, or carnivorous (meat-eating) animals in general. To avoid confusion, the former are referred to as *Carnivores*, with a capital C, and the latter as *carnivores*.

Distribution and Classification

Worldwide, the order Carnivora contains about 240 species in seven families. This does not include the 34 species of seals, sea lions, and the walrus, which are descended from the Carnivora and used to be included in the same order, but are now placed in their own order, the Pinnipedia. Native Carnivores are found everywhere except Antarctica, Australia, New Zealand, New Guinea, and a few other islands. Two of the seven families—the mongoose or civet (Viverridae) and hyena (Hyaenidae) families—are not found in the New World. Roughly 85 species from the other five families—the dog, raccoon, cat, bear, and weasel families—occur in the New World, of which 22 species and all but the bear family (the Ursidae) are represented in Costa Rica. The only South American ursid, the spectacled bear (*Tremarctos ornatus*), inhabited Costa Rica until about 10,000 years ago, but is now found only in the Andes and adjacent foothills.

The spectacled bear (*Tremarctos ornatus*).

Evolution

The Carnivora descended from ancient Insectivores, first appearing as small, weasellike creatures known as miacids about 55 million years ago, in the Paleocene period. For a long time—up until the end of the Eocene period, about 35 million years ago—the Carnivora was a relatively small order, overshadowed by several other groups of carnivorous mammals. The predominant carnivores during that period belonged to a now-extinct order, the Creodonta. Even though the Creodonta are now known to have evolved completely independently of the Carnivora, many creodonts were remarkably similar to modern-day Carnivores, resembling bears, foxes, cats, civets, dogs, or sea otters. The creodonts resembled Carnivores still further in that they possessed carnassial teeth (specialized teeth for cutting meat—see below), although the creodonts' carnassials evolved in a different part of the mouth. Some other, mostly omnivorous lineages around during that period also had some carnivorous members. There were carnivorous marsupials (which also evolved carnassial teeth, several times), and two groups of carnivorous condylarths, one containing small animals known as arctocyonids that resembled primitive Insectivores, the other containing large, hoofed predators up to the size of a grizzly bear known as mesonychids.

By the end of the Eocene, most of the creodonts and all of the carnivorous condylarths had become extinct, for unknown reasons, and the miacids started to radiate into numerous forms. Even though carnivores of similar shape and dentition to modern species had already been "invented" independently, this was the true beginning of the Carnivore families we know today. The miacids split into two major groups, one containing the ancestors of dogs, bears, raccoons and weasels, the other containing the ancestors of cats, mongooses, and hyenas. These two fundamental groups are recognized even today, as the suborders Caniniformia and Feliformia respectively.

Teeth, Skull, and Feeding

As the order name implies, all members of the Carnivora eat at least some meat, but many species are omnivorous and some are even mainly frugivorous. Most species have well-developed carnassial teeth—cheek teeth with bladelike edges for cutting meat. When the jaw closes, these teeth—the rearmost upper premolar and the front-most lower molar on each side of the mouth—work against each other like scissor blades. Correspondingly, the jaws of the Carnivora are structured in such a way that they cannot move from side to side as freely as those of most herbivorous mammals. Most of the Carnivora also have large canine teeth and a powerful bite for seizing and killing prey. Many of them have a short, somewhat curved lower jaw, and a large temporalis jaw muscle, both features that allow more force to be exerted at the front of the mouth. To accommodate the large temporalis muscle, members of the Carnivora tend to have wide zygomatic arches (through which the muscle passes) and a pronounced sagittal crest (a bony ridge along the top of the skull that provides the muscle extra anchorage). These and other generalisations about skull design are discussed in more detail on pp. 22-23.

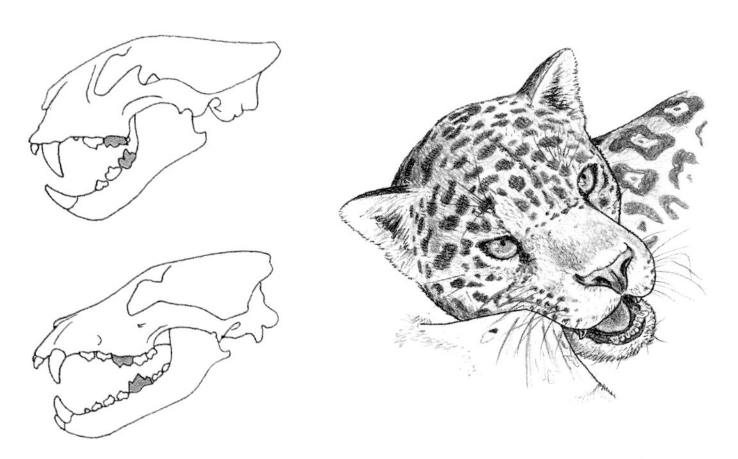

Shaded in red are the carnassial teeth of a cat (top left) and a dog (bottom left). The last upper premolar and the first lower molar of most Carnivores have a bladelike design, and work against each other like a pair of scissors. These so-called carnassial teeth allow Carnivores to slice off meat with the side of the mouth after they have made a kill (right).

Body Design

Carnivores range in size from the 15 cm (6 in.)-long, 50 g (2 oz) least weasel (*Mustela nivalis*) to the up to almost 3 m (10 ft)-long, 780 kg (1,700 lbs) grizzly bear (*Ursus arctos*).

Several features of Carnivore skeletons, other than the skull and teeth, are distinctive. Members of this order have small collar bones, or, in the case of some canids, no collar bones at all. The collar bones of the remaining canids, as well as those of felids, are tiny slivers suspended within the shoulder muscles. In other mammals, such as humans, the collar bones are attached firmly to other bones, and help both to hold the shoulder blade in place and to anchor muscles that control side-to-side movement of the limbs. The Carnivore design may have evolved because it improved the ability of primitive species—and many modern species—to chase prey on the ground: side-to-side limb movement is not important for running, while flexible shoulder blades permit a longer stride. Another aspect of limb structure peculiar to the Carnivora is the fusion of several wrist bones. This design makes the wrists firmer and may help various Carnivora run fast, climb or leap from trees, or subdue prey.

A skeletal feature that characterizes the males of all Carnivores except hyenas is a penis bone, also known as the *os penis* or baculum. This bone is not unique to the Carnivora—it is also found, for example, in most male rodents and some male Insectivores—but it is absent in the males of most other mammal orders. The baculum is usually long and thin in dogs, weasel relatives, and racoon relatives (that of the northern raccoon, for example is about 9 cm long) and short and broad in cats (that of the jaguar, for example, is less than 1 cm long). As compensation for this inferiority, the penises of felids are covered in tough spines.

BACULA (PENIS BONES) OF SELECTED CARNIVORES

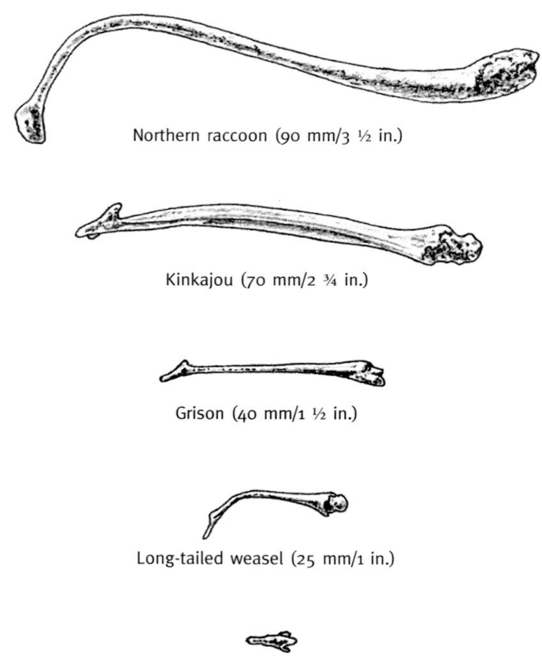

The baculum can be helpful in distinguishing species. Some Carnivores that are very similar in external anatomy, such as the northern and crab-eating raccoons, have quite different baculum designs. The baculum has several possible functions. It may aid penetration of the vagina, especially in mammals such as weasels in which males can be considerably larger than females. It may prevent the sperm duct, which runs down a groove in the baculum, from being blocked, especially in mammals such as canids that perform lengthy copulations in contorted body positions. And it may help increase the duration of copulation and stimulation to the female. Many female Carnivores are induced ovulators, which means that they ovulate only when stimulated by the action of mating, rather than when their estrous cycle dictates, as in humans and most other mammals. The baculum, the penile spines in cats, and the presence of both features in a few Carnivores such as raccoons, may be necessary to stimulate ovulation in induced ovulators.

Conservation

Since members of the Carnivora tend to be near the top of the food pyramid—in which the smallest, most abundant organisms at the bottom of the pyramid get eaten by larger, less numerous creatures above them, and those creatures in turn by yet larger and scarcer animals, and so on—they tend to be relatively scarce. It has been estimated that, while members of the Carnivora make up roughly 10% of mammalian genera, they represent only about 2% of mammalian biomass. Because of this natural rarity, they are easily put in jeopardy by human activities; 10 of Costa Rica's 22 Carnivora species are already listed as endangered.

Since the Carnivora is a large and diverse group, the four families found in Costa Rica are discussed independently over the following pages.

References

Ewer 1973; Kitchener 1991; Macdonald 1984b; Nowak 1991; Savage 1977; Turner & Antón 1997.

Dog Family (Canidae)

Distribution and Classification

The Canidae contains about 35 species worldwide and includes dogs, foxes, wolves, coyotes, and jackals. Native members of the dog family are found on all the continents except Antarctica and Australia. One species, the gray wolf (*Canis lupus*), has (or had) the biggest natural range of any terrestrial mammal other than *Homo sapiens*, inhabiting almost the entire Northern Hemisphere. The domestic dog (*Canis familiaris*) is, of course, found worldwide, but only because it is *Homo sapiens*' best friend. The name *C. familiaris* covers all 400 or so breeds of domestic dog, as well as the dingoes (wild domestic dogs) of Australia and New Guinea. Nineteen canid species are native to the New World. Eleven of these can be found south of Mexico, but only two, the gray fox and the coyote, are native to Costa Rica.

Evolution

The oldest-known canid fossils are from North America and Europe, and date back about 40 million years to the late Eocene period. Fossils show that the gray fox has inhabited North America for at least one million years. The coyote is thought to have first appeared as a distinct species about two to three million years ago, when it and the gray wolf may have diverged from common ancestors. Interestingly, despite millions of years as distinct species, wolves, coyotes, jackals, and domestic dogs (the latter two also believed to be descended from ancient forms of the gray wolf), can still mate with one another successfully.

Teeth, Skull, and Feeding

The dental formula of most canids, including the two species found in Costa Rica, is I 3/3, C 1/1, P 4/4, and M 2/3 = 42. The upper fourth premolar and the lower first molar are sharp, carnassial teeth. The canine teeth are, of course, prominent (the word *canine* comes from the Latin word *caninus*, which means pertaining to a dog), and are important weapons for wounding and holding on to prey. Ridges along the top and back of the skull and well-developed zygomatic arches provide anchorage for large jaw muscles. The teeth and skulls of canids are not as specialized as those of many other carnivores. The canine teeth are not particularly sharp, and canids do not use a precise, powerful killing bite as do felids and mustelids; large prey is simply mauled to death, and is only taken by species that hunt in packs. Indeed, canids are by no means exclusively carnivorous. They have large incisors and broad molars, and their long skulls and jaws provide room for a generous compliment of cheek teeth, all features that are typical of omnivorous mammals.

Body Design

Wild canids range in size from 35 cm (14 in.) and 1 kg (2 lbs and 3 oz) in the fennec fox (*Fennecus zerda*) of the deserts of Africa and the Middle East to up to 150 cm (60 in.) and 80 kg (175 lbs) in the gray wolf (*Canis lupus*) of the Northern Hemisphere. Domestic dogs can be heavier, the record being a 150 kg (330 lbs) Saint Bernard.

Canids are fairly uniform in shape, although the proportional length of their limbs varies. Most of the smaller canid species, principally the foxes, tend to have relatively short legs and prefer stalking and pouncing on their prey, or chasing it over short distances. Larger species such as the coyote, which tend to have relatively long legs, are very good runners. They may travel vast distances each day, and some use their stamina or speed to run down large prey. Coyotes, along with several other canids and other members of the Carnivora that are good runners, have a particularly resilient forearm design; the two forearm bones are bound together tightly with fibrous tissues, preventing the forearm from rotating. The small collar bones and fused wrist bones that are found in all Carnivores also contribute to the running ability of species like the coyote (see p. 199).

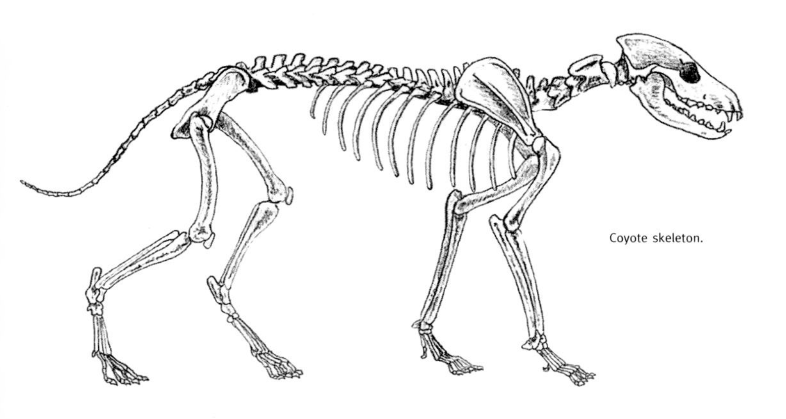

Coyote skeleton.

Canids are digitigrade. All but one African species have five toes on the forefeet and four on the hindfeet. The first digit on the forefoot, the dewclaw, does not touch the ground unless the foot sinks, so forefoot tracks reveal only four digits.

Dog and Cat Tracks

Canid tracks are easily confused with cat tracks, but they differ in several ways. The best-known distinction is that most canids have nonretractable claws, so claw marks are usually visible in their tracks, while cats almost always retract their claws while walking and very seldom leave claw marks. But beware: one of Costa Rica's species, the gray fox, is an exception to this rule. It has semiretractable claws and often leaves tracks with no claw marks. Other distinctions are the size of the heel pads, which are proportionately smaller in canids than in felids, and the position of the third and fourth digits (the middle two toes in tracks), which in canids tend to be farther in front of the outer two toes than they are in cats.

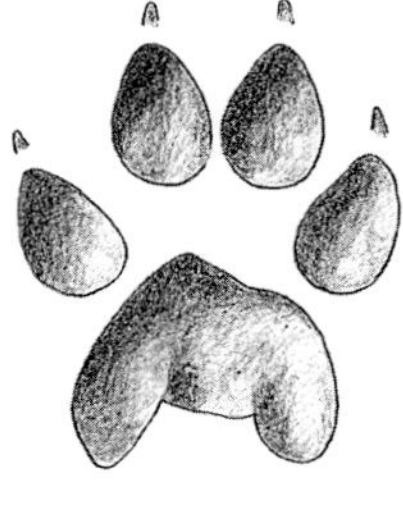
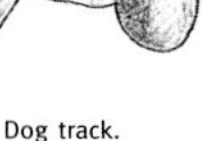

Dog track.

Cat track.

Senses

Most canids have good vision. This sense is most important at close range, both for following prey and for communicating with fellow dogs. Studies have shown that canids, especially those that live in packs, rely on a complex array of visual signals such as facial expressions and body postures to communicate when close together. Experiments suggest that dogs, like cats, have very limited color vision, and that colors are generally irrelevant to untrained canids.

Canids have excellent hearing. The upper limit of their hearing range is more than twice our own. Of course, such hearing ability did not evolve so that domestic dogs could respond to their owners' whistles. Rather, it probably prevails because it helps canids detect rodents, which communicate using ultrasound, and which are the principal prey of many species.

Canids' best sense, as any dog owner knows, is smell. Experiments have shown, for example, that they can detect the smell trace left on a glass by human fingerprints as long as six weeks after the glass has been handled. Each mammal has a corrugated membrane in its nose that is crudely indicative of its ability to smell. If one were to flatten out a dog's membrane, it would cover about 200 cm^2 (31 $in.^2$), while that of a person would cover only about 5 cm^2 (¾ $in.^2$). Canids use smell to hunt, to recognize their canine (and human) companions, and to communicate. Canids release odors with urine and feces, or by rubbing objects with scent glands located in areas such as at the base of the tail or between the toes. Marking strategies vary from species to species. The gray fox, for example, favors rubbing objects with its caudal (tail) gland which, running along almost half the length of the tail, is the largest caudal gland of any canid.

Social Organization and Reproduction

A number of features of canid social organization and reproduction are unusual among the Carnivora, and even among mammals. Lasting social bonds other than that between mother and offspring, for example, are rare in the Carnivora but common among canids. Both coyotes and gray foxes form pair bonds that can last for years, and some coyotes live in packs.

Canids also put more energy into raising young than do most other mammals. Compared to other carnivores, canids have the shortest gestation periods but the heaviest litter weights. In other words, they have more, but less developed young. In addition, canid pups tend to be dependent on their parents much longer than the young of most other mammals. This demanding reproductive style may help explain why many canids form long-term social bonds: mothers are often aided substantially by fathers and fellow pack members.

Reproduction in canids is also curious from an anatomical perspective. During mating, the tip of the penis of male canids swells up such that it cannot be withdrawn. This copulatory tie can keep male and female locked together for half an hour or more, usually with the couple standing side by side or in opposite directions. Prolonged mating is a common trait within the Carnivora. Many Carnivores are induced ovulators, and may need such prolonged mating bouts in order to stimulate the female enough that she ovulates. Canids, however, are not induced ovulators—they ovulate simply when their estrous cycle dictates, like humans and most other mammals—so provoking ovulation is not the function of the copulatory tie. Rather, the tie may increase a male's chances of impregnating a female by buying time for his sperm: if the female should subsequently happen to mate with another male, as many do, the first male's sperm would have a significant head start in the race to fertilize the female's eggs. The copulatory tie may have evolved specifically in canids because many species (or, perhaps, their ancestors) live in groups, where male competitors are more likely to mate with the same female in quick succession.

Gray foxes in copulatory tie.

Conservation

The gray fox and the coyote are among a fortunate few species of wild canids whose populations are still quite healthy; of the 19 canids native to the New World, 10 are already included on endangered species lists. Another species, the Falkland Island fox (*Dusicyon australis*), was poisoned into extinction at the end of the nineteenth century because it preyed on sheep. Canids are persecuted by farmers as killers, or potential killers, of livestock and poultry, and sometimes by hunters for sport or for pelts. Numerous studies have revealed that, while canids can cause serious losses to farmers in some areas, they generally cause much less damage than assumed. Most pest control policies recommend killing only problem individuals rather than eliminating entire populations. One reason canids should not be killed indiscriminately is that they can do people more benefit than harm by eating rodents, insects, and other pests. This was demonstrated on a large scale at the beginning of this century when a drastically successful campaign to remove coyotes from the western United States resulted in a plague of ground squirrels, causing massive damage to crops.

References

Bekoff 1977; Bekoff 1978; Berta 1987; Ewer 1973; Fox 1975; Gier 1975; Langguth 1975; Macdonald 1984c, 1995; Nowak 1991; Savage 1977; Sheldon 1992; Wayne & O'Brien 1987.

SPECIES ACCOUNTS

O: Carnivora F: Canidae

GRAY FOX

Urocyon cinereoargenteus
(plate 29)

Names: Also known as the tree fox. Genus from the Greek words *oura*, meaning tail, and *cyon*, meaning dog, because the gray fox has a tail with stiff hairs and no underfur, a feature more typical of dogs than foxes. The species name comes from the Latin words *cinereus*, meaning ash-colored, and *argenteus*, meaning silvery. There is only one other species in the genus, the similar but slightly smaller *U. littoralis*, or island gray fox, which occurs on a handful of islands off southern California. **Spanish**: *Zorro gris, tigrillo.*

Range: Southern edge of Canada to northwestern South America in Venezuela and Colombia; from sea level to at least 2,600 m (8,500 ft); mostly in relatively open habitats such as deciduous or partially disturbed forests. It is not known to occur in the wet, lowland Caribbean forests of Costa Rica or of other parts of Central America.

Size: 50 cm, 4 kg (20 in., 9 lbs); smaller here than in North America; males slightly larger than females.

Similar species: This is the only fox in Costa Rica. The somewhat similar cacomistle (pl. 31) has bold rings on its tail and lacks the fox's orangey markings. The fox's dorsal fur varies from bluish to brownish gray.

Natural history: Gray foxes are mostly crepuscular and nocturnal, but sometimes are active in broad daylight. They are omnivores, feeding on fruit, arthropods, and small vertebrates such as rodents, rabbits, birds, and reptiles, as well as on nuts, grains, grasses, and carrion. Occasionally they bury meal leftovers for subsequent retrieval. The diet probably varies with habitat and season. A study in Belize during the dry season found fruit and arthropods to be staples. In a study in Mexico, the gray fox, the coyote, and the coati, all of which are somewhat omnivorous, were found to partition food resources, with the fox favoring fruit, the coyote rodents, and the coati arthropods.

The gray fox is the only member of the dog family that readily climbs trees, a skill that can help it reach fruit or escape predators such as coyotes or domestic dogs. It can even ascend vertical trunks by grasping the tree with its front legs and pushing up using its hindlegs. It descends vertical trunks the same way, moving backward. Two anatomical features that are unusual among canids might be adaptations to this partly arboreal lifestyle: the gray fox has forearms with relatively wide rotational mobility, and semiretractable claws (which can be partially withdrawn when not needed and thus kept sharp). Gray foxes from forested regions have been found to have sharper and more curved claws than those from more sparsely vegetated regions.

Gray foxes may sleep in hollow trunks, and dens have been found up to 9 meters (30 ft) off the ground. They also den on the ground beneath piles of brush or rocks, or in tunnels. They may use tunnels abandoned by other animals or dig their own. Territorial and reproductive patterns have been studied in North America. Most documented home ranges have covered between about 75 and 140 ha (185 and 350 A). Males tend to have larger ranges than females. Males and females unite at least during the breeding season, and each pair usually has an exclusive territory. Pairs are generally monogamous and may stay together for years.

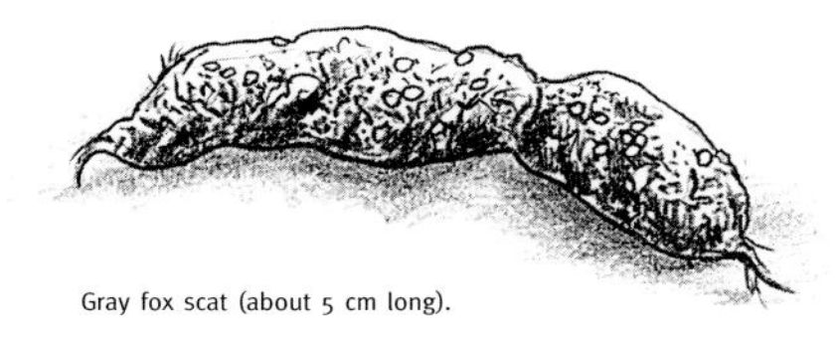

Gray fox scat (about 5 cm long).

Gestation lasts about two months. In North America, there is one litter a year, usually containing three to seven pups. In Monteverde, pups are born in the dry season. Males often help raise the young, but they spend less time at the den than females. Rarely, two females may rear young in the same den. Pups first venture from their den at the age of about five weeks, start to forage with their parents after about three months, and are independent at about four months. A parent was once observed trying to divert the attention of a potential predator away from the den by feigning an injury, a trick that is used often by certain ground-nesting birds. Pups tagged in North America have stayed in the parental home range until the following breeding season, then dispersing as far as 84 km (52 miles). Females are sexually mature at the age of 10 months. Gray foxes can live more than 13 years in captivity, but few more than five years old have been found in the wild.

Sounds: Mews, snarls, growls, barks, and screams.

Conservation: The gray fox's fondness for fruit might make it an important seed disperser for some plants. The gray fox may rarely eat poultry or crops such as corn, but is not a major pest. Occasionally it is hunted for sport. The pelt of Costa Rican gray foxes is rather thin and wiry and has never been of great value (although the thicker-pelted gray foxes of North America were hunted quite heavily in the 1970s). The gray fox can be fairly common in the seasonally dry habitats of northwestern Costa Rica, or in montane areas, such as Monteverde. It habituates easily to people, but tends to keep its distance and remain wary, unlike, say, some members of the raccoon family.

References

Banfield 1974; Cohen & Fox 1976; Delibes et al. 1989; Fox 1969; Fritzell 1987; Fritzell & Haroldson 1982; Hilje & Monge 1988; Jones 1982; Novaro et al. 1995; Reid 1997; Sheldon 1992; Timm & Laval 2000a; Trapp 1978; Trapp & Hallberg 1975; Turkowski 1971.

O: Carnivora F: Canidaee

COYOTE

Canis latrans

(plate 29)

Names: The scientific name comes from the Latin for barking dog. *Coyote* is derived from the species' Aztec name *coyotl.* **Spanish**: *Coyote.*

Range: Alaska to western Panama; from sea level to more than 3,400 m (11,200 ft) in Costa Rica; mostly in fairly open habitats such as dry forest, páramo (the scrubby habitat above the treeline), and agricultural areas. The coyote has expanded its range substantially since colonial times, probably in large part because deforestation has opened up and connected vast areas of suitable coyote habitat. In North America, it may also have benefited from humans' elimination of wolves, its main potential competitor.

Popular accounts suggest that the coyote inhabited the dry forests of northwestern Costa Rica by colonial times, but apparently it has spread south and into Panama, extending its range by some 16,000 km^2 (6,200 mi^2), since 1960. It is likely to continue to spread south into South America.

Size: 80 cm, 10 kg (31 in., 22 lbs); smaller in Costa Rica than in North America—18 kg (40 lbs) is the average in Alaska, where they are largest, and 34 kg (75 lbs) is the record; males larger than females.

Similar species: Coyotes are easily confused with domestic dogs, but can be distinguished from most by their long, slender legs, large, pointed ears, long muzzle, and bushy, black-tipped tail. Dog-coyote hybrids, known as coydogs or doyotes, have not been recorded in Costa Rica.

Natural history: Coyotes are mostly crepuscular and nocturnal. Animal prey constitutes more than 90% of the diet in most areas where coyotes have been studied, including Costa Rica. While the coyote of cartoon fame is perpetually obsessed with Beep-Beep the roadrunner, and consistently unsuccessful, wild coyotes vary their diets considerably according to habitat and season, and this flexibility has helped them establish themselves over a broad geographic range that continues to expand. In North America, the coyote's summer mainstay is rabbits and rodents, while the winter staple is large ungulates such as deer and elk, taken mostly as carrion after they have succumbed to the season's hardships.

Ctenosaur (*Ctenosaura similis*).

In one of the only studies on coyotes in the tropics, Christopher Vaughan and Miguel Rodríguez compared the diets of coyotes in Costa Rica's Cerro de la Muerte and in Palo Verde National Park. Cerro de la Muerte is a cold, highland habitat characterized by a blend of scrubby subalpine vegetation, oak forests, and farms, while Palo Verde is a hot, lowland habitat consisting principally of deciduous forest, marshes, and mangroves. At Cerro de la Muerte, Dice's rabbits, cattle, armadillos, small rodents, brocket deer, chickens, band-tailed pigeons, and grass prevailed in the coyote's diet. In Palo Verde, it was more diverse, including ctenosaurs (illustrated above) and iguanas (illustrated on p. 276), collared peccaries, cattle, hispid cotton rats, white-tailed deer, and a variety of other mammals, as well as locusts, blue-winged teal, purple gallinules (illustrated on p. 212), guácimo (*Guazuma ulmifolia*) fruits (illustrated on p. 310), and grass. The researchers presumed the larger prey such as cattle and deer had been eaten as carrion rather than killed by the coyotes themselves. The relative proportions of the different prey items changed seasonally at both sites. In Palo Verde, for example, coyotes took teal and gallinules only between February and April, as marshes dried up. Other known prey includes turtle eggs, snakes, and fish.

A pair of purple gallinules at Palo Verde. As wet areas shrink toward the end of the dry season, coyotes are able to prey on these birds.

Coyotes catch most live prey by stalking and pouncing, but are also capable of running prey down. They have been clocked at speeds of up to 64 km/h (40 mph) and have been known to chase prey for almost 5 km (3 miles). They are renowned for their cunning; Mexicans use the expression *muy coyote* to describe people who act shrewdly. Coyotes may play dead to lure in carrion feeders, and, in North America, they have even been seen pairing up with badgers to hunt burrowing animals: the coyote does the sniffing, the badger the digging, and both share the prey.

The social and reproductive biology of coyotes has been studied in North America. Coyotes may live alone, in pairs, or in packs typically containing no more than eight individuals. Pack members are often related, and there is usually a well-defined hierarchy, with one adult pair dominating. Coyotes habitually travel several kilometers in a night of hunting, and home ranges vary in size from about 1,000 to 10,000 ha (2,500 to 25,000 A). Among solitary individuals, males tend to have large, overlapping home ranges, while females have smaller, non-overlapping home ranges. Pairs are monogamous for at least one breeding season and often for several years.

Gestation lasts about two months. Litters most often containing four to seven pups are raised in burrows or other cavities. Coyotes modify dens abandoned by other animals or dig their own, creating burrows that are about 30 cm (1 foot) wide and up to 11 meters (36 feet) long, sometimes with more than one entrance. The den may be reused in successive years. Occasionally two mothers raise litters simultaneously in the same den. Pups are weaned at about a month and a half. They reach adult size at about nine months. Most pups leave their parents within their first year, but some may remain for up to three years. Young coyotes seeking a home range have been captured 160 km (100 miles) away from their natal ranges just months after leaving their parents. Coyotes can live 21 years in captivity but studies suggest that few live more than eight in the wild.

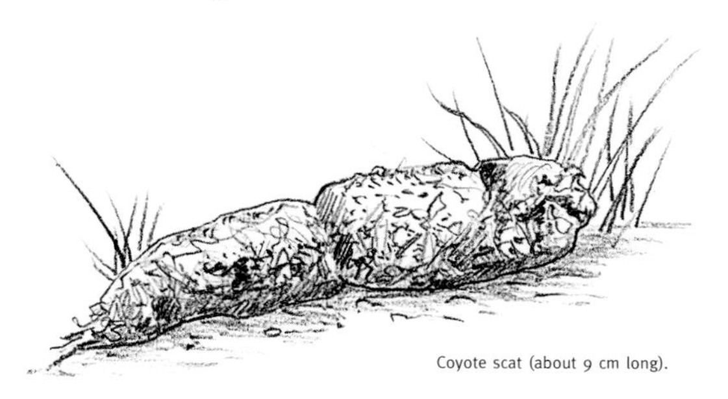

Coyote scat (about 9 cm long).

Sounds: A wide repertoire of sounds similar to those made by domestic dogs. The most conspicuous calls are whining howls emitted simultaneously by several individuals, and a series of short barks followed by a prolonged howl. Such calls are audible several kilometers away.

Conservation: In Costa Rica, coyotes have been known to damage crops of melon and corn, and to take poultry and livestock, and are thus persecuted by some farmers. Numerous studies have demonstrated that coyotes seldom do significant damage to livestock, however, for they generally feed only on animals that have been weakened or killed due other causes. The coyote seems to have benefited to some degree from human activities, judging from the expansion of its range through disturbed habitats in Costa Rica and Panama in recent years.

References

Arnaud 1993; Bekoff 1977, 1978; Bekoff & Wells 1980, 1986; Bowen 1984; Bowyer 1987; Delibes et al. 1989; Hilje & Monge 1988; Huxley & Servín 1995; Janzen 1983b; Jones 1982; Laundré & Keller 1984; Monge-Nájera & Morera 1986, 1987; Nowak 1991; Servin & Huxley 1993; Sheldon 1992; Springer 1980; Vaughan 1983b; Vaughan & Rodríguez 1986.

Raccoon Family (Procyonidae)

Distribution and Classification

Cacomistles, raccoons, coatis, olingos, and the kinkajou constitute the Procyonidae family. Authors disagree on how many of the numerous races of raccoons, coatis, and olingos should be considered separate species, so the family may contain anywhere from eight to 20 species. Seven species are currently recognized for Costa Rica: two raccoons, two olingos, one coati, one cacomistle, and the kinkajou. Procyonids are found only in the New World. Until recently, the world's two panda species (the lesser or red panda and the giant panda, both from Asia) were also included in this family. The pandas' placement in the Procyonidae was always controversial, however, and mounting evidence suggests that pandas are more closely related to bears (family Ursidae) than to procyonids. In more recent taxonomic lists, the pandas are placed (not always together) either with the bears or in their own families, the Ailuridae or Ailuropodidae.

Evolution

Procyon means doglike, and the procyonids, along with pandas and bears, are indeed thought to have originated from ancient members of the dog family. The oldest procyonid fossils are from Europe and date back about 50 million years to the Eocene period. Some procyonids stayed in Europe, but died out by about two million years ago. Procyonids have inhabited North America for at least 35 million years. As islands that now form southern Central America rose out of the sea and joined with North America, some procyonids spread south, eventually dispersing into South America following the completion of the land bridge some seven to two million years ago. A peculiar exception are primitive raccoons, whose fossils prove that they had arrived in South America by about 20 million years ago. The first members of the Carnivora known to have lived there, they are thought to have arrived by island-hopping from the north, either through what is now Central America or via the Antilles.

Teeth and Feeding

The procyonid dental formula is I 3/3, C 1/1, P 4/4 (3/3 in the kinkajou only), and M 2/2 = 40. Procyonids are among the least carnivorous of the Carnivora and, consequently, the teeth of many species lack some of the meat-eating specializations found in other members of the order. They have wide, low-crowned molars that lack the sharp, pointed cusps found on the molars of most carnivores, and only the cacomistles, the most carnivorous of the family, have carnassial teeth.

Body Design

Procyonids are plantigrade, and have five toes on both feet and hands, the design of which varies according to the locomotion and feeding habits of each species. Like some other highly arboreal creatures (such as squirrels and margays), kinkajous, olingos, and cacomistles have joints in their ankles that allow the hindfeet to be rotated 180°, a feature that is particularly useful when they are descending tree trunks headfirst. The kinkajou also has dexterous, almost primatelike hands with which to climb through treetops and reach fruit. Procyonids that spend more time on the ground, such as raccoons or coatis, have hindfeet that are only partially reversible. They are also capable of coming down tree trunks headfirst, but their descents are distinctly less graceful. Raccoons rely heavily on their sensitive hands to locate and then manipulate food, but they don't employ the converging grip used by kinkajous, and generally need to use both hands to grasp an object. Coatis eat food straight off the ground. Their undexterous but strong, clawed forepaws are designed for scraping, digging, and climbing trees.

Procyonids have long tails that help them keep their balance while moving about in trees. Here again the kinkajou has gone to the greatest extreme, for it has evolved a prehensile tail. Many procyonid tails have a ringed pattern that may help in communication. Bobbing tails in the understory are sometimes all that is visible of groups of coatis. The tails, which coatis tend to hold erect, may serve as banners that help group members stay together.

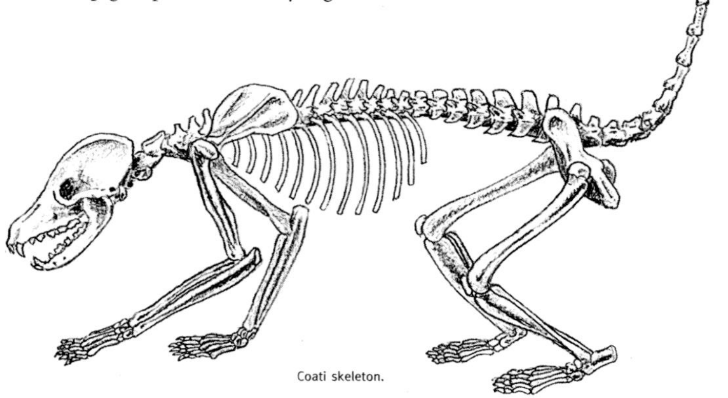

Coati skeleton.

Senses

Procyonids have good night vision, little or no ability to see color, good hearing, and an excellent sense of smell. The latter is important both in feeding and communication. Raccoons can detect food buried in 5 cm (2 in.) of sand. Likewise, coatis will immediately detect a banana suspended off the ground, but won't see it until they have walked around it a few times. If a fake (odorless) banana is suspended, they seldom notice it.

All procyonids except kinkajous have anal scent glands, which they use to mark territories, recognize other individuals, or mark arboreal runways to feeding sites. Kinkajous use glands located on their throat, chest, and abdomen. Like members of the weasel family, procyonids may also use scents from anal glands as a form of defense; cacomistles and olingos emit foul-smelling secretions when threatened.

Raccoons are famous for their sense of touch. Sensitive whiskers and highly developed nerves in their hands enable them to catch and eat food, sometimes without ever looking at it. The sense is also important to cacomistles, which have touch-sensitive hairs on their forearms to compliment their facial whiskers.

Conservation

Compared to many other Carnivores, most procyonids are quite common. Being omnivorous, they tend to occur lower down the food pyramid and have smaller home-range requirements than more carnivorous members of the order. They seldom cause serious damage to livestock, poultry, or crops, although raccoons and coatis can become a nuisance in populated areas by ripping apart bags of garbage. Procyonids are hunted in some parts of Costa Rica for their porklike meat. Raccoon pelts have never been highly sought after in Costa Rica as they have in North America, in part because they are of poor quality, but kinkajou fur is used for a variety of purposes. All procyonids are at least somewhat forest-dependent, but most seem to do well in fragmented areas where hunting is not too severe. In forest ecosystems, they are important as seed dispersers, as controllers of insect and rodent populations, and as prey for large predators.

References

Decker 1991; Dücker 1965; Ewer 1973; Ford & Hoffmann 1988; Ingles 1965; Janson et al. 1981; Jenkins & McClearn 1984; Lemoine 1977; Linares 1981; McClearn 1992a; Michels et al. 1960; Nowak 1991; Poglayen-Neuwall 1962, 1966; Poglayen-Neuwall & Toweill 1988; Pruitt & Burghardt 1977; Schaller et al. 1985; Stains 1984; Trapp 1972.

SPECIES ACCOUNTS

O: Carnivora F: Procyonidae

CENTRAL AMERICAN CACOMISTLE

Bassariscus sumichrasti

(plate 31)

Names: Also known as the Central American ringtail or bassarise. The common name is derived from the Aztec word *cacomixtle*. Genus from the Greek *-isc-*, meaning little, and *bassar*, meaning fox. The genus was formerly listed as *Jentinkia*. This species differs from the North American cacomistle or ringtail, *B. astutus* (found from southwestern Oregon and eastern Kansas to southern Mexico), in having dark rather than pale feet, nonretractable rather than semiretractable claws, triangular rather than round ears, and often lacking pale rings toward the tip of the tail. Since cacomistles differ from all other procyonids in that they have carnassial (meat-shearing) teeth, the two species are sometimes placed in their own family, the Bassariscidae. **Spanish**: *Cacomistle.*

Range: Southern Mexico to western Panama; from sea level to at least 2,800 m (9,200 ft); mostly in wet, evergreen forests and disturbed areas. In Costa Rica, this species has been recorded only in a handful of localities at mid- to high elevations in the Central and Talamancan mountain ranges.

Size: 45 cm, 1 kg (18 in., 2 lbs 3 oz).

Similar species: Raccoons (pl. 30, pp. 220 and 222) are larger, have a black mask around the eyes, and have shorter tails. Olingos (pl. 31, p. 232) are slightly larger; they have a face that is less foxlike in shape and lacks markings, and a tail that is less bushy and only faintly banded. Gray foxes (pl. 29, p. 208) have orangey markings and lack bands on the tail.

Natural history: Little is known of the natural history of the Central American cacomistle. It is generally solitary, although groups of up to nine have been recorded in fruiting trees. While the northern variety likes rocky, sometimes fairly open areas, and frequently travels along the ground, *B. sumichrasti* is mostly arboreal. It is mainly nocturnal. Six radio-collared cacomistles at Barva Volcano that were studied by Nélida García, Christopher Vaughan, and colleagues peaked in activity at around 10:00 PM and again at about 4:00 AM. Occasionally the cacomistles became active in the afternoon. For unknown reasons, afternoon activity was most common in May.

Cacomistles are omnivores. Possessing carnassial teeth, they are particularly well equipped to feed on meat. Indeed, the North American cacomistle is the most carnivorous of all procyonids, for almost 80% of its diet can be made up of vertebrates. Curiously, however, vertebrates seem to contribute only a relatively small part of the Central American cacomistle's diet—about 10% according to an estimate for cacomistles in Mexico. The favorite food of Central American cacomistles appears to be fruit, including figs (*Ficus* spp.), breadnuts (*Brosimum* spp., illustrated on p. 158), sapotes (*Manilkara* and *Poulsenia* spp.), and cecropias (*Cecropia* spp., illustated on p. 62). The most abundant remains visible in 10 scats collected at Barva Volcano were seeds of varablanca (*Hedyosmum mexicanum*), a relative of cloranthus. Cacomistles are also known to eat other plant parts, including flowers, on occasion.

Cacomistle scat (about 7 cm long).

Invertebrates appear to follow fruits in importance in cacomistles' diets. The Barva scats contained the remains of beetles, spiders, and ichneumon wasps. At Barva, cacomistles probably find at least some of their invertebrate prey in bromeliads, for they stop to check bromeliads frequently.

Cacomistles seldom spend long in any one place during a night of foraging. The six cacomistles at Barva used home ranges of 19 to 32 ha (50 to 80 A) during the seven months they were followed. The range of each male overlapped with that of one of the females, but there was no overlap in the ranges of individuals of the same sex. In a more seasonally variable habitat in Mexico, cacomistles were found to undertake short migrations just before winter, shifting their home ranges to lower elevations.

The Central American cacomistle has a gestation period of just over two months. Litters of two young have been recorded, but this is probably variable; the North American cacomistle produces litters of one to five young. Other life-history information is available only for the North American cacomistle. In that species, a mother may change dens frequently while raising young, and eats any excrement she finds around the den sites. These habits may help avoid the buildup of odors that could make a den obvious to predators. Young of *B. astutus* start to forage with the mother at the age of about two months, and are weaned after about four months. They leave their mother and reach sexual maturity by the time they are 10 months old. Captives can live over 16 years.

Sounds: Coughing barks, metallic chirps, growls, hisses, whimpers, and loud, slurred, two- or three-syllable birdlike cries. Due to the latter, this species is known as the *uyo* in El Salvador. *Uyo* calls are often answered immediately by neighboring cacomistles.

Conservation: Cacomistles can survive in partly disturbed habitats. The individuals studied at Barva actually spent more time foraging in secondary forests and shrubby pastures than in primary forest. Cacomistles are usually close to relatively undisturbed forest, however, which they probably depend on for certain foods and denning sites. In Mexico, cacomistles have been found to avoid secondary forests and pastures altogether.

Cacomistles can be beneficial to people because they catch rodents, and they are sometimes kept as pets for that purpose. The northern species used to be hunted quite heavily for its fur, known as California mink, and the Costa Rican variety may occasionally be hunted for its pelt or for sport. Although this species is common only in a handful of areas in the northern part of its range, Costa Rica is the only country in which it is legally protected.

References

Coates-Estrada & Estrada 1986; Emmons et al. 1997; Estrada & Coates Estrada 1985; García 1996; Poglayen-Neuwall 1973, 1992; Poglayen-Neuwall & Poglayen-Neuwall 1980, 1995; Poglayen-Neuwall & Toweill 1988; Trapp 1978; Vaughan et al. 1994.

O: Carnivora F: Procyonidae

CRAB-EATING RACCOON

Procyon cancrivorus

(plate 30)

Names: *Raccoon* comes from the North American indigenous word *arakun*, meaning scratching hands. *Procyon* means doglike, and *cancrivorus*, crab-eating. Genus previously listed as *Euprocyon*. **Spanish**: *Mapache* (or *mapachín*) *cangrejero*.

Range: Southwestern Costa Rica (up to at least Orotina and the Central Valley), and through South America east of the Andes to Uruguay and Argentina; from sea level to at least 1,200 m (4,000 ft); in forested or disturbed habitats near fresh- or saltwater. This species replaces the northern raccoon in South America; the ranges of the two species overlap only in Costa Rica and Panama.

Size: 60 cm, 5 kg (24 in., 11 lbs); males larger than females.

Similar species: *P. cancrivorus* is hard to distinguish from the northern raccoon (*P. lotor*) in the field. At close range, *P. cancrivorus* is said to give off a strong odor, which *P. lotor* does not. *P. cancrivorus* is slightly larger than *P. lotor*, with a less bushy and proportionately longer tail. Its fur tends to be shorter, stiffer, and less dense than that of the northern raccoon, for there is usually no underfur. The underparts and pale tail bands are often more orangey than those of *P. lotor*. On the back of the crab-eating raccoon's neck, the fur slants forward, starting with a whorl between the shoulders and meeting the regular, backward-facing fur along a V-shaped line between the ears. The northern raccoon has only backwards-facing fur on its neck. The feet and legs of *P. cancrivorus* are dark brown or blackish, whereas in *P. lotor* the forelimbs, and usually the hindlimbs, are pale gray. Note that habitat and crab-eating alone are not helpful in distinguishing the two species: the northern raccoon also frequents water, is fond of crabs and the like, and is abundant along the coast, while the crab-eating raccoon can also be found far inland.

Natural history: Although this is the only raccoon species other than *P. lotor* that is widespread (there are five other raccoon species that are confined to certain islands, and some or all of them may be no more than subspecies of the northern raccoon), *P. cancrivorus* has barely been studied in the wild and very little is known of its natural history.

Like the northern raccoon, the crabeater is largely solitary and nocturnal. In a study in Argentina, it was found to use secondary forests more than other habitats. Like *P. lotor*, it forages on the ground, but apparently always in or around water. It eats freshwater crabs (which can be common in forests far away from the coast), mollusks, fish, frogs, other small vertebrates, insects, and possibly fruits. It is a good climber, and dens in hollow trees. Captives have produced litters of two to six cubs, whose development is much like that of the northern raccoon. Crab-eating raccoons in Manuel Antonio National Park are reported to give birth in July and August. Two radio-collared females there were found to use home ranges of about 4 to 6 ha (10 to 15 A) for much of the year, but only about 1 to 2 ha (2 ½ to 5 A) during the season of cub birth.

Conservation: An unusual study in Manuel Antonio showed that people can affect crab-eating raccoons (and other mammals) in subtle ways, even in protected areas. Researchers Eduardo Carrillo and Christopher Vaughan found that, on weekdays, when few people visited the park, raccoons would be most active in the late afternoon, while on the more crowded weekends, they would be most active after dark. Such a shift is significant because it can alter, for example, what kinds of foods are available to an animal and what other species it has to compete with. The researchers also found that their raccoons used the park's camping areas more extensively during weekends because there was more trash to feed on. They speculated that the availability of trash might cause raccoons to become more abundant than they would be in a pristine environment, perhaps to the detriment of other, less-adaptable animals.

Beyond these interesting observations, the status of the crab-eating raccoon is unknown, in part because *P. cancrivorus* is so hard to distinguish from *P. lotor.* Since coexisting species seldom share exactly the same niche (because if they did, one would tend to out-compete and replace the other), it would be interesting to ascertain the relative abundance of *P. cancrivorus* and *P. lotor* in the several-hundred-kilometer stretch where their ranges overlap in Panama and southwestern Costa Rica, and any ecological differences that may exist between them.

References

Bisbal 1986; Carrillo & Vaughan 1993; Emmons & Feer 1997; Löhmer 1976; Reid 1997; Rodríguez & Chinchilla 1996; Yanosky & Mercolli 1993.

O: Carnivora F: Procyonidae

NORTHERN RACCOON

Procyon lotor

(plate 30)

Names: Also known as the common raccoon. When first described, the species was listed as a bear (*Ursus lotor*). The species name comes from the Greek *lotor*, meaning washer, in reference to the habit of some captives of dousing food in water. Since captives both manipulate food in the absence of water and rinse wet, clean items, they may not really be washing. Wild raccoons haven't been observed dunking food, but they do often hunt in water. It has been suggested that the captives could be going through the motions of prey capture to entertain themselves, or that rinsed items can be better identified by the raccoon's unusually sensitive hands. **Spanish**: *Mapache* (or *mapachín*) *norteño.*

Range: Southern Canada to Chiriquí, Panama (as well as France, Germany, the Netherlands, and Russia, where it was introduced as a game animal); from sea level to at least 2,800 m (9,200 ft); in forested and disturbed habitats, especially near water.

Size: 50 cm, 4 kg (20 in., 9 lbs); smaller in Costa Rica than in North America; males larger than females.

Similar species: See crab-eating raccoon (pl. 30, p. 221). In Costa Rica, the two species' ranges overlap only in the Central Valley and in the southwest.

Natural history: Northern raccoons have been studied extensively in North America, but very little in the tropics, although studies done in Manuel Antonio National Park and Tapantí National Wildlife Refuge by Eduardo Carrillo and Christopher Vaughan have revealed some details of the species' natural history in Costa Rica.

Northern raccoons are mostly nocturnal; in some areas they also emerge by day. At Manuel Antonio, northern raccoons habitually start foraging before dark and continue after dawn, being most active from 5:00 to 10:00 PM and from 3:00 to 7:00 AM. Along the coast, they often forage during the day to take advantage of low tide. And a radio-collared female at Tapantí foraged as much by day as by night, apparently to take advantage of dry spells.

The varied diet of the northern raccoon includes crustaceans, fish, frogs, worms, bird or turtle eggs, fruits, nuts, insects, and rarely, small vertebrates or carrion. In Manuel Antonio, land crabs (*Gecarcinus quadratus* and *Cardiosma crasum*) are this raccoon's favorite food, and fruit makes up most of the rest of the diet.

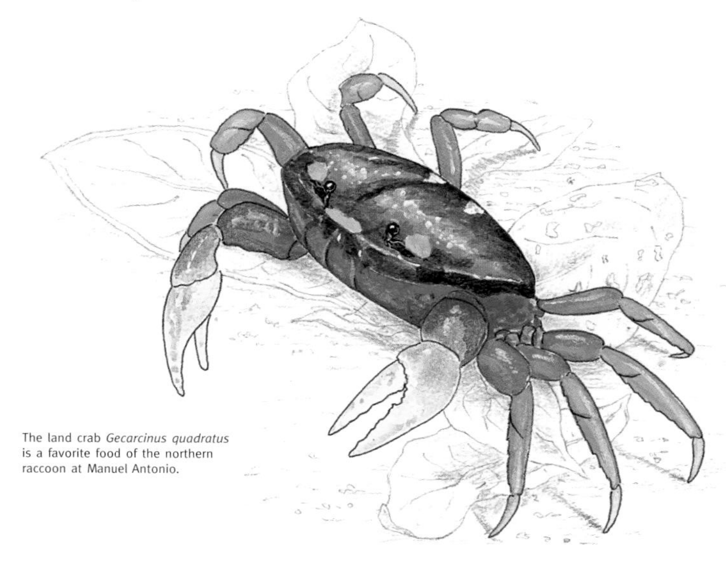

The land crab *Gecarcinus quadratus* is a favorite food of the northern raccoon at Manuel Antonio.

Northern raccoons den in cavities, preferring tree hollows well above the ground, but sometimes using other animals' ground burrows. In North America, home ranges have been found to vary from 0.2 ha to 4,900 ha (½ to 12,000 A), but average about 40 ha (100 A) for females and about 60 ha (150 A) for males. The downpour-avoiding female in Tapantí was found to move more quickly and over shorter distances than a typical North American raccoon, and radio-collared raccoons at Manuel Antonio also had somewhat smaller ranges, averaging 28 ha (70 A) for males and 15 ha (37 A) for females. The home ranges of males overlap those of one or more females, but not those of other adult males. The home ranges of neighboring females, who are often related, can overlap extensively.

During breeding season, a receptive female will allow a male to hang out and mate with her for up to a week or so before driving him away. A gestation period of about two months produces one to seven young. Males are not allowed near the young; they sometimes kill and eat them. Young nurse for two to four months. In North America, most leave their mothers at about nine months. Siblings may stay together for a few months after leaving their mothers. Females usually remain in their natal areas. Thus, mothers and daughters often use overlapping ranges, and some continue to associate with one another well into their adult lives. In one study, three generations of the same matriline reared litters simultaneously in the same brush pile. Adult males, on the other hand, disperse. Marked males have been recaptured as far as 275 km (170 mi) away from their birthplace. Raccoons become reproductively active at the age of one to two years. Northern raccoons can live 20 years in captivity, but seldom survive more than five in the wild.

Northern raccoon scat (the larger dropping is about 7 cm long).

Sounds: Yelps, hisses, and growls in antagonistic situations; whines and deep, bouncy churring sounds (somewhat resembling the call of a crake) made during peaceful social interactions, especially between mother and young.

Conservation: Northern raccoons sometimes eat corn and other crops, and turtle eggs. They may become a nuisance by ripping apart garbage bags, and they can carry diseases such as rabies. They are occasionally hunted for sport, food, or for their pelts (which are thinner and less valuable than those of northern raccoons in North America). They are less abundant in Costa Rica than they can be in North America, but do well in highly disturbed environments and are quite common in some areas.

References:
Carrillo 1990; Carrillo & Vaughan 1988, 1993; Emmons et al. 1997; Fritzell 1978; Gehrt & Fritzell 1997, 1998; Hilje & Monge 1988; Jones 1982; Lotze & Anderson 1979; Lyall-Watson 1963; Lynch 1967; Priewert 1961; Pruitt & Burghardt 1977; Sanderson 1983.

O: Carnivora F: Procyonidae

WHITE-NOSED COATI
Nasua narica
(plate 30)

Name: Also known as the northern coati. Lone, adult males are known as coatimundi, a name that originated with the Tupi of South America. Genus from the Latin *nasus*, meaning nose, and *naris*, meaning nostril. There are two other coati species: *N. nasua*, or the South American coati, found throughout most of South America east of the Andes, and *Nasuella olivacea*, or the mountain coati, of the Andes. Another small variety found only on Cozumel Island off Mexico used to be listed as a separate species, but is now generally considered a race of *N. narica*. **Spanish**: *Pizote* or, for lone males, *pizote solo*.

Range: Arizona and Texas in the United States to the northwestern tip of South America down to Ecuador; from sea level to at least 3,500 m (11,500 ft); mostly in or near forest.

Size: 55 cm, 5 kg (22 in., 11 lbs); males larger than females.

Similar species: Hard to confuse with any other mammal, but note that the base fur color, the extent and color of the frosting around the forelimbs, and the distinctness of the tail rings are all somewhat variable, even within geographic areas or litters.

Natural history: Since coatis habituate easily to people, they are among Costa Rica's most frequently seen wild mammals. They are almost exclusively diurnal, although they have been seen foraging at night on a handful of occasions. In hot areas such as Guanacaste, or Barro Colorado Island in Panama, they tend to rest during the midday heat, being most active from about 6:00 to 9:00 AM and from 3:00 to 5:00 PM.

Although coatis are quite at home in the canopy, sometimes climbing tall trees to get at fruit or to sleep, they spend most of their foraging time on the ground. They locate food principally by smell, and expose it by raking through the leaf litter, scratching into rotten trunks, or digging, or by plucking it off low vegetation. Relatively short limbs provide coatis leverage for digging (as do, for example, the short limbs of gophers or moles).

The staple foods of coatis are litter arthropods, other small animal prey, and fruit. In Guanacaste, favorite fruits include those of güiligüiste (*Karwinskia calderonii*), chicle (*Manilkara chicle*), the coffee relative crucillo (*Randia subcordata*), nance (*Brysonima crassifolia*), guácimo (*Guazuma ulmifolia*, illustrated on p. 310), and, in particular, figs (*Ficus* spp.). Favorite animal prey includes beetles and their larvae, crickets, caterpillars, millipedes, spiders, crabs (especially the land crabs *Cardiosma crasum* and *Gecarcinus quadratus* in Guanacaste—see illustration on p. 223), snails, and turtle eggs. At Nancite beach in Santa Rosa Park, they are the main predators of the eggs and hatchlings of olive ridley sea turtles. Occasionally coatis eat small vertebrates, such as lizards, rodents (including agoutis and spiny rats), frogs, and snakes; rarely, they take carrion. They also used to be seen grooming ticks off a pair of tapirs that frequented a Panamanian research station.

Coati scat (about 7 cm long).

The coati varies its diet seasonally according to the abundance of different foods. In the semideciduous forests of Barro Colorado Island in Panama, for example, researcher James Russell found that coatis dedicate almost 90% of their foraging time to animal foods during the wet season, but just over half their time during the dry season. This pattern seemed to reflect the abundance of arthropods in the leaf litter. Litter arthropods on the island are most abundant during the early wet season, when the thick layer of litter accumulated during the dry season gets soaked and starts to decompose. The arthropods decrease in abundance as the layer decomposes and thins, and become dramatically scarcer when the dry season sets in and the litter dries out. Regardless of the availability of arthropods, coatis can become almost entirely frugivorous when favorite fruits are abundant.

A noteworthy habit coatis in Panama have acquired is that of grooming themselves and fellow coatis with resin from the gumbo limbo relative *Trattinnickia aspera*. People use resin from various trees in this family (the Burseraceae)—Mexicans use the resin copal, for example, as cement and varnish, while the resins frankincense and myrrh, of biblical fame, are burned as incense—but its value to coatis is a mystery. Some of the possible benefits of fur-rubbing are discussed in the white-throated capuchin account on pp. 136-137.

Coatis have an interesting social arrangement. Females and young live in groups, known as bands, of up to 25 individuals, or occasionally more, while adult males are generally solitary. In times past, bands and individuals were thought to represent two different species, *N. sociabilis* and *N. solitaria*. While different bands share the peripheral parts of their 35 to 45 ha (85 to 110 A) ranges and tend to interact peacefully, encroaching lone males are usually driven off.

This unusual social arrangement seems to have several advantages. For one, band members probably gain safety in numbers. Researchers in Guanacaste found that members of larger bands enjoyed long, relaxed visits to a water hole, while members of small bands and solitary males drank more quickly, presumably because they were more vulnerable to predators. Adult males may be excluded in part because they are major predators on coati young. In one-on-one confrontations, females are unable to deter the larger males, but, as a group, they can prevail.

The exclusion of males also reduces competition for food among coatis. A study in Panama showed that adult males are more carnivorous than band members. This could be simply because vertebrates are not abundant enough to form a significant portion of the diets of entire bands, but are a worthwhile target for a solitary predator. Males may also eat more meat because bands exclude them from prime fruit- and insect-feeding locations. This dietary difference may help coatis survive when food is scarce, by reducing competition between bands and males.

One, or sometimes more, adult males are allowed to enter a band only during a two-week-long mating season, often around February. During this period, the lucky male or males take the leading role in driving off competitors, but they are still subordinate to the females. Competition among males is fierce. In one population, 30% of males were found to have wounds or damaged canine teeth.

Gestation lasts 10 or 11 weeks. Bands within an area produce young more or less in synchrony. In Panama and Santa Rosa, young are born around April or May—at the beginning of wet season, when the Panamanian study found arthropods and a number of fruits to be most dependable. In Santa Rosa, a second birthing period takes place from late July to early August, perhaps because most mothers there lose their young to white-throated capuchins and attempt second litters.

Pregnant females leave their bands a week or so before giving birth and build leaf nests in tree holes or amidst canopy branches. They return as soon as their two to seven young are strong enough to keep pace with the band. Thus, young abandon their nest for good at the age of only five to six weeks, a full month to two months earlier than do the young of other procyonids. Mothers benefit from leaving the nest as soon as possible because other band members then help protect the young, fanning out in an elliptical shape around the young while foraging, moving slower than normal to allow the young to keep up, and cooperating to fend off predators. Synchronous breeding may have evolved because it minimizes the period during which band members have to sacrifice foraging efficiency for the sake of young.

Females seem to have difficulty defending their young against males or other predators alone. A lone mother in Panama was seen to lose four of her five young in as many days, at least two of them to an adult male. Other predators on young include boa constrictors, raptors, other members of the Carnivora, and white-throated capuchins. In parts of Guanacaste, capuchins have become so proficient at raiding coati nests that almost no young survive beyond the age of two weeks.

A boa constricting a young coati.

Young coatis nurse until they are about four months old, and reach adult size at about 15 months. Females can bare young when they are two years old, but some do not reproduce until their fourth year. Russell found that the age at which females first reproduce was strongly influenced by year-to-year variations in food availability. The fruit crops his coatis relied on in the dry season were inconsistent, and females that had not bred previously would not produce their first litters in bad years. As females get older, they stop breeding but continue to associate with bands and may play an important role in helping their less-experienced companions find trees that fruit only periodically. Males leave their natal bands voluntarily around the age of two and have their first chance to breed just before they turn three. Since there tend to be more males than bands, males may not breed until they are four or five years old, and many may not breed at all. Captive coatis can live 17 years.

Sounds: Soft whines between band members; squeals, grunts, and thumpy chucking sounds during antagonistic encounters and in alarm.

Conservation: Coatis are probably important seed dispersers, as researcher Joel Sáenz demonstrated in a study in Guanacaste. Sáenz found that, in his region, coatis ate the fruits of 33 different plant species. They defecated almost all the seeds of those fruits intact, and they deposited the seeds far from the parent plants—from 250 to 1,120 meters (270 to 1,200 yards) away. By comparison, agoutis, which disperse many of the same seeds, seldom transport them further than 50 meters (55 yards).

Coatis, especially the fatter males, are sometimes hunted for their porklike meat. When they are hunted extensively, they tend to become more nocturnal. Sometimes they irk people by raiding garbage bins, or crops such as corn and cardamom. They can do well in disturbed habitats and are common in many areas.

References

Bonaccorso et al. 1980; Burger & Gochfeld 1992; Cornelius 1986; Decker 1991; Delibes et al. 1989; Emmons et al. 1997; Fedigan 1990; Gompper 1995, 1996; Gompper & Hoylman 1993; Gompper & Krinsley 1992; Hallwachs 1986; Hilje & Monge 1988; Howe 1996; Janzen 1970; Kaufmann 1962, 1983, 1987; Levings & Windsor 1996; McClearn 1990, 1992a,b; Newcomer & DeFarcy 1985; Overall 1980; Perry & Rose 1994; Russell 1979, 1981, 1983, 1984, 1996; Sáenz 1994a; Smythe 1970a.

O: Carnivora F: Procyonidae

KINKAJOU
Potos flavus
(plate 31)

Names: Also known as the honey bear. The name *kinkajou* originated with indigenous peoples in Brazil. The scientific name comes from the Latin *flavus*, meaning yellow, and the Greek *potes*, meaning drinker. One can understand how the kinkajou, with its arboreal habits, prehensile tail, gripping hands, rounded head, and short muzzle, was thought to be a primate (*Lemur flavus*) when first described. **Spanish**: *Martilla or marta.*

Range: Southeastern Mexico to South America east of the Andes, down to central Brazil; from sea level to at least 2,200 m (7,200 ft); in mature and disturbed forest.

Size: 50 cm, 2.5 kg (20 in., 5 ½ lbs); males larger than females.

Similar species: See olingo (pl. 31, p. 233).

Natural history: The kinkajou is mostly nocturnal, spending the daytime in tree hollows or dense tangles of vegetation in the upper levels of the forest. Although it belongs to the Carnivora, the kinkajou feeds mainly on fruit. In a recent study in Panama, kinkajous were found to dedicate about 90% of their feeding time to fruit, taking at least 78 different fruit species at that site alone. Favorites included figs (*Ficus* spp.), legumes (*Inga* sp.), caimito star apples (*Chrysophyllum cainito*), and the fruits of the borage relative muñeco blanco (*Cordia panamensis*), of the palms *Astrocaryum standleyanum* and *Scheelea zonensis*, and of a liana related to the sea grape, *Coccoloba parimensis*. Kinkajous also enjoy wild nutmegs (*Virola* spp., illustrated on pl. 22), hogplums (*Spondias mombin*, illustrated on p. 162), ojoches (*Brosimum* spp., illustrated on p. 158), sapotes (*Pouteria* spp.), almendros (*Dipteryx panamensis*), and cecropia fruits (*Cecropia* spp., illustrated on p. 62). Most of the fruit kinkajous select is ripe.

The rest of the kinkajou's diet consists of flowers and nectar, young leaves and buds, insects, small vertebrates, honey, and birds' eggs. The Panamanian kinkajous ate seven species of flowers, including those of a philodendron and the female flowers of various palms. They took nectar from balsa (*Ochroma pyramidale*, illustrated on p. 38), the kapok relative ceibo (*Pseudobombax septenatum*), and the flacourtia relative *Tetrathylacium johansenii*. They are also known to take nectar from *Quararibea* trees, which are also kapok relatives.

In most studies to date, insects have represented only a tiny fraction of the diet, possibly having been ingested incidentally with fruit and nectar. Nonetheless, six stomachs analyzed in the Bolivian Amazon contained almost 90% ants and ant-nest carton material. Likewise, kinkajous feed on vertebrates only rarely, although they have been seen catching and eating bats that visit fruiting trees in Mexico, and some captives in Peru were found to relish mice.

Several anatomical specializations seem to reflect the kinkajou's arboreal, fruit-eating habits. Flexible knee and ankle joints allow the hindfeet to be rotated 180° backward, enabling the kinkajou to descend or hang from tree limbs headfirst. The prehensile tail (a feature that, among the world's 240 members of the Carnivora, is present only in the kinkajou and the binturong, a 14 kg (30 lb), arboreal, bearlike frugivore found in Asia) is of course also an adaptation for life in the trees. The kinkajou's hands resemble our own—the fingers can be splayed and converge toward the center of the palm when the hand is closed. This enveloping grip helps kinkajous grip thin branches in the outer parts of trees where fruits often grow as well as pluck and manipulate the fruits. And an exceptionally long, narrow tongue helps the kinkajou probe fruits, flowers, and insect nests.

Kinkajous have very long tongues.

Curiously, kinkajous are more lethargic than other procyonids; they have unusually low muscle masses, metabolic rates, heart pulses, and body temperatures when compared to other mammals their size.

The social arrangement of kinkajous is not well understood. They usually travel alone, and sometimes exclude others of the same sex from their 10 to 40 ha (25 to 100 A) home ranges, behavior that is typical of the Carnivora. At other times, however, kinkajous congregate peacefully in fruiting trees and even groom and play with each other. Some groups have been recorded foraging and sleeping together for months at a time. It has not been determined whether members of such groups are related. Kinkajous den in hollow trees. Gestation lasts just under four months, producing one or, rarely, two young, which are born in the dry season. Males are sexually mature after a year and a half, and females in their third year. Captives can live 23 years.

Sounds: Slurred, two-syllable chirps, the second syllable lower in pitch than the first, somewhat resembling the call of the olingo but more whistly and higher pitched, and with the greatest emphasis at the beginning of the first syllable. The volume, pitch, and frequency of the chirps rise when the animal is alarmed or excited, turning into a hiccuping "wick-a-wick-a-wick" series. Also barks and shrill, penetrating screams; hisses during antagonistic encounters; and tongue clicks, made by females in heat.

Mythology: Indigenous Colombians believe that if a kinkajou barks during the daytime, a member of their family will die.

Conservation: Kinkajous are probably important seed dispersers for many plant species. They seldom spend long in a given feeding tree, so seeds, many of which are swallowed whole and passed intact, are typically carried some distance before they are defecated. Kinkajous may also pollinate flowers, for they often pick up pollen on their faces as they drink nectar.

In some areas, kinkajous are captured for the pet trade or killed for their meat or pelts. Their soft, dense, attractive fur has been used to make wallets and belts, and to line horse saddles and tail harnesses. Kinkajous cannot survive in deforested areas. Of 162 individuals collected in a study in Venezuela, for example, only 3% were found in open habitats. In forested areas where hunting pressures are not too severe, however, they can be common; they are among the most frequently seen nocturnal mammals in Costa Rica.

References

Bisbal 1986; Charles-Dominique et al. 1981; Crandall 1971; Emmons & Gentry 1983; Emmons et al. 1997; Estrada & Coates-Estrada 1985; Ford & Hoffmann 1988; Forman 1985; Handley 1976; Janson et al. 1981; Julien-Laferrière 1993; Kays 1999; Kays & Gittleman 1995; Leopold 1959; MacClintock 1985; McNab 1989; Müller & Kulzer 1977; Müller & Rost 1983; Poglayen-Neuwall 1962, 1966; Pruitt & Burghardt 1977; Redford 1989; Reid 1997; Skutch 1960; Timm & Laval 2000a; Walker & Cant 1977.

O: Carnivora F: Procyonidae

OLINGOS

Bassaricyon gabbii and *B. lasius*
(plate 31)

Names: *Bassar* means fox and *cyon* means dog. The species name is after William More Gabb, a nineteenth-century British researcher who conducted some of the first studies of the geology, paleontology, and zoology of the Talamanca region in southern Costa Rica. A second species of olingo, *B. lasius*, is known only from Estrella de

Cartago and is endemic to Costa Rica. It has longer and thicker fur than *B. gabbii* (*lasius* is the Greek for woolly) and slightly different skull and teeth. As more specimens are collected, *B. lasius*, as well as four other olingo species in South America, may turn out to be no more than subspecies of *B. gabbii*. **Spanish**: *Olingo*. Locals seldom distinguish the olingo from the kinkajou by name, often using the kinkajou's name, *martilla*, to refer to both.

Range: Nicaragua through the coastal forests of northwest South America to Ecuador and the western edge of the Amazon basin to Bolivia; from sea level to at least 2,000 m (6,600 ft); in mature and disturbed forest.

Size: 43 cm, 1.5 kg (17 in., 3 ½ lbs).

Similar species: Kinkajous (pl. 31, p. 229) are larger. In some areas—notably in dry habitats—kinkajous may have the same drab-brown coloration as olingos, but they are usually more orangey. The olingo's tail differs from the kinkajou's in that it is more bushy, broadens rather than tapers at the tip, has faint rings that kinkajous' tails lack, and is not prehensile (gripping). While kinkajous use their tail as a "fifth limb" to clamber between branches, olingos use their tail only for balance, and tend to jump around more than kinkajous. Even in dim light, it is usually possible to discern which of these styles of locomotion an animal uses.

Natural history: Olingos are mostly nocturnal and arboreal. They run and jump with great agility through the branches, sometimes clearing 3 meters (10 feet) in a single leap. Like kinkajous, they feed on fruit, nectar, invertebrates, and small vertebrates, and the two species can sometimes be found feeding in the same trees. Subtle morphological differences between olingos and kinkajous, however, suggest that the two species, whose geographic ranges and habitat preferences are also very similar, occupy somewhat different feeding niches.

Olingos lack prehensile tails, which may make them less proficient than kinkajous at reaching the extremities of fruit and flower-laden trees. They also have sharper, less flattened cheek teeth than kinkajous, and may be more carnivorous. Olingos have been studied very little in the wild, but there are a number of anecdotal reports of them feeding on mice and other vertebrates. One in Monteverde was seen chasing down a variegated squirrel and killing it with a series of bites to the back of the head. Another Monteverdean individual, which frequently arrives to steal sugar water from feeders at the Hummingbird Gallery, also snags the occasional hummingbird, a demonstration that has horrified more than one group of feeder spectators.

Olingos den in tree cavities. They usually sleep and travel alone, and studies on captives suggest they are less social than kinkajous. Adult males fight when kept together, and may therefore maintain exclusive territories in the wild. Gestation lasts two and a half months. Single young are born in the dry season. They reach sexual maturity at end of their second year. Captives can live at least 17 years.

Sounds: When alarmed, two-syllable, sneeze-like coughs, the first syllable rising in pitch and longer than the short, lower-pitched second syllable, somewhat similar to the call of the kinkajou, but more hoarse, less whistly, and lower-pitched, with the greatest emphasis at the end of the first syllable.

Conservation: Since, like kinkajous, olingos seldom spend long in a single fruiting tree, they may be good seed dispersers. They visit and may pollinate flowers of balsa (*Ochroma* spp.) and kapok (*Quararibea* spp.) trees. They are hunted little but are vulnerable to deforestation. They are usually less common than kinkajous, and are legally protected within Costa Rica.

References

Decker & Wozencraft 1991; Eisenberg 1989; Emmons et al. 1997; Cristina Hansen, pers. comm.; Janson et al. 1981; Patricia Maynard, pers. comm.; Poglayen-Neuwall & Poglayen-Neuwall 1965; Reid 1997.

Weasel Family (Mustelidae)

Distribution and Classification

There are about 65 species of mustelids, distributed almost worldwide. They are absent only from the continents of Antarctica and Australia. The family is split into five subfamilies, three of which are represented in Costa Rica. The largest subfamily, with about 34 species, is the Mustelinae, which includes Costa Rica's long-tailed weasel, tayra, and grison. The Mephitinae subfamily contains the world's nine skunk species, all of which are found in America, three of them in Costa Rica. The world's 13 otter species, one of which occurs in Costa Rica, comprise the subfamily Lutrinae. The other two subfamilies are the Mellivorinae (which contains just one species, the honey badger, which resembles a grison and inhabits Africa, India, and the Middle East) and the Melinae (with eight species of badgers). The scientific family name comes from the Latin words *mus*, meaning mouse, and *telum*, meaning spear: many weasels, especially those in the genus *Mustela* (such as Costa Rica's long-tailed weasel), like to eat rodents and have spearlike body shape and speed.

Evolution

The ancestors of Costa Rica's mustelids originated in the north. The oldest fossils are from Europe and date back about 50 million years to the Eocene period. Roughly 35-million-year-old fossils have been found in North America. As chunks of what is now southern Central America joined up with North America, completing a land bridge to South America between seven and two million years ago, Mustelids were able to spread south through Costa Rica and into South America.

Teeth, Skull, and Feeding

The mustelid dental formula is I 3/2-3, C 1/1, P 2-4/2-4, M 1/1-2 = 28 to 38. Although some species are omnivorous, all mustelids have teeth that are well adapted for meat-eating. Mustelids' large canines are important for seizing and killing prey. Their rear premolars and front molars are carnassial teeth—teeth with blade-like edges for slicing through meat. Their rear molars are the only teeth designed for crushing, so mustelids, like many carnivores, chew at bones with the rear corner of the mouth. Mustelids' incisors are less important and occupy a narrow space at the front of the mouth. Two of the lower incisors are set back behind the others, forming a second, space-saving row. Because the incisors take up little space, and because mustelids have fewer molars and premolars than many other mammals, the muzzle is relatively short and the head, wedge-shaped. This design could be advantageous to species that forage in narrow tunnels. The rear part of mustelids' skulls, on the other hand, is relatively long, and the space beneath the zygomatic arches (the cheekbones), wide, providing room for large temporal muscles. Bony ridges at the top and back of the skull provide firm anchorage for these muscles. Large temporal muscles are responsible for mustelids' great biting strength.

Body Design

Most mustelids have relatively long bodies and short legs, and are thus well designed for fitting down burrows, where they often sleep or hunt. Long necks help some species carry large prey without tripping the forefeet. Mustelids have five-toed, plantigrade feet. The toes have curved, nonretractable claws that are useful for capturing prey, climbing, and digging. Species that dig a lot have enlarged claws on their forepaws. Body size ranges from 20 cm (8 in.) and 25 g (1 oz) in the least weasel to up to a meter (3 feet) and 40 kg (90 lbs) in the wolverine, sea otter, and giant otter. Mustelids can be very aggressive, and successfully attack or defend themselves against animals much larger than themselves.

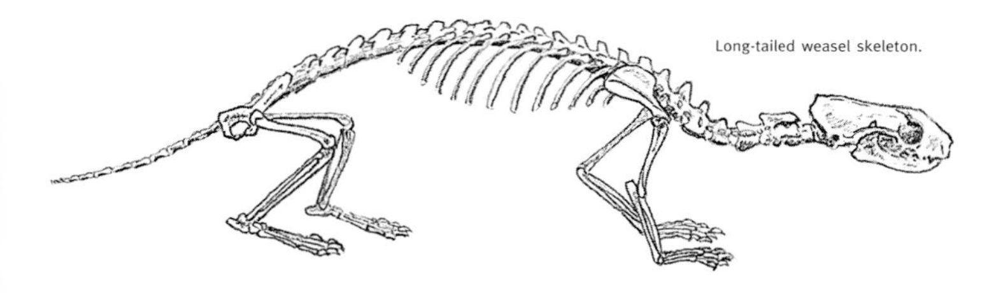
Long-tailed weasel skeleton.

Senses and Scent Glands

Mustelids see poorly, sometimes failing to notice motionless, quiet observers, but have good senses of hearing and smell. As in many other mammals, a pair of scent glands at the base of the tail helps individuals recognize one another and mark territories. In mustelids, however, these glands are unusually large and, in a number of species, produce a remarkably foul-smelling liquid that is used in defense. In America, skunks are the most famous offenders. Their glands, which resemble small hose nozzles, can squirt musk a distance of 3 meters (10 feet). Humans can smell skunks' musk—a thick, yellow fluid known as n-butyl mercaptan—from over a mile away. At close range, the smell of the musk causes nausea. If the musk makes contact with skin, it burns like acid; if it gets into the eyes of a potential predator, it causes temporary blindness (lasting 15 to 20 minutes in people). For those unfortunate enough to come into direct contact with the musk, tomato juice, ammonia, vinegar, or gasoline are reputed to be effective in removing the odor.

Skunks generally do not discharge unless they are highly provoked. Their distinctive coloration, which they flaunt with a variety of threat displays, helps warn or remind predators not to approach. (Warning, or aposematic, coloration is well known in creatures such as coral snakes or monarch butterflies, but is rare among mammals.) Nonetheless, skunks can produce almost one third of a liquid ounce of musk per week, and can squirt five times in succession if a predator is particularly stupid.

A number of other mustelids have similar capabilities. The European polecat, the African zorilla and striped weasel, the Middle Eastern marbled polecat, and the Asian stink badgers, ferret badgers, and hog badger, among others, are all capable of stinking their predators into retreat. Some, including Costa Rica's long-tailed weasel, are incapable of squirting scent like skunks do, but can secrete enough musk to be quite unappetizing.

Reproduction

Perhaps the most interesting aspect of reproduction in mustelids is the ability of many species to control the duration of their pregnancies. To date, this is known to occur only in temperate zones, where such species mate at the end of breeding season and, by delaying the implantation of fertilized eggs in the uterus, do not begin embryonic development until the following breeding season. Weasels have thus been recorded to have gestation periods of over a year, although the period of actual embryonic development usually lasts less than two months. Of the 37 or so mustelid species whose reproductive biology has been studied (the life histories of the remaining 28 are unknown), 17 species use delayed implantation. Mustelids are not unique in this regard, but worldwide only 47 mammals (in the orders Xenarthra, Insectivora, Chiroptera, Carnivora, Pinnipedia, and Artiodactyla) are known to delay implantation.

Mustelids give birth in underground dens or other cavities that are often lined with dry leaves and other material the female pushes in with her head. Underground dens are typically situated a little above the level of the tunnels so they don't fill with water when it rains. Young are born naked and blind but grow quickly. In most species, the young are independent by the age of two months. The age at which they reach sexually maturity varies from two and a half weeks (for the Northern Hemisphere's stoat, *Mustela erminea*, in which females mate with an adult male while still in their nest, blind and helpless, and delay implantation until they are fully grown) to two years.

Conservation

Many members of the this family (mink, for example) have soft, durable fur that is highly prized. Although demand has dropped now, hunting in decades past has wiped out numerous populations and pushed several species to the verge of extinction. Skunks (whose fur is marketed as

Alaskan Sable or Black Marten) and river otters are occasionally hunted for their pelts in Costa Rica. Mustelids may also be hunted for sport or as predators of poultry and eggs. In North America, skunks are the principal vectors of rabies, representing up to two thirds of the wild animals found with rabies each year. Skunks have long incubation periods, and thus ample opportunity to spread the disease by biting fellow skunks or other animals. Transmission of the disease to domestic animals or humans, however, is exceedingly rare. Skunks and other mustelids can benefit humans by eating insects and rodents, and they are left alone by farmers in many parts of the world for that reason. Skunks and tayras can also eat a lot of fruit, and it is likely they help disperse some seeds. Some mustelids, such as skunks, do well in disturbed areas, but many other species, such as the river otter, require more pristine environments and have suffered enormously from loss of habitat.

References

Barker 1956; Ewer 1973; Howard & Marsh 1982; King 1989; Macdonald 1995; Mead 1989; Nowak 1991; Pruitt & Burghardt 1977; Verts 1967; Voigt 1984.

SPECIES ACCOUNTS

O: Carnivora F: Mustelidae

LONG-TAILED WEASEL
Mustela frenata
(plate 33)

Names: The origins of the word *Mustela* are explained on p. 235. *Frenata* comes from the Latin *fraenum*, meaning bridle, in reference to a bridle-like pattern of white stripes around the head that characterizes long-tailed weasels from Mexico (where the type specimen was caught) to Nicaragua. **Spanish**: *Comadreja.*

Range: Southern British Columbia in Canada to Bolivia; from sea level to at least 4,000 m (13,000 ft), especially above 1,000 m (3,300 ft); in forest or partially disturbed habitats, often near water.

Size: 20 cm, 150 g (8 in., 5 oz); males larger than females.

Similar species: The long-tailed weasel is longer, thinner, and quicker than squirrels and lacks a bushy tail. In the northern parts of its range, the longtail turns white during winter to match the snow, but no seasonal color change takes place in the tropics. Central American longtails do, however, have geographically variable facial patterns. Rather than the aforementioned bridle pattern typical of animals from northern Central America, Costa Rican longtails have white spots or stripes around their eyes, or no white facial markings at all.

Natural history: This weasel is active by night or day, but it is shy and rarely seen. When not breeding, it is generally solitary. It eats vertebrates such as rodents, rabbits, birds, bats, frogs, lizards, and snakes, some of which may be several times its size. It also takes bird eggs, carrion, insects, and worms. Young sometimes eat their own siblings.

The longtail hunts by zigzagging along the ground, frequently investigating tunnels, and using its keen senses of smell and hearing to locate prey. With its long, thin body and short legs, it is well suited for searching in burrows. An extremely flexible backbone enables it to turn around in a narrow burrow by rolling upside down and walking back over its body. The longtail can also climb trees by leaping on to and then spiraling up trunks. In Monteverde, it is one of the main predators on the eggs and hatchlings of quetzals, which nest in cavities about 10 meters (33 feet) off the ground. Longtails generally kill prey captured aboveground with a bite to the nape that pierces the base of the skull or severs the spinal cord, but they dismiss prey taken underground with a suffocating bite to the throat. Longtails are voracious, eating as much as a third of their body weight or more each day. Such feeding fuels the longtail's unusually high metabolic rate, which in turn is necessary to keep it warm: being long, thin, and short-furred, the longtail loses heat relatively quickly.

Occasionally longtails perform frenzied dances, a trait that has intrigued biologists for decades. Some have suggested that the dance serves to mesmerize rabbits and other dim-witted prey, enabling the weasel to approach close enough to pounce on them. The longtail dances with or without an audience, however, and its convulsions may instead be a response to the discomfort caused by parasitic worms lodged in the weasel's nasal sinuses. This condition, known as Skrjabingylosis, is common in small mustelids. Weasels become worm hosts when they eat mice or, especially, shrews that have eaten snails, if those snails have been burrowed into by worm larvae that left their original weasel host in the weasel's feces. The worms get from the weasel's stomach to its nasal sinuses by burrowing along the spinal cord.

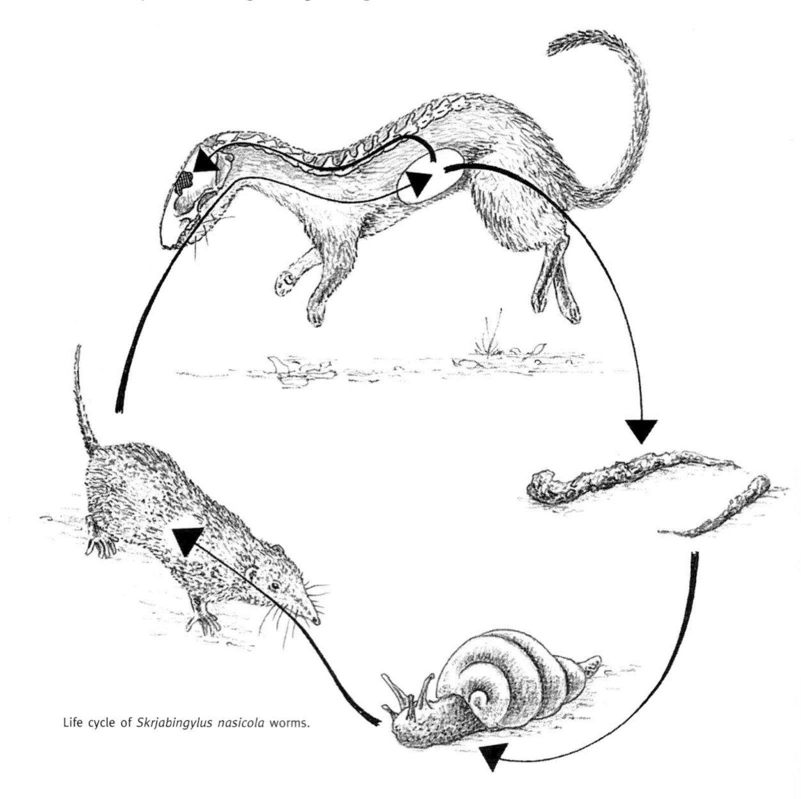

Life cycle of *Skrjabingylus nasicola* worms.

The home range size of longtails (in North America) has been found to vary substantially—from 4 to 160 ha (10 to 395 A)—depending on habitat, season, and the availability of prey. Males tend to have larger ranges than females. The ranges of neighboring males overlap little, but each male's range typically overlaps that of at least one female and expands during the breeding season as the male scouts for mates. Longtails den in rock piles, hollow logs, or in the burrows of their victims. The main chamber, which is typically lined with dry grass and fur from the weasel's victims, is about 3.5 to 5 cm (1 ½ to 2 in.) wide, and is connected to other passages that may serve as bathrooms or food-storage areas.

Temperate-zone longtails have gestation periods of up to 11 months, although embryonic development lasts only about three and a half weeks (see p. 237). Their litters contain three to 12 young that are weaned within a month and can hunt for themselves within two months. The mother raises the young alone. Temperate zone females mate at just three to four months of age, when they are still with their mother and not quite fully grown. They mate with the local male who, due to a high turnover of resident males, is unlikely to be their father. Males do not mate until their second year. Longtails may live nine years in captivity, but field studies suggest that few live more than two years in the wild.

Sounds: Series of short, low-pitched trills, especially during social interactions and while hunting; foot stamping and high-pitched screeches, often repeated several times, as threat; squeals in distress.

Conservation: The longtail occasionally raids chicken coops, where it may kill several birds without eating them. This "surplus killing" is also perpetrated by other mustelids, foxes, and wild cats. Presumably such carnivores kill available prey instinctively; since they are seldom, if ever, presented the opportunity to kill many animals at a time in the wild, they have not evolved any way to control reflexes that are wasteful in unnatural circumstances. Perhaps because the longtail's canines leave a distinct pair of holes in each dead bird, it is falsely reputed to be a blood-sucker. Since it preys heavily on rodents, it may do farmers more good than harm. Although inconspicuous, this species may be quite common at higher elevations.

References

Brown & Lasiewski 1972; DeVan 1982; DeVos 1960; Dougherty & Hall 1955; Fagerstone 1987; King 1977, 1989; Linduska 1947; Nowak 1991; Sheffield & Thomas 1997; Svendsen 1976, 1982; Weber & Mermod 1983, 1985; Wheelwright 1983.

O: Carnivora F: Mustelidae

GREATER GRISON

Galictis vittata
(plate 32)

Names: Also known as the greater huron (as opposed to the lesser grison or huron, *G. cuja*, which is similarly colored but smaller and replaces it through central and southern South America). Genus from either the Latin *galea*, meaning helmet, or the Greek *gale*, meaning weasel, and from the Greek *iktis*, also meaning weasel; *vittata* is the Latin for striped. Also listed as *Grison vittata*. *Grison* comes from the Latin for gray. **Spanish**: *Grisón*.

Range: Southern Mexico to Peru, Bolivia, and southeastern Brazil; from sea level to at least 1,500 m (5,000 ft); in forest or open habitats. Most records in Costa Rica are from the lowlands, below about 500 m (1,650 ft), especially on the Atlantic side of the country and in the fields around Carara Biological Reserve. There have also been several recent sightings, including one of a family group, in the Monteverde area between 1,300 and 1,500 m (4,300 and 5,000 ft).

Size: 50 cm, 3 kg (20 in., 6 ½ lbs); males larger than females.

Natural history: This distinctive mammal is rare throughout its range, and little is known of its natural history. One unfortunate researcher spent months setting traps for grisons in the vicinity of Carara. More than 6,000 trap hours yielded opossums, iguanas, vultures, and even a tortoise, but not a single grison. In addition to being hard to observe or catch, grisons are adept at shedding radio-transmitters, in part because, unlike most other mammals, they are thicker at the neck than at the head, and easily slip out of radio-collars. The following information is based on random observations in the wild, and studies on captives.

Grisons may be active by day or night. One that was radio-collared briefly in Venezuela was active about 10 to 12 hours a day, mostly at night. Grisons travel alone, in pairs, or in groups that probably consist of family members. When active, they stop frequently to check their surroundings, extending their necks and heads and sniffing and peering in different directions. They seem to have poor eyesight but an excellent sense of smell; they sometimes fail to notice people standing still right next to them, but smell them and flee quickly when downwind.

Grisons are good climbers, but forage mostly at ground level. They feed principally on small vertebrates, including rodents, opossums, reptiles, and birds. They also take fruit, arthropods, and eggs. They can kill prey larger than themselves; an individual in Monteverde was observed attacking and killing a young, roughly 4 kg (9 lbs) paca with a bite to the throat. They are particularly adept at catching snakes and have been observed killing large, aggressive species such as the mussurana (*Clelia clelia*) that most predators leave alone.

Grisons will kill and eat even large, aggressive snakes such as the mussurana (shown here eating a Fer-de-lance).

Grisons probably spend part of their time hunting in water. They have a small amount of webbing between their toes, are excellent swimmers, and have often been observed in the vicinity of rivers or in flooded rice fields. Captives sometimes retrieve prey from underwater; aquatic prey, including frogs and fish, have been found in the stomachs of wild grisons.

With their short legs, narrow heads, and elongate bodies, grisons are also able to pursue prey underground. In South America, they are major predators on cavies (wild guinea pigs), which live down tunnels. During the nineteenth-century in Chile, grisons were domesticated and used to flush game from burrows, just as their cousins, the ferrets, are used in Europe.

If cornered, grisons can defend themselves vigorously. When startled or captured, they secrete a strong-smelling, yellowish liquid from anal glands. This musk may serve as a further deterrent, rather like the musk of the grison's relatives, the skunks. Like skunks, grisons have a bold, distinctive color pattern that might help warn or remind predators to stay clear.

The territorial arrangement of this species is unknown. The radio-collared individual in Venezuela, a female, traveled 2 to 3 km (1 ¼ to 2 mi) each night and used a home range of 415 ha (1,025 A) during the month she was followed. Grisons den in hollows amidst tree roots or rocks, or in other animals' burrows. They may also excavate their own holes.

They have a gestation period of about 40 days, which results in two to four young. Random observations of adult pairs and of cubs suggest that grisons pair up and mate at the end of the dry season, in late April or May, and give birth in June or July. Young may be weaned at about three months and leave their mother one or two months after that, by which time they are almost her size. Captives can live at least 10 years.

This hole in the bank of a trail in Monteverde was used by a grison for several weeks.

Sounds: Snorts, low-pitched growls, high-pitched barks, and screams in alarm; nasal "anh-anh" sounds by young.

Conservation: Grisons sometimes take poultry and may be killed as pests. Although they are rare everywhere, Costa Rica is the only country in which they are legally protected.

References

Dalquest & Roberts 1951; Elizondo 1987; Michael Fogden, pers. comm.; Denis Gómez, pers. comm.; Jones 1982; Kaufmann & Kaufmann 1965; Malavassi 1991; Moojen 1943; Nowak 1991; Osgood 1943; Reid 1997; Rood 1970; Sunquist et al. 1989; Timm et al. 1989; Sergio Vega, pers. comm.

O: Carnivora F: Mustelidae

TAYRA
Eira barbara
(plate 32)

Names: *Eira* comes from the South American Guaraní language and means honey-eater; tayras occasionally raid bee hives. *Barbara* is the Greek for strange. **Spanish**: *Tolomuco, cholomuco, tejón.*

Range: Central Mexico to northern Argentina; from sea level to at least 2,400 m (7,900 ft); in or near forest.

Size: 65 cm, 5 kg (26 in., 11 lbs).

Similar species: Best distinguished from the jaguarundi (pl. 35, p. 269) and from the river otter (pl. 32, p. 253) by bushy tail. The tayra's coat color is variable. Over much of its range it has a blond head and neck that contrasts sharply with its dark body. Some individuals in the savannas of Guyana and Bolivia are entirely blond. In Costa Rica, however, the tayra typically has no more than a blond patch on the throat, and often no blond at all, although the head and neck may be somewhat paler than the rest of the body. This dark color form is rare outside Costa Rica and Panama.

Natural history: The tayra is seen more often than many other members of its family or even order, yet it has been studied little in the wild. It is more conspicuous in part because it is active during the daytime. Three tayras radio-collared in Belize were consistently active between about 6:00 AM and 4:00 PM, with a peak in activity in the early morning, while one followed in Venezuela was most active at dawn and dusk. The tayra travels in small family groups, pairs, or, most often, alone, moving swiftly along the ground in search of a variety of foods.

Its diet is comprised of fruit, arthropods, and meat. Fruits eaten by the tayra include guavas (*Psidium guajava*), cecropias (*Cecropia* spp., illustrated on p. 62), hogplums (*Spondias mombin*, illustrated on p. 162), palm fruits (*Astrocaryum standleyanum*), and coffee fruits (*Coffea* spp.). Its scat sometimes consists entirely of seeds. Vertebrate prey includes rabbits, opossums, rodents (including agoutis and pacas), birds (as well as bird eggs and nestlings), iguanas and other lizards, and snakes (including pit vipers, which are venomous). Sometimes the tayra takes on prey much larger than itself. It has been known to attack tamarin, squirrel, and even capuchin monkeys, two-toed sloths, and brocket deer. It also takes honey from the nests of stingless bees (*Trigona* spp.) and other species. Although tayras do much of their foraging on the ground, they are excellent climbers, a skill they use to reach fruit and other foods or to escape danger. Naturalist Alexander Skutch described seeing one climb 30 meters (100 ft) up a clean, vertical trunk to get at a laughing falcon nest in San Isidro el General.

Tayra scat (about 8 cm long).

Radio-collared tayras have been found to roam large distances, travelling an average of almost 7 km (4 ½ miles) each day, and covering home ranges of between 900 and 2,400 ha (2,200 and 5,900 A). Tayras den in hollow logs or trees, beneath rocks, or in other animals' burrows. The entrance to a tayra burrow at the base of a tree in Santa Rosa National Park was about 12 cm (5 in.) wide. Gestation lasts a little over two months. One to four young (often two) nurse for two to three months. Captives can live 18 years.

Sounds: Growls, pants, snorts, hisses, and clicking sounds during confrontations or in alarm; catlike yowls during courtship.

Conservation: Tayras sometimes take poultry and can damage crops such as corn or sugar cane, but they are not serious pests. Their pelts are not highly sought after. They may help control rodent populations, and in some areas they have been tamed by indigenous peoples for that purpose. Tayras probably help disperse the seeds of some plants; intact scats are often visible in their scats.

References

Bisbal 1986; Marc Egger, pers. comm.; Elizondo 1987; Emmons & Feer 1997; Emmons et al. 1997; Enders 1935; Galef et al. 1976; Grzimek 1975; Hall & Dalquest 1963; Janzen 1983e; Kaufmann & Kaufmann 1965; Konecny 1989; Nowak 1991; Poglayen-Neuwall 1975, 1978; Poglayen-Neuwall & Poglayen-Neuwall 1976; Skutch 1992; Sunquist et al. 1989; Vaughn 1974; Wille 1987.

O: Carnivora F: Mustelidae

STRIPED HOG-NOSED SKUNK

Conepatus semistriatus

(plate 33)

Names: The word *skunk* comes from the North American Algonquian name *Seganku.* The scientific name comes from the Mexican Aztec word *conepatl*, meaning digger, and the Latin *semi*, meaning half, and *striatus*, meaning striped. There are four other species of hog-nosed (*Conepatus*) skunks outside Costa Rica, distributed from the southern United States to northern Argentina. Some believe that the five forms of hognose represent races of a single species. **Spanish**: *Zorro hediondo rayado.*

Range: Veracruz, Mexico, to the Andes and Pacific slope of Peru, and in eastern Brazil; from sea level to at least 4,100 m (13,500 ft); mostly in relatively open habitats such as pastures, secondary growth, and dry forests.

Size: 40 cm, 2.5 kg (16 in., 5 ½ lbs); males larger than females.

Similar species: This species is similar in coloration to the predominant form of the North American striped skunk (*Mephitis mephitis*). Distinguished from the spotted skunk (pl. 33, p. 249) by its larger size and white V pattern (two longitudinal stripes, one on each side of the body, that come together on the top of the head); from the hooded skunk (pl. 33) by broader stripes higher up on the sides of the body and by its tail, which has no black on it except at the very base; and from both of these species by its unusually long snout, which extends well beyond the lower lip and which is naked for the first three centimeters. Note that in Costa Rica the spotted skunk is found only in the northwest and the central valley, and the hooded skunk only in the northwest (where it is the commonest of the three species); the hognose is the only species of skunk in the rest of the country.

Natural history: The striped hog-nosed skunk has been studied little. It is mostly nocturnal and solitary. It forages on the ground, using its flexible snout and long, sharp front claws to root for small insects and the like, small vertebrates, and fruit. Hog-nosed skunks in the Andes are resistant to pit-viper venom and may feed on them; the same might be true of Costa Rican hog-nosed skunks (as well as spotted skunks—see p. 249).

Hognoses can leave distinctive areas of churned up topsoil where they have been foraging. Their unmistakable odor can also reveal their presence, sometimes hours after they have passed.

A radio-collared female in Venezuela was active for about six hours and traveled 1 to 2 km (⅔ to 1 ¼ mi) each night. Her home range incorporated 53 ha (130 A), but she only used 18 ha (45 A) of the range during the wet season. Striped hog-nosed skunks den in hollow logs, down the abandoned burrows of armadillos and other animals, and under buildings. They may also excavate tunnels themselves. Gestation lasts about two months, resulting in litters of four or five young. Captive hognoses can live at least six and a half years.

Conservation: Hognoses have short, coarse fur that is not as valuable as that of the *Mephitis* skunks, but they are hunted heavily in some parts of their range. An estimated 155,000 pelts were being traded annually during the 1970s. Like other skunks, hognoses are important vectors of rabies. They do well in disturbed areas and persist unless hunting is very intensive. *C. semistriatus* is fairly common in many parts of Costa Rica.

References

Broad et al. 1988; Eisenberg 1989; Grimwood 1969; Howard & Marsh 1982; Janzen & Wilson 1983; Jones 1982; Nowak 1991; Sunquist et al. 1989.

O: Carnivora F: Mustelidae

SPOTTED SKUNK

Spilogale putorius

(plate 33)

Names: Also known as the tree skunk or hydrophobia cat (hydrophobia is another name for rabies, a disease it has been known to carry), among other names. Genus from the Greek *spilos*, meaning spot, and *gale*, meaning weasel; *putorius* is the Latin for putrid. Some authorities consider spotted skunks in Central America and those west of the Continental Divide in North America to be a separate species—*S. gracilis* or the western spotted skunk—and use the name *S. putorius* or eastern spotted skunk only for those occurring east of the Continental Divide in North America, from the southern tip of Canada to the northern tip of Mexico. The two are separated because they are largely isolated from one another geographically, and, in North America at least, have different breeding strategies (see below). There is one other species of spotted skunk, *S. pygmaea*, which is restricted to a small stretch of the Pacific coast of Mexico. **Spanish**: *Zorro hediondo manchado.*

Range: Southern British Columbia in Canada to the Valle Central of Costa Rica; from sea level to 3,000 m (10,000 ft), but not known to occur above about 1,550 m (5,100 ft) in Costa Rica; in open areas with some cover, brush, and relatively dry forests.

Size: 25 cm, 500 g (10 in., 1 lb); males larger than females.

Similar species: Distinguished from the other two skunk species in Costa Rica by smaller size and by white stripes and spots, which are arranged differently in each individual but never form unbroken lines as in the other species. The patch on the forehead is consistent, and absent in the other skunks.

Natural history: Spotted skunks are mostly nocturnal and solitary. They are primarily insectivorous, but they also take small vertebrates such as rodents, lizards, snakes, and birds, both alive and as carrion. Vertebrate prey is dispatched with a piercing bite to the back of the neck. The spotted skunk is considered more carnivorous than other skunks, and this may be reflected in the fact that it has proportionately longer carnassial teeth. Rattlesnake remains have been found in spotted skunk scats, and it is conceivable that spotted skunks are important predators on rattlers, given that most other predators leave them alone. There is some evidence that spotted skunks are immune to rattler venom; moreover, rattlers become visibly agitated when they detect skunk must, reacting the same way they do when in the presence of king snakes, one of their main predators. *Spilogale* also enjoys fruit and eggs. Rather like the hooded skunk, it breaks eggs too big to be crushed in its mouth by kicking them backward beneath its body.

Like other skunks, *Spilogale* warns (or reminds) potential predators of its putrid weaponry by erecting its tail. If the warning is not heeded, this species is apt to raise its tail even higher by doing a handstand, and may even advance toward the foe in this unusual position. *Spilogale* may arch its hindquarters over its head and spray unimpressed challengers while still in a handstand, or adopt the more conventional horseshoe position, pointing its rear end toward the target and arching its head around to see where it is aiming.

The spotted skunk also makes use of its acrobatic capabilities to climb trees, where it may escape predators, forage, or sleep. The spotted skunk's ability to use trees is one of several differences between it and Costa Rica's other two skunk species that may help the three species coexist. Other important distinctions may be its more carnivorous diet and its smaller size, the latter enabling it to pursue prey down narrower burrows. The spotted skunk also has proportionately longer claws, which may have prevailed because they improve its ability to climb trees, or to dig up and then snare rodents and other animals.

Little is known of the movement patterns of this species, in part because spotted skunks are adept at squeezing out of traps and pulling off ear tags and radio transmitters. A male in North America had a large home range of 4,359 ha (10,800 A), and traveled farther during breeding season than at other times of year. Spotted skunks den in cavities amidst rocks, in hollow logs or trees (sometimes several meters off the ground), in buildings, or in any other hole that is secluded, dark, and dry. They line their dens with dry grass. They change dens periodically, and a den may be used by more than one skunk at different times. They are shy and sometimes abandon dens disturbed by humans.

The reproductive cycle of spotted skunks has not been studied in the tropics. Curiously, in North America, the eastern, *putorius* race of the spotted skunk does not delay implantation of the fertilized egg in the uterus, has a gestation period of about two months, and usually gives birth around June, while the western, *gracilis* race does delay implantation (like many other temperate-zone weasels), consequently has a gestation period of seven months or more, and usually gives birth in April. Two to nine young are weaned after about two months, by which time their anal scent glands are ready for action. They reach adult size at about four months. Captives can live more than nine years.

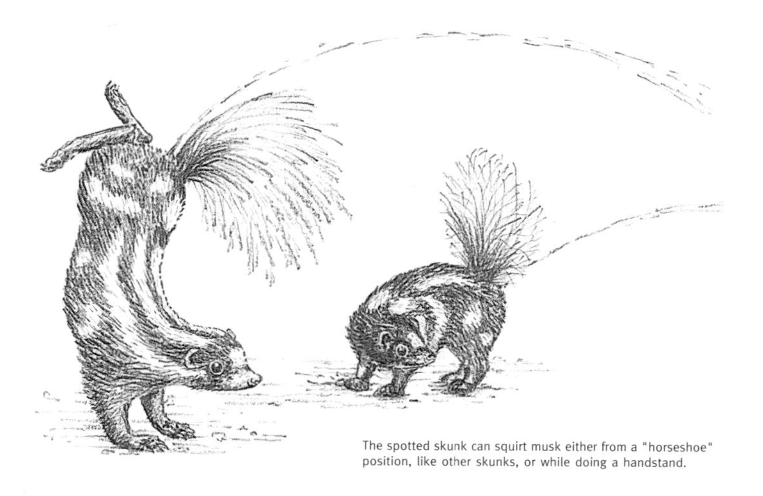

The spotted skunk can squirt musk either from a "horseshoe" position, like other skunks, or while doing a handstand.

Sounds: Foot stamping as threat; guttural grunts; raspy, high-pitched screeches resembling the calls of the white-throated magpie jay.

Conservation: In some areas the spotted skunk is persecuted as a poultry and egg thief, or for its soft pelt, while in others it is considered beneficial since it eats insect and rodent pests. It is listed as endangered or vulnerable in many parts of its range, but not in Costa Rica. It is much less common here than Costa Rica's other two skunk species.

References

Baker & Baker 1975; Barker 1956; Crabb 1944, 1948; Egoscue et al. 1970; Howard & Marsh 1982; Kaplan & Mead 1991; Kinlaw 1995; Lowery 1974; Manaro 1961; McCullough & Fritzell 1984; Mead 1968a,b, 1981, 1989; Nowak 1991; Rosatte 1987; Van Gelder 1953, 1959; Zeiner 1975.

O: Carnivora F: Mustelidae

HOODED SKUNK

Mephitis macroura

(plate 33)

Names: Hooded skunk is somewhat of a misnomer for this species in Costa Rica. Hooded forms, which have a broad white band running from on top of the head down the back, and which predominate through most of this species' range, are rare or absent in Costa Rica. *Mephitis* is the Latin for noxious odor; *macroura* derives from the Greek words *makros*, meaning large, and *oura*, meaning tail. There is only one other species in this genus, the striped skunk (*M. mephitis*), well known in the United States and Canada, where it is the predominant—or in the northern part of its range, the only—skunk species. **Spanish**: *Zorro hediondo encapuchado.*

Range: Arizona and southwestern Texas in the United States to northwestern Costa Rica; from sea level to at least 2,400 m (7,900 ft), but known only from the lowlands in Costa Rica; mostly in open habitats and dry forests. Officially reported from Costa Rica for the first time in 1982.

Size: 35 cm, 2 kg (14 in., 4 ½ lbs); males larger than females.

Similar species: Distinguished from the striped hog-nosed skunk (pl. 33, p. 247) by lack of hoglike nose, longer tail, and color. The color of *M. macroura* is variable. In Costa Rica, the body may be all black or have a white stripe running along each side. The striped form differs from the striped hog-nosed skunk in having usually thinner stripes (no more than 4 cm or 1 ½ in. wide at the broadest point) that are lower down on the sides of the body. The stripes do not join at the shoulders to form a V shape as in the hognose. The tail is typically 50 to 100% black, and any white hairs are usually mixed in with the black ones. The tail of the hog-nosed skunk is clean white, sometimes with some black near the base.

Natural history: Like other skunks, *M. macroura* is mostly solitary and nocturnal. It is omnivorous, feeding on insects, fruits, small vertebrates, and garbage. An individual that lived near a house in Santa Rosa National Park also relished chicken eggs and was adept at breaking them. Since the eggs were too large to be broken in the skunk's small mouth, the skunk picked them up with its forepaws and repeatedly hurled them back between its hindlegs until they cracked. Similar behavior has been observed in North American skunks. These observations, as well as a Costa Rican study of the predation of eggs on the ground in which eaten and uneaten eggs were found scattered around nest sites, suggest that skunks might be major predators on the eggs of ground-nesting birds.

Hooded skunk scat (about 6 cm long).

Details of the reproductive biology of this genus are based principally on studies of the northern species, *M. mephitis*. Gestation lasts 6 months or a little more, resulting in litters of one to 10 (usually four or five) kittens, which are weaned after about two months and are independent after about four months. Females can breed in their second year. The egg-eating individual in Guanacaste produced a litter of at least four toward the end of the rainy season. Northern *Mephitis* skunks can live more than 12 years in captivity.

Sounds: Low churrings, birdlike twitters, squeals, and loud screeches have been described for this genus; foot stamping as threat.

Conservation: *Mephitis* skunks can be a nuisance to poultry farmers by eating chickens or eggs. They can also be important vectors of rabies. On the other hand, they can benefit farmers by eating rodents and insects. In some areas, they are hunted for their pelts; in Guatemala, their scent glands are used in folk remedies; and their gall bladders are used in Chinese medicines. The hooded skunk is hunted little in Costa Rica, and is common in forested and disturbed areas alike in the northwest.

References

Barker 1956; Janzen 1978; Janzen & Hallwachs 1982; Jones 1982; Lowery 1974; Reid 1997; Verts 1967; Zeiner 1975.

O: Carnivora F: Mustelidae

NEOTROPICAL RIVER OTTER

Lutra longicaudis
(plate 32)

Names: Sometimes referred to as the southern river otter, but this name is confusing since it is also used for *L. provocax*, a similar species found in Chile and Argentina. A third similar species, *L. canadensis* or the northern river otter, replaces *L. longicaudis* throughout North America. Scientific name from the Latin words for otter and long-tailed. Genus sometimes listed as *Lontra*. **Spanish**: *Nutria*; *perro de agua*.

Range: Northwestern Mexico to Peru and Uruguay; from sea level to 3,000 m (10,000 ft); in or near rivers, lakes, or swamps.

Size: 60 cm, 9 kg (24 in., 20 lbs); males larger than females.

Natural history: River otters are most often seen alone, but sometimes in pairs or family groups. They are active by day or night. Although they have been known to wander several miles over land, they generally stay close to water. Their coat, consisting of two types of fur, is designed for water: dense underfur traps air and serves as an insulating layer that is kept dry, even while the otter is swimming, by an outer layer of long, oily guard hairs.

River otters do most of their foraging underwater, where they can stay submerged for up to eight minutes. They catch food with their mouths. Their diet includes crustaceans (such as shrimp, crabs, and crayfish), mollusks and other aquatic invertebrates, fish, frogs, and birds. By studying several hundred feces collected along the Puerto Viejo and Sarapiquí Rivers at La Selva, biologist Romeo Spínola found that otters there feed principally on shrimp in the genera *Atya* and *Machrobrachium*, clingfish (*Gobiesox nudus*), cichlid fish (Cichlidae), and catfish (*Rhamdia* spp.). All these animals are relatively sedentary and

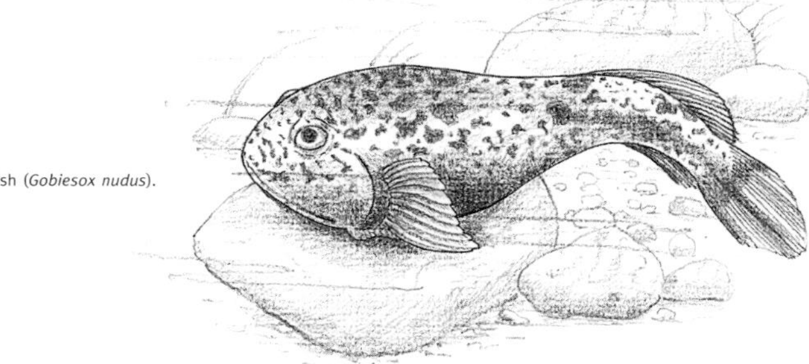

A clingfish (*Gobiesox nudus*).

slow-moving, as are the favorite prey of river otters studied elsewhere. The otter gets at shrimp and clingfish by overturning underwater stones, beneath which they like to hide. It has been calculated that otters eat between 15 and 20% of their weight each day, which would be equivalent to more than 1 kg (2 ¼ lbs) a day for this species.

River otters like to play. A favorite game for family groups is to toboggan down mud banks into the water. They may scurry back up the bank to repeat the slide dozens of times, sometimes coming down in unison. Otters are shy, however, so often the only indication of their presence are riverbed tracks and piles of crustacean shell-filled scats on rocks and logs. At La Selva, scats were found most frequently next to relatively deep water. Scat sites may be used repeatedly by several individuals. Scat odors help neighbors recognize one another and probably let males know when females are in heat.

Neotropical river otter scat; the larger pellet is about 8 cm long.

In order to meet their considerable dietary needs, otters in this genus use home ranges stretching anywhere from about 7 to 80 km (4 to 50 mi) of river. Territorial arrangement in this genus is variable, but neighboring males tend to have mostly non-overlapping ranges that are larger than those of females. In some areas there is a male hierarchy, with the top-ranking individuals occupying the best stretches of river.

Researchers in Brazil recently studied the denning habits of this species for the first time. Over the course of a year, they managed to locate 108 otter shelters. Most of the shelters were located along river banks in natural cavities. The cavities ranged from deep caves to shallow recesses beneath overhanging rocks or trees. The dens of other otter species often have an entrance underwater, with a tunnel that slopes up to a leaf-lined chamber above the high-water level, but no such entrances were found in Brazil. Sometimes the otters also constructed their own shelters by digging shallow (no more than 50 cm or 20 in. deep) recesses along several meters of river bank, or simply by clearing a small space amidst dense grass on top of a river bank. These two types of shelters built by the otters themselves were found mostly in deforested areas, where natural cavities were scarce. They tended to be less stable than the natural cavities and were often destroyed by floods.

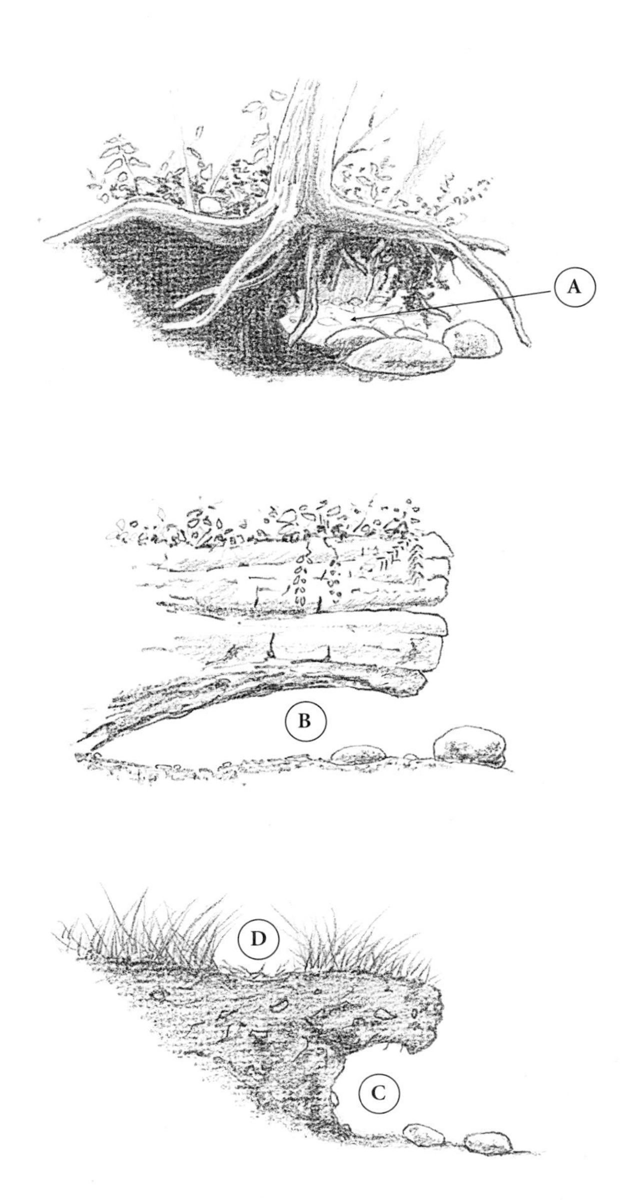

Researchers in Brazil have found that river otters prefer to shelter in natural cavities beneath tree roots (A), rock overhangs (B), or boulders. In deforested areas, natural cavities are scarce, so otters excavate shelters in mud banks (C), or clear resting spaces amidst thick grass above river banks (D). The latter shelters are less stable than the natural cavities, and more vulnerable to flooding. (Drawings after Pardini and Trajano 1999.)

Reproduction in Neotropical river otters has not been studied. The northern species mates at the end of breeding season, but females delay implantation of the fertilized egg in the uterus until the beginning of the following breeding season. After implantation, embryonic development lasts about two months. Litters of one to five young can swim after two months. The mother carries cubs that are reluctant to learn to the middle of a pool and dives. The male sometimes helps with the cubs' training. Cubs stop nursing after three or four months, leave the mother after eight months to a year, and are sexually mature at two or three years. The northern species can live 23 years in captivity.

Sounds: Snorts, like those of a swimmer trying to remove water from nostrils (although otters are able to close their nostrils when underwater); purrs, whistles, screeches, and staccato chuckles during social interactions; grunts and hisses when threatened.

Conservation: The Neotropical river otter has become extinct or exceedingly rare over much of its former range. Pelt hunters took a huge toll in eras past. For example, an estimated 30,000 were being killed annually in Peru and Colombia alone during the early 1970s. Even as late as 1980, there was a recorded worldwide trade of some 37,443 Neotropical otter skins. Although fur-trading has since lessened, habitat destruction has not; clean, unpolluted rivers and lakes with undisturbed bankside cover and adequate food supply are increasingly hard to find.

References

Barker 1956; Broad 1987; Bussing 1998; Chanin 1985; Duplaix 1984; Emmons et al. 1997; Estes 1989; Foster-Turley et al. 1990; Gallo 1989; Grzimek 1975; Helder & Andrade 1997; Macdonald & Mason 1992; Mason & Macdonald 1986; Melquist 1984; Melquist & Dronkert 1987; Melquist & Hornocker 1983; Nowak 1991; Olimpio 1992; Pardini 1998; Pardini & Trajano 1999; Parera 1992, 1993; Spínola 1994; Spínola & Vaughan 1995a,b; Van Zyll de Jong 1972, 1987.

Cat Family (Felidae)

Distribution and Classification

The natural distribution of the world's 37 or so cat species includes all the continents except Australia and Antarctica. All races of the domestic cat belong to the same species, *Felis catus*, which is of course found worldwide. Twelve species of wild felids are native to the New World, six of which inhabit Costa Rica. Until recently, most of the world's cats, and all Costa Rica's species except the jaguar, were usually listed in the genus *Felis*. Recent genetic studies have given taxonomists a clearer understanding of the phylogeny (the evolutionary history) of cats, however, so cats are now separated into more diverse genera that correspond to their different evolutionary lineages.

Evolution

The oldest fossils of catlike creatures are of saber-toothed cats that date back to the early Oligocene period, about 35 million years ago. Sabertooths (the Machairodontinae) were common for millions of years, disappearing only recently, at the end of the Pleistocene some 10,000 years ago.

Sabertooth cats inhabited the Americas until just 10,000 years ago. Illustrated here is *Smilodon populator*, which measured about 120 cm (4 ft) at the shoulders. It is thought to have used its large size and proportionately massive forelimbs to pull down large herbivores, which were abundant during the Pleistocene period.

A second lineage of cats (the Felinae), the ancestors of today's species, are known to have inhabited France by the Miocene period, some 25 million years ago. There are numerous theories about possible lines of descent between these primitive felines and the cats we know today. Recent molecular comparisons suggest that, about 12 million years ago, the felines split into two main groups. Interestingly, three of Costa Rica's cats—the ocelot, margay, and oncilla—are thought to be derived from one group (known as the South American lineage), while the remaining three—the jaguarundi, puma, and jaguar—are thought to be derived from the other (known as the Old World lineage). The ancestors of Costa Rica's cats thus traveled a long way to get here. Jaguars, for example, are thought to have descended from primitive leopards, possibly in Africa. Jaguars are known from fossils in Europe, from where they could have spread into North America via the Bering land bridge and down to South America over the Panamanian land bridge. South American jaguars known from fossils dating back to the Pleistocene (the Ice Ages, some two million to 10,000 years ago) were larger and longer-legged than today's variety. Smaller jaguars may have prevailed because they were better suited to the forests that replaced open habitats at the end of the Ice Ages.

Teeth and Feeding

The dental formula for the Felidae is I 3/3, C 1/1, P 2/2 or 3/2 and M 1/1 = 28 to 30. Cats are the most exclusively meat-eating members of the order Carnivora and, correspondingly, have the most specialized dentition. They kill with a bone-splintering bite to their prey's head or neck, for which the canine teeth—the broadest and strongest of any carnivore—are essential. (A radio-collared jaguar in Belize that lost its canines in an accident died of starvation within a few weeks.) A gap between the canines and the cheek teeth allows the canines to sink deep into prey. Cats' carnassials (cheek teeth adapted for meat-slicing) are the largest and sharpest of any carnivore. Any meat the carnassials leave behind is scraped off bones by sharp papillae on the upper surface of felids' tongues, or by the incisors. The incisors are also used for grabbing hold of prey and removing feathers or fur. Meat can be swallowed in chunks and is easy to digest, so felids have no grinding teeth and a short digestive tract. The lack of grinding teeth gives felids relatively short jaws. Shorter jaws mean more biting power at the front of the mouth, for they give one of the main biting muscles, the temporalis, a greater mechanical advantage.

Body Design

Cats are digitigrade. They have four toes on their hindpaws and five on their forepaws. The first digit on the forefoot, the dewclaw, does not touch the ground unless the foot sinks, so forefoot tracks usually reveal only four digits. All cats except the cheetah have retractable claws that are generally withdrawn during walking, and thus kept sharp for prey-catching or tree-climbing. Thus, claw marks are almost never visible in felid tracks. This feature distinguishes cat tracks from the otherwise similar tracks of all dogs except the gray fox. This and other distinctions between cat and dog tracks are discussed on p. 205.

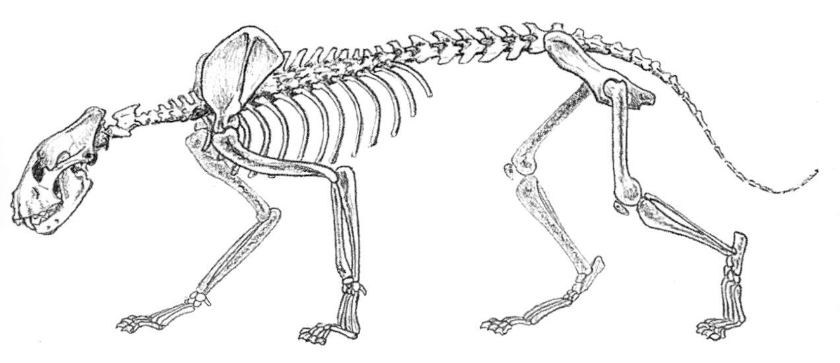

Jaguar skeleton.

If one were to line up Costa Rica's six cat species in order of size—or, perhaps more easily, if one were just to visualize such a line-up—one would find an almost perfect size gradient from the smallest, the oncilla, to the largest, the jaguar. This gradient is probably not mere coincidence. Competition generally prevents very similar species from coexisting. Since cats are otherwise fairly uniform in design, the size gradient may reflect a precise partitioning of food resources. In other words, different-sized cats prefer different-sized food, and can thus live together without competing to such a degree that any of them are unable to exist. This of course is a simplification, because there are numerous other factors that influence competition. The similar-sized margay and jaguarundi, for example, appear to differ principally in habitat use (see pp. 265 and 270). But the studies described in the species accounts do generally support the idea that a dietary, size-dependent gradient exists between Costa Rica's cats.

Melanism

The famous black panther of the Old World tropics is not a distinct species, as was once thought, but rather, a melanistic (dark) form of the leopard. Costa Rica has its own black *Panthera*, in the form of melanistic jaguars, although they are rare. Indeed, melanistic individuals have been recorded for all Costa Rica's spotted cats and for the puma. Studies on melanistic leopards and domestic cats revealed that, in those species, melanism is usually caused by a recessive gene. Recessive genes must be in both parents in order to influence offspring. Thus, if recessive genes and their traits aren't either beneficial, not detrimental, or linked to another, more important trait, they tend to disappear. Black panthers, which represent up to 50% of the leopard population in one part of Asia, are thought to have persisted because their dark coloration is well suited to dark tropical forests. It used to be presumed that melanism in the cats of tropical America was also caused by recessive genes, but, surprisingly, melanism in jaguars was recently found to be caused by a dominant gene. Why a feature controlled by a dominant gene should be so rare in the wild is a mystery.

Senses

Short muzzles and forward-facing eyes allow felids extensive stereoscopic vision (see p. 24). Hence they have excellent depth perception, which is important for pouncing on prey or moving about in trees. Felids also have excellent night vision; domestic cats need only 17% of the amount of light that humans do to make out objects in the dark. Like many other nocturnal mammals, cats possess a reflective layer behind the retina known as the tapetum lucidum, and retinas with more rods (the receptors that work in low light) and fewer cones than our own. Cats in captivity can learn to distinguish colors, but only after extensive training, so color is thought to be virtually meaningless to cats in the wild.

Nonetheless, daytime vision is important to felids, few of which are exclusively nocturnal. To protect their sensitive retinas from bright light, cats are able to narrow their pupils to the size of a pinhead. While our pupils are controlled by a circle of muscles that can contract only partially, the pupils of most felids are controlled by muscles that cross each other, forming an elliptical shape that can be pulled shut. In Costa Rica, jaguars and jaguarundis have round pupils, while the remaining cats have elliptical pupils.

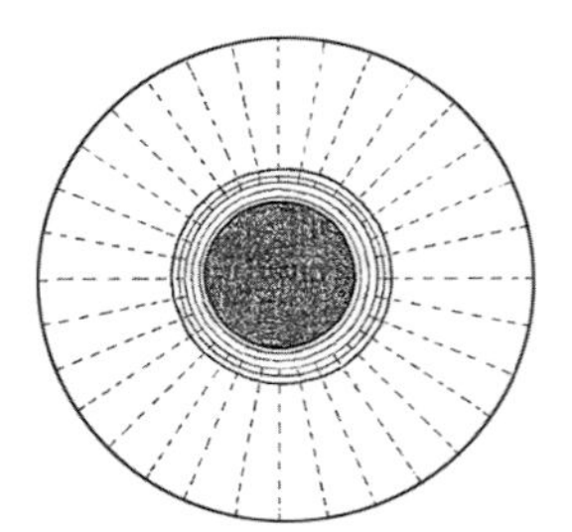

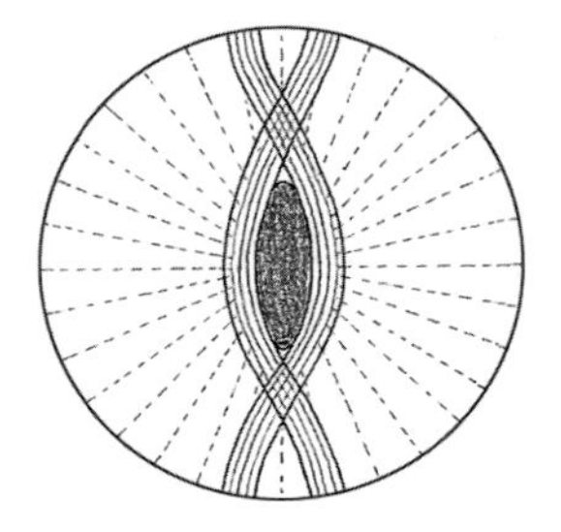

In most mammals, the muscles that control the aperture of the pupil are circular (left) and, like rubber bands, cannot contract beyond a certain point. By contrast, most cats have interlacing pupil muscles (right) that can close the pupil to the size of a pinhead and protect the cat's sensitive eyes from bright light.

Cats also have excellent hearing. They can hear sounds at frequencies three times higher than our own upper limit (although their ability to hear high-pitched sounds, like our own, declines with age). This skill helps them detect the ultrasonic sounds made by rodents, the staple prey of many species. Ultrasound doesn't travel very far, so cats' large, rotatable ears are useful as amplifiers.

Cats' sense of smell, on the other hand, is not as well-developed as in most other members of the Carnivora. It is seldom used to locate prey, but is important in communication. Cats release scents with urine and feces or rub and scratch scents onto tree trunks and into the ground from glands in their cheeks, rump, anus, and between the toes. Cats leave marks most frequently in areas where their ranges border or overlap those of their neighbors. Scent marks tell cats how often their neighbors are using certain areas and convey information about the neighbors themselves. Radio-collared jaguars in Belize, for example, were found to mark much more frequently when they were close to one another than when they were far apart. The same jaguars would move into areas of territorial overlap as soon as their neighbors were elsewhere, and if a cat died, its territory would be reoccupied within weeks. It has also been demonstrated that chemical changes in females' scents let males know when the females are ready to mate.

Conservation

Since cats are at the very top of the food pyramid, they are, even in pristine conditions, among the rarest of Costa Rica's mammals. They are thus particularly vulnerable to deforestation and hunting, and all six of Costa Rica's wild cat species are listed as endangered. The fur trade took a huge toll in the 1960s, 1970s, and early 1980s. Fur trading is minimal now, but cats are still hunted, for sport or because they are perceived as a threat to livestock and poultry. This threat is greatly exaggerated. Studies have shown that, where cats have access to sufficient healthy forest, they very seldom prey on livestock, even if it is available. Hunters increase the likelihood that cats will take livestock by killing the cats' natural prey, or by injuring the cats and thus hindering their ability to catch wild animals. Sadly, the appearance of one problem animal often leads to the persecution of an entire population. And by using dogs trained to track down cats and corner them in trees where they can be shot, hunters can locate and kill cats with shocking ease.

As cat populations decline and habitat is carved up into ever-smaller islands, the potential for inbreeding increases. Probably the best studied remnant cat population is one of some 30 to 50 pumas isolated in Florida that are the only remaining representatives of a puma subspecies known as the Florida panther. Millions of dollars spent, too late, to try to save them have revealed many signs of inbreeding—including a crook at the end of the tail that characterizes the subspecies—and computer predictions suggest they will be extinct by the year 2015. Costa Rica's cats may not be far from a similar point of no return; many respected biologists believe that the jaguar is there already. Cats may have a better chance of surviving in Costa Rica than in most other Central American countries, but if large, pristine areas continue to be cut down and current hunting practices are not stopped, the ultimate forms of tropical animal life will be among the first to disappear.

References

Belden 1988; Collier & O'Brien 1985; Ewer 1973; Fegus 1991; Hemmer 1976; Hughes 1977, 1985; Kiltie 1984; Kitchener 1991; Kurten 1973a,b; O'Brien et al. 1987; Peters & Hast 1994; Rabinowitz 1986a,b; Rabinowitz & Nottingham 1986; Radinsky 1975; Robinson 1970, 1976; Savage 1977; Turner & Antón 1997; Van Valkenburgh & Ruff 1987; Werdelin 1985.

SPECIES ACCOUNTS

O: Carnivora F: Felidae

ONCILLA

Leopardus tigrina
(plate 34)

Names: Also known as the little spotted cat or tiger cat. Genus listed as *Felis* until recently. Scientific name from the Greek words *leon*, meaning lion, and the Latin *pardus*, meaning leopard, and *tigris*, meaning tiger. Oncilla is derived from *onca*, a Native American name for the jaguar. **Spanish**: *Tigrillo*. (Costa Ricans seldom distinguish between this and the similar margay and use the names *tigrillo* and *caucel* interchangeably for either.)

Range: Costa Rica to northern Argentina (although it has not been recorded from Panama nor many parts of South America); from sea level to at least 3,200 m (10,500 ft); in forest or thick brush.

Size: 50 cm, 2 kg (20 in., 4 ½ lbs); tail 28 cm (11 in.)—55% of head and body length; males larger than females.

Similar species: The oncilla can be difficult to distinguish from the margay (pl. 34) in the field, and indeed, the two were originally considered one species. The oncilla is the size of a house cat or a little larger. The margay averages larger, although there can be size overlap in extreme cases. Margays with a head and body length of as short as 43 cm (17 in.) and oncillas as long as 59 cm (23 in.) have been recorded. The margay has a bushier and proportionately longer tail, larger eyes, and a more bulging muzzle around the whiskers. The margay usually has large, closed rosettes along its sides, while the oncilla usually has proportionately smaller and more numerous spots or open rosettes, but the pattern in both can vary. And finally, the hair on the nape of the margay grows forward, while oncilla nape hair is directed backward.

Natural history: Almost nothing is known of the natural history of the oncilla in the wild. It is probably mostly nocturnal. It is a good climber but is apparently less arboreal than the margay. The stomachs of two individuals from Costa Rica's Talamanca range contained rodents—naked-footed mice (*Peromyscus nudipes*) and a spiny pocket mouse (*Heteromys desmarestianus*)—a shrew (*Cryptotis* sp.), and a large-footed finch (*Pezopetes capitalis*). Gestation lasts about two and a half months, resulting in one or two young.

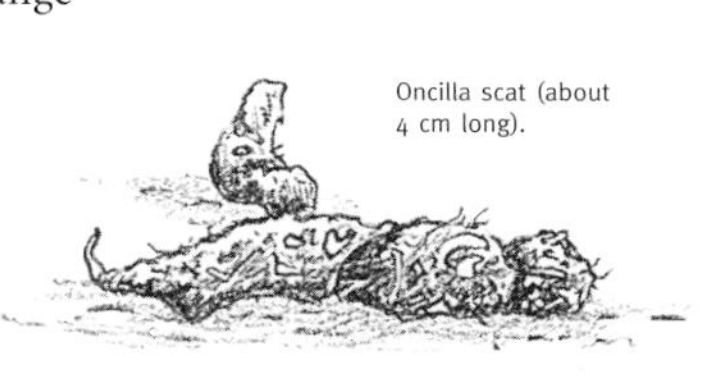

Oncilla scat (about 4 cm long).

An astonishing one fifth of oncilla specimens and pelts are reported to be melanistic (black or mostly black). Melanistic populations tend to be concentrated in certain areas, such as northern Venezuela and southeastern Brazil in the case of the oncilla. It is not known how common they are in Costa Rica, although one of the few specimens recorded here, from the Tapantí area, was melanistic. Note that the spotted coat pattern remains visible in melanistic oncillas, as it does in other melanistic spotted cats.

From afar, melanistic forms of spotted cats such as this oncilla appear jet black. At close range and from certain angles, however, their spots are still faintly visible.

Conservation: The oncilla is one of the least frequently observed mammals in Costa Rica. Habitat loss and hunting have undoubtedly contributed to its rarity. International traders, however, managed to locate some 84,500 pelts in 1983 alone. About 24 oncilla skins are needed to make a fur coat.

References

Bewick 1807; Broad 1987; Emmons & Feer 1997; Gardner 1971a; Leyhausen & Falkena 1966; McMahan 1986; Mondolfi 1986; Morúa 1986; Nowak 1991; Oliveira 1998b; Thornback & Jenkins 1982.

O: Carnivora F: Felidae

MARGAY

Leopardus wiedii
(plate 34)

Names: Genus listed as *Felis* until recently. See p. 262 for the origin of *Leopardus*. Species name after German naturalist Prince Maximillian zu Wied, whose mammal collection provided the type specimen—the specimen used to describe the species. The English name is thought to be derived from the Guaraní word *mbaracayá*, which means wild cat. **Spanish**: *Caucel*. (Costa Ricans seldom distinguish between this and the similar oncilla, and use the names *caucel* and *tigrillo* interchangeably for either.)

Range: Mexico to Argentina and Uruguay (now extinct in the southern United States); from sea level to at least 3,000 m (10,000 ft); mostly in pristine or old secondary forest.

Size: 55 cm, 3.5 kg (22 in., 8 lbs); tail 39 cm (15 in.)—70% of head and body length; males usually larger than females.

Similar species: Easy to confuse with the oncilla (see pl. 34, p. 262). The ocelot (pl. 34, p. 266) has a proportionately shorter tail and smaller eyes, and averages larger. Infrequently, ocelots with a head and body length of as little as 66 cm (26 in.) and margays as long as 72 cm (28 in.) have been recorded, but this overlap is rare. In one analysis of a couple of hundred specimens, only 2% of the margays were larger than the smallest ocelot. In ocelots, the rosette markings fuse together, forming broad, uneven bands along the side of the body. In margays, each rosette is usually distinct.

Natural history: The margay is mostly solitary and nocturnal. Wild individuals in Belize and captives are most active between about 1:00 and 5:00 AM. It is the most arboreal of the New World cats, and this is reflected in its anatomy. Its tail is proportionately longer than that of most other cats, and may therefore be more effective as a balancing tool or as a counterweight for springing off branches. Its claws are proportionately longer, which may improve the margay's ability to grip trunks and branches. When leaping between branches, a fully outstretched margay can catch itself with a single paw. The margay's leaping ability is phenomenal: it can jump vertically 2.5 m (8 ft) in a single bound. And its hind ankles are unusually flexible, thanks to a peculiar arrangement of its metatarsal bones. This feature, shared with squirrels and some raccoon relatives but unique in the cat world, allows the hindfeet to supinate (rotate backward) 180° so that the margay can grip tree limbs firmly, even when coming down tree trunks headfirst. (No doubt many a domestic cat owner would wish their pet had the same ability.)

The margay eats small vertebrates, arthropods, fruit, and, on occasion, leaves. In accordance with its arboreal habits, the margay takes more arboreal food than other cats. In a study of margays, jaguarundis, and ocelots occurring in the same area in Belize, for example, margays were found to eat more arboreal mammals, birds, and fruit than the other species. Vertebrate prey includes mice, squirrels, mouse opossums, rabbits, young agoutis and pacas, and various amphibians and reptiles. Margays take birds of all sizes, from small passerines to tinamous and resplendent quetzals; a margay was once observed pulling the latter out of its tree-hole nest in Monteverde.

Margay scat (about 5 cm long).

Two radio-collared males in Belize and Brazil had home ranges of 1,100 and 1,600 ha (2,700 and 4,000 A). Unlike ocelots, margays typically defecate in inconspicuous places and conceal their scats with debris. They rest in trees or vine tangles, often 7 to 10 m (23 to 33 ft) above the ground. Gestation lasts about two and a half months. Unlike oncillas and ocelots, which have two pairs of nipples and sometimes give birth to two or more young, margays have only one pair of nipples and usually give birth to a single kitten. Small litter size is another feature typical of arboreal mammals. Blue-eyed young start to leave their den at five weeks, and stop nursing at about two months. They are similar to adults in size and coloration by the age of 10 months, and reach sexual maturity at about two years. Captive margays can live 20 years.

Sounds: According to one expert, margays have eight distinct vocalizations: purrs, meows, barking meows, moans, hisses, spits, growls, and snarls.

Conservation: Margays were hunted heavily for their fur through the 1960s, 1970s, and early 1980s. About 29,785 pelts (15 of which are needed to make a fur coat) were traded internationally in 1980 alone. Fur trading is minimal now, but margays are still pursued by indiscriminate hunters and poultry farmers. Margays do not seem to do well in disturbed habitats, and may thus be particularly affected by deforestation.

References

Azevedo 1996; Emmons & Feer 1997; Kitchener 1991; Konecny 1989; McMahan 1986; Mellen 1993; Mondolfi 1986; Oliveira 1994, 1998b; Petersen 1979; Petersen & Petersen 1978; Prator et al. 1988; Tewes & Schmidly 1987; Wheelwright 1983.

O: Carnivora F: Felidae

OCELOT

Leopardus pardalis

(plate 34)

Names: Genus listed as *Felis* until recently. See p. 262 for the origin of *Leopardus*. *Pardalis* is the Latin for leopardlike. The English name comes from the Mexican Aztec word *tlalocelotl*, meaning field tiger. **Spanish**: *Manigordo*, in reference to the ocelot's broad front paws; *ocelote*.

Range: Texas (although it is now almost extinct in the United States, and exceedingly rare in Mexico) to northern Argentina and Uruguay; from sea level to 3,800 m (12,500 ft); in forest or thick brush.

Size: 75 cm, 12 kg (30 in., 26 lbs); tail 35 cm (14 in.)—45% of head and body length; males usually larger than females.

Similar species: See margay (pl. 34, p. 264).

Natural history: Ocelots are mostly nocturnal, although they sometimes begin activity in the late afternoon. Individuals radio-tracked in Peru peaked in activity just before and during the first hours of dark, while others in Venezuela were most active at the beginning and end of each night and still others followed in Belize were most active from 9:00 PM to midnight.

Although good climbers, ocelots are less arboreal than margays and feed mostly on ground-dwelling animals. Their primary hunting strategy is not to stalk prey, as many other cats do, but to roam around and pounce quickly on suitable prey as it is encountered. Sometimes they crouch on logs and swipe passing rodents. In Peru, ocelots take prey animals weighing less than 1 kg (2 ¼ lbs) in roughly the same proportion they occur in the wild. They take larger prey less frequently, although they have been known to kill animals as large as collared peccaries and brocket deer. In most places ocelots have been studied, including Corcovado, rodents, especially spiny rats (*Proechimys* spp.), are the staple. In Peru, three species of spiny rats comprised about a third of the diet, and other small rodents made up another third. In Belize, opossums (*Didelphis marsupialis* and *Philander opossum*) and armadillos (*Dasypus novemcinctus*) are favorites. Other food includes mouse opossums, agoutis, porcupines, rabbits, kinkajous, coatis, anteaters, sloths, monkeys, bats, small crocodilians, iguanas and smaller lizards, snakes, freshwater turtles, fish, land crabs, frogs, birds, arthropods, fruit, and grass. Ocelots carefully pluck feathers from birds before eating them. Like the large cats, ocelots will sometimes conceal a partially eaten kill and revisit it.

Ocelot scat (about 8 cm long). Ocelot scats measured by Federico Chinchilla in Corcovado ranged from 17 to 25 mm in diameter—consistently smaller than those of pumas and jaguars from the same area.

Mammalogist Louise Emmons and colleagues discovered that, in Peru at least, ocelots vary their hunting strategy according to the moon cycle. Small rodents—the staple prey of ocelots there—tend to be lunar phobic, which means they shy away from moonlight. Apparently (see p. 185), rodents stay fully active on moonlit nights, but simply avoid open areas, perhaps to escape the notice of owls, visually-oriented rodent specialists that are known to hunt most actively and most successfully on moonlit nights. The researchers found that ocelots, in turn, also avoid trails and other open areas on moonlit nights, presumably because their prey is scarce there, and/or because ocelots cannot approach prey from well-lit areas without being seen.

Radio-collared ocelots in various studies have tended to spend about 10 hours on the move and have covered between 3 and 7 km (2 and 4 ½ mi) each night. This is quite variable, however: ocelots have also been recorded walking for more than 12 hours without rest, or spending more than a day in the same spot. Recorded home ranges vary from 100 to 3,100 ha (250 to 7,700 A). Females and young males stop more often and cover less ground than adult males, apparently travelling mostly to find prey. Adult males have home ranges that are several times larger than those of females and that encompass or overlap those of several females. The ranges of adults of the same sex generally do not overlap.

Ocelots mark their ranges by spraying urine on vegetation and simultaneously scraping clear a roughly 10 by 15 cm (4 by 6 in.) patch of ground with their hindfeet. They also scratch logs and defecate in prominent places, frequently along the sandy borders of streams. Defecation spots may be used repeatedly, and by more than one individual. Ocelots rest in trees, between buttress roots, under tree falls, in caves, or in thick brush.

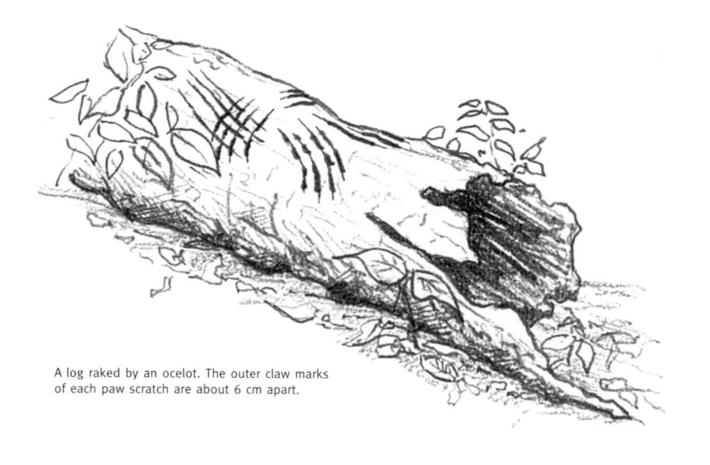

A log raked by an ocelot. The outer claw marks of each paw scratch are about 6 cm apart.

A gestation period of 70 to 80 days results in one, two, or rarely three or four kittens. Young start to leave the den and accompany the mother on hunts after about a month and begin eating solid food at about two months. Their eyes, which are bright blue at first, turn brown at about three months. The young are independent at about a year, but often stay in their mother's home range for another year or more. In Peru, females bare young every other year. Captives can live 20 years.

Sounds: Purrs; hisses when threatened.

Conservation: Ocelots can survive in disturbed habitats, but fragmentation of pristine forests has made them more vulnerable to hunting. They have beautiful pelts (13 of which are required to make a coat) and are relatively docile, and have thus been among the fur and pet trades' biggest victims. Having imported as many as 133,069 ocelot pelts in a single year, the United States banned their importation in 1972. Tens of thousands of pelts were still being traded on international markets well into the 1980s. Recently the demand for pelts has dropped, although they are still killed in Costa Rica by game hunters and farmers. Ocelots reproduce so slowly that, even in protected areas, they recover from such losses very slowly. According to one report, ocelots are unlikely to be able to sustain the annual removal of more than 2 to 3% of their population.

References

Bisbal 1986; Broad 1987; Chinchilla 1997; Clarke (J.) 1983; Dice 1945; Emmons 1982, 1987, 1988; Emmons et al. 1989; Kitchener 1991; Konecny 1989; Laack 1991; Ludlow & Sunquist 1987; McMahan 1986; Mondolfi 1986; Murray & Gardner 1997; Nowak 1991; Sunquist 1992; Sunquist et al. 1989; Taber et al. 1997; Tewes & Everett 1986; Timm et al. 1989.

O: Carnivora F: Felidae

JAGUARUNDI

Herpailurus yagouaroundi

(plate 35)

Names: Genus listed as *Felis* until recently. Genus from the Greek words *herpes*, meaning creeper, or *herpa*, meaning strange, and *ailouros*, meaning cat; *yagouaroundi* is derived from Guaraní natives' name for this cat. **Spanish**: *León breñero*; *yaguarundi*.

Range: Southwestern United States (as well as introduced, feral population in Florida) to northern Argentina; mostly in lowlands but up to at least 3,200 m (10,500 ft); more in scrubby areas with some forest, often near streams, than deep into large areas of wet forest.

Size: 65 cm, 5 kg (26 in., 11 lbs).

Similar species: The coloration of jaguarundis varies considerably, from black, to brownish gray, to sandy or reddish brown. This variation caused the jaguarundi to be listed as two distinct species until the early twentieth century, when different color phases were found together in the same litter. The fur of all color phases may be uniform in color or grizzled. Jaguarundis are weasel-like in form, and dark individuals (the most common) are easily confused with tayras (pl. 32, p. 245). They differ in having a long, thin tail, and lacking a blond patch on the chest.

Grayish individuals (found mainly in the dry northwest of Costa Rica) are distinguished from gray foxes (pl. 29, p. 208) by tail, larger size, small ears, and broader muzzle. The reddish form (also found mainly in the northwest) is distinguished from pumas (pl. 35, p. 272) by smaller size, usually reddish rather than usually tan color, lack of white around muzzle, and lack of dark tail tip.

Natural history: Despite being the most commonly seen wild cat in Central and South America, the jaguarundi has barely been studied in the wild. Unlike Costa Rica's other wild cats, the jaguarundi is active mostly by day. Three jaguarundis radio-collared in Belize started moving at around 4:00 AM each morning, becoming most active at around 11:00 AM. They decreased activity through the afternoon and returned to relative inactivity by sunset. The jaguarundi has round pupils, rather than the elliptical pupils of most other cats, and this may be a reflection of its more diurnal habits (see p. 260).

The jaguarundi forages mostly on the ground, but is an agile climber that will ascend trees when pursued or, occasionally, to get at food. It feeds mainly on small rodents (especially hispid cotton rats in Belize), rabbits, birds (especially ground-dwelling species such as tinamous, quail, and certain doves), whiptail lizards, iguanas, and arthropods. Other documented foods include opossums, small monkeys, fish, leaves, and fruit.

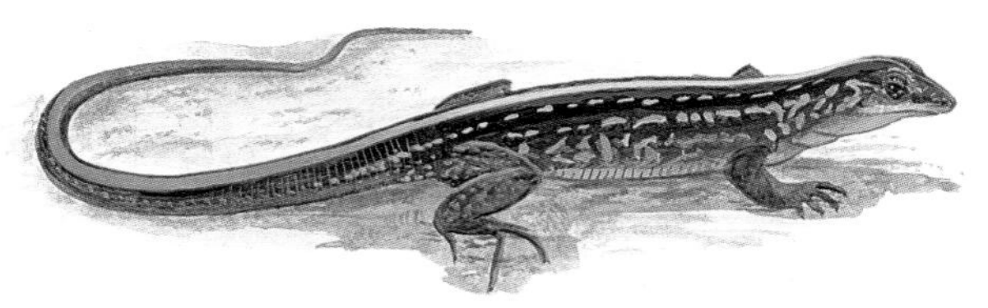

Central American whip-tailed lizard (*Ameiva festiva*).

Some biologists have wondered about the extent of competition between the jaguarundi and the similar sized margay. In other cats, differences in size—and perhaps especially in mouth gape—appear to cause a stratification of diets that helps the various species coexist (see, for example, the puma on p. 273). Since the jaguarundi and the margay are almost identical in size and gape, the jaguarundi's more diurnal activity and its preference for more open habitats and ground-dwelling prey may be critical in this regard.

The jaguarundi can cover great distances for its size. A male and a female in Brazil had home ranges of about 1,800 and 700 ha (4,400 and 1,700 A) respectively, while a female and two males followed in Belize wandered more than 6 km (4 miles) a day, the female covering a home range of 2,000 ha (5,000 A), and the males of 8,800 and 10,000 ha

(22,000 and 25,000 A). These home ranges are larger even than those used by many jaguars. Researchers are not sure why a cat that does not appear to have a particularly specialized diet would use such large areas. It may be that adults never settle permanently. Social interactions or competition with other carnivores could also be factors.

As in other cats, male jaguarundis mark their ranges by scraping a patch of ground with their hindfeet and urinating on or near the pile of accumulated earth, or by scratching fallen tree trunks. Jaguarundis den under tree falls, in dense thickets, or in hollow trees.

Jaguarundi scat (about 7 cm long).

A gestation period of two to two and a half months results in litters of one to four kittens. Kittens start leaving their den at about a month, and stop nursing at about two months. Females reach sexual maturity when they are about one and a half years old. Jaguarundis can live 15 years in captivity.

Sounds: Repeated short, ascending whistles, somewhat resembling the dusk call of the clay-colored robin; hisses and growls when threatened; loud screams by females while mating.

Conservation: The area necessary to support a viable population of a given species is difficult to estimate because it depends on so many variables. Cat expert Tadeu de Oliveira used diverse information to come up with a fascinating but sobering mean of 3,521 km^2 (352,100 ha or 880,000 A) as the area necessary for jaguarundis. Since the jaguarundi has never been hunted for its pelt, and since it can do well in disturbed habitats, it has more potential for maintaining a viable population than do Costa Rica's other native cats. Even so, according to one estimate, a mere 3 to 6% of the jaguarundi's range is currently protected. In the agricultural areas that form much of its present range, the jaguarundi is persecuted by indiscriminate game hunters and irate poultry farmers.

References

Bisbal 1986; Crawshaw 1995; Emmons et al. 1997; Hulley 1976; Kiltie 1984; Konecny 1989; Manzani & Monteiro-Filho 1989; McCarthy 1992; Mellen 1993; Mondolfi 1986; Nowell & Jackson 1996; Oliveira 1994, 1998a; Tewes & Schmidly 1987; Wemmer & Scow 1977.

O: Carnivora F: Felidae

PUMA

Puma concolor

(plate 35)

Names: Also known as the mountain lion, cougar, catamount, panther, painter, and purple feather, among many other names. *Puma* comes from the Peruvian Quecha language; *concolor* is the Latin for one-colored. **Spanish**: *León*; *puma*.

Range: The Yukon in Canada to the southern tip of the Americas in Chile and Argentina (the broadest latitudinal range of any wild New World mammal); from sea level to at least 3,300 m (11,000 ft) in Costa Rica (and to 5,300 m or 17,400 ft in the Andes); in almost any relatively undisturbed habitat.

Size: 110 cm, 50 kg (43 in., 110 lbs); smaller than in North America; males larger than females.

Similar species: See Jaguarundi (pl. 35, p. 269). Color varies from pale tan (especially in Guanacaste) to reddish. Melanistic individuals are very rare.

Natural History: The puma is active by day or night. It ambushes prey from the ground or from trees. To kill large prey, the puma stalks to within about 15 m (50 ft), and then rushes and jumps on the victim's back, breaking the victim's neck with a powerful bite to the back of the head. The puma is an excellent climber, and can leap to branches more than 5 m (16 ft) off the ground; in a study in Corcovado, four out of 11 puma scats contained monkey remains, from three different species. Other favorite foods in tropical habitats include spiny rats, agoutis, pacas, porcupines, opossums, rabbits, armadillos, anteaters, bats, peccaries, deer, iguanas, and snakes.

Puma attacks on humans in North America have received much press recently, although such attacks are exceedingly rare; one detailed study documented only nine fatal and 44 nonfatal attacks in the United States and Canada between 1890 and 1990. Biologist Charles Foerster recently published a dramatic account of how he was attacked by a puma while tracking tapirs alone one night in Corcovado. Yelling and shining a flashlight in the puma's eyes halted the animal's initial rush for about a minute. Since hurling radio-telemetry gear at it failed to halt a subsequent advance, in which the puma raised up on its hind legs and lashed out with its forepaws, Foerster had to defend himself by wheeling his arms to deflect blows downward to his thighs, hitting the puma with his flashlight, hiding behind trees, and continuing to yell. Despite tripping and falling on his back at one point, he was eventually able to scare off the unusually persistent puma. His (and others') advice to future victims is never to run away from a large, unfriendly cat, since this is likely to excite a feline still further, but rather, if necessary, to shout and be as assertive as possible.

Puma scat (about 16 cm long). Puma scats collected by Federico Chinchilla in Corcovado averaged slightly smaller than jaguar scats (measuring 27 to 29 mm in diameter as opposed to 28 to 34 mm in jaguar scats). Since both are quite variable and overlap in size, however, they can only be distinguished with certainty when tracks are visible alongside.

The puma seems at first glance to fill a similar feeding niche to that of the jaguar. When two apparently like animals occur together, however, there are generally subtle ecological differences between them. Otherwise, in theory, the better-adapted one would have replaced the other. In some regions, pumas seem to prefer higher, drier ground than do jaguars, or to avoid areas with high densities of jaguars. Several studies suggest that, where pumas and jaguars occur together, pumas tend to take less large prey. Similar partitioning has been observed in leopards and their larger cousins, tigers, in Nepal. In North America, where jaguars are absent and pumas are larger, big animals such as deer and elk are staples in pumas' diets.

Like jaguars, pumas frequently drag large prey to concealed eating sites. Prey may be partially eaten, concealed with leaves and other debris, and marked with strong-smelling urine. A puma may stick around to guard its cache, or revisit it on subsequent days. Pumas tend to feed first on the rump of large prey, while jaguars often start near the head.

Pumas must travel great distances to secure food. Stomach analyses in North America show that pumas there don't eat every day and suggest they eat on average only six days out of nine. Pumas can cover 65 km (40 mi) in a night, and their home ranges have been estimated at up to 7,500 ha (18,500 A) in Belize and an extraordinary 180,000 ha (440,000 A) in Texas.

Pumas mark their ranges by spraying urine, defecating, and scraping roughly 10 by 25 cm (4 by 10 in.) patches of ground with their hindfeet. They also scratch trees, perhaps to mark territories, or perhaps only to remove loose claw sheaths.

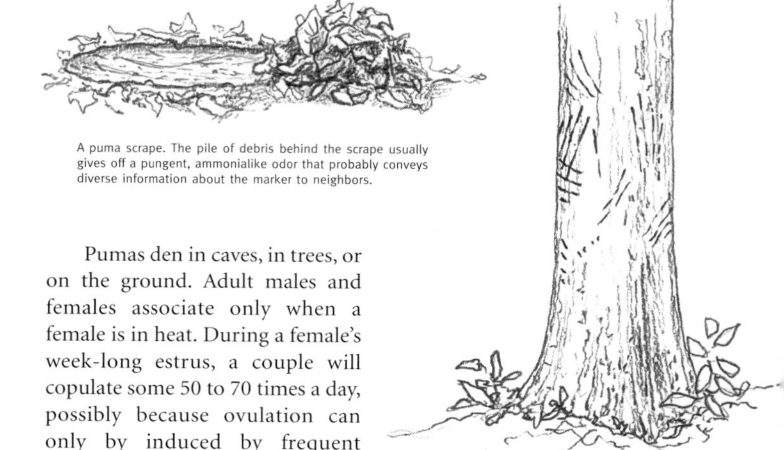

A puma scrape. The pile of debris behind the scrape usually gives off a pungent, ammonialike odor that probably conveys diverse information about the marker to neighbors.

A puma scratching post. Like ocelots, pumas also scratch horizontal logs, but the outer claw marks made by each puma paw are about 9 cm apart.

Pumas den in caves, in trees, or on the ground. Adult males and females associate only when a female is in heat. During a female's week-long estrus, a couple will copulate some 50 to 70 times a day, possibly because ovulation can only by induced by frequent mating (see p. 201). Gestation lasts about three months. One to six young (but most often three or four in North America, and perhaps less in the tropics) nurse for three or four months, by which time they are about half adult size. Their coat is spotted (perhaps a genetic holdover from spotted ancestors) and their eyes blue for at least three months. They stay with the mother until they are one and a half to two years old. Siblings often travel together for a few months after leaving their mother. Pumas reach sexual maturity at about two and a half years, although males may not mate until they are older, once they have established a territory. Pumas can live 20 years in captivity. A recent analysis of over 1,000 puma skulls, however, revealed that few pumas live longer than nine years in the wild.

Sounds: Hisses when threatening at close range; prolonged "yowling" by females in heat; short, very high-pitched whistles and repeated "chucking" noises by cubs.

Conservation: Pumas were never hunted for their pelts as heavily as the spotted cats, but they are still affected by habitat destruction, and they are persecuted by game hunters and farmers. In Costa Rica, they are rare outside protected areas.

Even within protected areas, the long-term viability of puma populations is questionable. To form an estimate of how much land pumas need to maintain viable populations, researcher Paul Beier pooled information from dozens of different studies on factors such as birth rates, litter size, longevity, and the carrying capacity of different habitats. He calculated that 100,000 to 220,000 ha (250,000 to 550,000 A), depending on the variables just mentioned, would be the minimum area needed to ensure the survival of an isolated population of pumas for 100 years.

Beier also calculated that, if a population were connected to another by just a single wildlife corridor, the minimum area estimate dropped significantly—to between 60,000 and 160,000 ha (150,000 and 400,000 A). In a country as small as Costa Rica, no single park could possibly meet these requirements. But if parks were connected by wildlife corridors—both within Costa Rica and beyond—the long-term outlook for pumas in this region would be much more optimistic.

References

Beier 1991, 1993; Chinchilla 1997; Currier 1983; Eaton 1976b; Emmons 1987; Emmons et al. 1997; Foerster 1996; Gay & Best 1996; Hornocker 1970; Iriarte et al. 1990; Kiltie 1984; Lindzey 1987; Nowak 1991; Robinette et al. 1959; Schaller & Crawshaw 1980; Seidensticker et al. 1973; Seidensticker & Lumpkin 1992; Sunquist 1981; Sunquist & Sunquist 1989; Taber et al. 1997.

O: Carnivora F: Felidae

JAGUAR

Panthera onca

(plate 35)

Names: English name from the native American word *yaguar*, which means pouncing killer; genus from the Greek word *panthera*, meaning panther; *onca* is a name for the jaguar used by native South Americans. The genus *Panthera* contains all the world's big cats—the jaguar, the leopard, the snow leopard, the tiger, and the lion. **Spanish**: *Tigre*; *jaguar*.

Range: Mexico to northern Argentina (now extinct in the southwestern United States, most of Argentina, and Uruguay); from sea level to about 3,500 m (11,500 ft) on Cerro de la Muerte; in almost any natural habitat, but mostly in large, protected expanses of lowland wet forest, often near water.

Size: 150 cm, 80 kg (60 in., 175 lbs); males larger than females.

Natural history: Costa Rica's largest carnivore is active by day or night. Although it is a good climber and swimmer, it hunts mainly on the ground. It catches prey by stalking and ambushing at close range. It appears to be an opportunistic feeder, taking whatever prey (larger than about 1 kg or 2 ¼ lbs) is most available in a given habitat. It favors capybaras in a swampy area of Brazil, for example, armadillos in the Cockscombe Basin of Belize, and collared peccaries—especially immature ones—in Calakmul, Mexico. In Costa Rica, sloths and iguanas are staples at La Selva, while white-lipped peccaries, olive ridley and Pacific green sea turtles, red brocket deer, sloths, and iguanas are favored in Corcovado. Other known prey includes tapirs, pacas, agoutis, opossums, anteaters, monkeys, procyonids, skunks, ocelots, pumas, fish, large birds, boas and other snakes, small crocodilians, fresh water turtles, and turtle eggs. The jaguar also eats grass.

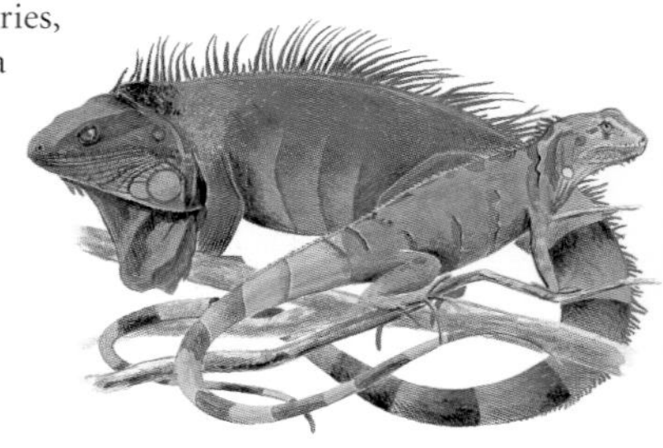

Green iguana (*Iguana iguana*); adult and juvenile.

An example of the opportunistic feeding behavior of jaguars was documented recently by Eduardo Carrillo in Corcovado. He found that jaguars there adjust their hunting according to the activity cycles of nesting sea turtles. During the last quarter and new moon phases, when most turtles come to nest, the jaguars would focus their hunting effort on patrolling beaches at night. During the rest of each month, they would spend more time back in the forest and would hunt more by day, when their main forest prey, white-lipped peccaries, were active. This cycle also affected how much the jaguars moved around, for they covered much larger distances when hunting in the forest. Carrillo made the extraordinary discovery that this hunting cycle was in turn influenced by the El Niño sea-temperature phenomenon. Apparently as a result of unusual conditions created by El Niño, during 1997 and part of 1998 relatively few sea turtles came to nest, and the jaguars turned their attention more permanently to the forest.

Olive ridley sea turtle (*Lepidochelys olivacea*).

Jaguars do not kill large mammals by biting through their throats or asphyxiating them as do several Old World cats, but rather by pouncing on their backs or sides and twisting their necks such that they fall to the ground. The victims' necks often break in the fall. Jaguars have large heads and tremendous jaw strength. It has been calculated that, while jaguars are only 1.2 times larger than pumas in body length, their relative bite force is 1.6 times greater. This enables them to kill even large prey with a splintering bite through the neck or skull. It also enables them to penetrate the armor of turtles and crocodilians, which pumas (and most other predators) ignore. In South America, jaguars bite through the carapace of land turtles weighing 8 kg (18 lbs) or more and then scoop out the meat. They treat smaller turtles like hamburgers, chewing them up whole, carapace and all. When feeding on really large turtles, such as sea turtles, jaguars simply bite off the head and neck, and insert their forepaws through the neck hole to scoop out the flesh from inside the carapace chamber.

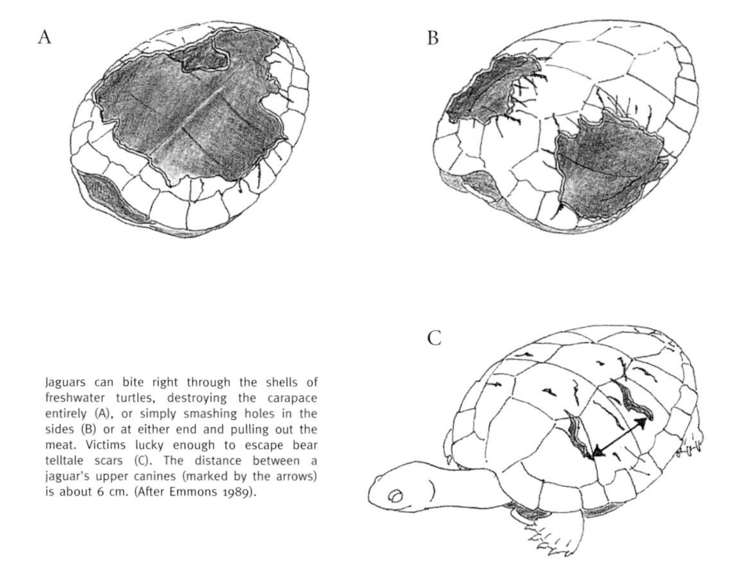

Jaguars can bite right through the shells of freshwater turtles, destroying the carapace entirely (A), or simply smashing holes in the sides (B) or at either end and pulling out the meat. Victims lucky enough to escape bear telltale scars (C). The distance between a jaguar's upper canines (marked by the arrows) is about 6 cm. (After Emmons 1989).

Large victims are frequently dragged a considerable distance to a concealed eating site, where they may be revisited for several days. A jaguar followed in Brazil dragged a partially eaten collared peccary 1 km (⅔ mi), while jaguars in Costa Rica will haul olive ridley and green sea turtles—the latter weighing up to 200 kg (440 lbs)—well into the forest behind beaches at Tortuguero, Corcovado, and Nancite. Feeding sites may be reused; the remains of 14 turtles were found at one site 150 m (165 yds) inland from Nancite beach. Unlike pumas, jaguars seldom conceal their kills with debris.

Jaguars eat sea turtles by biting off the head and scooping out flesh through the neck hole. Victims are often dragged to secluded spots behind the beach and revisited on following nights. Some spots, such as this one, with a fresh carcass in the foreground and old kills in the background, are used repeatedly.

Jaguars eat more than 1 kg (2 ¼ lbs) of meat daily in captivity, and typically roam several kilometers each day to find sufficient food and patrol home ranges in the wild. Studies in Brazil and Belize have found females to have home ranges of between 1,000 and 3,800 ha (2,500 and 9,400 A). Males have larger ranges that overlap those of several females. Home-range size estimates for some jaguars reach as high as 39,000 ha (96,000 A). Jaguars mark their ranges by spraying urine, defecating, and scraping roughly 10 by 37 cm (4 by 15 in.) patches of ground with their hindfeet. They also scratch trees, perhaps to mark territories, or perhaps only to remove loose claw sheaths.

Jaguars rest in trees, on the ground, or in caves. Adult males and females associate only when a female is in heat. During this period—which lasts about a week—jaguars will copulate some 100 times a day, probably because ovulation can only by induced by frequent mating (see p. 201). Gestation lasts three to three and a half months. Jaguars can give birth to up to four cubs at a time. In captivity, they usually produce two cubs per litter. There are few records of litters in the wild, but three litters observed in Corcovado were all of single cubs. Records from zoos indicate that cubs can walk within three weeks, nurse for five to six months, are independent after one to two years, and reach sexual maturity at two to four years. Young may stay near their mother until they are more than two years old. In a rare record from the wild in Corcovado, a cub left its mother when it was about 19 to 20 months old. The mother mated shortly thereafter and raised another single cub. At that pace, she would have been producing a cub about every 22 months. With such a low reproductive turnover, jaguars would be slow to recover from losses due to hunting. Captives can live 22 years.

Jaguar scat (about 18 cm long).

Sounds: Grunts; prolonged, deep, rumbles; loud "sawing" calls, each a series of repeated, hoarse, almost cowlike grunts that can last a minute or more. Hunters used to imitate jaguar calls by yanking a waxed cord attached through a drum-like instrument made of a gourd and deer skin or other similar materials. Call imitations are reputed to have lured hundreds of jaguars to their deaths.

Mythology: According to legend of the Bribri tribe of southern Costa Rica, life on earth was created when Sibú, the Bribri God-Creator, planted seeds in the blood of a baby jaguar. The blood was deposited on the earth's surface by a vampire bat (see p. 116).

Conservation: The future of the jaguar in Costa Rica and throughout its range is uncertain. According to very optimistic estimates, the jaguar today occupies less than one-third of its original range in Central America, and less than two-thirds of its original range in South America. The fur trade took its toll in decades past. An estimated 15,000 jaguars were being killed annually in the Brazilian Amazon through the 1960s, for example, and the United States imported 13,516 pelts in 1968 alone. Today, habitat loss and persecution by hunters and farmers are the greatest threats.

Jaguars sometimes kill livestock, but studies have shown that most jaguars leave livestock alone, even when ranches lie within their home ranges. The few individuals that do attack livestock tend to be old or injured, often by gunshot wounds, and presumably therefore have difficulty hunting wild prey. Seven out of eight cattle killers examined by jaguar expert Alan Rabinowitz in Belize, for example, had old shot gun wounds to the skull. Unfortunately, the acts of one problem animal often provoke the persecution of an entire population. There are very few records of jaguars attacks on people, and most of these are speculative.

The only place in Costa Rica where jaguars or their tracks are seen frequently is Corcovado. According to an estimate that came out of an international symposium of cat experts, an isolated jaguar population needs about 4,000 km^2 (400,000 ha or a million acres) to be sure of producing at least some successful offspring each year. There is no longer any stretch of contiguous jaguar habitat this large in Central America, and unless a major, multinational effort is made to link protected areas, the jaguar's days in this region are probably numbered.

References

Aguilar 1986; Aranda 1993; Bozzoli 1986; Braker & Greene 1994; Carr 1953; Carrillo 2000; Carrillo et al. 1994; Chinchilla 1997; Crawshaw & Quigley 1991; Eaton 1976b; Emmons 1987, 1989; Emmons et al. 1997; Greene & Losos 1988; Koford 1976, 1983a; Mondolfi & Hoogesteijn 1986; Nowak 1991; Quigley 1988; Rabinowitz 1986a,b; Rabinowitz & Nottingham 1986; Rodríguez & Chinchilla 1996; Schaller 1980; Schaller & Crawshaw 1980; Schaller & Vasconcelos 1978; Seymour 1989; Swank & Teer 1989; Taber et al. 1997; Thornback & Jenkins 1982; Vaughan 1983a; Watt 1987.

Manatees (order Sirenia)

Distribution and Classification

The Trichechidae contains only three species, one of which is found in Costa Rica. The West African manatee (*Trichechus senegalensis*) is found in coastal waters and rivers from Senegal to Angola and has been found 2,000 km up the Niger river. The Amazonian manatee (*T. inunguis*) is found along the Amazon river and its tributaries as far west as Ecuador and northern Peru, and in the Rupununi and Essiquibo rivers of Guyana. Costa Rica's species, the West Indian Manatee (*T. manatus*), is split into two subspecies. The subspecies known as the Florida manatee (*T. m. latirostris*) resides year-round in the coastal waters and rivers of Florida, Georgia, and the Bahamas, migrating north as far as Virginia and west as far as Louisiana when the water is warm enough in the summer. The subspecies known as the Antillean manatee (*T. m. manatus*) inhabits the Atlantic coast and rivers of Central and South America and some Caribbean islands, from Mexico (although populations north of Veracruz have been practically wiped out) to the mouth of the Amazon in Brazil.

Together with a similar-looking creature called the dugong, which is found in the Indian and western Pacific Oceans, the manatees comprise the only living members of the order Sirenia. The sirenians were so named because Christopher Columbus supposedly thought they were the mermaids, or sirens, famous in mythology for luring sailors onto rocks with their beauty. Sirenians are completely unrelated to other, apparently similar, marine mammals such as seals or walruses. Rather, they belong to a superorder known as the subungulates, which includes hyraxes (which look like rodents), the aardvark (which looks like an anteater), and elephants.

Evolution

Nails on the flippers and vestigial pelvic bones hint at manatees' terrestrial ancestors, which probably originated in the Old World at the beginning of the Eocene period, about 55 million years ago. It has been hypothesized that five or 10 million years later they dispersed to South America, where one of them slowly evolved into something resembling a manatee in order to feed on sea grasses that were then abundant in the western Atlantic ocean. By the Pliocene period, around five million years ago, that primitive manatee had split into the three groups that became the ancestors of today's species. One group had floated up the Amazon, another back to Africa, and the ancestors of the manatee now in Costa Rica, into the Caribbean. In the Caribbean, manatees quickly replaced a previously abundant dugong relative, possibly because their dental arrangement was better suited for coping with abrasive food plants and large quantities of silt that were in the water during this period. After the Pliocene, sirenians became less abundant and diverse, at first due to climate change and, recently and suddenly, due to human interference.

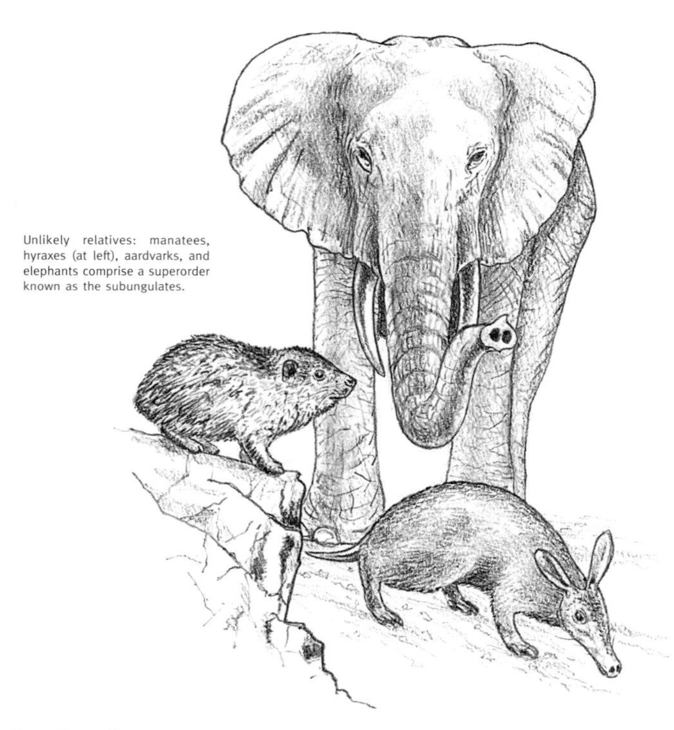

Unlikely relatives: manatees, hyraxes (at left), aardvarks, and elephants comprise a superorder known as the subungulates.

Teeth and Feeding

The only marine mammals that feed almost exclusively on plants, manatees have evolved in numerous ways to deal with the challenges of this diet. Since vegetation provides little protein, manatees must eat vast quantities of it to survive, and the plant fibers quickly wear down their teeth. Freshwater grasses are particularly abrasive, for they contain silica spicules, a defensive feature that wears down herbivores' teeth. Manatees chew with bony plates in the front of the mouth and a set of up to 40 molars that are replaced constantly. As in elephants, teeth are replaced horizontally—they emerge at the back of the mouth and are slowly pushed forward, becoming worn down and falling out as they reach the front. Manatees lose their four incisors (which are the tusks in elephants and the dugong) when still young.

Vegetation is not only hard to chew; it is also hard to digest. No mammalian stomach enzyme can break down cellulose, so manatees, like most other herbivorous mammals, digest their food through a process of bacterial fermentation. Manatees are not ruminants like, say, deer or cows, and they do not have a chambered stomach. Rather, they ferment food in the rear part of their 45 meter (150 foot) long digestive tract, mostly in the large intestine. As much as 90 kg (200 lbs) of vegetation is sent down this tract each day, and it takes about a week to be digested.

As a result of their low-energy diet, manatees have an unusually low metabolic rate—about a third of that of a typical mammal the same size. They save energy by moving slowly, and by living in warm tropical and subtropical waters, where they don't have to work too hard to stay warm.

Body Design

Manatees have two five-fingered flippers that they use for paddling, walking, or even trotting along river bottoms; helping stuff food into the mouth; scratching; and hugging other manatees. Their uniquely rounded tail is used primarily for propulsion. They cruise at between 4 and 10 km/h (2 and 6 mph), but can reach 25 km/h (16 mph) for distances of up to 100 meters (110 yards). This may help them evade sharks, crocodilians, or jaguars, although no predator other than people has ever actually been recorded. Another, less obvious use for the tail has been observed in male-weary females. They ground their front ends in shallows and slap over-enthusiastic suitors over the head as they approach.

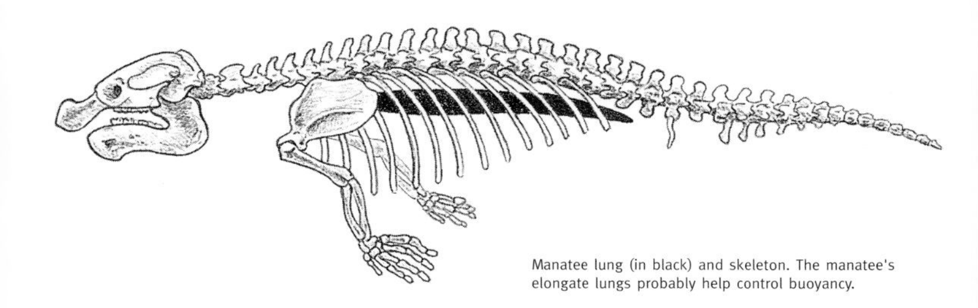

Manatee lung (in black) and skeleton. The manatee's elongate lungs probably help control buoyancy.

Manatees have a number of internal anatomical features that are unusual. While almost all mammals have seven neck vertebrae, manatees (along with two-toed sloths) have only six. Manatees also have uncommonly thick, heavy bones, most of which lack marrow cavities. A heavy skeleton may serve as ballast, helping compensate for flatulence caused by the fermentation of large quantities of vegetation and enabling manatees to stay submerged without having to expend too much energy paddling. Highly unusual lungs, which are housed separately in almost meter-long cavities known as hemidiaphragms that run along each side of the body, probably aid buoyancy control as well. Manatees might be able to control their depth in the water simply by contracting or expanding the cavities.

Finally, manatees differ from other marine mammals (such as dolphins, whales, and seals) in having proportionately small brains. A larger brain might be unnecessary because manatees eat a food that does not require complicated hunting techniques, do not have a particularly complicated vocal repertoire to emit or interpret, and do not maintain complex social relationships. A large brain could also be too great an energetic burden on a mammal with such low metabolic rates.

Senses

Manatees have good hearing, an important sense in their murky, aquatic environment, where the potential for visual or olfactory communication is poor. Anatomical evidence suggests that manatees absorb sound through their cheeks, which are adjacent to their large ear bones, rather than through the tiny ear openings just behind their eyes. Like their relatives, the elephants, manatees may use infrasound, which has a frequency too low to be audible to us. Low-frequency sounds travel farther than high-frequency sounds, and may help estrous females attract males, and mothers stay close to their calves. A mother can hear her calf call from at least 60 meters (200 feet) away.

Manatees have small eyes. Although they have reasonable binocular vision, or depth perception, they seem far-sighted, sometimes colliding with objects that are very close. The eyes have rods, which may provide reasonable night vision, and cones, which may provide color vision. A nictitating membrane (a third, transparent eyelid) protects the eyes, as in many aquatic creatures.

Since manatees have whiskers around the mouth and tend to kiss and nuzzle, they may be able to recognize chemical odors on one another, at least when they raise themselves out of the water. Females in heat may attract males by rubbing scents onto exposed rocks and other objects.

Conservation

In Rudyard Kipling's famous story *The White Seal*, a wise, old sea cow leads a seal to a secret lagoon where they and their friends will be safe forever from their human enemies. Unfortunately, the real sea cow, a 4 ton (9,000 lb) dugong relative called *Hydrodamalis gigas*, did not have such luck. Shipwrecked Russian sailors under the command of Vitus Bering discovered it in the Bering Sea in 1741 and survived off its meat for several months while they used its hide and bones to construct a new boat. When the sailors got home, word of the valuable and easy-to-kill creature spread quickly. It was extinct by about 1768.

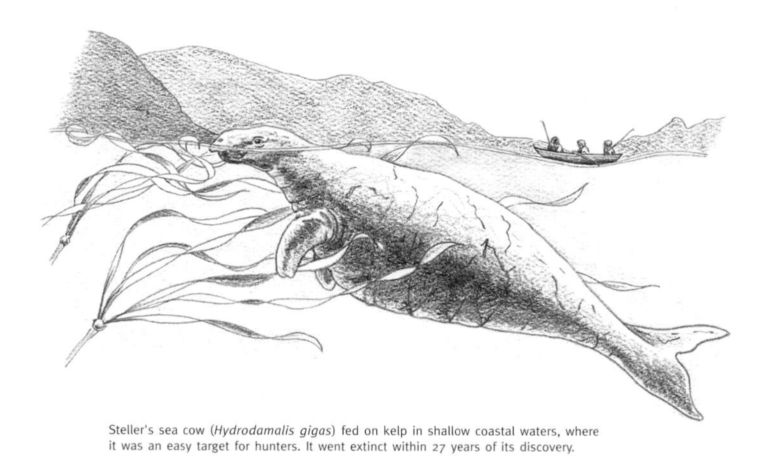

Steller's sea cow (*Hydrodamalis gigas*) fed on kelp in shallow coastal waters, where it was an easy target for hunters. It went extinct within 27 years of its discovery.

Like *Hydrodamalis*, manatees are peaceful and do not compete with people for any food resource. Some may even benefit us by controlling weed and mosquito populations. But since people also favor the coastline and rivers they inhabit, manatees will come into ever-increasing contact with the hunters, motorboats, pollution, and other alterations that have already killed most of their kind. And since the coastal river systems they inhabit often lie directly downstream from major agricultural areas or cities, even manatees in protected areas can be vulnerable to pollution and other environmental problems originating far away, as is the case with Costa Rica's manatee population (see p. 290). A slow reproductive turnover makes manatees all the more vulnerable to extinction. The few thousand manatees that are left in the world are protected under CITES Appendix 1, but they may need more than that to escape the fate of their Arctic cousin.

References

Caldwell & Caldwell 1985; De Jong et al. 1981; Domning 1977, 1981, 1982a, 1982b, 1989; Fischer & Tassey 1993; Hartman 1979; Husar 1978; Irvine 1983; Jiménez 2000; Lefebvre et al. 1989; MacLaren 1967; O'Shea & Salisbury 1991; Rathbun & O'Shea 1984; Reynolds 1979; Reynolds & Odell 1991; Scheffer 1973; Schevill & Watkins 1965; Stejneger 1887.

SPECIES ACCOUNT

O: Sirenia F: Trichechidae

WEST INDIAN MANATEE

Trichechus manatus
(plate 38)

Names: Also known as the Antillean manatee or Caribbean manatee. Genus from the Greek *echo*, meaning to have, and *trichion*, meaning hair, because the body is covered with hairs. Manatee comes from a Honduran Garifuna word meaning woman's breast, because manatees' nipples are located up near the forelimbs, as in humans. **Spanish**: *Manatí*.

Range: Saltwater (usually close to river mouths) or, especially, freshwater that is slow-moving and shallow, but not less than a meter deep; along the Atlantic coast of the Americas and around Caribbean islands, from Virginia (summer only) in the United States down to the Amazon river in Brazil. Although found mostly near the coast, manatees also show up far inland. They inhabit Lake Izabal in the interior of Guatemala, are found hundreds of kilometers up the Orinoco river in Venezuela, and used be seen in Lake Nicaragua, all the way up the San Juan River that marks part of Costa Rica's northern border. Manatees were introduced to the Panama Canal around 1960 to control weeds, and a few have crossed the canal to the Pacific.

Size: 300 cm, 500 kg (120 in., 1,100 lbs).

Natural history: North and Central America's only herbivorous, aquatic mammal has been studied extensively in Florida, but very little over the rest of its range. Basic details of the status and natural history of manatees in Costa Rica were revealed for the first time only in the late 1990s, thanks to a survey conducted by biologist Ignacio Jiménez.

Manatees are very hard to see. They have a low metabolic rate, which allows them to stay submerged for up to 20 minutes when resting. When active, they surface for air about every five minutes, but expose only their nostrils. Since they can renew about 90% of the air in their lungs in a single breath (a resting human renews only about 10% per breath), their visits to the surface can be brief. Thus, the only sign of a manatee might be a patch of clipped aquatic grass, or a stream of bubbles rising to the surface; manatees are full of hundreds of pounds of fermenting vegetation and are famous for their flatulence.

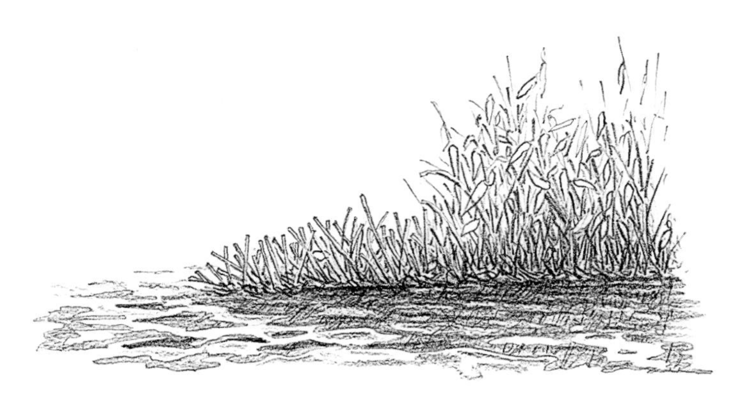

Manatee sign: clipped grass at the water's edge.

Manatees in Costa Rica are even harder to see than those in Florida, for the water here is murkier, and the manatees here are both rarer and mostly active at twilight and at night. Florida manatees, by contrast, alternate short periods of activity and rest throughout the day and night. Costa Rica's manatees may have become more nocturnal because they have been under greater pressure from hunters in recent decades than have Florida's manatees.

The activity patterns of manatees in Costa Rica are also influenced by tide and rainfall. Manatees appear to move around most when the tide is rising, perhaps because the incoming sea partly neutralizes the outgoing river current, enabling manatees to swim more easily. Manatees take advantage of heavy rains and the consequent swelling of rivers to move upstream and get at food in areas that would normally be too shallow to penetrate.

Manatees are essentially solitary, but herds can form at feeding areas or when a female is in heat. When together they can be playful. They nuzzle and kiss one another, and in Florida they have been seen playing follow the leader and body-surfing underwater currents in unison.

Manatees feed on at least 60 species of plants. Jiménez found that the favorite foods of manatees in Costa Rica were aquatic grasses such as a rice relative (*Oryza latifolia*), guinea grass (*Panicum maximum*), cola de zorro (*Hymenochne amplexicaule*), and, especially, the thinner and more tender *Brachiaria* grass; other aquatic plants such as water hyacinths (*Eichhornia crassipes*), hydrilla (*Hydrilla verticillata*), and an evening primrose relative (*Ludwigia* sp.); and forest vegetation overhanging bodies of water, including a number of vines, and a kapok relative, the provision tree (*Pachira aquatica*). All of these plants are common along the waterways of northeastern Costa Rica. In deforested areas, manatees in Costa Rica also pull themselves partially out of the water to feed on cattle-pasture grasses.

Plants fed upon by manatees in Costa Rica: *Eichhornia*, *Ludwigia*, and *Pachira* (from left to right).

Manatees do us a service by eating some of these plants. Water hyacinths and hydrillas, as well as another manatee food plant, the water lettuce (*Pistia stratoites*), often choke up waterways, making them unusable for many creatures, including humans. By opening up waterways, manatees also remove potential breeding sites for malaria-carrying *Anopheles* mosquitoes.

Although manatees are largely herbivorous, captives will eat dead fish, and wild manatees in parts of Jamaica appear to feed on fish tangled in nets. In both cases, they strip the flesh and leave behind the fish skeletons. It is not known whether manatess would ever be able to catch fish under more natural circumstances.

Manatees can breed at any time of year, although they do so mostly or only in certain seasons in some areas. In Florida, most (but not all) births occur in spring or summer, while in Tabasco, Mexico, all calves are born during the wet season, when water levels are at their highest and food is most abundant. Whether manatee reproduction is seasonal in Costa Rica is not known. Manatees are not monogamous; a female in heat may be courted simultaneously by a half dozen or more males. Gestation lasts 12 to 13 months. Manatees give birth to single calves or, rarely, twins, each weighing about 30 kg (66 lbs) and measuring about 120 cm (47 in.). Calves feed independently after a year, but stay with the mother for up to two years. Females are sexually mature by the time they are three or four years old. Sexually active females produce only one calf every two to five years. Manatees are long-lived, often surviving 30 or 40 years, and occasionally as long as 60 years.

Sounds: Chirp-squeaks, squeals, and screams, generally of short duration (less than half a second), some of which have been likened to the shrieks made by seagulls. Manatees seem to have a relatively limited vocabulary and are not thought to use sounds in echolocation (as do some dolphins).

Mythology: The manatee is prominent in Native American folklore throughout its range. Tribes along the Orinoco River in the Amazon, for example, believe that people who drown are carried to a submerged city, where they are reborn as manatees or river dolphins. According to a legend of the Bribri of southern Costa Rica, the manatee came into being when a tapir hunter was punished for being abusive toward his wife. (Tapirs themselves are reincarnations of a mistreated woman in Bribri legend—see p. 300.) Sibú, the Bribri God-Creator, caused the hunter to fall into a stream, where the hunter was condemned to spend the rest of his life in the form of a manatee.

Conservation: If the Bribri legend is true, the hunter has received ample punishment. The manatee's meat, oil, leather, and bones have been sought after for centuries. Meat from different parts of the manatee's body is said to vary in color and flavor—resembling veal, pork, chicken, or fish—apparently due to varying concentrations of hemoglobin in different muscles. Manatee oil is used for cooking and as lamp fuel. The manatee's strong, solid bones, which can be carved like ivory, were used traditionally to make various weapons and fish hooks. Manatee leather is exceedingly resilient, too. In the 1930s, 40s, and early 50s, before the advent of synthetic leather, about 4,000 to 7,000 manatee hides were being traded each year on world markets.

Various parts of the manatee's body are said to have medicinal properties that help cure ailments such as diarrhea, arthritis, impotence, menstrual pains, severe coughing, and eye problems. The most bizarre medicinal practice uses the manatee's earbone, which children wear around their necks or wrists as a sort of good luck charm, to ward off disease. (Curiously, the earbone is used in the same way by Costa Ricans and by various tribes in Venezuela, more than 1,000 km away.)

Interviews conducted by Jiménez revealed that manatee hunting in Costa Rica reached its peak during the 1950s and 60s, when manatees were still abundant here. By the 1970s, manatees had become rare, and remain so today. After more than three years of research in the late 1990s, Jiménez estimated the manatee population along the entire Caribbean coast and river systems north of Limón, where most of Costa Rica's manatees are found, as well as in the southeastern corner of Nicaragua, at just 30 to 100 individuals.

This precariously small population appears at first glance to be well protected. Most of the population's range lies within protected areas, and even outside parks the manatee is legally protected as an endangered species. The parks in turn have attracted ecotourism, which has provided an economic alternative for many exhunters. Since manatees are, in addition to all this, exceedingly hard to find, they are now hunted very seldom in Costa Rica.

Manatees in Costa Rica still face a variety of other threats, however. Illegal fishing nets stretched across lagoons, often within Costa Rica's protected areas, are known to have tangled and drowned several manatees over recent years. Ecotourism, while benefiting manatees in many ways, has also brought more motorboat traffic, a serious threat to the shallow water-loving manatees, although boat traffic is not nearly as great a threat here as it is in Florida.

Protected areas and the wildlife within them can also be affected by problems originating far away. Apparently pristine river systems can be seriously altered by deforestation and pollution upstream. Deforestation inevitably leads to massive erosion, which silts up rivers and changes their depth and flow patterns. As a result, rivers may become too shallow or quick for manatees, and they may be stripped of the aquatic and bankside vegetation that manatees depend on for food. This process may have caused manatees to disappear from a number of rivers they appear to have inhabited previously, including the Sarapiquí, San Carlos, Pacuare, Matina, Reventazón, Parismina, and upper San Juan rivers.

The rivers currently inhabited by manatees receive massive quantities of toxic agricultural runoff, mostly from the hundreds of square kilometers of banana plantations that cover the Atlantic lowlands. Such runoff has been blamed for the sporadic appearance of masses of dead fish in the region. A dead manatee was located during one such die off. In samples both outside and within the Tortuguero river system, Jiménez has found high concentrations of organophosphates, agricultural chemicals that are known to be lethal to mammals. He found the chemicals not only in water and sediment, but also in some of the manatee's aquatic food plants. (Fortunately, Chiquita Brands and various smaller companies have recently begun to reduce the use of toxic chemicals through the Banana OK and other programs.) Open-pit gold mines proposed along some 700,000 ha (1,700,000 A) that drain into the San Juan river would increase the threat of pollution to the region's manatees still further, for such mining produces vast quantities of cyanide and heavy metals.

Today, the manatee is as important a symbol as it has ever been. The fact that manatees still exist throughout a large stretch of Costa Rica, and that the population has a good chance of recovering, is a credit to Costa Rica's extensive network of parks. But the fact that the manatee is still at risk of extinction in the country is also an important reminder that we are still creating environmental problems that can jeopardize the health of ecosystems even within park boundaries. The recovery or demise of the manatee over the coming years will thus symbolize much of what is right or wrong with the environment in Costa Rica.

References

Bozzoli 1986; Colmenero 1986; Domning 1982a; Domning & Hayek 1986; Frantzius 1869; Hartman 1979; Husar 1978; Irvine 1983; Jiménez 1995, 1998, 2000; Jiménez & Altrichter 1998; Lefebvre et al. 1989; Ligon 1983; MacLaren 1967; Odell et al. 1978; O'Donnell 1981; O'Shea & Salisbury 1991; Powell 1978; Reynolds 1979; Reynolds & Odell 1991; Reynolds et al. 1995; Schevill & Watkins 1965; Stewart 1995; Thornback & Jenkins 1982.

KPCOFGS

Tapirs (order Perissodactyla)

Distribution and Classification

The order Perissodactyla contains just three families and 16 species worldwide: seven horses, five rhinoceroses, and four tapirs. Only three perissodactyls occur naturally in the New World—the Brazilian tapir (*Tapirus terrestris*), which inhabits northern and central South America east of the Andes; the mountain tapir (*T. pinchaque*), found at high elevations in the Andes; and Baird's tapir (p. 297), the only odd-toed ungulate native to Costa Rica. The fourth species, the Malayan tapir (*T. indicus*), inhabits Asia. Perissodactyls differ from the artiodactyls, or even-toed ungulates, in having a proportionately large third digit on each foot upon which they place most or all of their weight (see illustration on p. 306). All ungulates (from the Latin for hoofed ones) belong to one of these two orders.

Evolution

Perissodactyls first appeared about 60 million years ago and, until relatively recently, were much more abundant and widespread than they are today. Although tapirs are now the only perissodactyls native to the New World and are restricted to the tropics, horses, rhinos, and tapirs all used to inhabit North America. Biologist Daniel Janzen believes that some of Guanacaste's best known plants—such as gourd trees (*Crescentia alata*, illustrated below, and *C. cujete*), guácimos (*Guazuma ulmifolia*, illustrated on p. 310), guapinols (*Hymenaea courbaril*, illustrated on p. 298), and Guanacaste trees (*Enterolobium cyclocarpum*)—may owe their existence to

The gourd tree is one of several Costa Rican plants that may have been dispersed originally by native perissodactyls, which were abundant in the Americas until the end of the Pleistocene, just 10,000 years ago. Today, the gourd tree is dispersed mostly by domestic horses, and perhaps by Costa Rica's only remaining native perissodactyl, the tapir.

ancient perissodactyls. These trees lack obvious native dispersers today, but their fruits are relished and efficiently dispersed by domestic horses, which are similar to horses that roamed the region until close to the end of the Ice Ages some 10,000 years ago. Such plants may only be here today because livestock (and possibly tapirs) have partially assumed the role of the extinct native perissodactyls.

Tapirs appear to have originated in North America, where their fossils date back 50 million years. From there, they presumably spread to Europe, where fossils date back 35 million years. As continents drifted apart and climates changed, tapir populations became separated, which would explain how the Malayan tapir got to the other side of the world. North American tapirs spread south once North and South America became reconnected, and subsequently disappeared from North America about 10,000 years ago.

Teeth and Feeding

The dental formula of tapirs is I 3/3, C 1/1, P 4/3, and M 3/3 = 42. The back three premolars are enlarged, resembling molars, and help crush food. The teeth are exceedingly durable; tapirs can exert 140 to 230 kg (300 to 500 lbs) of pressure when biting a hard seed and yet, even in old individuals, tooth-wear is reported to be minimal.

The foliage that forms the basis of a tapir's diet is full of cellulose, which cannot be digested by any mammalian stomach enzyme. Like other herbivorous mammals, tapirs have microorganisms in their stomachs that help digest plant material through a process of fermentation. Perissodactyls lack the complex, four-chambered stomachs found in most artiodactyls, however, and do not ferment food in the foregut or ruminate (see pp. 304-305). Rather, they ferment food toward the rear of their guts, in the cecum and in the colon. This hindgut fermentation process deals with leafage more quickly but less efficiently than the foregut fermentation of ruminants. Ruminants can take about 80 hours to digest cellulose and end up absorbing about 60% of the carbohydrates they eat, while perissodactyls finish after about 48 hours, but absorb only about 45% of the carbohydrates. As a result, deer droppings usually have a fine consistency, while tapir droppings are much coarser, containing clearly visible undigested plant fragments.

There are other pros and cons to the two fermentation strategies. The faster digestion process gives perissodactyls less time to break down the toxins that are present in many tropical plants. One way they may get around this is by eating a "mixed salad" of leaves in order to avoid suffering too much from the toxins of any single plant. Studies in Guanacaste, Corcovado, and Panama's Barro Colorado Island have demonstrated that Baird's tapir will eat a huge variety of plants. Janzen discovered that, in Guanacaste alone, foliage of at least 150 native plant species was acceptable to a captive tapir, while researcher Charles

Foerster observed a wild tapir eat 126 plant species over the course of a year in Corcovado. Of course, some plants are more acceptable than others, and many cannot be eaten by tapirs at all. The Guanacaste captive had elaborate food preferences, and completely rejected leaves from over 300 plant species.

On the other hand, hindgut fermentation allows perissodactyls to make better use of another type of food—fruit. While even-toed ungulates ferment their food before it reaches the stomach and thus lose much of the nutritional value of any fruit they eat, odd-toed ungulates like the tapir ferment food only after it has passed through the stomach. Although tapirs may not know this, they clearly relish fruits. When available, sonzapote (*Licania platypus*) fruits were the center of attention of five tapirs studied by Foerster in Corcovado (see p. 298). The Guanacaste captive was so fond of guapinol (*Hymenaea guapinol*) fruit pods that Dr. Janzen's feeding experiments would have to be abandoned if there were any pods in vicinity of the tapir's corral, since the animal would be too distracted trying to get at them; and if a limitless quantity of bananas was supplied, the captive tapir sometimes ate until it fell asleep, with a banana still hanging out of its mouth.

Body Design

Tapirs have three individually hoofed toes on their hindfeet and four on their forefeet. The fourth forefoot toe is a little higher than the others and is not always visible in tracks. Tapirs move easily through the muddy areas they like to frequent. They splay their toes as their feet sink, then bring the toes together as the feet are withdrawn. The narrower foot is easily and silently extracted.

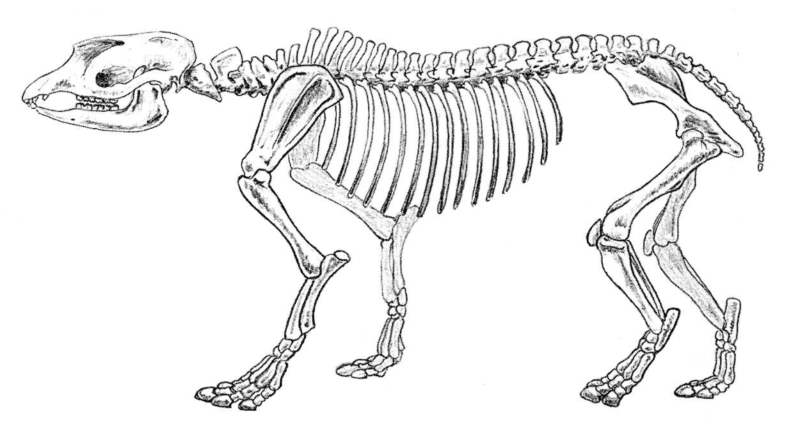

Tapir skeleton.

Senses

Tapirs have small eyes and appear to rely relatively little on vision. They do, however, use some basic facial expressions in communication. Like horses, they flatten their ears and flare their lips when irritated, and hold their ears still and upright when nervous. As in cats, white markings on the ears may accentuate such expressions. Tapirs may also have good night vision, for they move with apparent ease through dense vegetation in the dark. The eyes of their cousins the horses, at least, extend backward deep into the skull, creating light-gathering tubes that provide horses with night vision rivaling that of owls.

Tapirs have good hearing. They are easily startled by noises and use a small repertoire of sounds in communication.

Their keenest sense seems to be smell. They smell the air constantly when nervous and can seldom be observed for long from downwind. Smell is also important in communication. Tapirs mark certain spots by scraping with their hindfeet and urinating, and thus probably convey chemical messages to neighbors. Distinctive, white urine crystals can collect at favorite marking spots. Tapirs also use smell to find and identify food, passing their snouts over any food item before eating it. If their sense of smell lets them down, an apparently delicate sense of taste may pick up the error; the Guanacaste captive would quickly spit out unwanted leaves that had been handed to it inside sandwiches of palatable leaves, or seeds that had been imbedded in favorite fruits.

Conservation

Perissodactyls are important both as plant predators and plant dispersers, and as food for large carnivores. Persecuted by hunters and usurped by people and non-native ungulates, 14 of the world's 16 perissodactyls, including the four tapirs, currently appear on endangered species lists.

References

Foerster 1998; Freeland & Janzen 1974; Horwich & Lyon 1993; Janzen 1982d,e, 1983c,f,g,h,k; Janzen & Martin 1982; Klingel 1977; Macdonald 1995; Nowak 1991; Terwilliger 1978.

SPECIES ACCOUNT

O: Perissodactyla F: Tapiridae

BAIRD'S TAPIR

Tapirus bairdii
(plate 36)

Names: The word *tapir* is thought to be derived from the native Brazilian word *tapy*, meaning thick, in reference to tapirs' 3 cm (1 ¼ in.)-thick hide. Spencer Baird was a prominent nineteenth-century zoologist and ornithologist, and the founder of the famous National Museum in the United States. **Spanish**: *Danta*; *tapir*.

Range: Southern Mexico to Ecuador; from sea level to bamboo thickets near the summit of Chirripó, Costa Rica's tallest mountain at 3,880 m (12,730 ft); generally in habitats with cover and few people, usually near water.

Size: 200 cm, 250 kg (80 in., 550 lbs); males slightly smaller than females.

Natural history: Tapirs are generally solitary, and are active by day or night. Within a given area, however, they tend to have distinct activity patterns that may be influenced by the intensity of hunting and other human activities, and by climate. In Santa Rosa, two radio-collared tapirs were most active between 6:00 and 8:00 PM and between 4:00 and 7:00 AM, while in Corcovado, five radio-collared tapirs showed peak activity at around 7:00 PM and from 3:00 to 4:00 AM.

Costa Rica's largest land mammal is more athletic than one might expect. Its tapered form allows it to push silently through vegetation or crash away from predators at high speed. It can ascend almost vertical terrain, walk along river bottoms with only the top of its head above water, swim, and even dive for periods of 15 to 30 seconds. When fleeing, a tapir will flatten anything in its path; its thick skin helps protect it against thorns, splintered branches, and the claws of large cats. Tapirs spend most of their active time meandering slowly through the forest, however, sniffing and eating plants as they go. A prehensile snout helps them pluck vegetation. If choice leaves are out of normal reach, tapirs will raise up on their hindlegs, or use their noses and teeth to bend over or snap saplings with stems less than about 2.5 cm (1 in.) thick.

Three impressively extensive studies have provided detailed information on the diet of Baird's tapirs in Costa Rica. In Guanacaste, Daniel Janzen conducted hundreds of feeding experiments on a captive; in Corcovado, Eduardo Naranjo documented the diet of tapirs in the wild through direct observation, and, especially, by managing to find and analyze 136 scats; and also in Corcovado, Charles Foerster was able to habituate a wild tapir to his presence and spend an astonishing 286 hours observing her from a distance of just a few meters.

Both the Corcovado studies found more than ¾ of the tapir's diet to consist of leaves and stems, and most of the remainder, of fruit. Occasionally, the tapirs also fed on bark and flowers. Foerster calculated that his tapir spent more than 70% of her active time eating, averaging about 50 mouthfuls every 10 minutes, and consuming an average of over 15 kilos of food per day! All three studies, as well as another direct observation study done in Panama, show that tapirs feed on a vast variety of plants (see pp. 294-295), but tend to have favorites.

The Guanacaste captive's favorites included pulp from guapinol (*Hymenaea courbaril*, illustrated below) seed pods (so relished that if guapinol was placed between the tapir's lips, the tapir would eat it while sound asleep!), fruits of the gourd tree (*Crescentia alata*, illustrated on p. 293), foliage of balsa (*Ochroma pyramidale*) and madroño (*Alibertia edulis*), and both foliage and fruits of guácimo (*Guazuma ulmifolia*, illustrated on p. 310). In Corcovado, 40% of the mouthfuls taken by Foerster's tapir were of just three foods—leaves and stalks of the Swiss cheese philodendron (*Monstera* sp.) comprised 23%, the fruit of sonzapote (*Licania platypus*, illustrated below) 10%, and the leaves and stalks of a coffee relative (*Psychotria* sp.) 7%. (The fact that a philodendron is the top food choice is curious, since members of this family contain crystals of calcium oxalate, a caustic substance that deters most herbivores.) And in Monteverde, indirect evidence suggests that tapirs there snack on shrimp plant relatives (members of the Acanthaceae such as *Hansteinia*, *Justicia*, *Razisea*, and *Poikilacanthus*), various coffee relatives (Rubiaceae), palms, bamboos, and a number of tree saplings.

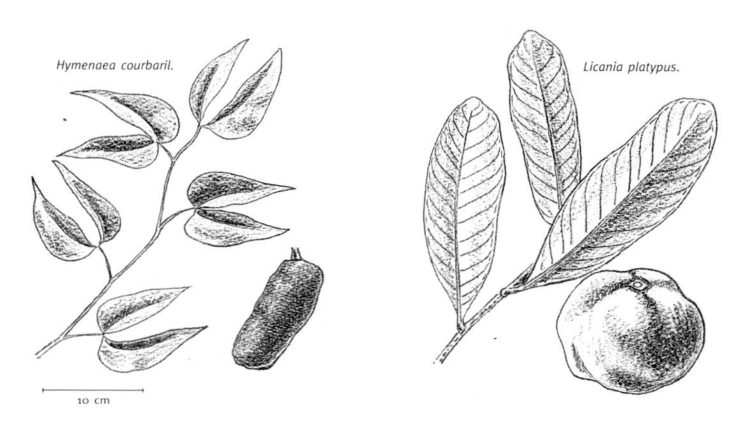

Two of the tapir's favorite foods: the fruits of guapinol (*Hymenaea courbaril*) and of sonzapote (*Licania platypus*).

The two Corcovado studies also revealed that the tapir's diet varies seasonally. When certain fruits were available, the tapirs ate more fruit and less leaves and stems. Foerster's tapir ate the most fruit in February and March, and especially between May and July, when *Licania* fruits were available. During the *Licania* months, she was also more choosy with foliage, eating almost no stems and selecting only the newest leaves (which tend to be less toxic and more nutritious).

Seasonal variation in diet can also influence habitat use. The Corcovado studies, as well as observations made in Panama and Belize by other researchers, have revealed that tapirs spend much of their time in secondary forest or light gaps, perhaps because such habitats offer more vegetation at a height accessible to tapirs. Foerster, who radio-tracked four other tapirs in addition to the habituated female, found that all five spent more time in secondary forest than in primary. When the fruits of *Licania*, a primary forest species, were available, however, the tapirs tended to spend more time in that habitat. And while tapirs usually forage alone, Foerster observed as many as four individuals feeding under the same *Licania* tree.

One habitat component that seems to be essential to tapirs is fresh water. Tapirs usually rest in or near water. They tend to flee there when disturbed by predators, and they may find relief in water from insects, too. Tapirs also seem to require water to defecate; captives usually defecate in water if it is available, and 96% of the feces Naranjo found in Corcovado were in clear, shallow, slow-moving river pools.

Tapirs deposite roughly 6 cm wide balls of dung in piles (left), rather like their cousins, horses. They usually defecate in water, however, where fresh dung floats to the surface (right).

Home-range sizes for Foerster's five tapirs in Corcovado, as well as two radio-collared in Guanacaste, varied from 62 to 232 ha (153 to 573 A). In any given month, however, the Corcovado tapirs only used from 15 to 37 ha (37 to 91 A). Gestation lasts 13 months and usually results in a single calf that, as in deer, is left in a secluded spot and visited periodically for nursing during the first week or so of its life. A streaked coloration, which is lost after four to eight months, makes the calf less conspicuous to predators. Calves leave their mothers after 10 to 11 months but may not reach full size and sexual maturity until they are three or four years old. Females have young about every other year. Captives can live 35 years.

Sounds: Snorts, grunts, and squeaks in alarm; hiccuping click sounds (made by the tongue and the palate) during social interactions; and loud, descending whistles that are sometimes answered by other, distant tapirs.

Mythology: For the Bribri natives (and other related indigenous groups of southern Costa Rica), the tapir is an animal to be feared and respected and, as we might interpret it today at least, a somewhat feminist symbol. For the Bribris, tapirs on earth are spirits of a superior Tapir, who is the sister of Sibú, the Bribri God. According to Bribri legend, Sibú sent Tapir to earth so that he could marry her off in exchange for a wife of his own. Tapir is very wise, however, and recognized her brother's selfish motives. Tapir is also able to see into the future, and saw that if she got married, her husband would be unfaithful to her. Being stubborn and independent, Tapir refused to get married, so Sibú allowed the Bribris to hunt tapirs, her spirits on earth.

Only certain Bribri clans are allowed to hunt tapirs, and those must do so with extensive ritual. The hunter who kills the tapir must then live in isolation for two weeks or more; if he is visited by his wife during this period, his wife is likely to fall sick. Only one woman in each clan is trained to cook the tapir appropriately, without cutting it up too much. Each clan member receives one piece of meat and must not ask for more, and girls must receive their meat from the hands of an elder. Abusing such traditions is said to bring terrible punishment from Tapir, who is vengeful of Sibú and men, and jealous of women.

Conservation: Because of these beliefs, tapirs were not hunted extensively in southern Costa Rica until the early 1900s, with the establishment of large banana plantations, and with them, nonindigenous hunters. For most of the 1900s, tapirs were hunted heavily all over Costa Rica for their fatty meat and thick hide, and for sport. Today they are hunted much less, for they are rare outside protected areas, are legally protected, do not have great economic worth, and seldom damage agriculture. Even within protected areas, however, tapir hunting continues. In Corcovado, tapirs are killed frequently around the edges of the park, and sometimes even in its interior: Foerster found one victim less than 3 km (2 mi) from the park headquarters at Sirena.

Tapirs mature and reproduce so slowly that, even in areas that have been effectively protected for years, populations still have not recovered from past hunting. Since tapirs frequently forage in areas of secondary forest, moderate habitat disturbance in and of itself may not be detrimental to them, but generally tapirs are found only in remote regions. The species may number no more than a few hundred in Costa Rica. In Corcovado, for example—which at about 40,000 ha (100,000 A) is the largest area of protected lowland forest along the Pacific coast of Central America, and one of the areas where tapirs seem most common—the aforementioned studies produced estimates of between 150 and 500 tapirs for the entire park. According to one calculation, 200 is the minimum number of tapirs required to reduce the risk of inbreeding enough to ensure the long-term survival of an isolated population.

The presence or absence of tapirs probably has a large impact on ecosystems. On one hand, tapirs predate many plants by eating and destroying seeds and by snapping or defoliating saplings. Biologist Daniel Janzen has observed that, where tapirs have reappeared in Santa Rosa, favorite food plants less than a couple of meters tall are exceedingly scarce. On the other hand, although tapirs can crack even hard seeds in their mouths and destroy many more in their stomachs, they also swallow and pass some seeds whole. This, as well as their habit of defecating in water, may make them important dispersers for some plants.

References

Bozzoli 1986; Carter 1984; Dirzo & Miranda 1990; Foerster 1998; Fragoso 1991; Horwich & Lyon 1993; Janzen 1981, 1982d,e, 1983k; Jones 1982; Lawton 2000; Naranjo 1995a,b; Terwilliger 1978; Vaughan 1978; Williams 1984.

Peccaries and Deer (order Artiodactyla)

Distribution and Classification

The Artiodactyla, or even-toed ungulates, are so called because, with the exception of the two peccary species found in Costa Rica, they have two or four toes on all feet. The order is split into three suborders—the Suiformes includes peccaries, true pigs, and hippopotamuses; the Tylopoda includes llamas, camels, and their relatives; and the Ruminantia includes mouse deer, deer, giraffes, pronghorns, antelopes, cattle, sheep, and goats.

The order contains nine families and 180 or so species. Only two families—the peccary and deer families—and four species occur in Costa Rica. Costa Rica's two peccary species belong to the family Tayassuidae. The family contains only one other species, the Chacoan peccary (*Catagonus wagneri*), which was known only from a fossil until a living population was found in Paraguay in 1972. The species is now known to inhabit Paraguay and neighboring eastern Bolivia and northern Argentina. There are nine other piglike animals in the world, all of which belong to the Old World family, the Suidae. Features that distinguish peccaries from these "true pigs" include a large scent gland on the rump, upper incisors (the tusks) that point downward rather than outward or upwards, and several aspects of their internal anatomy. While true pigs can give birth to as many as 12 young, peccaries typically produce only two, which, unlike those of true pigs, are not born in a nest and can stand and run just a day after birth.

Costa Rica's two deer species belong to the family Cervidae. The family contains some 40 to 45 species, about a third of which occur naturally in America and Europe, and the rest in Asia. Some have been introduced to islands such as Australia, New Zealand, New Guinea, and Cuba. All mammals with antlers—about 39 species—belong to this family.

Evolution

Artiodactyls—along with other ungulates and whales (see pp. 327-328)—are descended from condylarths, primitive herbivorous mammals that inhabited the earth from around the time dinosaurs became extinct, about 65 million years ago.

Peccaries first appeared in North America, at the beginning of the Oligocene period, about 40 million years ago. Since then, many species have appeared and gone extinct. Peccaries about twice the size of the species around today, mostly in the genus *Platygonus*, used to live as far north as the Yukon in Canada, and some persisted until just 12,000 years ago. One subfamily—the Old World peccaries—inhabited Europe, Africa, and Asia until about five million years ago. Deer ancestors inhabited Asia about 40 million years ago, but no fossils older than about 20 million years have been found in North America. The artiodactyls, including *Platygonus*, reached Costa Rica and subsequently South America from the north as the Panamanian land bridge was formed.

Teeth, Skull, and Feeding

The dental formulae are I 2/3, C 1/1, P 3/3, and M 3/3 = 38 for peccaries, and I 0/3, C 0/1, P 3/3, and M 3/3 = 32 for Costa Rica's deer.

Peccaries rake in food with their hooves and with their hard but flexible nose disk (known as the rhinarium), seize it with their incisors, and crush it with their molars. Their most distinctive teeth, however, are the canines, which grow continually until the peccary is four or five years old. The upper and lower canines rub against each other, and are thus worn down and sharpened. Since peccaries eat very little meat, their canines are not used to catch or kill prey, but they may still be important in feeding. The canines interlock, preventing the jaw from moving sideways, so peccaries crush food with an up and down movement rather than with the more typical sideways chewing motion that is used by, for example, true pigs and deer. This design may help keep the jaw from slipping or dislocating when the peccaries smash hard seeds. The logic of this theory is easily understood by anyone who has tried to use a nutcracker with loose hinges. The canines are also important in defense and in social interactions. They are used both as weapons and as acoustic instruments. By clacking the canines together, peccaries produce sounds that help intimidate predators or fellow herd members. Peccaries seldom bite one another, but within herds they frequently use an open-mouthed, canine-revealing threat display that is an important back-off signal.

Costa Rica's two deer species, meanwhile, have lower canines that are greatly reduced, resembling incisors, and they lack upper canines and upper incisors. They snip off vegetation by closing their sharp lower incisors against a callous pad on the upper gum or they simply yank off vegetation using their tongues.

The skulls of male deer are unmistakable because they have antlers. Antlers serve as ornaments with which to woo females and intimidate rival males, and as weapons with which to headbut or "rut" with stubborn male competitors. Deer lose and regrow their antlers annually in accordance with their breeding cycle. The breeding and antler cycles of white-tailed deer in northwestern Costa Rica are described on pp. 322-324. Those of the red brocket deer have not been documented in Costa Rica.

Stomach—What is a Ruminant?

Deer, like most artiodactyls, are ruminants. Vegetation, which comprises the bulk of a ruminant's diet, contains much cellulose, which cannot be broken down by any mammal's digestive enzymes. Ruminants first pass food to a large pouch in the stomach known as the rumen. The rumen contains bacteria and protozoa that break down the cellulose through fermentation. Ruminants then regurgitate the fermented food, known as cud, to the mouth for further chewing, or ruminating. When the cud is swallowed for a second time, it bypasses the rumen, moving straight through three more stomach chambers and then the rest of the digestive tract. By swallowing first and chewing later, once they are in a secluded spot, ruminants also reduce their exposure to predators. Other herbivorous mammals also use bacteria to break down cellulose, but the bacteria are housed further along in the digestive tract, usually in an appendage called the cecum that is situated between the short and large intestines, and food is never ruminated.

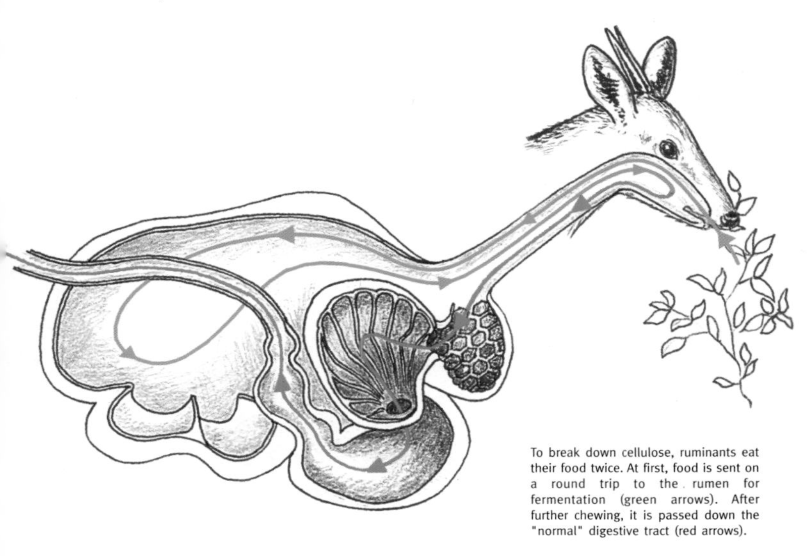

To break down cellulose, ruminants eat their food twice. At first, food is sent on a round trip to the rumen for fermentation (green arrows). After further chewing, it is passed down the "normal" digestive tract (red arrows).

Body Design

Peccaries have four toes on their forefeet and three on their hindfeet, while deer have four toes on all feet. Like perissodactyls, artiodactyls are unguligrade, placing their weight entirely on their nails (the hooves), but unlike perissodactyls, they place their weight on the hooves of the third and fourth digits (see illustration on p. 306). The second and fifth toes are also hooved, but are small and do not touch the ground unless the foot sinks deep.

A couple of peculiarities in foot design help artiodactyls run efficiently, which they must do to avoid being eaten. The bones behind the third and fourth toes are fused, forming a single, more lightweight element known as the cannon bone. Strung alongside the cannon bone and over the ankle joint are a set of long ligaments. When the animal places its weight on the foot and the ankle bends, the ligaments stretch like rubber bands, giving the animal extra spring when the foot is lifted.

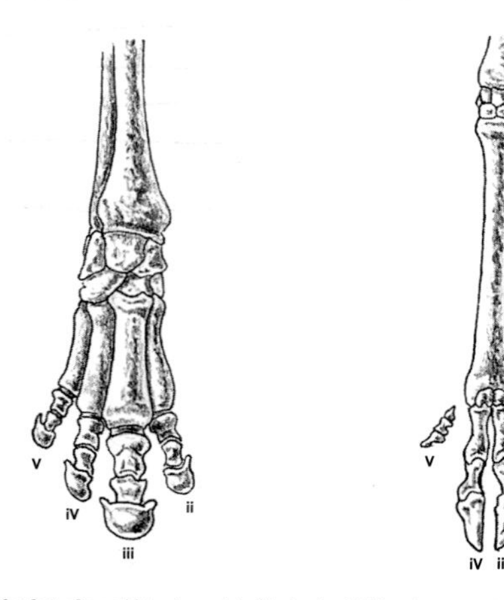

The forefeet of an odd-toed ungulate (the tapir, at left) and an even-toed ungulate (a deer, at right). Odd-toed ungulates place their weight either on digits ii, iii, and iv (in the case of tapirs and rhinos) or solely on digit iii (in the case of horses); even-toed ungulates place their weight on digits iii and iv. (Drawings after Beddard, in Nowak 1991).

Senses

Artiodactyls have poor eyesight but excellent senses of smell and hearing. Being highly social, peccaries depend on sounds and odors to stay close to one another or to locate their herds when separated. Peccary scent is strong and unmistakable. Even humans can detect peccaries by smell before hearing or seeing them, and the musky, sweaty odor can linger for hours. The odor comes from fluid secreted by a visible 5 by 8 cm (2 by 3 in.) gland (which resembles a nipple and was mistaken for the navel by early naturalists) situated on the rump about 16 cm (6 in.) in front of the tail. The fluid, which apparently can be released at will and even squirted a short distance, is amber in color, but turns jet black when exposed to the air. It is rubbed on to trees and rocks, and forms dark, oily stains on favorite marking spots. When peccaries interact, they often rub their cheeks against one another's scent gland.

Although deer are not as smelly as peccaries, they are no less dependent on their sense of olfaction. A pocket just in front of the eye is their only conspicuous scent gland. During breeding season, males mark by rubbing saplings and overhanging branches with these glands and another located on the forehead. Whitetails also have glands on their feet and ankles. Dominant males in particular use these glands, along with urine or feces, to mark scrapes during the breeding season.

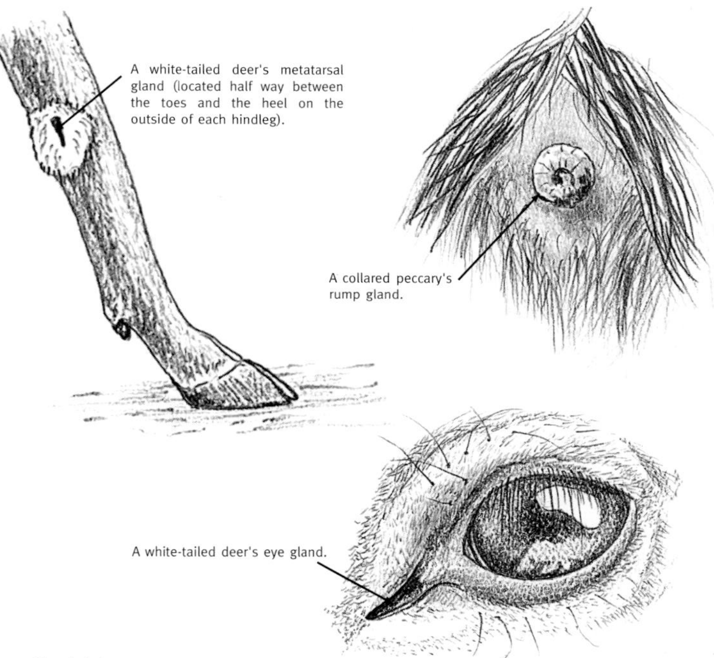

Social Arrangement

The social arrangements of collared peccaries and white-tailed deer have been studied in North America, and are interesting to compare.

Collared peccaries live in herds in which no clear hierarchical systems have been determined. Biologists recorded thousands of interactions between members of wild herds in Arizona and found that only a small percentage of interactions were unfriendly. Even when females were in heat, there was little or no antagonism between males. Furthermore, adults tolerated and even helped feed and defend young that were not their own. Young were tolerated when they shouldered adults away from food sources, stole food from adults' mouths, and clambered over bedded-down adults in order to wedge themselves between the warm bodies. Females nursed others' young, and all adults clustered around the young when certain predators came near.

This cooperation and apparent lack of competition is also reflected in the composition of the herds. Herds have been found to contain adult sex ratios of about 1:1, and there seems to be little movement of individuals from one herd to another. These things would be unusual in a hierarchical society in which a dominant adult was driving off his or her competitors. This social arrangement is unusual among mammals. Some experts think there is more competition between collared peccaries than is apparent, and that it is concealed by a well-established dominance hierarchy. Another possibility is that competition is minimal because herds contain closely related individuals, but the relatedness of herd members has never been studied.

Within groups of white-tailed deer, on the other hand, there are clear hierarchies. Males are generally dominant over females (except their own mothers), and larger, older individuals are generally dominant over smaller, younger ones. Subordinates typically avoid eye contact with, and step out of the way of, their superiors. If necessary, males competing for females will resolve hierarchies with head-butting and shoving contests. As a result, male white-tailed deer differ from the females in being larger and in possessing antlers; both features have been selected because they improve the males' chances of intimidating or fighting off rivals, and of wooing females. In peccaries, on the other hand, there are no such differences between males and females.

Conservation

Artiodactyls are important both as seed predators and dispersers, and as food for large carnivores. They are among Costa Rica's most important game species, and have the potential to continue to be if well managed. In many parts of the country, native artiodactyl populations have been decimated by unsustainable levels of hunting and habitat destruction, and usurped by their non-native cousins. The collared peccary and both deer species are still common in many areas; the white-lipped peccary, which is more vulnerable for a number of reasons that are discussed in the species account, is now extinct over most of its former range in Costa Rica, and its future here is precarious.

References

Altrichter 1997; Atkeson & Marchinton 1982; Byers & Bekoff 1981; Halls 1984; Kile & Marchinton 1977; Kiltie 1981c; Moore & Marchinton 1974; Nowak 1991; Rodríguez & Solís 1994; Sowls 1997; Wetzel 1977; Woodburne 1968.

SPECIES ACCOUNTS

O: Artiodactyla F: Tayassuidae

COLLARED PECCARY

Tayassu tajacu
(plate 36)

Names: Also known as the javelina, because of its javelinlike tusks. *Tayassu* comes from the Brazilian Tupí language and means taya gnawer. Taya are bulbs of plants in the philodendron family (see pp. 310-311). *Tajacu* may be derived from the same. **Spanish**: *Saíno.*

Range: Arizona and Texas in the United States to northern Argentina; from sea level to over 3,000 m (10,000 ft) at Cerro de la Muerte in Costa Rica; in or near forest or dense second growth, often near water.

Size: 90 cm, 19 kg (35 in., 40 lbs).

Similar species: See white-lipped peccary (pl. 36, p. 313). The coloration of collared peccaries can vary with place and season. In Arizona, for example, the darker outer portion of their banded hair falls off at the beginning of summer, leaving the coat thinner and paler and thus more appropriate for the heat.

Natural history: Collared peccaries are active by day or night. In hot areas such as Guanacaste, or in areas where they have been hunted, they tend to be more active from late afternoon to early morning.

Their diet includes fruits, seeds, roots, tubers, leaves, and some animal food. The diet of collared peccaries in Costa Rica has been studied most extensively in the seasonally dry forests of Palo Verde by Michael McCoy and colleagues. Analysis of more than 100 scats showed that the diet of collared peccaries at Palo Verde varies dramatically over the course of a year. During the late wet season, there is an abundance of fruits and seeds relished by the peccaries, such as hogplums (*Spondias mombin*, illustrated on p. 162), moridero (*Eugenia salamensis*), and nances (*Byrsonima crassifolia*). In the early dry season, fruits become scarce and the peccaries switch to a diet apparently consisting mostly of roots, tubers, and vines, such as vines of the wild grape *Cissus rhombifolia*. The fruit famine is broken in February with the appearance of guácimo fruits (*Guazuma ulmifolia*, illustrated below), which the peccaries eat almost exclusively during that month. In the late dry season, guácimo fruits become scarcer, and the peccaries start eating others, especially those of raintrees (*Pithecellobium saman*) and Guanacaste trees (*Enterolobium cyclocarpum*, illustrated pl. 22). In the early wet season, the peccaries start to eat more fibrous foods such as roots and tubers again, and rummage around the bases of guácimo and rain trees for old seeds left over from the dry-season crop.

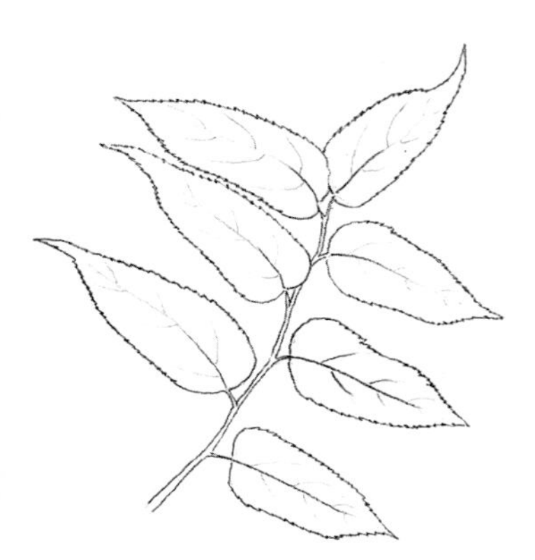

Guácimo (*Guazuma ulmifolia*): leaves (left), new fruits (right), and close-up of an old fruit on the ground (center). In Guanacaste, guácimo fruits are a seasonal staple for collared peccaries—as well as for many other mammals, including variegated squirrels, agoutis, coyotes, coatis, white-tailed deer, and tapirs.

In wetter habitats, important foods include fruits and seeds in the palm, sapote, and pea families. A peculiar food used by collared peccaries in wet habitats in Costa Rica is the terrestrial philodendron known as dumb cane (*Dieffenbachia*, illustrated on opposite page). Almost no other vertebrates, or even insects, feed on dumb cane, for its sap contains high quantities of toxic calcium oxalate crystals, which affect metabolism and can destroy mammals' kidneys. Peccaries probably break down such toxins by fermenting them in their digestive tracts. (The peccary's ability to stomach such chemicals may have played a key role in enabling the species to establish itself in the arid southwest of the United States, where the prickly pear cactus *Opuntia*, another oxalic acid-laden plant, forms a major part of the peccary's diet.) Peccaries may select such plants because they are rich in certain minerals (see pp. 315-316). The peccary and dumb cane may even enjoy a mutually beneficial relationship. Dumb cane reproduces vegetatively, so even small pieces of stem that are broken off and trampled can reroot and grow into new plants; the dense clumps of this plant that are common in wet, lowland forests may be created in part by marauding peccaries.

Dumb cane (*Dieffenbachia*).

Animal prey has not been recorded as an important food in most areas where collared peccaries have been studied, including Guanacaste. In the Amazon, however, collared peccaries have been found to eat significant quantities of insects, snails, worms, and millipedes, and even some small lizards, frogs, birds, and rodents; and in captivity, they readily accept meat. Since collared peccaries dig a lot, they might feed more on worms than is apparent from direct observation or fecal and stomach analyses, as white-lipped peccaries do in Corcovado (see p. 315).

Collared peccaries live in herds containing from a few to 30 individuals. The herds studied by McCoy in Guanacaste used home ranges averaging 118 ha (290 A). They tended to stay in smaller areas for much of the year, however, ranging further only when food was scarce in the early wet season and, especially, in the early dry season. During these periods of fruit scarcity, the peccaries would also take fewer and shorter rests. Preferred resting sites in Guanacaste were cool spots in rock or dirt caves or beneath fallen trees. In some areas, collared peccary herds split into smaller groups during periods of food scarcity in order to spread out and make use of smaller food patches. Unlike white-lipped peccaries, collared peccaries are territorial, defending home ranges that overlap little, if at all, with those of neighboring herds. Their much smaller home ranges probably wouldn't provide for more than one herd, but are also easier to defend.

Collared peccaries are generally shy and unlikely to attack people. When startled by humans or other potential predators, they charge away and often try to hide in logs or holes. If cornered, they can inflict deep bites.

One to five, but usually two chestnut- or tan-colored young are born after a gestation period of just under five months. In Guanacaste, most young are born in March and April. Young can run after a few hours and join the herd a day after birth. If the herd is startled and there is a stampede, young not strong enough to keep up sit motionless and are left behind. Once danger has passed, they call back their mother. Young nurse for six to eight weeks and become independent after two to three months. Once weaned, they may habitually follow an adult other than their mother. Females reach sexual maturity when they are a little less than one year old. Captives can live 24 years, but of over 400 wild peccaries examined in Arizona, less than 5% were older than eight, and none were older than 16.

Collared peccary scat. The main clump is about 8 cm long.

Sounds: Purrs (mostly by young), low grunts, and doglike barks help herd members stay together or find one another when separated; growls and tooth clacks are threats; loud "woofs" are alarm calls, typically made as the herd flees.

Conservation: Collared peccaries travel in single file, leaving trails that are conspicuous to hunters. They are hunted for their meat and hides, and sometimes as agricultural pests. They can damage melon, rice, corn, and root crops such as yuca, camote, and tiquisque. They do not roam far from suitable habitat to do so, however. At La Selva, researcher Isa Torrealba found that yuca crops bordering the protected area are pillaged by collared peccaries frequently, but those more than 400 m (440 yards) away are untouched. Collared peccaries are important seed predators; they may also disperse some seeds by spitting them out or passing them whole. Collared peccaries do not need such large expanses of pristine habitat as do white-lipped peccaries, and since they travel in smaller herds, they are less vulnerable to hunting; they are common in most protected areas.

References

Babbitt 1988; Barreto et al. 1997; Bodmer 1989, 1991; Byers & Beckoff 1981; Croat 1983; Enríquez et al. 1991; Fragoso 1999; Gottdenker & Bodmer 1998; Kiltie 1981b; McCoy 1985; McCoy et al. 1990; McCoy et al. 1983; Oldenburg et al. 1985; Olmos 1993; Peres 1996; Robinson & Eisenberg 1985; Rodríguez & Chinchilla 1996; Sowls 1997; Timm 1994a; Timm & Laval 2000a; Timm et al. 1989; Torrealba 1993; Zervanos & Hadley 1973.

O: Artiodactyla F: Tayassuidae

WHITE-LIPPED PECCARY

Tayassu pecari

(plate 36)

Names: Sometimes listed under the genus *Dicotyles.* Both parts of the scientific name come from the Brazilian Tupi language. *Tayassu* means taya gnawer. Taya are bulbs of plants in the philodendron family. *Pecari* means path maker. **Spanish**: *Chancho de monte.*

Range: Southern Mexico to northeastern Argentina; from sea level to about 1,900 m (6,200 ft); mostly in large areas of pristine, wet forest, often near water. It has also been recorded in seasonally dry habitats and savannas in Guanacaste (although it is now extinct in that region) and parts of South America.

Size: 100 cm, 30 kg (40 in., 66 lbs).

Similar species: The collared peccary (pl. 36, p. 309) is smaller, has a pale shoulder collar, and lacks stark white fur on its chin and lower cheek. Note, however, that the collared peccary can have pale, grizzled fur around its face, and that the amount of white on white-lipped peccaries can vary—a few have virtually none, while others have white stretching down to their legs and belly. The collared peccary generally travels in smaller herds, is less noisy, and, where the two species occur together, tends to frequent drier, higher ground.

Natural history: Until recently, almost all that was known of the behavior of white-lipped peccaries in the wild was based on studies done in the Amazon. Over the last few years, however, research conducted by Mariana Altrichter and colleagues in Corcovado has revealed details of the species' natural history in Costa Rica.

White-lipped peccaries are active by day or night. In Corcovado, they are mostly diurnal, usually foraging in the early morning and late afternoon and resting between about 10:00 AM and 2:00 PM.

They feed on fruits, seeds, stalks, leaves, insects and other small animals, roots, tubers, and, occasionally, fungi and flowers, all of which they either take off the ground or dig up. Both in the Peruvian Amazon and in Corcovado, Costa Rica, whitelip feces contain about two-thirds fruit and seed remains and one-third vegetation remains. Favorite fruits and seeds in Corcovado include those of the Moraceae family such as milk trees (*Brosimum* spp., illustrated on p. 158) and figs (*Ficus* spp.), as well as hogplums (*Spondias* spp., illustrated p. 162), the kapok relative *Quararibea asterolepis*, wild cashew (*Anacardium excelsum*), camarón (*Licania operculipetala*), legumes (*Inga* spp.), and palms, including coyolillos (*Astrocaryum* spp.) and yolillos (*Raphia taedigera*).

Although Altrichter found the *Astrocaryum* palm seeds to be the peccaries' favorite food during March and April, Corcovado's whitelips do not feed nearly as heavily on palms as do whitelips in the Amazon. This is curious, because the whitelips' association with palms (such as *Iriartea*, *Astrocaryum*, and *Euterpe*) in the Amazon is considered fundamental to their ecology there.

Iriartea palm nuts.

Palm seeds are eaten by relatively few vertebrates because they are extremely hard and because they appear in massive but patchy and seasonal crops that few animals can exploit efficiently. The design and behavior of white-lipped peccaries enable them to deal with both problems much more effectively than other fruit- and seed-eating mammals, including even their close relatives, the collared peccaries. Peccary expert Richard Kiltie has calculated that a different jaw design gives whitelips a biting force at least 1.3 times that of collared peccaries, and thus the ability to eat harder seeds. And unlike collared peccaries (or any other Neotropical mammal) whitelips travel in huge herds—typically containing between 50 and 100, and occasionally as many as 300 individuals—that seldom stay long in one place, wandering up to 10 km (6 mi) each day and covering vast home ranges of between 4,000 and 20,000 ha (10,000 and 50,000 A). This behavior is well suited for exploiting the massive but scattered palm crops.

Since palms seeds are tough (and not necessarily very nutritious either), it is possible that whitelips only exploit them heavily in the face of competition with certain other animals. Perhaps, as Altrichter has speculated, whitelips eat fewer palm seeds in Corcovado because competition is less intense than in the Amazon. Collared peccaries, for example, appear much less common than whitelips in Corcovado.

Animal matter appears to be eaten by whitelips only occasionally in some areas, but is a major component of the diet in others. Turtle scientist Archie Carr claimed that "where (white-lipped) peccaries pass there are almost no small ground-dwelling animals to be found," and that herds regularly plundered beaches of turtle eggs, and, especially, hatchlings waiting to emerge just beneath the surface. In Peru, whitelips were found to feed on significant quantities of snails, millipedes, worms, and, occasionally, birds. In Brazil, a herd spent two months foraging almost exclusively on eels and other fish that were exposed as a pond dried up, and one individual even snaffled up a snake. And in Corcovado, Altrichter discovered that whitelips spend about one-third of their time digging, mainly to feed on earthworms.

It's hard to tell if a peccary eats earthworms, because the worms may be indistinguishable to anyone watching peccaries feed, and are not detectable in stomach or fecal samples. Only by comparing soil samples was Altrichter able to gauge the importance of earthworms in the diet of her peccaries. Soil right next to where the peccaries excavated contained far more earthworms than soil collected at random from elsewhere, and also far more than the actual spot where the peccaries had been digging. In other words, the peccaries selected spots with lots of worms and ate the worms as they dug.

Earthworms are not the only food that whitelips dig for. In some dry areas in South America, whitelips feed heavily on roots. They also unearth seeds and seedlings. Richard Kiltie has noticed that whitelips often dig around the bases of plants and logs and has suggested that such areas would be a good place to look for small stashes of seeds hoarded by rodents. In light of Altrichter's new discovery, one wonders if such spots aren't also good for worms.

White-lipped peccary scat. The main clump is about 9 cm long.

Many of the white-lipped peccary's foods are seasonal. Most plants only fruit once a year, and, in Corcovado, relatively few do so in the late wet season. Worms, on the other hand, become virtually inaccessible during the driest months, because the ground becomes too hard for the peccaries to rake, and because the worms are deeper in the soil. Vegetation, however, is available all the time. In Corcovado, whitelips feed on the leaves and stalks of heliconias and philodendrons throughout the year, and especially in the late wet season, when fruits and seeds are most scarce. Philodendrons are an interesting choice, because they are unpalatable to most other animals (see p. 310). Peccaries may select them because they are rich in certain minerals. Nutritional analyses have shown, for example, that whitelips in Corcovado greatly increase their intake of calcium during the wet season by feeding on dumb cane philodendrons (*Dieffenbachia* spp.). This period coincides with their reproductive season, when pregnant or lactating females are most in need of calcium. This is also the time of the year when peccaries eat the most worms. Unlike most other whitelip foods, worms are full of protein, which is also especially important to mothers and their growing young.

White-lipped peccaries eat the leaves and stalks of heliconia plants, especially when fruits and seeds are scarce.

The foods available to whitelips in Corcovado vary not only through time but also in location, so whitelips need to move around with the different seasons. Altrichter found that, during the dry season, when the forest there offers the most fruits and seeds, whitelips stay inland, mostly in primary forest. As the wet season progresses and fruits and seeds become scarcer, they spend more time in secondary habitats (where heliconia leaves and stalks are always available), and along the coast (where, at that time of year, camarón fruits are available). As fruits and seeds become scarcer, whitelips also have to move around more to find enough food, much as collared peccaries do in Guanacaste (see p. 311). At the end of the wet season, Altrichter's peccaries left her study area entirely.

Whitelips have an exaggerated reputation for ferocity. The large herds can make terrifying noises and may charge if threatened, but months of observations by biologists have revealed that they generally run away from perceived threats. Lyle Sowls, who has spent nearly 40 years studying peccaries, noted that during extensive observations at a bait station in Brazil, individuals occasionally threatened him by jumping a short distance in his direction, but invariably backed down if he walked purposely toward them. In the case of a stampede, climbing a tree is the safest escape strategy. If a climbable tree is not at hand, standing still may help since peccaries have poor vision. The only person (as far as is known) who has tested this theory, however, was bitten.

Whitelips in Corcovado mate between January and March. A gestation period of about five months results in one to three young, but most often twins. Young are reddish brown at birth. They gradually darken during their first year, turning black like the adults by the time the time they are about one-third grown, in their second year. Females and young tend to travel in the center of herds, while males spread out along the edges. Females reach sexual maturity at about a year and a half. Genetic studies suggest that females tend to stay in their natal herds, while males leave and join other herds. These herds may not be far away, since whitelip herds have broadly overlapping home ranges and occasionally even forage within sight of one another. Captive whitelips can live at least 13 years.

Sounds: Retching noises by young; low rumbles, loud barks, and whines help herd members stay together or find one another when separated; grumbles, snorts, and tooth clacks are alarm noises or threats. While collared peccaries emit series of three to eight tooth clacks, white-lipped peccaries usually emit single clacks.

Conservation: White-lipped peccaries are indicator species, being among the first mammals to disappear when pristine habitats are disturbed by humans. Their foraging style requires large areas of forest, their meat is highly prized, and their large herd size makes it easy for hunters to kill many at a time. They may also be considered pests in some areas, for they occasionally raid crops of yuca, camote, corn, bananas, and sugar cane. The species is already extinct through most of its former range in Costa Rica, and is now seen regularly only in Corcovado National Park. Even there, whitelips are hunted extensively, especially in the late wet season when their migrations take them to the edges of the park and beyond.

The absence of such a large and formerly numerous mammal undoubtedly has tremendous repercussions on ecosystems. Whitelips were found to be the most important single food of the jaguar in Corcovado, for example, and their elimination from other areas has probably contributed to the jaguar's demise. The absence of whitelips must also greatly affect the plant composition of a forest. Whitelips are major seed predators on palms and other plants, and possibly dispersers of seeds they spit out or pass whole. The only intact seeds Altrichter and colleagues found in whitelip scats were smaller than 3 mm in diameter (such as those of figs and guavas), but some of those seeds subsequently germinated. Since whitelips travel such vast distances, they could be important dispersers for small-seeded plants.

References

Altrichter 1997; Altrichter et al. 1999; Barreto et al. 1997; Bodmer 1989 & 1991; Campero 1999; Carrillo 2000; Carrillo et al. 1997a,b; Chinchilla 1997; Crandall 1971; Donkin 1985; Emmons et al. 1997; Fragoso 1998; Gottdenker & Bodmer 1998; Hernández et al. 1995; Kiltie 1981a,b, 1982; Kiltie & Terborgh 1983; López 1999; Mayer & Wetzel 1987; McHargue & Hartshorn 1983b; Olmos 1993; Peres 1996; Sowls 1997; Vaughan 1983a; Wille 1987.

O: Artiodactyla F: Cervidae

RED BROCKET DEER

Mazama americana

(plate 37)

Names: The common name distinguishes this species from three other species of brocket deer, although none of the others is found in Costa Rica—the gray brocket deer (*Mazama gouazoubira*) inhabits San José Island off Panama and South America down to Uruguay, and two highland species (*M. rufina* and *M. chunyi*) inhabit the Andes. The genus is derived from one of the species' native Mexican names, *mazame*. **Spanish**: *Cabro de monte.*

Range: Eastern Mexico to northern Argentina; from sea level to about 2,800 m (9,300 ft); in or near forest or dense scrub.

Size: 110 cm, 25 kg (43 in., 55 lbs).

Similar species: White-tailed deer (pl. 37, p. 321) are larger, have white facial markings, paler underparts, and pale brown rather than dark reddish upperparts. Some red brocket deer have very dark faces and legs. White-tails tend to inhabit relatively open, sparsely vegetated habitats, while brockets favor dense, evergreen forest. Correspondingly, white-tails move with their heads held high, and males have long, branched antlers, while brockets move with their heads held low, and males have short, unbranched, backward-pointing antlers. Both characteristics enable brockets to push through heavy vegetation without getting entangled.

Natural history: This quiet, solitary animal is active by day or night. It often feeds along trails or in open areas bordering forest, where it can be approached closely by a cautious observer. Sometimes it wades into shallow streams to urinate.

One might expect this deer, like most other ruminants, to feed predominantly on leaves, but the red brocket prefers other foods. In Venezuela, it eats more fruit and seeds than leaves, particularly in the wet season. In Suriname, the red brocket favors fruits, seeds, and fungi when these are available. When these foods become scarce at the end of the wet season and during the dry season, the red brocket eats more leaves, twigs, vines, shoots, and flowers. Likewise, fruits and seeds (especially palm seeds) were found to comprise 81% of the diet of the red brocket in the Peruvian Amazon.

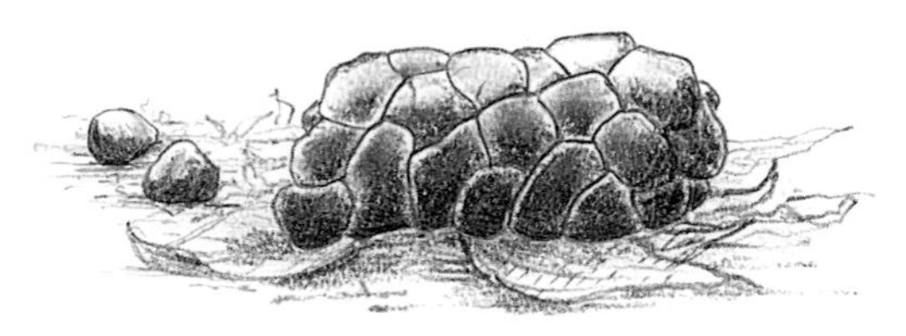

Red brocket deer scat. The main clump is about 5 cm long.

By fermenting food before absorbing any of its nutrients (see p. 305), ruminants digest leafage more efficiently, but they lose much of the nutritional value of rich foods such as fruit. Brocket deer, however, demonstrate that foregut fermentation can be an effective way to deal with certain fruits and seeds. Since brockets, like other ungulates, are strictly terrestrial, they have "second choice" at fruits—arboreal animals have the first pick, before fruits have fallen. Arboreal creatures tend to leave seeds alone in favor of fruit pulp, which is more nutritious and easier to digest, and they can more easily afford to avoid fruits or seeds that are well defended with hard casings or toxins. Thus, the fruits and seeds that are less desirable are the ones that are most abundant on the forest floor. Ungulate expert Richard Bodmer has speculated that the ruminant stomach enables red brockets to take advantage of abundant but tough or poisonous foods. He found that, lacking the strong jaws and teeth that peccaries use to smash hard seeds in their mouths, brocket deer swallow seeds whole and wear down hard casings through fermentation. Fermentation may also detoxify seeds by breaking down compounds before they move further down the digestive tract for absorption.

Red brockets have a gestation period of about seven and a half months. Females sometimes mate and become pregnant while still nursing their previous fawn, and may reproduce continuously. In several parts of South America at least, red brockets can breed at any time of year, but they often mate during the wet season and give birth in the dry season. The males' antler-growing cycle is also variable, being aseasonal in some areas, annual in others, and taking up to two years in others. The reproduction and antler cycle of red brockets has not been studied in Costa Rica.

The red brocket usually gives birth to a single fawn, which is left in a concealed spot and visited periodically by the mother for nursing until it can follow her. The white spotting on fawns starts to disappear by the time the fawn is about a month old. Spots along the ribs disappear by about two months, and spots on the back of the neck by about three months. Females are sexually mature by the time they are a year old. Captives can live over 13 years. Studies in Venezuela suggest that, in the wild, very few survive more than eight years, and most less than four years.

Sounds: Stamps ground with forefoot when nervous; emits a sneezelike snort when startled, often as it takes flight.

Conservation: The red brocket is hunted for its meat and sometimes as an agricultural pest. Its habit of fleeing only a short distance and then freezing makes it vulnerable to dog hunting, and its tendency to come out to forest edges to feed makes it an easy gun target. In pristine areas, however, its ability to disappear into areas of difficult access to hunters, and its apparently fast reproductive turnover, enable it to persist longer than many other large mammals. It does well in high elevation forests, which are relatively intact in Costa Rica. It is important as food for large carnivores (it was found to be a staple of both jaguars and pumas in Corcovado, for example), and as a seed predator. It may disperse some seeds by spitting them out or passing them whole, but destroys most; only 6% of stomach samples analyzed in Peru contained any whole seeds.

References

Barrette 1987; Bisbal 1994; Bodmer 1989, 1991; Branan & Marchington 1987; Branan et al. 1985; Chinchilla 1997; Dietrich 1993; Eisenberg & McKay 1974; Emmons & Feer 1997; Emmons et al. 1997; Gardner 1971b; Jones 1982; MacNamara & Eldridge 1987; Timm et al. 1989; Roberto Wesson, pers. comm.

O: Artiodactyla F: Cervidae

WHITE-TAILED DEER

Odocoileus virginianus
(plate 37)

Names: The species was given its scientific name by eccentric nineteenth-century naturalist Constantine Rafinesque. Rafinesque is famous for the zeal with which he found and named new species—on one occasion, for example, he named 12 new "species" of thunder and lightening above the Ohio river. The names he invented were sometimes inaccurate. Having found a hollow white-tailed-deer tooth in a cave in Virginia, he named the genus *Odocoileus*, when the correct Greek for hollow tooth is *Odontocoelus*. **Spanish**: *Venado*; *venado colablanca.*

Range: Southern Canada to northern South America as far as Bolivia and Brazil; from sea level to at least 1,300 m (4,300 ft) in Costa Rica; mostly in deciduous forest and other relatively open habitats, rarely in mature wet forest.

Size: 130 cm, 30 kg (51 in., 65 lbs); smaller in Costa Rica than in North America; males larger than females.

Similar species: See red brocket deer (pl. 37, p. 318).

Natural history: White-tailed deer forage by day or night, singly or in small groups. Groups typically consist either of does and their fawns and yearlings, or of adult males. Whitetails feed mainly on leaves, supplemented with fruit, seeds, flowers, fungi, twigs, and bark. Favorite foods in northwestern Costa Rica include the leaves of silk cotton trees (*Cochlospermum vitifolium*, illustrated at right), yellow hogplum trees (*Spondias mombin*, illustrated on p. 162), red hogplum trees (*S. purpurea*), wild hibiscus (*Malvaviscus arboreus*), vines in the poinsettia (*Dalechampia scandens*) and morning glory (*Ipomoea* spp.) families, and various types of grass (such as *Cyperus*, *Brachiaria*, and *Oplismenus*).

Leaf of the silk cotton tree (*Cochlospermum vitifolium*).

Whitetails in Mexico are very fond of mistletoes (*Phoradendron* spp.); according to one hypothesis, whitetails may be accountable for the lack of these common parasites on branches less than about 2 m (7 ft) off the ground in Guanacaste. Whitetails in northwestern Costa Rica also enjoy fruits and seeds, such as acorns (*Quercus oleoides*), panama seeds (*Sterculia apetala*), guácimos (*Guazuma ulmifolia*, illustrated on p. 310), figs (*Ficus* spp.), and nances (*Brysonima crassifolia*).

White-tailed deer droppings (each about 1 to 2 cm long).

Whitetails are frequent visitors to salt licks. Studies in Palo Verde (and other parts of the world) have shown that these patches of exposed earth, typically located on exposed river banks or in dried-up swamps, are unusually rich in sodium. By gnawing at such patches (a habit known as pedophagy, from the Greek *pedon*, meaning ground, and *phago*, meaning to eat), whitetails may compensate for sodium deficiencies in their largely herbivorous diet.

Whitetails in North America use overlapping home ranges of anywhere from 24 to several hundred hectares. A few whitetails radio-tracked in Palo Verde had home ranges of under 23 ha (57 A). Males use larger ranges than do females, especially during the breeding season. Whitetails do not form lasting pair bonds, and both bucks and does are polygamous. Bucks compete with one another for breeding rights mostly through displays (see p. 308). Usually one buck will cede before a conflict escalates to physical confrontation, but if not, head-butting and pushing contests ensue. Such belligerence can cost bucks broken antlers or, rarely, the bucks can become permanently entangled and die of starvation.

In northwestern Costa Rica, the rutting and breeding season lasts from July to October. Bucks shed their antlers in January or February, and immediately start growing new ones. As the antlers develop, they are soft and covered with a velvety skin that supplies blood for growth. By June or July, the antlers are fully grown. Circulation in the velvet stops and the layer dies and peels away, revealing the hard antlers beneath. Bucks hasten the shedding of velvet by scraping their antlers against trees and shrubs.

After the velvet has fallen off, and as the breeding season gets under way, bucks continue to rub trees, but in a more vigorous fashion. These later tree rubs serve as visual and scent marks; bucks tear away bark with their antlers to expose conspicuous, roughly 20 cm (8 in.)-long stretches of pale inner wood, and rub against the mark with enlarged scent glands that they develop on their foreheads during the breeding season. Studies in the United States have shown that bucks there have preferences when choosing rubbing trees. They tend to select roughly 2 m (6 ft)-tall saplings with trunks about 2.5 cm (1 in.) in diameter, and prefer those that have smooth, aromatic bark and few low branches.

A white-tailed deer rub.

Another marking strategy used by bucks (and occasionally by does) is ground-scraping. The deer rake roughly 30 to 75 cm (1 to 2 ½ ft) long depressions with their hooves, usually in areas free of ground vegetation. Sometimes they defecate or urinate in the scrape. Bucks often lick or rub their forehead against branches that overhang scrape marks.

A white-tailed deer scrape.

Gestation lasts six and a half to seven months, such that fawns are born in the dry season, mostly in February and March. While older females in North America frequently give birth to three fawns in a litter, Costa Rican whitetails usually give birth to single fawns, or occasionally to twins. Newborns are left in a concealed spot and visited periodically for nursing. The fawns start following the mother after 10 days to a month. They can eat leaves by the time they are two months old. Most fawns in northwestern Costa Rica reach this age right around the onset of the rainy season in May, when there is a flush of new vegetation. They are usually weaned shortly after they start to eat vegetation, ceasing to nurse completely by the time they are four months old. They lose their spotted coat at the age of four to six months. They may stay with the mother for a year or more. Females have been known to breed after just six months but seldom do so until their second year. Males are sexually mature at about one and a half years of age. White-tails can live 16 years in the wild, although very few survive that long.

Reproduction and antler cycle of the white-tailed deer in northwestern Costa Rica. Males are depicted in green, females in red, and fawns in orange (adapted from Solís et al. 1986).

Sounds: Forefoot stamping when nervous; whistling snorts in alarm; guttural grunts by males when rutting; lamblike bleats by fawns.

Mythology: The Bribri of Costa Rica believe that the deer was sent by their God, Sibú, to help form and shape land and sea. For many cultures, the annually regrowing antlers of whitetails symbolize regeneration.

Conservation: Regeneration has indeed been the story of the whitetail in Costa Rica over recent years. As a source of meat, leather, and other products, the white-tailed deer has been one of the most important game animals in the Americas. To a lesser degree, it is also hunted as a pest, for it can damage crops such as corn and beans where farms border suitable cover. Like many of the earth's resources, whitetails started to be exploited in a nonsustainable fashion in the twentieth century. In 1944 alone, a reported 60,000 whitetail hides were exported to South America from Costa Rica, and the deer's meat was so abundant during that period that it was cheaper than beef. Such massive hunting pressure, coupled with the destruction of their principal habitat, the dry forests of the northwest, had made whitetails rare in Costa Rica by the 1960s.

Since then, the creation of hunting laws and, especially, protected areas has enabled whitetails to recuperate, at least in the northwestern parks. The increasing scarcity of jaguars, pumas, and other deer predators has probably aided this recovery. White-tailed deer in turn probably contribute to the process of forest regeneration. In deciduous areas at least, they forage extensively both in scrubby, regrowing areas and in forest, and probably defecate or spit out primary forest plants into regenerating areas.

References

Atkeson & Marchinton 1982; Bozzoli 1986; Dimare 1994; Emmons & Stark 1979; Gavin et al. 1984; Halls 1984; Hilje & Monge 1988; Jaeger 1978; Janzen 1983i, 1986; Kile & Marchinton 1977; Marchinton & Hirth 1984; Mena 1978; Rodríguez & Solís 1994; Rodríguez et al. 1985a,b; Rue 1968; Sáenz 1994b; Smith 1982; Teer 1994; Vaughan & Rodríguez 1994.

Dolphins (order Cetacea)

Distribution and Classification

The order Cetacea (from the Greek *ketos*, which means sea monster) is the largest of the three orders of marine mammals, with about 76 species of dolphins, porpoises, and whales. The other two marine orders are the Sirenia, with four species (pp. 281 to 291), and the Pinnipedia, with 34 species of seals and sea lions.

The order is typically divided into two suborders. Cetaceans that do not echolocate and have baleen plates (see p. 328) instead of teeth, a double blow hole, and a symmetrical skull comprise the suborder Mysticeti. There are only 10 species in this suborder, known collectively as the baleen whales. Cetaceans that echolocate and have teeth, a single blowhole, and (with a few exceptions) an asymmetrical skull comprise the suborder Odontoceti. There are 66 species in this suborder, known collectively as the toothed whales. The dolphin family (Delphinidae) belongs to the toothed whale suborder. It contains 32 species, distributed almost worldwide. In addition to dolphins, the family includes the melon-headed whale, pilot whales, and killer whales.

Some 25 species of cetaceans occur or are expected to occur in Costa Rican waters, but all except the tucuxi are essentially marine, and therefore fall outside the scope of this book. For those interested, Reid 1997 includes descriptions and illustrations of Central America's marine mammals.

Evolution

Cetaceans have a bizarre history. Their roots can be traced back to a group of terrestrial, hoofed, herbivorous mammals known as condylarths that inhabited the earth around the time dinosaurs finally disappeared, about 65 million years ago. Some descendents of the condylarths—either from a group of predators known as the mesonychids or from the line that gave rise to modern artiodactyls—started to feed around water. These animals were furred, hoofed, and carnivorous, and decidedly unwhalelike. One of these, named *Ambulocetus natans* (illustrated overleaf), inhabited the earth about 50 million years ago, and was likened in a recent National Geographic article to a "shaggy crocodile."

The descendents of *Ambulocetus* became steadily less amphibious and more aquatic. Through countless generations of natural selection, their nostrils moved back and higher up the skull to facilitate breathing at the water's surface. Their limbs became ever shorter and more streamlined, their body hair shortened and disappeared, and their hoofs gradually morphed into flippers. By about 46 million years ago, they had evolved webbed feet. By about 35 million years ago, they had lost all external traces of hindlimbs, but had developed large, muscular tails for propulsion. By this time, they looked much like modern cetaceans, and had already split into the two groups that are recognized today—the Mysticeti and the Odontoceti. Dolphins appear to be a relatively recent offshoot of the Odontoceti, since the oldest delphinid fossils date back only about 15 million years.

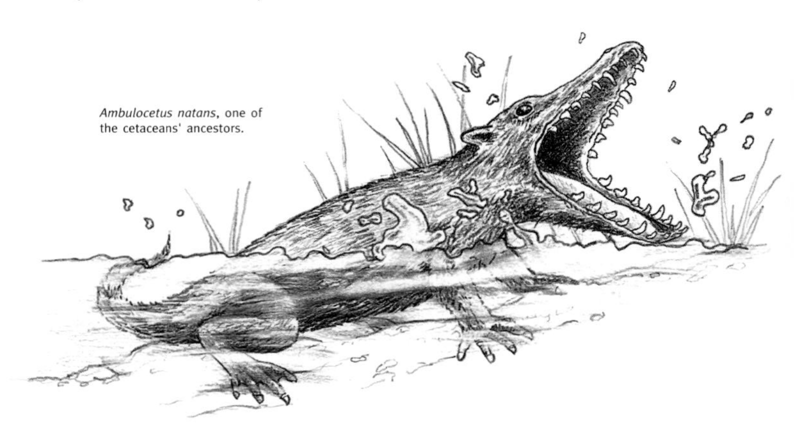

Ambulocetus natans, one of the cetaceans' ancestors.

Teeth, Skull, and Feeding

The Mysticeti feed by swimming with their mouths open, and then using their huge tongues to force water out through their baleen plates. The plates—horny, comblike structures that evolved from the ridges that arch across the palate in the roof of the mouth—arch down from the upper jaw and act as giant sieves. As water is expelled, the plates trap mostly small food, such as plankton, inside the mouth for swallowing. As a curious holdover from the group's past, Mysticeti fetuses develop teeth, but reabsorb them before birth in favor of the baleen plates.

Dolphins and other Odontoceti, on the other hand, tend to catch larger prey—mainly fish and squid—which they pursue actively and grasp with their teeth. One of them, the killer whale, is the world's largest carnivore, sometimes taking other cetaceans or seals. Toothed whales tend to swallow prey whole and seldom use their teeth for chewing. The teeth are of a uniform, conical shape, and cannot be distinguished into incisors, canines, premolars, or molars.

Body Design

Since they never have to support their weight on land, cetaceans do not face the same size limitations as terrestrial mammals. Among them is the largest animal ever to inhabit the earth, the up to 33 m (110 ft) and 190,000 kg (200 ton) blue whale (a resident of Costa Rican waters off the Pacific coast), which weighs more than 30 times the weight of largest living land mammal, the elephant. Large size is especially important to cetaceans because water absorbs heat much more effectively than air; large creatures have less surface area relative to their volume than smaller animals, and thus retain heat longer. Delphinids are among the smaller cetaceans, ranging in size from the tucuxi to the up to 10 m (33 ft)-long killer whale.

The flippers of cetaceans are modified forelimbs, while the tail fluke and the dorsal fins have no skeletal support. What remains of the hindlimbs—tiny vestiges of their terrestrial past—is now used only to anchor the muscles of the genitalia (see illustration overleaf). (Ancient genes still express themselves from time to time, however, causing a few aberrant whales to be born with visible hindlegs.) In males, the genitalia muscles control the movement of the penis in and out of the abdomen, where the penis is housed when not in use in order to improve streamlining. Likewise, the teats of females are housed in slits on either side of the abdomen. The streamlining of delphinids and other cetaceans is further enhanced by the absence of external ears. Instead, they have a tiny hole on each side of the head that connects directly to the inner ear.

Dolphins propel themselves with up and down movements of the tail fluke, and steer with their three fins. They are faster than any other marine mammals, with some species capable of swimming at 30 km/h (20 mph) for sustained periods. They use their speed and agility to escape predators or, occasionally, to attack them; groups of dolphins have been known to kill sharks by ramming them.

Of course, unlike most other marine creatures (such as, for example, sharks and other fish), marine mammals must come to the surface to breathe. Dolphins and other cetaceans have a number of interesting adaptations to help them deal with this problem. Their nostrils (the blowholes) are situated on their backs, a feature any human swimmer would envy. As they start a dive, they exhale, in order to reduce buoyancy and sink faster. While underwater they reduce their oxygen needs by slowing their heartbeat and the rate of blood supply to their muscles. They can tolerate high levels of carbon dioxide, repaying their oxygen debt upon resurfacing. Like manatees, they can exchange 90% of the air in their lungs with a single breath. That single breath is enough to enable some whales to stay submerged for as long as an hour.

Senses

Dolphins have no sense of smell and lack sweat and scent glands. They have good vision, especially underwater, and can also see objects several meters above water. Many dolphins, including the tucuxi, use visual displays during courtship. Their eyes are protected from saltwater by an oily fluid secreted by tear glands.

The delphinids' most developed sense by far is hearing. They use a diverse and complex language of clicks and whistles to navigate, communicate, and find prey. Clicks are used mostly in echolocation, while whistles are used in social interactions. In at least some dolphin species, individuals produce unique patterns of whistles that may help them recognize one another. Dolphins lack true vocal chords and are thought to produce sounds by moving air from their lungs into nasal sacs situated near the blowhole, which they can do without coming to the surface to breathe. The sounds are reflected along the top of the skull to the forehead, where a pocket of fat known as the melon focuses and directs the sounds outward.

The need to interpret these complicated sounds may explain why dolphins have evolved such large brains. The only mammals with brains larger than our own are some cetaceans and elephants, and dolphins' large brains are particularly impressive because of their relatively small body size. Some have speculated that, since bats also interpret elaborate sonic information but have relatively small brains, dolphins' large brains may have some other, as yet not understood function. Whatever the reason, dolphins have been brainy for a long time—fossils show that while humans acquired their current brain size only about 100,000 years ago, cetaceans have had large brains for millions of years.

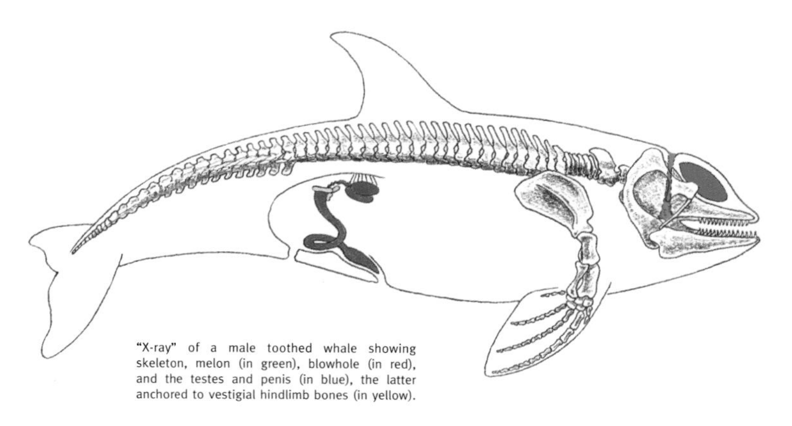

"X-ray" of a male toothed whale showing skeleton, melon (in green), blowhole (in red), and the testes and penis (in blue), the latter anchored to vestigial hindlimb bones (in yellow).

Conservation

Cetacean populations have probably suffered more over the last two centuries than at any other time in their history. The baleen plates of the large whales have been used for a variety of purposes, such as to stiffen shirt collars; oil from cetaceans' blubber and other parts of the body has been used for lamp fuel and in the production of margarine, soaps, explosives, and lubricants for small clocks, among many other things; cetaceans' bones have been ground to make glue, gelatin, and fertilizer; and some cetaceans have been hunted for their meat. Dolphins have been used for most of these purposes on occasion, although they have never been persecuted as actively as the whales.

Cetacean hunting has slowed significantly over recent years, but dolphins still fall victim to irresponsible fishing practices. Despite dolphin-friendly tuna campaigns, as many as 100,000 dolphins a year died in nets intended for tuna and other fish through the 1980s. Another growing threat is, of course, pollution. Toxins from industrial, agricultural, and domestic waste tend to work their way up the food pyramid, becoming most concentrated in the ocean's top carnivores. Since the late 1980s, tens of thousands of large marine mammals, including dolphins, have suddenly been dying in European waters due to viral infections. Since the deaths have been focused around polluted areas, and since many of the pollutants are known to hinder mammals' immune systems, some believe that pollution is the underlying cause of these epidemics. Since some pollutants also cause infertility, there is also concern about how effectively depleted populations will be able to recover.

References

Brownell et al. 1989; Chadwick 2001; Harrison 1972, 1974, 1977; Janzen & Wilson 1983; Leatherwood & Reeves 1983; Lilly 1977; Macdonald 1995; Nowak 1991; Rice 1984; Terry 1983.

SPECIES ACCOUNT

O: Cetacea F: Delphinidae

TUCUXI

Sotalia fluviatilis

(plate 38)

Names: Also known as the gray or river dolphin. The name *tucuxi*, which originated with the Tupi natives of the Amazon, is pronounced "tookooshee." The genus is derived from the Latin *sote*, meaning to keep, and the Greek *alia*, meaning gathering, since pods of tucuxis maintain tight formations as they swim. *Fluviatilis* means of rivers. There are some subtle differences between tucuxis from the Amazon river basin and those found everywhere else, and experts often distinguish the two by referring to the former as riverine and the latter as marine, although the marine tucuxi also ascends rivers. The two populations are sometimes listed as different subspecies: *S. f. fluviatilis* and *S. f. guianensis*, respectively. Features that set the marine tucuxi—the one found in Costa Rica—apart from the riverine tucuxi include larger body size, more teeth in the upper jaw, and a longer gestation period. **Spanish**: *Delfín tucuxi*; *delfín de río*.

Range: Honduras to southern Brazil, along the Atlantic coast and inland up large rivers, notably the Amazon and its tributaries, where some populations live as far west as the base of the Andes, some 2,500 km (1,500 miles) from the ocean. In South America, tucuxis seldom frequent sections of river that are less than 6 m (20 ft) deep. The presence of this species in Costa Rica was not confirmed officially until 1997, when experts, following tips from locals, found a group of 10 off the coast of the Gandoca Manzanillo area. Locals had been noticing groups of tucuxis in salt and fresh water since at least the early 1990s.

Size: 170 cm, 50 kg (67 in., 110 lbs).

Similar species: At least six other dolphin species are found in Costa Rican waters, but the tucuxi is the only one that forages extensively in freshwater. When in saltwater, the tucuxi stays within a few kilometers of the coast. The only other Costa Rican dolphin species that enters river mouths to feed is the bottle-nosed dolphin (*Tursiops truncatus*). At the Sixaola river, Costa Rica's southeastern border with Panama, bottle-nosed dolphins have never been observed more than 500 m (550 yds) inland from the mouth, while tucuxis have been seen several kilometers upstream. The tucuxi has similar coloration to the bottlenose, but is distinguished by its much smaller size (bottlenoses can reach body lengths of up to 4 m or 13 ft); its triangular, only slightly curved dorsal fin; and its longer, thinner snout. The two species sometimes swim together. All other Costa Rican dolphins, except juvenile Atlantic spotted dolphins (*Stenella frontalis*), have distinctive, bold patterns. The young spotted dolphins are distinguished by their more curved dorsal fins, thinner bodies and flippers, and slightly longer beak.

The bottle-nosed dolphin (top) is the only dolphin species other than the tucuxi that enters Costa Rican river mouths. The tucuxi (bottom) is similar to the bottlenose in color, but can be distinguished by its smaller size; shorter, less curved dorsal fin; and longer, narrower snout.

Natural history: Tucuxis forage by day, especially in the early morning and late afternoon. They feed mostly on fish, such as members of the herring (Clupeidae), croaker and seatrout (Sciaenidae), coney (Serranidae), and anchovy (Engraulidae) families, and also enjoy shrimp and crabs. Riverine tucuxis take fish up to about 37 cm (15 in.) in length. Tucuxis, even young ones, tend to have very worn-down teeth, presumably as a result of chewing on hard or silt-covered prey.

Tucuxis travel in pods of up to 30 (but most often just two to four) tightly clustered individuals that move in close synchrony. Pod members have been seen cooperating to catch food by forming circles around schools of fish and then converging in unison, or by herding fish up against banks or even fishing nets. Pods tend to be consistent in daily movement patterns. Tucuxis typically come to the surface to breathe at least once a minute, often arching their heads and upperparts above the surface of the water. They are less curious than other dolphins and are quick to swim away if disturbed. They also tend to swim more slowly and leap out of the water less often than other dolphins, but occasionally wild tucuxis surf waves and perform leaping somersaults.

A gestation period of an estimated 11 to 12 months results in single, roughly 65 cm (26 in.) long young. Calves may stay with their mothers for several years, as do the calves of most dolphin species; there is evidence that some groups of three or four tucuxis consist of adult pairs and their young. Captive tucuxis can live at least 16 years. The tucuxi's skin color is thought to become lighter with age.

Sounds: Short puffs upon surfacing; whistles, pops, and clicks underwater. In other dolphins at least, clicks are used in echolocation to navigate and find food as well as to herd schools of fish. For a species that lives in often murky waters, and in tight social groups, sounds may be particularly important.

Mythology: Tucuxis are considered sacred by some South American natives, who believe tucuxis return the bodies of drowned people to shore.

Conservation: Recently, tucuxis have become economically valuable in Costa Rica, for they are one of the main attractions on boat tours offered at Manzanillo. At sea, tucuxis are vulnerable to irresponsible fishing practices. Inland, much of their potential freshwater habitat has become developed and polluted.

References

Best & da Silva 1984; Borobia & Barros 1989; Borobia et al. 1991; Bössenecker 1978; Caldwell & Caldwell 1970; Carr 1994; Carwardine 1995; da Silva 1994; da Silva & Best 1996; Diaz 1997; Geise 1989; Leatherwood & Reeves 1983; Magnusson et al. 1980; Nowak 1991; Perrin & Reilly 1984.

ACKNOWLEDGMENTS

My first thank you must go to my publishers, John K. McCuen and Marc Roegiers of Zona Tropical Publications, for giving me the opportunity to create this book. I feel very fortunate that their unwavering patience and encouragement endured as our original concept of a relatively simple publication evolved into an eight-year-long project.

Much of the information presented in this book might appear to be quite straightforward, but figuring out, for example, what an animal eats, how long it lives, or how far it wanders in fact requires countless hours of challenging research. None of that research was done by me; it was conducted by the hundreds of authors listed in the bibliography and the thousands of nameless assistants who helped them. I would like to apologize to them for any injustices I may have committed in reporting their work and to express my deepest gratitude and admiration to them, for this book is a fruit of their labor.

Tracking down all these researchers' publications was not an easy task, and I could not have done it without the help of many people. I am indebted to the staff of the Biodoc library at the Universidad Nacional in Heredia and the Biblioteca Nacional in San José for hours of shelf-pacing in search of journal back issues, as well as to the unfortunate photocopier operators at both institutions. For help in exploiting the resources of the Organization for Tropical Studies library at La Selva on several occasions, I am grateful to Ruth Rodríguez and Orlando Vargas. Richard Laval and Frank Joyce generously allowed me to pillage their extensive personal libraries. Valuable source literature was also loaned to me by Federico Chinchilla, David Cordero, Marc Egger, Mark Fenton, Ricardo Guindon, Bill Haber, Cristina Hansen, Julia Lowther, Trudy Millerstrom, Alan Pounds, Bernal Rodríguez, Roberto Wesson, and Willow Zuchowski. Annie Wainwright kindly sought after and sent me some of the more obscure references from libraries in the United States.

For helping me acquire the tracks, skeletons, dead mammals, and other objects I used as models for the illustrations in this book, I am grateful to Denis Calderón, Eladio Cruz, Jim Daniels, Mercedes Díaz, Mark Fenton, Jimmy Flott of the La Salle Natural History Museum in San José, Bill Haber, Werner Hagnauer and the late Lily Bodmer of the Las Pumas wildcat sanctuary at Finca La Pacífica in Guanacaste, Laura May, mammal skeleton expert Trudy "Bones" Millerstrom, Alexander Molina, Luis Obando, Lucas Ramírez, Pedro Rosafa, and Willow Zuchowski. For sharing anecdotes of past mammal abundance and hunting lore, I thank Don Ebelio González, Don Wilford "Wolf" Guindon, Don Miguel Salazar, and Don Angel Villegas. And for protecting the immeasurably beautiful forests where I reside, and where I have done most of my observation of Costa Rican mammals and their signs, I thank the Tropical Science Center, the Monteverde Conservation League, and the handful of individuals whose foresight enabled the protection of the land these two organizations now manage.

A special acknowledgment must go to Bob Law for miraculously salvaging an early draft of the book—months of work—from the clutches of a computer virus that surreptitiously ate its way through my hard drive and floppy disks. A number of individuals read and commented on subsequent drafts, or parts thereof, and I am greatly indebted to the following for this time- and energy-consuming favor: Federico Chinchilla, Marc Egger, Deborah Fitzpatrick, Valerie Giles, Bill Haber, Cristina Hansen, Frank Joyce, Richard Laval, Alan Pounds, Andrew Vallely, and Willow Zuchowski. For constructive criticism of the illustrations, I thank Orlando Calvo, Mark Fenton, Larry Landstrom, Richard Laval, Katherine Stoner, and Roberto Wesson.

Designing this book was a considerable challenge, and could not have been pulled off without the tremendous technical expertise and patience of Adrián Soto of Servigráficos.

For their indirect but equally important involvement with this book, or what led up to it, I am eternally grateful to my parents, Dod and Annie Wainwright, as well as to John Newmark, Mi Wainwright, Captain John and Penelope Garrett, Sarah DiPerna, Valerie Austin, Martin King, Anthony Holcombe, Susan and André Roegiers, Jim Butler, Don Mitchell, Ricardo Soto, Beatriz Rodríguez and Rolando Herrera and their children Fabiola and Alonso, Jim Wolfe, Luz Delgado, Dina Fernández, and to my colleagues, the guides and park guards of Monteverde's various reserves.

My final and most profound thank you goes to my wife, Patricia Maynard, and sons, Mark and Kyle, for all their support, companionship, patience, and love.

Mark Wainwright

BIBLIOGRAPHY

Acha, P. N., and A. M. Alba. 1988. Economic losses due to *Desmodus rotundus*. In: Greenhall & Schmidt '88. Pp. 207-14.

Adams, J. K. 1989. *Pteronotus davyi*. Mamm. Species 346: 1-5. Am. Soc. Mammalogists.

Adler, G. H., M. Endries, and S. Piotter. 1997. Spacing patterns within populations of a tropical forest rodent, the spiny rat, on islands in Panama. J. Zool. London 241: 43-53.

Aguilar, C. H. 1986. Religión y magia entre los indios de Costa Rica de origen sureño: chamanismo. Tercera edición. Editorial Univ. de Costa Rica, San José.

Aiello, A. 1985. Sloth hair: unanswered questions. In: Montgomery '85a. Pp. 213-18.

Allen, C. H., R. L. Marchinton, and W. M. Lentz. 1985. Movement, habitat use and denning of opossums in the Georgia Piedmont. Am. Midl. Nat. 113: 408-12.

Almeda, F., and C. M. Pringle (eds.). 1988. Tropical rainforests: diversity and conservation. Calif. Acad. Sc. and Amer. Assoc. for the Adv. of Sc.

Alonso-Mejía, A., and R. A. Medellín. 1992. *Marmosa mexicana*. Mamm. Spec. 421. Am. Soc. Mammalogists.

Altenbach, J. S. 1989. Prey capture by the fishing bats *Noctilio leporinus* and *Myotis vivesi*. J. Mamm. 70: 421-24.

Altrichter, M. 1997. Estrategia de alimentación y comportamiento del chancho cariblanco *Tayassu pecari* en un bosque húmedo tropical de Costa Rica. Masters Thesis. Univ. Nacional, Heredia.

Altrichter, M., J. Sáenz, and E. Carrillo. 1999. Chanchos cariblancos (*Tayassu pecari*) como depredadores y dispersores de semillas en el Parque Nacional Corcovado, Costa Rica. Brenesia 52: 53-59.

Alvarez, J., M. R. Willig, J. K. Jones, Jr., and W. D. Webster. 1991. *Glossophaga soricina*. Mamm. Species 379. Am. Soc. Mammalogists.

Anderson, P. K. 1984. Sea cows and manatees. In: Macdonald '84a. Pp. 292-99.

Anderson, S., and J. K. Jones, Jr. (eds.). 1984. Orders and families of Recent mammals of the world. John Wiley & Sons, New York.

Antikatzides, T., S. Erichsen, and A. Spiegel (eds). 1975. The laboratory animal in the study of reproduction. Gustav Fischer Verlag, Stuttgart.

Aranda, J. M. 1981. Rastros de los mamíferos silvestres de Chiapas. Inst. Nac. Invest. Rec. Bióticos, México.

Aranda, J. M. 1993. Habitos alimentarios del jaguar (*Panthera onca*) en la Reserva de la Biósfera de Calakmul, Campeche. In: Medellín & Ceballos '93. Pp. 231-38.

Aranda, J. M., and I. March. 1987. Guía de los mamíferos silvestres de Chiapas. Inst. Nac. Invest. Rec. Bióticos. Xalapa, Veracruz, México.

Arauz, J. 1993. Estado de conservación del mono tití (*Saimiri oerstedi citrinellus*) en su área de distribución original, Manuel Antonio, Costa Rica. Masters Thesis. Univ. Nacional, Heredia.

Arellano-Sota, C. 1988. Biology, ecology and control of the vampire bat. Rev. Infect. Dis. 10(4): S615-19.

Arita, H. T., and K. Santos-del-Prado. 1999. Conservation biology of nectar-feeding bats in Mexico. J. Mamm. 80(1): 31-41.

Arnaud, G. 1993. Alimentación del coyote (*Canis latrans*) en Baja California Sur, México. In: Medellín & Ceballos '93. Pp. 205-15.

Arroyo-Cabrales, J., and J. K. Jones, Jr. 1988. *Balantiopteryx plicata*. Mamm. species 301. Am. Soc. Mammalogists.

Association for Tropical Biology, and the Organization for Tropical Studies (eds.). Symposium and annual meeting: tropical diversity origins, maintenance, and conservation. ATB and OTS, San José.

Atkeson, T. D., and R. L. Marchinton. 1982. Forehead glands in white-tailed deer. J. Mamm. 63: 613-17.

Atramentowicz, M. 1982. Influence du milieu sur l'activité locomotrice et la reproduction de *Caluromys philander*. Rev. Ecol. (Terre et Vie) 36: 375-95.

Atramentowicz, M. 1986. Dynamique de population chez trois marsupiaux didelphidés de Guyane. Biotropica 18: 136-49.

Azevedo, F. C. C. de. 1996. Notes on the behavior of the margay *Felis wiedii* (Schinz, 1821), (Carnivora, Felidae), in the Brazilian Atlantic Forest. Mammalia 60: 325-28.

Babbitt, K. 1988. Preliminary examination of collared peccary foraging on *Deiffenbachia*. OTS 88-3: 345-47.

Bakarr, M. J. 1990. Rodents as dispersers of VA mycorrhizal fungus (VAMF) spores: evidence from a Costa Rican highland forest. In: Loiselle '90. Pp. 233-37.

Baker, H. G. 1983. *Ceiba pentandra*. In: Janzen '83c. Pp. 212-15.

Baker, M. 1996. Fur rubbing: use of medicinal plants by capuchin monkeys. Am. J. Primatol. 38: 263-70.

Baker, R. H. 1983. *Sigmodon hispidus*. In: Janzen '83c. Pp. 490-92.

Baker, R. H., and M. W. Baker. 1975. Montane habitat used by the spotted skunk (*Spilogale putorius*) in Mexico. J. Mamm. 56: 671-73.

Baker, R. J., and C. L. Clark. 1987. *Uroderma bilobatum*. Mamm. Species 279. Am. Soc. Mammalogists.

Baker, R. J., J. K. Jones, Jr., and D. C. Carter (eds). 1977. Biology of bats of the New World family Phyllostomatidae, Part II. Spec. Publ. Mus. Texas Tech Univ. 13.

Baker, R. J., J. C. Patton, H. H. Genoways, and J. W. Bickham. 1988. Genic studies of *Lasiurus* (Chiroptera: Vespertilionidae). Occas. Paper Mus. Texas Tech Univ. 117: 1-15.

Baldwin, J. D. 1970. Reproductive synchronization in squirrel monkeys (*Saimiri*). Primates 11: 317-26.

Baldwin, J. D., and J. I. Baldwin. 1971. Squirrel monkeys (*Saimiri*) in natural habitats in Panama, Colombia, Brazil, and Peru. Primates 12: 45-61.

Baldwin, J. D., and J. I. Baldwin. 1972a. Population density and use of space in howling monkeys (*Alouatta villosa*) in south-western Panama. Primates 13: 371-79.

Baldwin, J. D., and J. I. Baldwin. 1972b. The ecology and behavior of squirrel monkeys (*Saimiri oerstedii*) in a natural forest in western Panama. Folia Primatol. 18: 161-84.

Baldwin, J. D., and J. I. Baldwin. 1976. Primate populations in Chiriquí Panama. In: Thorington & Heltne '76. Pp. 20-31.

Baldwin, J. D., and J. I. Baldwin. 1981. The squirrel monkeys, genus *Saimiri*. In: Coimbra-Filho & Mittermeier '81. Pp. 277-330.

Banfield, A. W. F. 1974. The mammals of Canada. Univ. Toronto Press, Buffalo.

Barbour, R. W., and W. H. Davis. 1969. Bats of America. University Press of Kentucky, Lexington.

Barbour, R. W., and W. H. Davis. 1974. Mammals of Kentucky. University Press of Kentucky, Lexington.

Barker, W. 1956. Familiar animals of North America. Harper Row.

Barnard, C. J. 1984. Shrews. In: Macdonald '84a. Pp. 758-63.

Barnes, R. D., and Barthold, S. W. 1969. Reproduction and breeding behaviour in an experimental colony of *Marmosa mitis* Bangs (Didelphidae). J. Reprod. Fert., suppl. 6:477-82.

Barreto, G. R., O. E. Hernández, and J. Ojasti. 1997. Diet of peccaries (*Tayassu tajacu* and *T. pecari*) in a dry forest of Venezuela. J. Zool. London 241: 279-84.

Barrette, C. 1987. The comparative behavior and ecology of chevrotains, musk deer, and morphologically conservative deer. In: Wemmer '87. Pp. 200-213

Bateman, G. C., and T. A. Vaughan. 1974. Nightly activities of mormoopid bats. J. Mamm. 55: 45-65.

Bauchop, T. 1978. Digestion of leaves in vertebrate arboreal folivores. In: Montgomery '78. Pp. 193.

Beier, P. 1991. Cougar attacks on humans in the United States and Canada. Wildlife Society Bulletin 19: 403-12.

Beier, P. 1993. Determining minimum habitat areas and habitat corridors for cougars. Conserv. Biol. Vol. 7, No. 1: 94-108.

Bekoff, M. 1977. *Canis latrans*. Mamm. Species 79. Am. Soc. Mammalogists.

Bekoff, M. (ed.). 1978. Coyotes: biology, behavior and management. Academic Press, New York.

Bekoff, M., and M. C. Wells. 1980. The social ecology of coyotes. Sci. Am. 242: 130-48.

Bekoff, M., and M. C. Wells. 1986. Social ecology and behavior of coyotes. In: Rosenblatt et al. '86. Pp. 251-338.

Belcher, A., G. Epple, L. Greenfield, I. Kunderling, L. Soolnick, and A. Smith, III. 1989. Social communication in callitrichids: complexity of scent images. Am. J. Primatol. 18.

Belden, R. C. 1988. The Florida panther. Audubon Wildlife Report. Academic Press, Florida. Pp. 515-32.

Bennett, S., L. J. Alexander, R. H. Crozier, and A. G. Mackinlay. 1988. Are megabats flying primates? Contrary evidence from a mitochondrial DNA sequence. Austral. J. Biol. Sci. 41: 327-32.

Berta, A. 1987. Origin, diversity, and zoogeography of the South American Canidae. Fieldiana Zool., n.s. 39: 455-71.

Best, R. C., and V. M. F. Da Silva. 1984. Preliminary analysis of reproductive parameters of the boutu, *Inia geoffrensis*, and the tucuxi, *Sotalia fluviatilis*, in the Amazon River system. In: Perrin et al. '84. Pp. 361-69.

Best, R. C., and A. Y. Harada. 1985. Food habits of the silky anteater (*Cyclopes didactylus*) in the central Amazon. J. Mamm. 66(4):780-81.

Best, T. L. 1995a. *Sciurus variegatoides*. Mamm. Species 500. Am. Soc. Mammalogists.

Best, T. L. 1995b. *Sciurus deppei*. Mamm. Species 505. Am. Soc. Mammalogists.

Bewick, T. 1807. A general history of the quadrupeds. Bewick and Hodgson, Newcastle upon Tyne.

Biggers, J. D. 1967. Notes on reproduction of the woolly opossum (*Caluromys derbianus*) in Nicaragua. J. Mamm. 48: 678-80.

Bisbal, F. J. 1986. Food habits of some neotropical carnivores in Venezuela (Mammalia, Carnivora). Mammalia 50: 329-39.

Bisbal, F. J. 1994. Biología poblacional del venado matacán (*Mazama* spp.) (Artiodactyla: Cervidae) en Venezuela. Rev. Biol. Trop. 42(1/2): 305-314.

Blest, A. D. 1964. Protective display and sound production in some New World Arctiid and Ctenuchid moths. Zoologica 49: 161-81.

Bloedel, P. 1955a. Hunting methods of fish-eating bats, particularly *Noctilio leporinus*. J. Mamm. 36: 390-99.

Bloedel, P. 1955b. Observations on life histories of Panama bats. J. Mamm. 36: 232-35.

Bodmer, R. E. 1989. Frugivory in Amazonian ungulates. Ph.D. diss., Clare Hall, Cambridge Univ., England.

Bodmer, R. E. 1991. Strategies of seed dispersal and seed predation in Amazonian ungulates. Biotropica 23(3): 255-61.

Boinski, S. 1985. Status of the squirrel monkey (*Saimiri oerstedi*) in Costa Rica. Primate Conserv. 6: 15-16.

Boinski, S. 1987a. Birth synchrony in squirrel monkeys (*Saimiri oerstedi*): a strategy to reduce neonatal predation. Behav. Ecol. Sociobiol. 21: 393-400.

Boinski, S. 1987b. Habitat use by squirrel monkeys (*Saimiri oerstedi*) in Costa Rica. Folia Primatol. 49: 151-67.

Boinski, S. 1987c. Mating patterns in squirrel monkeys (*Saimiri oerstedi*): implications for seasonal sexual dimorphism. Behav. Ecol. Sociobiol. 21: 13-21.

Boinski, S. 1987d. The status of *Saimiri oerstedi citrinellus* in Costa Rica. Primate Conserv. 7: 69-72.

Boinski, S. 1988. Use of a club by a wild white-faced capuchin (*Cebus capucinus*) to attack a venomous snake (*Bothrops asper*). Am. J. Primatol. 14: 177-9.

Boinski, S. 1989. Why don't *Saimiri oerstedii* and *Cebus capucinus* form mixed-species groups? Int. J. Primatol. 10: 103-14.

Boinski, S., and C. L. Mitchell. 1994. Male residence and association patterns in Costa Rican squirrel monkeys (*Saimiri oerstedi*). Am. J. Primatol. 34: 157-69.

Boinski, S., and J. D. Newman. 1988. Preliminary observations on squirrel monkey (*Saimiri oerstedi*) vocalizations in Costa Rica. Am. J. Primatol. 14: 329-43.

Boinski, S., and P. E. Scott. 1988. Association of birds with monkeys in Costa Rica. Biotropica 20: 136-43.

Boinski, S., and R. M. Timm. 1985. Predation by squirrel monkeys and double-toothed kites on tent-making bats. Am. J. Primatol. 9: 121-27.

Bonaccorso, F. J. 1979. Foraging and reproductive ecology in a Panamanian bat community. Bull. Florida State Mus. 24: 359-408.

Bonaccorso, F. J., W. E. Glanz, and C. M. Sandford. 1980. Feeding assemblages of mammals at fruiting *Dipteryx panamensis* (Papilionaceae) trees in Panama: seed predation, dispersal, and parasitism. Rev. Biol. Trop. 28: 61-72.

Bonino, N. 1990. Historia natural, evaluación del daño y combate de la taltuza *Orthogeomys heterodus* (Rodentia: Geomyidae) en una zona hortícola de Costa Rica. Masters Thesis. Univ. Nacional, Heredia.

Bonino, N. 1993. Características físicas y reproductivas de la taltuza *Orthogeomys heterodus* (Rodentia: Geomyidae) en Costa Rica. Brenesia 39-40: 29-35.

Bonino, N. 1994. Ambito de acción, uso del habitat y actividad diaria de la taltuza *Orthogeomys heterodus* (Rodentia: Geomyidae) en una zona hortícola de Costa Rica. Rev. Biol. Trop. 42(1/2): 297-303.

Bonino, N., and L. Hilje. 1992. Estimación de la abundancia de la taltuza *Orthogeomys heterodus* (Rodentia: Geomyidae) y el daño producido en una zona hortícola de Costa Rica. Manejo Integrado de Plagas (Costa Rica) 23: 26-31.

Borobia, M., and N. B. Barros. 1989. Notes on the diet of marine *Sotalia fluviatalis*. Marine Mamm. Sci. 5: 395-99.

Borobia, M., S. Siciliano, L. Lodi, and W. Hoek. 1991. Distribution of the South American dolphin *Sotalia fluviatalis*. Can. J. Zool. 69: 1025-39.

Bössenecker, P. G. 1978. The capture and care of *Sotalia guianensis*. Aquatic Mammals 6: 13-17.

Boucher, D. H. 1981. Seed predation by mammals and forest dominance by *Quercus oleoides*, a tropical lowland oak. Oecologia 49: 409-14.

Bowen, W. D. 1984. Coyote. In: Macdonald '84a. Pp. 62-3.

Bowyer, R. T. 1987. Coyote group size relative to predation on mule deer. Mammalia 51(4): 515-26.

Boyce, M. S. (ed.). 1988. Evolution of life histories of mammals. Yale Univ. Press, New Haven.

Bozzoli, M. E. 1986. El nacimiento y la muerte entre los bribris. Primera reimpresión. Editorial Univ. de Costa Rica, San José.

Bradbury, J. W. 1983a. *Saccopteryx bilineata*. In: Janzen '83c. Pp. 488-9.

Bradbury, J. W. 1983b. *Vampyrum spectrum*. In: Janzen '83c. Pp. 500-1.

Bradbury, J. W., and L. H. Emmons. 1974. Social organization of some Trinidad bats. I. Field Studies. Emballonuridae. Z. Tierpsychol. 45: 225-55.

Bradbury, J. W., and S. L. Vehrencamp. 1976a. Social organization and foraging in emballonurid bats. I. Field Studies. Behav. Ecol. Sociobiol. 1: 337-81.

Bradbury, J. W., and S. L. Vehrencamp. 1976b. Social organization and foraging in emballonurid bats. II. A model for the determination of group size. Behav. Ecol. Sociobiol. 1: 383-404.

Bradbury, J. W., and S. L. Vehrencamp. 1977a. Social organization and foraging in emballonurid bats. III. Mating systems. Behav. Ecol. Sociobiol. 2: 1-17.

Bradbury, J. W., and S. L. Vehrencamp. 1977b. Social organization and foraging in emballonurid bats. IV. Parental investment patterns. Behav. Ecol. Sociobiol. 2: 19-29.

Braker, E. H., and H. W. Greene. 1994. Population biology: life histories, abundance, demography, and predator prey interactions. In: McDade et al. '94. Pp. 244-55.

Branan, W. V., and R. L. Marchington. 1987. Reproductive ecology of white-tailed and red brocket deer in Suriname. In: Wemmer '87. Pp. 344-51.

Branan, W. V., M. Werkhoven, and R. L. Marchinton. 1985. Food habits of brocket and white-tailed deer in Suriname. J. Wildl. Mgmt. 49: 972-6.

Brandon, C. 1983. *Noctilio leporinus*. In: Janzen '83c. Pp. 480-1.

Bravo, S. P., M. M. Kowalewski, and G. E. Zunino. 1995. Dispersión y germinación de semillas de *Ficus monckii* por el mono aullador negro (*Alouatta caraya*). Boletín primatológico latinoamericano 5: 25-27.

Bravo, S. P., and G. E. Zunino. 1998. Effects of black howler monkey (*Alouatta caraya*) seed ingestion on insect larvae. Am. J. Primatol. 45: 411-15.

Broad, S. 1987. The harvest of and trade in Latin American spotted cats (Felidae) and otters (Lutrinae). Wildlife Trade Monitoring Unit, Cambridge.

Broad, S., R. Luxmoore, and M. Jenkins. 1988. Significant trade in wildlife: a review of selected species in CITES appendix II. Volume I: Mammals. Internatl. Union Conserv. Nat. Gland, Switzerland.

Broadbooks, H. E. 1952. Nest and behavior of a short-tailed shrew, *Cryptotis parva*. J. Mamm. 33: 241-3.

Bronson, F. H. 1979. The reproductive ecology of the house mouse. Quart. Rev. Biol. 54: 265-99.

Brooke, A. P. 1990. Tent selection, roosting ecology and social organization of the tent-making bat, *Ectophylla alba*, in Costa Rica. J. Zool. Lond. 221: 11-19.

Brooke, A. P. 1994. Diet of the fishing bat, *Noctilio leporinus* (Chiroptera: Noctilionidae). J. Mamm. 75: 212-18.

Brown, J. H., and R. C. Lasiewski. 1972. Metabolism of weasels: the cost of being long and thin. Ecology 53: 939-43.

Brown, P., T. W. Brown, and A. D. Grinnell. 1983. Echolocation, development, and vocal communication in the lesser bulldog bat, *Noctilio albiventris*. Behav. Ecol. Sociobiol. 13: 287-98.

Brownell, R. L., Jr., K. Ralls, and W. F. Perrin. 1989. The plight of the "forgotten" whales. Oceanus 32(1): 5-11.

Buchanan, O. M., and T. R. Howell. 1965. Observations on the natural history of the thick-spined rat, *Hoplomys gymnurus*, in Nicaragua. Ann. Mag. Nat. Hist., ser. 13, 8: 549-59.

Bucher, J. E. 1975. Studies of behavior and locomotion in some species of Neotropical marsupials. M. S. Thesis. Wright State Univ., Dayton, Ohio.

Bucher, J. E., and I. Fritz. 1977. Behavior and maintenance of the woolly opossum (*Caluromys*) in captivity. Lab Anim. Sci. 27: 1007-12.

Bucher, J. E., and R. S. Hoffmann. 1980. *Caluromys derbianus*. Mamm. Species 140. Am. Soc. Mammalogists.

Burger, J., and M. Gochfeld. 1992. Effect of group size on vigilance while drinking in the coati, *Nasua narica* in Costa Rica. Anim. Behav. 44: 1053-57.

Bussing, W. A. 1998. Freshwater fishes of Costa Rica. Second edition. Editorial de la Univ. de Costa Rica.

Byers, J. A., and M. Bekoff. 1981. Social, spacing and cooperative behaviour of the collared peccary *Tayassu tajacu*. J. Mamm. 62: 767-85.

Caldwell, D. K., and M. C. Caldwell. 1970. Echolocation-type signals by two dolphins, genus *Sotalia*. Quart. J. Florida Acad. Sci. 33: 124-31.

Caldwell, D. K., and M. C. Caldwell. 1985. Manatees—*Trichechus manatus*, *Trichechus senagalensis*, and *Trichechus inunguis*. In: Ridgeway & Harrison '85. Pp. 33-66.

Cameron, G. N., and P. A. McClure. 1988. Geographic variation in life history traits of the hispid cotton rat (*Sigmodon hispidus*). In: Boyce '88. Pp. 33-64.

Cameron, G. N., and S. R. Spencer. 1981. *Sigmodon hispidus*. Mamm. Species 158. Am. Soc. Mammalogists.

Campero, H. 1999. Variación y estructura genética dentro y entre grupos de chanchos de monte (*Tayassu pecari*) en el Parque Nacional Corcovado, Costa Rica. Masters Thesis. Univ. Nacional, Heredia.

Cantero, M. 2000. ¿Hasta cuándo sobrevivirán? La Nación, 8 nov. 2000. Seccion Viva.

Carr, A. 1953. High jungles and low. Univ. Florida Press, Gainesville.

Carr, T. 1994. The manatees and dolphins of the Miskito Coast Protected Area, Nicaragua. Caribbean Conservation Corporation. Contract no. T94070376 to the Marine Mammal Commission.

Carrillo, E. 1990. Patrones de movimientos y hábitos alimentarios del mapachín (*Procyon lotor*) en el Parque Nacional Manuel Antonio, Costa Rica. Masters Thesis. Univ. Nacional, Heredia.

Carrillo, E. 2000. Ecology and conservation of white-lipped peccaries and jaguars in Corcovado National Park, Costa Rica. PhD. Thesis. Univ. Massachusetts Amherst.

Carrillo, E., R. Morera, and G. Wong. 1994. Depredación de tortuga lora (*Lepidochelys olivacea*) y de tortuga verde (*Chelonia mydas*) por el jaguar (*Panthera onca*). Vida Silv. Neotrop. 3(1): 48-49.

Carrillo, E., J. Sáenz, T. K. Fuller, and M. Altrichter. 1997a. Size and stability of white-lipped peccary (*Tayassu pecari*) herds in Corcovado National Park, Costa Rica. In: Association.... '97.

Carrillo, E., J. Sáenz, T. K. Fuller, and M. Altrichter. 1997b. Home range and activity patterns of white-lipped peccaries (*Tayassu pecari*) in Corcovado National Park, Costa Rica. In: Association.... '97.

Carrillo, E., and C. Vaughan. 1988. La influencia de la lluvia sobre los movimientos de un mapachín en un bosque nuboso de Costa Rica. Rev. Biol. Trop. 36(2B): 373-6.

Carrillo, E., and C. Vaughan. 1993. Variación en el comportamiento de *Procyon* spp. (Carnivora: Procyonidae) por la presencia de turistas en un área silvestre de Costa Rica. Rev. Biol. Trop. 41(3): 843-48.

Carrillo, E., and G. Wong. 1992. Registro y medidas de restos de un *Cabassous centralis* (Edentata: Dasypodidae) en el Parque Nacional Manuel Antonio, Costa Rica. Brenesia 38: 153-54.

Carter, D. C. 1984. Perissodactyls. In: Anderson & Jones '84. Pp. 549-62.

Carwardine, M. 1995. Whales, dolphins, and porpoises. Dorling Kindersly.

Casinos, A., F. Bisbal, and S. Boher. 1981. Sobre tres ejemplares de *Sotalia fluviatalis* del Lago de Maracaibo (Cetacea-Delphinidae). Publ. Dept. Zool. Univ. Barcelona 7: 93-6.

Ceballos, G., and R. A. Medellín. 1988. *Diclidurus albus*. Mamm. Species 316. Am. Soc. Mammalogists.

Ceballos, G., and A. Miranda. 1987. Los mamíferos de Chamela, Jalisco. Instituto de Biología, Univ. Nac. Autón. México.

Chacón, M. 1996. Manejo en cautiverio y evaluación económica de la reproducción del tepezcuintle (*Agouti paca*) en la región atlántica de Costa Rica. Masters Thesis. Univ. Nacional, Heredia.

Chadwick, D.H. 2001. Evolution of whales. National Geographic 200(5). November issue. Pp. 64-77.

Chanin, P. 1985. The natural history of otters. Facts on File, New York.

Chapman, C. A. 1986. *Boa constrictor* predation and group response in white-faced *Cebus* monkeys. Biotropica 18(2): 171-72.

Chapman, C. A. 1988a. Patch use and patch depletion by the spider and howling monkeys of Santa Rosa National Park, Costa Rica. Behaviour 150: 99-116.

Chapman, C. A. 1988b. Patterns of foraging and range use by three species of Neotropical primates. Primates 29(2): 177-94.

Chapman, C. A. 1990a. Ecological constraints on group size in three species of Neotropical primates. Folia Primatol. 55: 1-9.

Chapman, C. A. 1990b. Association patterns of spider monkeys: the influence of ecology and sex on social organization. Behav. Ecol. Sociobiol. 26: 409-14.

Chapman, C. A., and L. J. Chapman. 1990. Reproductive biology of captive and free-ranging spider monkeys. Zoo Biol. 9: 1-9.

Chapman, C. A., and L. Lefebvre. 1990. Manipulating foraging group size: spider monkey food calls at fruiting trees. Anim. Behav. 39: 891-96.

Chapman, C. A., and D. M. Weary. 1990. Variability in spider monkeys' vocalizations may provide basis for individual recognition. Am. J. Primatol. 22: 279-84.

Chapman, J. A. 1983. *Sylvilagus floridanus*. In: Janzen '83c. Pp. 492-4.

Chapman, J. A. 1984. Latitude and gestation period in New World rabbits (Leporidae: *Sylvilagus* and *Romerolagus*). Am. Nat. 124: 442-45.

Chapman, J. A., and G. Ceballos. 1990. The cottontails. In: Chapman & Flux '90. Pp. 95-110.

Chapman, J. A., and G. A. Feldhamer (eds.). 1982. Wild mammals of North America: biology, management, and economics. Johns Hopkins Univ. Press, Baltimore.

Chapman, J. A., and J. E. C. Flux (eds.). 1990. Rabbits, hares and pikas. Status survey and conservation action plan. IUCN, Gland, Switzerland.

Chapman, J. A., A. L. Harman, and D. E. Samuel. 1977. Reproductive and physiological cycles in the cottontail complex in western Maryland and nearby West Virginia. Wildl. Monogr. 56: 1-73.

Chapman, J. A., J. G. Hockman, and M. M. Ojeda C. 1980. *Sylvilagus floridanus*. Mamm. Species 136. Am. Soc. Mammalogists.

Charles-Dominique, P. 1983. Ecology and social adaptations in didelphid marsupials: comparison with eutherians of similar ecology. In: Eisenberg & Kleiman '83. Pp. 395-422.

Charles-Dominique, P. 1986. Inter-relations between frugivorous vertebrates and pioneer plants: *Cecropia*, birds, and bats in French Guyana. In: Estrada & Fleming '86. Pp. 119-35.

Charles-Dominique, P., M. Atramentowicz, M. Charles-Dominique, H. Gerard, A. Hladik, C. M. Hladik, and M. F. Prevost. 1981. Les mamifères frugivores arboricoles noctunes d'une forêt guyanaise: interrelations plantes-animaux. Rev. Ecol. (Terre et Vie) 35: 341-435.

Checklist of CITES species. CITES Secretariat/World Conservation Monitoring Centre. 1996.

Cheney, D. L., and R. M. Seyfarth. 1990. How monkeys see the world. Univ. of Chicago Press.

Chevalier-Skolnikoff, S. 1990. Tool use by wild *Cebus* monkeys at Santa Rosa National Park, Costa Rica. Primates 31(3): 375-84.

Chinchilla, F. A. 1997. La dieta del jaguar (*Panthera onca*), el puma (*Felis concolor*), el manigordo (*Felis pardalis*) (Carnivora: Felidae) en el Parque Nacional Corcovado, Costa Rica. Rev. Biol. Trop. 45(3): 1223-29.

Choate, J. R. 1970. Systematics and zoogeography of Middle American shrews of the genus *Cryptotis*. Univ. Kansas Publ. Mus. Nat. Hist. 19: 195-317.

Ciochon, R. L., and A. B. Chiarelli (eds). 1980. Evolutionary biology of the New World monkeys and continental drift. Plenum Press, New York.

Clarke, J. A. 1983. Moonlight's influence on predator / prey interactions between short-eared owls (*Asio flammeus*) and deermice (*Peromyscus maniculatus*) Behav. Ecol. Sociobiol. 13: 205-09.

Clarke, M. R. 1981. Aspects of male behavior in the mantled howler monkeys (*Alouatta palliata*) in Costa Rica. Am. J. Primatol. 3: 1-22.

Clarke, M. R. 1982. Socialization, infant mortality, and infant-nonmother interactions in howling monkeys (*Alouatta palliata*) in Costa Rica. PhD Thesis. Univ. California, Davis.

Clarke, M. R. 1983. Infant-killing and infant disappearance following male takeovers in a group of free-ranging howling monkeys (*Alouatta palliata*) in Costa Rica. Am. J. Primatol. 5: 241-47.

Clutton-Brock, T. H. (ed.). 1977. Primate ecology: studies of feeding and ranging behaviour in lemurs, monkeys and apes. Academic Press, London.

Clutton-Brock, T. H. 1984. Primates. In: Macdonald '84a. Pp. 306-17.

Coates-Estrada, R., and A. Estrada. 1986. Manual de identificación de campo de los mamíferos de la estación de biología "Los Tuxtlas". Univ. Nac. Aut. México.

Cockerill, R. A. 1984. Deer. In: Macdonald '84a. Pp. 520-27.

Cody, M., and J. Smallwood (eds.). 1996. Long-term studies of vertebrate communities. Academic Press, New York.

Cohen, J. A., and M. W. Fox. 1976. Vocalizations in wild canids and possible effects of domestication. Behav. Proc. 1: 77-92.

Coimbra-Filho, A. F., and R. A. Mittermeier (eds.). 1981. Ecology and behavior of Neotropical primates. Vol. 1. Acad. Brasil. de Cienc., Rio de Janeiro.

Collett, S. F. 1981. Population characteristics of *Agouti paca* (Rodentia) in Colombia. Michigan State Univ. Mus. Publ. Biol. Ser. 5: 485-602.

Collier, G. E., and S. J. O'Brien. 1985. A molecular phylogeny of the Felidae immunological distance. Evolution 39: 473- 87.

Collins, L. R. 1973. Monotremes and marsupials: a reference for zoological institutions. Smithsonian Inst. Publ. 4888: 1-323.

Colmenero, L. C. 1986. Aspectos de la ecología y comportamiento de una colonia de manatíes (*Trichechus manatus*) en el municipio de Emiliano Zapata, Tabasco. An. Inst. Biol. Univ. Nac. Autón. Méx 56: 589-602.

Corbet, G. B. 1984a. Cavy-like rodents. In: Macdonald '84a. Pp. 684-85.

Corbet, G. B. 1984b. Squirrel-like rodents. In: Macdonald '84a. Pp. 604-5.

Cornelius, S. 1986. Turtles of Santa Rosa National Park. Editorial Incafo, Madrid.

Costello, R. K., C. Dickinson, A. L. Rosenberger, S. Boinski, and F. S. Szalay. 1993. Squirrel monkey (genus *Saimiri*) taxonomy: a multidisciplinary study of the biology of species. In: Kimbal & Martin '93. Pp. 177-210.

Crabb, W. D. 1944. Growth, development and seasonal weights of spotted skunks. J. Mamm. 25: 213-21.

Crabb, W. D. 1948. The ecology and management of the prairie spotted skunk in Iowa. Ecol. Monogr. 18: 201-32.

Crandall, S. L. 1971. The manangement of wild mammals in captivity. Univ. Chicago Press.

Crawshaw, P. G. 1995. Comparative ecology of ocelot (*Felis pardalis*) and jaguar (*Panthera onca*) in a protected subtropical forest in Brazil and Argentina. Ph.D. diss. Univ. Florida, Gainesville.

Crawshaw, P. G., Jr., and H. B. Quigley. 1991. Jaguar spacing, activity, and habitat use in a seasonally flooded environment in Brazil. J. Zool. 223: 357-70.

Crerar, L. M., and M. B. Fenton. 1984. Cervical vertebrae in relation to roosting posture in bats. J. Mamm. 65: 395-403.

Cresticelli, F. (ed.). 1977. The visual system in vertebrates. Springer, Berlin.

Croat, T. B. 1983. *Dieffenbachia*. In: Janzen '83. Pp. 234-36.

Currier, M. J. P. 1983. *Felis concolor*. Mamm. Species 200. Am. Soc. Mammalogists.

Dalquest, W. W., and J. H. Roberts. 1951. Behavior of young grisons in captivity. Amer. Midl. Nat. 46: 359-66.

Dare, R. J. 1974. The social behavior and ecology of spider monkeys, *Ateles geoffroyi*, on Barro Colorado Island. PhD diss. Univ. Oregon.

da Silva, V. M. F. 1994. Aspects of the biology of the Amazonian dolphins genus *Inia* and *Sotalia fluviatalis*. Ph.D. diss. Univ. Cambridge, England.

da Silva, V. M. F., and R. C. Best. 1996. *Sotalia fluviatalis*. Mamm. Spec. 527. Am. Soc. Mammalogists.

Decker, D. M. 1991. Systematics of the coatis, genus *Nasua* (Mammalia: Procyonidae). Proc. Biol. Soc. Wash. 104: 370-86.

Decker, D. M., and W. C. Wozencraft. 1991. Phylogenetic analysis of recent procyonid genera. J. Mamm. 72(1): 42-55.

De Jong, W., A. Zweers, and M. Goodman. 1981. Relationship of aardvarks to elephants, hyraxes, and sea cows from alpha-crystallin sequences. Nature 292(5823): 538-40.

De Jong, W., A. Zweers, K. A. Joysey, J. T. Gleaves, and D. Boulter. 1985. Protein sequence analysis applied to xenarthran and pholidote phylogeny. In: Montgomery '85.

Delgado, R. 1990. Costrucción de túneles y ciclo reproductivo de la taltuza *Orthogeomys cherriei* (Allen) (Rodentia: Geomyidae). Rev. Biol. Trop. 38(1): 119-27.

Delgado, R. 1992. Ciclo reproductivo de la taltuza *Orthogeomys cherriei* (Rodentia: Geomyidae). Rev. Biol. Trop. 40(1): 111-15.

Delibes, M., L. Hernández, and F. Hiraldo. 1989. Comparative food habits of three carnivores in Western Sierra Madre, Mexico. Z. Säugetierk. 54: 107-10.

DeVan, R. 1982. The ecology and life history of the long-tailed weasel. Ph.D. diss. Univ. Cincinnati.

DeVos, A. 1960. *Mustela frenata* climbing trees. J. Mamm. 41: 520.

Diaz, D. 1997. Los pequeños del Caribe. In: Viva: Revista Diaria de La Nación, 5 octubre, 1997. San José. Pp 1-2.

Dice, L. R. 1945. Minimum intensities of illumination under which owls can find dead prey by sight. Am. Nat. 79: 385-416.

Dickman, C. R. 1984a. Anteaters. In: Macdonald '84a. Pp. 772-5.

Dickman, C. R. 1984b. Armadillos. In: Macdonald '84a. Pp. 780-83.

Dickman, C. R. 1984c. Edentates. In: Macdonald '84a. Pp.770-1.

Dickman, C. R. 1984d. Sloths. In: Macdonald '84a. Pp. 776-8.

Diersing, V. E. 1981. Systematic status of *Sylvilagus brasiliensis* and *S. isonus* from North America. J. Mamm. 62: 539-556.

Diersing, V. E. 1984. Lagomorphs. In: Anderson and Jones '84. Pp. 241-54.

Dietrich, J. R. 1993. Biology of the brocket deer (Genus *Mazama*) in northern Venezuela. PhD Thesis. Univ. Basilea.

Dimare, M. I. 1994. Hábitos alimentarios del venado cola blanca en la Isla San Lucas, Puntarenas, Costa Rica. In: Vaughan & Rodríguez '94. Pp. 73-102.

Dinerstein, E. 1985. First records of *Lasiurus castaneus* and *Antrozous dubiaquercus* from Costa Rica. J. Mamm. 66: 411-12.

Dinerstein, E. 1986. Reproductive ecology of fruit bats and the seasonality of fruit production in a Costa Rican cloud forest. Biotropica 18: 307-18.

Dinerstein, E. 2000. The influence of fruit-eating bats on the dynamics and composition of neotropical premontane cloud forests. In: Nadkarni & Wheelwright 2000. Pp. 434-35.

Dirzo, R., and A. Miranda. 1990. Altered patterns of herbivory and diversity in the forest understory: a case study of the possible consequences of contemporary defaunation. In: Price et al. '90. Pp. 273-87.

Domning, D. P. 1977. An ecological model for late Tertiary sirenian evolution in the North Pacific Ocean. Systematic Zoology 25(4): 352-362.

Domning, D. P. 1981. Sea cows and sea grasses. Paleobiology 7(4): 417-20.

Domning, D. P. 1982a. Commercial exploitation of manatees *Trichechus* in Brazil c. 1785-1973. Biol. Cons. 22: 101-26.

Domning, D. P. 1982b. Evolution of manatees: a speculative history. J. Paleontol. 56: 599-619.

Domning, D. P. 1989. Kelp evolution: a comment. Paleobiology 15(1): 53-56.

Domning, D. P., and L. C. Hayek. 1986. Interspecific and intraspecific morphological variation in manatees (Sirenia: *Trichechus*). Marine Mamm. Sci. 2: 87-144.

Donkin, R. A. 1985. The peccary - with observations on the introduction of pigs to the New World. Trans. Amer. Philos. Soc. 75: 1-152.

Dougherty, E. C., and E. R. Hall. 1955. The biological relationships between American weasels (genus *Mustela*) and nematodes of the genus *Skrjabingylus* Petrov 1927 (Nematoda: Metastrongylidae), the causative organisisms of certain lesions in weasel skulls. Rev. Ibérica de Parasit., Tomo Extraord., Granada, Spain.

Dresser, B., R. Reece, and E. Maruska (eds.). 1988. Proceedings of the fifth world conference on breeding endangered species in captivity.

Dubock, A. C. 1984. Rodents as pests. In: Macdonald '84a. Pp. 600-3.

Dücker, G. 1965. Colour vision in mammals. J. Bombay Nat. Hist. Soc. 61: 572-86.

Dunning, D. C. 1968. Warning sounds of moths. Z. Tierpsychol. 25: 129-38.

Duplaix, N. 1984. Otters. In: Macdonald '84a. Pp. 124-29.

Durant, P. 1983. Estudios ecológicos del conejo silvestre *Sylvilagus brasiliensis meridensis* (Lagomorpha: Leporidae): en los páramos de los Andes venezolanos. Caribbean J. Sci. 19(1-2): 21-29.

Durant, P. 1984. Estudios ecológicos y necesidad de protección del conejo de páramo. Rev. Ecol. Conserv. Ornit. Latinoamer. 1: 35-37.

Eaton, R. L. (ed.). 1976a. The World's Cats, Vol. 3. Carnivore Research Institute, Washington D. C.

Eaton, R. L. 1976b. Why some felids copulate so much. In: Eaton '76a. Pp. 74-94.

Egoscue, H. J., J. G. Bittmenn, and J. A. Petrovich. 1970. Some fecundity and longevity records for captive small mammals. J. Mamm. 51: 622-23.

Eisenberg, J. F. 1961. The nest-building behavior of armadillos. Proc. Zool. Soc. Lond. 137: 322-24.

Eisenberg, J. F. 1973. Reproduction in two species of spider monkeys, *Ateles fusciceps* and *A. geffroyi*. J. Mamm. 54: 955-57.

Eisenberg, J. F. 1974. The function and motivational basis of hystricomorph vocalizations. Symp. Zoolog. Soc. London 34: 211-47.

Eisenberg, J. F. 1976. Communication mechanisms and social integration in the black spider monkey, *Ateles fusciceps robustus*, and related species. Smithson. Contrib. Zool. 213: 1-108.

Eisenberg, J. F. (ed.). 1979. Vertebrate ecology in the northern Neotropics. Smithsonian Institution Press, Washington.

Eisenberg, J. F. 1983a. *Ateles geoffroyi*. In: Janzen '83c. Pp. 451-53.

Eisenberg, J. F. 1983b. The mammalian radiations: an analysis of trends in evolution, adaptation, and behavior. Univ. Chicago Press.

Eisenberg, J. F. 1988. Reproduction in polyprotodont marsupials and simlar-sized eutherians with a speculation concerning the evolution of litter size in mammals. In: Boyce '88. Pp. 291-311.

Eisenberg, J. F. 1989. Mammals of the Neotropics: the northern Neotropics. Volume I. Univ. Chicago Press.

Eisenberg, J. F., L. R. Collins, and C. Wemmer. 1975. Communication in the Tasmanian devil (*Sarcophilus harrisii*) and a survey of auditory communication in the Marsupialia. Z. Tierpsychol. 37: 379-99.

Eisenberg, J. F., and D. G. Kleiman. 1977. Communication in lagomorphs and rodents. In: Sebeok '77. Pp. 634-54.

Eisenberg, J. F., and D. G. Kleiman (eds.). 1983. Advances in the study of mammalian behavior. Special Pub. no. 7. Am. Soc. Mammalogists. Shippensburg, Pa.

Eisenberg, J. F., and I. Golani. 1977. Communication in Metatheria. In: Sebeok '77. Pp. 575-99.

Eisenberg, J. F., and E. Maliniak. 1967. Breeding the murine opossum *Marmosa* in captivity. Intl. Zoo Yearb. 7: 78-79.

Eisenberg, J. F., and E. Maliniak. 1985. Maintenance and reproduction of the two-toed sloth *Chloepus didactylus* in captivity. In: Montgomery '85a. Pp. 327-32.

Eisenberg, J. F., and G. M. McKay. 1974. Comparison of ungulate adaptations in the new and old world tropical forests with special reference to Ceylon and the rain forests of Central America. In: Geist & Walther '74. Pp. 585-602.

Eisenberg, J. F., and R. W. Thorington. 1973. A preliminary analysis of a Neotropical mammal fauna. Biotropica 5: 150-61.

Elizondo, L. 1987. *Eira barbara* y *Gallictis vittata* (Familia Mustelidae). Fundacion Neotropica – Fundacion de Parques Nacionales. Programa de de Patrimonio de Costa Rica. Mimeograph.

Emmons, L. H. 1982. Ecology of *Proechimys* (Rodentia, Echimyidae) in southeastern Peru. Trop. Ecol. 23: 280-90.

Emmons, L. H. 1987. Comparative feeding ecology of felids in a neotropical rainforest. Behav. Ecol. Sociobiol. 20: 271-83.

Emmons, L. H. 1988. A field study of ocelots (*Felis pardalis*) in Peru. Rev. Ecol. (Terre et Vie) 43: 133-57.

Emmons, L. H. 1989. Jaguar predation on chelonians. J. Herpetology 23: 311-14.

Emmons, L. H., and F. Feer. 1997. Neotropical rainforest mammals: a field guide. Second edition. Univ. Chicago Press.

Emmons, L. H., and A. H. Gentry. 1983. Tropical forest structure and the distribution of gliding and prehensile-tailed vertebrates. Am. Nat. 121: 513-24.

Emmons, L. H., P. Sherman, D. Bolster, A. Goldizen, and J. Terborgh. 1989. Ocelot behavior in moonlight. In: Redford & Eisenberg '89. Pp. 233-42.

Emmons, L. H., and N. Stark. 1979. Elemental composition of a natural mineral lick in Amazonia. Biotropica 11(4): 311-13.

Emmons, L. H., B. M. Whitney, and D. L. Ross. 1997. Sounds of neotropical rainforest mammals: an audio field guide. Compact disk. Cornell Laboratory of Ornithology, Ithaca, New York.

Enders, R. K. 1935. Mammalian life histories from Barro Colorado Island, Panama. Bull. Mus. Comp. Zool. 78: 385-502.

Enders, R. K. 1966. Attachment, nursing, and survival of young in some didelphids. In: Rowlands '66. Pp. 195-203.

Enders, R. K. 1980. Observations on *Syntheosciurus*: taxonomy and behavior. J. Mamm. 61: 725-27.

Engelmann, G. F. 1985. The phylogeny of the Xenarthra. In: Montgomery '85a. Pp. 51-64.

Engstrom, M. D., B. K. Lim, and F. A. Reid. 1994. Two small mammals new to the fauna of El Salvador. Southwest. Nat. 39: 281-306.

Engstrom, M. D., F. A. Reid, and B. K. Lim. 1993. New records of two small mammals from Guatemala. Southwest. Nat. 38: 80-82.

Enríquez, A., M. Guerrero et al. 1991. Efecto de la pertubación por *Tayassu tajacu* sobre la población de *Dieffenbachia* sp. (Araceae). OTS 91-2: 293-97.

Estes, J. A. 1989. Adaptations for aquatic living by carnivores. In: Gittleman '89. Pp. 242-82.

Estrada, A. 1984. Resource use by howler monkeys (*Alouatta palliatta*) in the rain forest of Los Tuxtlas, Mexico. Intl. J. Primatol. 5: 105-31.

Estrada, A., A. Anzures, and R. Coates-Estrada. 1999. Tropical rain forest fragmentation, howler monkeys (*Alouatta palliatta*), and dung beetles at Los Tuxtlas, Mexico. Am. J. Primatol. 48: 253-62.

Estrada, A. and R. Coates-Estrada. 1984. Fruit eating and dispersal by howling monkeys (*Alouatta palliatta*) in the tropical rain forest of Los Tuxtlas, Mexico. Am. J. Primatol. 6: 77-91.

Estrada, A. and R. Coates-Estrada. 1985. A preliminary study of resource overlap between howling monkeys (*Alouatta palliatta*) and other mammals in the tropical rain forest of Los Tuxtlas, Mexico. Am. J. Primatol. 9: 27-37.

Estrada, A. and R. Coates-Estrada. 1986. Frugivory in howling monkeys (*Alouatta palliatta*) at Los Tuxtlas, Mexico: dispersal and fate of seeds. In: Estrada & Fleming '86. Pp. 93-105.

Estrada, A. and R. Coates-Estrada. 1991. Howling monkeys (*Alouatta palliatta*), dung beetles (Scarabaeidae) and seed dispersal: ecological interactions in the tropical rain forest of Los Tuxtlas, Veracruz, Mexico. J. Trop. Ecol. 7: 459-74.

Estrada, A. and R. Coates-Estrada. 1996. Tropical rain forest fragmentation and wild populations of primates at Los Tuxtlas. Intl. J. Primatol. 5: 759-83.

Estrada, A. and R. Coates-Estrada. 1988. Tropical rain forest conversion and perspectives in the conservation of wild primates (*Alouatta* and *Ateles*) in Mexico. Am. J. Primatol. 14: 315-27.

Estrada, A. and T. H. Fleming (eds.). 1986. Frugivores and seed dispersal. Dordrecht: Junk.

Estrada, A., S. Juan-Solano, T. Ortíz, and R. Coates-Estrada. 1999. Feeding and general activity patterns of a howler monkey (*Alouatta palliatta*) troop living in a forest fragment at Los Tuxtlas, Mexico. Am. J. Primatol. 48: 167-83.

Evans, P. G. H. 1984a. Toothed whales. In: Macdonald '84a. Pp. 176-77.

Evans, P. G. H. 1984b. Whales and dolphins. In: Macdonald '84a. Pp. 162-75.

Evans, P. G. H. 1987. The natural history of whales and dolphins. Facts on File, New York.

Evans, S. 1999. The green republic: a conservation history of Costa Rica. Univ. Texas Press.

Ewer, R. F. 1973. The carnivores. Cornell Univ. Press, New York.

Fagerstone, K. A. 1987. Black-footed ferret, long-tailed weasel, short-tailed weasel, and least weasel. In: Novak et al. '87. Pp. 548-73.

Fedigan, L. M. 1990. Vertebrate predation in *Cebus capucinus*: meat-eating in a neotropical monkey. Folia Primatol. 54: 196-205.

Fedigan, L. M. 1993. Sex differences and intersexual relations in adult white-faced capuchins (*Cebus capucinus*). Intl. J. Primatol. 14(6): 853-77.

Fedigan, L. M., and L. M. Rose. 1995. Interbirth interval in three sympatric species of neotropical monkey. Am. J. Primatol. 37: 9-24.

Fegus, C. 1991. The Florida panther verges on extinction. Science 251: 1178-80.

Fenton, M. B. 1992. Bats. Facts on File, New York.

Fenton, M. B., and D. R. Griffin. 1997. High-altitude pursuit of insects by echolocating bats. J. Mamm. 78(1): 247-50.

Fenton, M. B., P. Racey, and J. M. V. Rayner (eds.). 1987. Recent advances in the study of bats. Cambridge Univ. Press.

Ferrari, S. F. 1991. An observation of western black spider monkeys, *Ateles paniscus chamek*, utilzing an arboreal water source. Biotropica 23(3): 307-08.

Findley, J. S. And D. E. Wilson. 1974. Observations on the neotropical disk-winged bat, *Thyroptera tricolor* Spix. J. Mamm. 55: 562-71.

Fischer, M. S., and P. Tassy. 1993. The interrelation between Proboscidea, Sirenia, Hyracoidea, and Mesaxonia: the morphological evidence. In: Szalay et al. '93.

Fishkind, A. S., and R. W. Sussman. 1987. Preliminary survey of the primates of the Zona Protectora and La Selva Biological Station, northeast Costa Rica. Primate Conserv. 8: 63-66.

Fitch, H. S., and H. W. Shirer. 1970. A radiotelemetric study of spatial relationships in the opossum. Am. Midl. Nat. 84: 170-86.

Fleming, T. H. 1970. Notes on the rodent faunas of two Panamanian forests. J. Mamm. 51: 473-90.

Fleming, T. H. 1971a. *Artibeus jamaicensis*: delayed embryonic development in a neotropical bat. Science 171: 402-04.

Fleming, T. H. 1971b. Population ecology of three species of neotropical rodents. Misc. Publ. Mus. Zool. Univ. Michigan 143: 1-77.

Fleming, T. H. 1972. Aspects of the population dynamics of three species of opossums in the Panama Canal Zone. J. Mamm. 53: 619-23.

Fleming, T. H. 1973. The reproductive cycles of three species of opossums and other mammals in the Panama Canal Zone. J. Mamm. 54: 439-55.

Fleming, T. H. 1974a. The population ecology of two species of Costa Rican heteromyid rodents. Ecology 55: 493-510.

Fleming, T. H. 1974b. Social organization in two species of Costa Rican heteromyid rodents. J. Mamm. 55: 543-61.

Fleming, T. H. 1977. Growth and development of two species of tropical heteromyid rodents. Am. Midl. Nat. 98: 109-23.

Fleming, T. H. 1983a. *Carollia perspicillata*. In: Janzen '83c. Pp. 457-58.

Fleming, T. H. 1983b. *Heteromys demarestianus*. In: Janzen '83c. Pp. 474-75.

Fleming, T. H. 1983c. *Liomys salvini*. In: Janzen '83c. Pp. 475-77.

Fleming, T. H. 1988. The short-tailed fruit bat: a study in plant-animal interactions. Univ. Chicago Press.

Fleming, T. H., and G. J. Brown. 1975. An experimental analysis of seed hoarding and burrowing behavior in two species of Costa Rican heteromyid rodents. J. Mamm. 56: 301-15.

Fleming, T. H., E. R. Heithaus, and W. B. Sawyer. 1977. An experimental analysis of the food location behavior of frugivorous bats. Ecology 58: 619-28.

Fleming, T. H., E. T. Hooper, and D. E. Wilson. 1972. Three Central American bat communities: structure, reproductive cycles, and movement patterns. Ecology 53: 555-69.

Flux, J. E. C., and P. J. Fullagar. 1983. World distribution of the rabbit (*Oryctolagus cuniculus*). Acta Zool. Fennica 174: 75-77

Foerster, C. 1996. Researcher attacked by puma in Corcovado National Park, Costa Rica. Vida Silv. Neotrop. 5(1): 57-58.

Foerster, C. 1998. Ecología de la danta centroamericana *Tapirus bairdii* en un bosque húmedo tropical de Costa Rica. Masters thesis. Univ. Nacional, Heredia.

Fontaine, R. 1980. Observations on the foraging association of double-toothed kites and white-faced capuchin monkeys. Auk 97: 94-98.

Ford, L. S., and R. S. Hoffmann. 1988. *Potos flavus*. Mamm. Species 321. Am. Soc. Mammalogists.

Forman, L. 1985. Genetic variation in two procyonids: phylogenetic, ecological and social correlates. Ph.D. diss.: New York Univ.

Forsyth, A. 1990. Portraits of the rainforest. Firefly Books, Ontario.

Foster, M. S., and R. M. Timm. 1976. Tent-making by *Artibeus jamaicensis* (Chiroptera: Phyllostomatidae) with comments on plants used by bats for tents. Biotropica 8: 265-69.

Foster-Turley, P., S. Macdonald, C. Mason (eds.). 1990. Otters: an action plan for their conservation. Intl. Union for Conservation of Nature and Natural Resources. Kelvyn Press, Inc., Illinois.

Fowler, J. M., and J. B. Cope. 1964. Notes on the harpy eagle in British Guiana. Auk 81: 257-73.

Fox, M. W. 1969. Ontogeny of prey-killing behavior in Canidae. Ibid. 35: 242-58.

Fox, M. W. 1975. The wild canids. Nostrand Reinhold, New York.

Fox, M. W. and J. A. Cohen. 1977. Canid communication. In: Sebeok '77. Pp. 728-48.

Fragoso, J. M. 1991. The effects of hunting on tapirs in Belize. In: Robinson & Redford '91. Pp. 154-62.

Fragoso, J. M. 1998. Home range and movement patterns of white-lipped peccary (*Tayassu pecari*) herds in the northern Brazilian Amazon. Biotropica 30: 458-69.

Fragoso, J. M. 1999. Perception of scale and resource partitioning by peccaries: behavioral causes and ecological implications. J. Mamm. 80(3): 993-1003.

Francq, E. N. 1969. Behavioral aspects of feigned death in the opossum, *Didelphis marsupialis*. Am. Midl. Nat. 81: 556-68.

Frantzius, A. von. 1869. Die Säugethiere Costaricas. Archiv. Naturgesch. 35: 247-325.

Freeland, W. J. and D. H. Janzen. 1974. Strategies in herbivory by mammals: the role of secondary compounds. Am. Nat. 108(961): 269-89.

Freese, C. H. 1976a. Censusing *Alouatta palliata*, *Ateles geoffroyi* and *Cebus capucinus* in the Costa Rican dry forest. In: Thorington & Heltne '76. Pp. 4-9.

Freese, C. H. 1976b. Predation on swollen-thorn acacia ants by white-faced monkeys *Cebus capucinus*. Biotropica 8: 278-81.

Freese, C. H. 1977. Food-habits of white-faced capuchins *Cebus capucinus* (Primates: Cebidae) in Santa Rosa National Park, Costa Rica. Brenesia 10/11: 43-56.

Freese, C. H. 1978. Behavior of white-faced capuchins *Cebus capucinus* at a dry-season waterhole. Primates 19: 275-86.

Freese, C. H. 1983. *Cebus capucinus*. In: Janzen '83c. Pp. 458-60.

Freese, C. H., and J. Oppenheimer. 1981. The capuchin monkeys, genus *Cebus*. In: Coimbra-Filho & Mittermeier '81. Pp. 331-90.

Freiheit, C. 1966. Courtship activity of the paca. J. Mamm. 46: 707.

Fritzell, E. K. 1978. Aspects of raccoon (*Procyon lotor*) social organization. Can. J. Zool. 56: 260-71.

Fritzell, E. K. 1987. Gray fox and island gray fox. In: Novak et al. '87. Pp. 408-20.

Fritzell, E. K., and K. J. Haroldson. 1982. *Urocyon cinereoargenteus*. Mamm. Species 189. Am. Soc. Mammalogists.

Froehlich, J. W., and R. W. Thorington, Jr. 1996. The genetic structure and socioecology of howler monkeys (*Alouatta palliata*) on Barro Colorado Island. In: Leigh et al. '96. Pp. 291-305.

Froehlich, J. W., R. W. Thorington, Jr., and J. S. Otis. 1981. The demography of howler monkeys (*Alouatta palliata*) on Barro Colorado Island, Panama. Intl. J. Primatol. 2(3): 207-36.

Fullard, J. H., M. B. Fenton, and J. A. Simmons. 1979. Jamming bat echolocation: the clicks of Arctiid moths. Can J. Zool. 57: 647-49.

Galef, B. G., Jr., R. A. Mittermeier, and R. C. Bailey. 1976. Predation by the tayra (*Eira barbara*). J. Mamm. 57: 760-61.

Gallo, J. P. 1989. Distribución y estado actual de la nutria o perro de agua (*Lutra longicaudis annectens* Major, 1897) en la Sierra Madre del Sur, México. Tesis de Maestría. Univ. Nac. Autón. de México.

Garber, P. A., and R. W. Sussman. 1984. Ecological distinctions between sympatric species of *Saguinus* and *Sciurus*. Am. J. Phys. Anthropol. 65: 135-46.

García, N. E. 1996. Ecología del cacomistle (*Bassariscus sumichrasti*) en un bosque nuboso tropical, Costa Rica. Masters thesis. Univ. Nacional, Heredia.

Gardner, A. L. 1971a. Notes on the little spotted cat *Felis tigrina oncilla* in Costa Rica. J. Mamm. 52: 461-465.

Gardner, A, L. 1971b. Postpartum estrus in a red brocket deer, *Mazama americana*, from Peru. J. Mamm. 52: 623-24.

Gardner, A. L. 1977. Feeding habits. In: Baker et al. '77. Pp. 293-350.

Gardner, A. L. 1982. Virginia opossum. In: Chapman & Feldhamer '82. Pp. 3-36.

Gardner, A. L. 1983a. *Didelphis marsupialis*. In: Janzen '83c. Pp. 468-69.

Gardner, A. L. 1983b. *Oryzomys caliginosus*. In: Janzen '83c. Pp. 483-85.

Gardner, A. L., and G. K. Creighton. 1989. A new generic name for Tate's *Microtarsus* group of South American mouse opossums (Marsupialia: Didelphidae). Proc. Biol. Soc. Wash. 102: 3-7.

Gardner, A. L., and D. E. Wilson. 1970. A melanized *subcutaneous* covering of the cranial musculature in the phyllostomid bat, *Ectophylla alba*. J. Mamm. 52(4): 854-855.

Gavin, T. A., L. H. Suring, P. A. Vohs, Jr., and E. C. Meslow. 1984. Population characteristics, spatial organization, and natural mortality in the Columbian white-tailed deer. Wildlife Monographs 91: 1-49.

Gay, S. W., and T. L. Best. 1996. Age-related variation in skulls of the puma (*Puma concolor*). J. Mamm. 77(1): 191-98.

Gehrt, S. D., and E. K. Fritzell. 1997. Sexual differences in home ranges of raccoons. J. Mamm. 78(3): 921-31.

Gehrt, S. D., and E. K. Fritzell. 1998. Duration of familial bonds and dispersal patterns for raccoons in south Texas. J. Mamm. 79(3): 859-72.

Geise, L. 1989. Estrutura social, comportamental e populacional de *Sotalia* sp. (Gray, 1886) (Cetacea, Delphinidae) na região estuarino-lagunar de Cananéia, SP e na Baía de Guanabara, RJ. M.Sc. thesis, Univ. São Paulo.

Geist, V., and F. Walther (eds.). 1974. The behaviour of ungulates and its relation to management. IUCN Publication, n.s., no. 24, 2 vols. Morges, Switzerland.

Genoways, H. H. (ed.). 1987. Current mammalogy, volume 1. Plenum Press, New York.

Genoways, H. H. and J. H. Brown (eds.). 1993. Biology of the Heteromyidae. Spec. Publ. 10. Am. Soc. Mammalogists.

Giacalone, J., N. Wells, and G. Willis. 1987. Observations on *Syntheosciurus brochus* in Volcán Poás National Park, Costa Rica. J. Mamm. 68: 145-46.

Gier, H. T. 1975. Ecology and behavior of the coyote (*Canis latrans*). In: Fox '75. Pp. 247-62.

Gillette, D. D., and C. E. Ray. 1981. Glyptodonts of North America. Smithsonian Contributions to Paleobiology 40. Smithsonian Inst. Press, Washington D.C.

Gilmore, D. P. 1977. The success of marsupials as introduced species. In: Stonehouse & Gilmore '77. Pp. 169-78.

Gittleman, J. L. (ed.). 1989. Carnivore behavior, ecology, and evolution. Cornell Univ. Press.

Glander, K. E. 1974. Baby-sitting, infant sharing, and adoptive behavior in mantled howling monkeys. Am. J. Phys. Anthrop. 41: 482.

Glander, K. E. 1975. Habitat and resource utilization: an ecological view of social organization in mantled howling monkeys. Ph.D. diss., Univ. of Chicago.

Glander, K. E. 1978a. Howling monkey feeding behavior and plant secondary compounds: a study of strategies. In: Montgomery '78. Pp. 561-78.

Glander, K. E. 1978b. Drinking from arboreal water sources by mantled howling monkeys (*Alouatta palliata* Gray). Folia Primatol. 29: 206-17.

Glander, K. E. 1979. Feeding associations between howling monkeys and basilisk lizards. Biotropica 11(3): 235-36.

Glander, K. E. 1980. Reproduction and population growth in free-ranging mantled howling monkeys. Am. J. Phys. Anthropol. 53: 25-36.

Glander, K. E. 1983. *Alouatta palliata*. In: Janzen '83c. Pp. 448-49.

Glander, K. E. 1996. The howling monkeys of La Pacifica. Published independently.

Glanz, W. E. 1984. Food and habitat use by two sympatric *Sciurus* species in central Panama. J. Mamm. 65: 342-46.

Glanz, W. E. 1996. The terrestrial mammal fauna of Barro Colorado Island: censuses and long-term changes. In: Leigh et al. '96. Pp. 445-68.

Glanz, W. E., R. W. Thorington, Jr., J. Giacalone-Madden, and L. R. Heaney. 1996. Seasonal food use and demographic trends in *Sciurus granatensis*. In: Leigh et al. '96. Pp. 239-52.

Glass, B. P. 1985. History of classification and nomenclature in Xenarthra (Edentata). In: Montgomery '85a. Pp. 1-3.

Gliwicz, J. 1984. Population dynamics of the spiny rat *Proechimys semispinosus* on Orchid Island (Panama). Biotropica 16(1): 73-78.

Goffart, M. 1971. Function and form in the sloth. Pergamon Press, New York.

Golley, F. B., and E. Medina (eds.). 1975. Tropical ecological systems. Springer-Verlag, New York.

Gómez, L. D. 1983. Variegated squirrels eat fungi, too. Brenesia 21: 458-60.

Gompper, M. E. 1995. *Nasua narica*. Mamm. Species 487. Am. Soc. Mammalogists.

Gompper, M. E. 1996. Sociality and asociality in white-nosed coatis (*Nasua narica*): foraging costs and benefits. Behav. Ecol. 7(3): 254-63.

Gompper, M. E., and A. M. Hoylman. 1993. Grooming with *Trattinnickia* resin: possible pharmaceutical plant use by coatis in Panama. J. Trop. Ecol. 9: 533-40.

Gompper, M. E., and J. S. Krinsley. 1992. Variation in social behavior of adult male coatis. Biotropica 24(2a): 216-19.

González, A., and M. Alberico. 1994. Relaciones competitivas entre *Proechimys semispinosus* and *Hoplomys gymnurus* en el Occidente Colombiano. Caldasia 17: 313-24.

González, L. 1992. Dominancia jerárquica de machos aulladores (*Alouatta palliata*) en un grupo de monos de la Estación Biológica La Selva. OTS 92-2: 203-13.

Goodwin, G. G., and A. M. Greenhall. 1961. A review of the bats of Trinidad and Tobago. Bull. Am. Mus. Nat. Hist. 122: 191-301.

Gottdenker, N., and R. E. Bodmer. 1998. Reproduction and productivity of white-lipped and collared peccaries in the Peruvian Amazon. J. Zool. London 244.

Gould, E. 1969. Communication in three genera of shrews (Soricidae): *Suncus*, *Blarina*, and *Cryptotis*. Communications in behavioral biology, pt. A, vol. 3: 11-31. Academic Press, New York.

Grand, T. I. 1978. Adaptations of tissue and limb segments to facilitate moving and feeding in arboreal foliage. In: Montgomery '78. Pp. 231-41.

Grand, T. I. 1983. Body weight: its relationship to tissue composition, segmental distribution of mass, and motor function. 3. The Didelphidae of French Guyana. Aust. J. Zool. 31: 299-312.

Gray, T. J. 1967. Man and rat: and investigation of the age old fight. World Health Magazine, April 1967 issue.

Greene, H. W. 1988. Species richness in tropical predators. In: Almeda & Pringle '88. Pp. 259-80.

Greene, H. W. 1989. Agonistic behaviour by three-toed sloths, *Bradypus variegatus*. Biotropica 21: 369-72.

Greene, H.W. and J.B. Losos. 1988. Systematics, natural history, and conservation. BioScience 38: 458-62.

Greene, H. W. and C. M. Rojas. 1988. *Orthogeomys underwoodii* (Rodentia: Geomyidae) on the Osa Peninsula, Costa Rica, with comments on the biological significance of pelage markings in tropical pocket gophers. Brenesia 29: 95-99.

Greene, H. W. and M. Santana. 1983. Field studies of hunting behavior by bushmaster. Am. Zool. 23: 897.

Greenhall, A. M. 1968. Notes on the behavior of the false vampire bat. J. Mamm. 49: 337-40.

Greenhall, A. M. 1988. Feeding behavior. In: Greenhall & Schmidt '88. Pp. 111-31.

Greenhall, A. M., G. Joermann, U. Schmidt, and M. R. Seidel. 1983. *Desmodus rotundus*. Mamm. Species 202. Am. Soc. Mammalogists.

Greenhall, A. M., and U. Schmidt (eds.). 1988. Natural history of vampire bats. CRC Press, Florida.

Greenhall, A. M., U. Schmidt, and G. Joermann. 1984. *Diphylla ecaudata*. Mamm. Species 227. Am. Soc. Mammalogists.

Greenhall, A. M., U. Schmidt, and W. Lopez-Forment. 1971. Attacking behavior of the vampire bat, *Desmodus rotundus*, under field conditions in Mexico. Biotropica 3(2): 136-41.

Greenhall, A. M., and W. A. Schutt, Jr. 1996. *Diaemus youngi*. Mamm. Species 533. Am. Soc. Mammalogists.

Greenlaw, J. S. 1967. Foraging behavior of the double-toothed kite in association with white-faced monkeys. Auk 84: pp. 596-97.

Gribel, R. 1988. Visits of *Caluromys lanatus* (Didelphidae) to flowers of *Pseudobombax tomentosum* (Bombacaceae): a probable case of pollination by marsupials in central Brazil. Biotropica 20: 344-47.

Grimwood, I. R. 1969. Notes on the distribution and status of some Peruvian mammals. Spec. Publ. Amer. Comm. Internatl. Wildl. Protection, no. 21.

Grzimek, B. (ed.). 1975. Grzimek's animal life encyclopedia. Mammals, I - IV. Van Nostrand Reinhold, New York. Vols. 10-13.

Haber, W. A., W. Zuchowski, and E. Bello C. 1996. An introduction to cloud forest trees. La Nación S. A., San José.

Hafner, M. S. 1991. Evolutionary genetics and zoogeography of Middle American pocket gophers, genus *Orthogeomys*. J. Mamm. 72: 1-10.

Hafner, M. S., and D. J. Hafner. 1987. Geographic distribution of two Costa Rican species of *Orthogeomys*, with comments on dorsal pelage markings in the Geomyidae. Southwest. Nat. 32: 5-11.

Hall, E. R., and W. W. Dalquest. 1963. The mammals of Veracruz. Univ. Kansas Publ., Mus, Nat. Hist. 14: 165-362.

Hall, E. R., and K. R. Kelson. 1959. The mammals of North America. The Ronald Press Co., New York.

Halls, L. K. (ed.). 1984. White-tailed deer: ecology and management. Stackpole Books, Harrisburg, Pa.

Hallwachs, W. 1986. Agoutis (*Dasyprocta punctata*), the inheritors of guapinol (*Hymenaea courbaril* Leguminosae). In: Estrada & Fleming '86.

Hamilton, R. B., and D. T. Stalling. 1972. *Lasiurus borealis* with five young. J. Mamm. 53: 190.

Handley, C. O. Jr. 1976. Mammals of the Smithsonian Venezuelan project. Brigham Young Univ. Sci. Bull. Biol. Ser. 20(5): 1-90.

Handley, C. O., Jr., D. E. Wilson, and A. L. Gardner, eds. 1991. Demography and natural history of the common fruit bat, *Artibeus jamaicensis*, on Barro Colorado Island, Panamá. Smithsonian Contrib. Zool. 511: 1-173.

Happel, R. 1983. *Saimiri* as a probable pollinator of *Passiflora*. Brenesia 21: 455-6.

Happel, R. 1986. Socioecology and conservation of the red-backed squirrel monkey (Primates, Cebidae, *Saimiri oerstedii*). Brenesia 25-26: 245-50.

Harrison, R. J. 1972, 1974, and 1977. Functional anatomy of marine mammals (3 vols.). Academic Press, London.

Hartenberger, J. -L. 1985. The order Rodentia: major questions on their evolutionary origin, relationships and suprafamilial systematics. In: Luckett & Hartenberger '85. Pp. 1-33.

Hartman, D. S. 1979. Ecology and behavior of the manatee *Trichechus manatus* in Florida. Special Publication 5. Am. Soc. Mammalogists.

Hayes, M., and R. K. Laval. 1989. The mammals of Monteverde: an annotated checklist, second edition. Tropical Science Center, San José.

Heaney, L. R. 1983. *Sciurus granatensis*. In: Janzen '83c. Pp. 489-90.

Heaney, L. R., and R. S. Hoffmann. 1978. A second specimen of the Neotropical montane squirrel, *Syntheosciurus poasensis*. J. Mamm. 59: 854-5.

Heaney, L. R., and R. W. Thorington, Jr. 1978. Ecology of Neotropical red-tailed squirrels, *Sciurus granatensis*, in the Panama Canal Zone. J. Mamm. 59: 846-51.

Heithaus, E. R., and T. H. Fleming. 1978. Foraging movements of a frugivorous bat, *Carollia perspicillata* (Phyllostomatidae). Ecol. Monogr. 48: 127-43.

Heithaus, E. R., T. H. Fleming, and P. A. Opler. 1975. Foraging patterns and resource utilization in seven species of bats in a seasonal tropical forest. Ecology 56: 841-54.

Helder, J., and H. K. Andrade. 1997. Food and feeding habits of the neotropical river otter *Lutra longicaudis* (Carnivora, Mustelidae). Mammalia 61: 193-203.

Hemmer, H. 1976. Fossil history of living Felidae. In: Eaton '76. Pp. 1-14.

Herd, R. M. 1983. *Pteronotus parnellii*. Mamm. Species 209. Am. Soc. Mammalogists.

Hernández, O. E., G. R. Barreto, and J. Ojasti. 1995. Observations of behavioral patterns of white-lipped peccaries in the wild. Mammalia 59: 146-8.

Hershkovitz, P. 1955. On the cheek pouches of the tropical American paca, *Agouti paca*. Saugetierk. Mitteil. 3: 66-70.

Hershkovitz, P. 1977. Living New World monkeys (Platyrrhini). Vol. I. Univ. Chicago Press.

Hilje, L., and J. Monge. 1988. Diagnóstico preliminar acerca de los animales vertebrados que son plagas en Costa Rica. Univ. Nacional, Heredia, Costa Rica.

Hladik, C. M. 1978. Adaptive strategies of primates. In: Montgomery '78.

Hladik, A., and C. M. Hladik. 1969. Rapports trophiques entre végétación et primates dans la forêt de Barro Colorado (Panama). Terre et Vie 23: 25-117.

Hood, C. S., and J. K. Jones, Jr. 1984. *Noctilio leporinus*. Mamm. Species 216. Am. Soc. Mammalogists.

Hood, C. S., and J. Pitochelli. 1983. *Noctilio albiventris*. Mamm. Species 197. Am. Soc. Mammalogists.

Hoogmoed, M. S. 1983. The occurrence of *Sylvilagus brasiliensis* in Surinam. Lutra 26(1): 34-45.

Hooper, E. T. 1968. Habitats and food of amphibious mice of the genus *Rheomys*. J. Mamm. 49: 550-53.

Hooper, E. T., and J. H. Brown. 1968. Foraging and breeding in two sympatric species of Neotropical bats, genus *Noctilio*. J. Mamm. 49: 310-12.

Hooper, E. T., and M. D. Carlton. 1976. Reproduction, growth and development in two contiguously allopatric rodent species, genus *Scotinomys*. Misc. Publ. Mus. Zool. Univ. Michigan, no. 51.

Hornocker, M. G. 1970. An analysis of mountain lion predation upon mule deer and elk in the Idaho Primitive Area. Wildl. Monogr. 21.

Horwich, R. H., and J. Lyon. 1993. A Belizean Rain Forest. Second edition. Orang-utan Press, Gays Mills, WI.

Howard, W. E., and J. N. Amaya. 1975. European rabbit invades western Argentina. J. Wildl. Mgmt. 39: 757-61.

Howard, W. E., and R. E. Marsh. 1982. Spotted and hog-nosed skunks: *Spilogale putorius* and allies. In: Chapman & Feldhamer '83. Pp. 664-73.

Howe, H. F. 1996. Fruit production and animal activity at two tropical trees. In: Leigh et al. '96. Pp. 189-200.

Howell, A. B. 1930. Aquatic mammals, their adaptation to life in the water. Springfield, Baltimore.

Howell, D. J. 1974. Feeding and acoustic behavior in glossophagine bats. J. Mamm. 55: 263-76.

Howell, D. J. 1983. *Glossophaga soricina*. In: Janzen '83c. Pp. 472-4.

Howell, D. J., and P. Burch. 1974. Food habits of some Costa Rican bats. Rev. Biol. Trop. 21: 281-94.

Howell, D. J., and R. C. Hodgkin. 1976. Feeding adaptations in the hairs and tongues of nectar-feeding bats. J. Morph. 148: 329-36.

Hoyt, R. A., and J. S. Altenbach. 1981. Observations on *Diphylla ecaudata* in captivity. J. Mamm. 62: 215-16.

Hughes, A. 1977. The topography of vision in mammals of contrasting lifestyle: comparative optics and retinal organisation. In: Cresticelli '77. Pp. 613-756.

Hughes, A. 1985. New perspectives in retinal organisation. Prog. Ret. Res. 4: 243-314.

Hulley, J. T. 1976. Maintenance and breeding of captive jaguarundis *Felis yagouaroundi* at Chester Zoo and Toronto. Intl. Zoo Yearb. 16: 120-22.

Hunsaker, D., II. 1977a. The ecology of New World marsupials. In: Hunsaker '77b. Pp. 95-156.

Hunsaker, D., II. 1977b. The biology of marsupials. Acad. Press, New York.

Hunsaker, D., II., and D. Shupe. 1977. Behavior of New World marsupials. In: Hunsaker '77b. Pp. 279-347.

Husar, S. L. 1978. *Trichechus manatus*. Mamm. Species 93. Am Soc. Mammalogists.

Huxley, C., and J. Servín. 1995. Estimación del ámbito hogareño del coyote (*Canis latrans*) en la Reserva de la Biosfera la Michilía, México. Vida Silv. Neotrop. 4(2): 98-106.

Ingles, L. G. 1965. Mammals of the Pacific states: California, Oregon, and Washington. Stanford Univ. Press.

Iriarte, J. A., W. L. Franklin, W. E. Johnson, and K. H. Redford. 1990. Biogeographic variation of food habits and body size of the American puma. Oecologia 85: 185-90.

Irvine, A. B. 1983. Manatee metabolism and its influence on distribution in Florida. Biological Conservation 25: 315-34.

Izor, R. J. 1985. Sloths and other mammalian prey of the harpy eagle. In: Montgomery '85. Pp. 343-46

Jaeger, E. C. 1978. A source book of biological names and terms. Third edition. Charles C. Thomas Publisher, Springfield, Illinois.

Janis, C. 1984. Odd-toed ungulates. In: Macdonald '84a. Pp. 480-81.

Janis, C., and P. J. Jarman. 1984. The hoofed mammals. In: Macdonald '84a. Pp. 468-79.

Janos, D. P., C. T. Sahley, and L. H. Emmons. 1995. Rodent dispersal of vesicular-arbuscular mycorrhizal fungi in Amazonian Peru. Ecology 76: 1852-58.

Janson, C. H. 1984. Capuchin-like monkeys. In: Macdonald '84a. Pp. 352-7.

Janson, C. H., J. Terborgh, and L. H. Emmons. 1981. Non-flying mammals as pollinating agents in the Amazonian forest. Biotropica 13 (Repro. Bot. Suppl.): 1-6.

Janzen, D. H. 1970. Altruism by coatis in the face of predation by *Boa constrictor*. J. Mamm. 51: 387-89.

Janzen, D. H. 1978. Predation intensity on eggs on the ground in two Costa Rican forests. Am. Midl. Nat. 100: 467-70.

Janzen, D. H. 1981. Digestive seed predation by a Costa Rican Baird's tapir. Biotropica 13: 59-63.

Janzen, D. H. 1982a. Attraction of *Liomys* mice to horse dung and the extinction of this response. Anim. Behav. 30: 483-89.

Janzen, D. H. 1982b. Fruit traits, and seed consumption by rodents, of *Crescentia alata* (Bignoniaceae) in Santa Rosa National Park, Costa Rica. Am. J. Botany 69: 1240-50.

Janzen, D. H. 1982c. Removal of seeds by tropical rodents from horse dung in different amounts and habitats. Ecology 63: 1887-1900.

Janzen, D. H. 1982d. Seed removal from fallen Guanacaste fruits (*Enterolobium cyclocarpum*) by spiny pocket mice (*Liomys salvini*). Brenesia 19/20: 425-9.

Janzen, D. H. 1982e. Seeds in tapir dung in Santa Rosa National Park, Costa Rica. Brenesia 19/20: 129-35.

Janzen, D. H. 1982f. Wild plant acceptability to a captive Costa Rican Baird's tapir. Brenesia 19/20: 99-128.

Janzen, D. H. 1983a. *Canis latrans*. In: Janzen '83c. Pp. 456-7.

Janzen, D. H. 1983b. *Coendou mexicanum*. In: Janzen '83c. Pp. 460-1.

Janzen, D. H. (ed.). 1983c. Costa Rican natural history. Univ. Chicago Press, Chicago.

Janzen, D. H. 1983d. *Crescentia alata*. In: Janzen '83c. Pp. 222-24.

Janzen, D. H. 1983e. *Eira barbara*. In: Janzen '83c. Pp. 469-70.

Janzen, D. H. 1983f. *Enterolobium cyclocarpum*. In: Janzen '83c. Pp. 241-43.

Janzen, D. H. 1983g. *Guazuma ulmifolia*. In: Janzen '83c. Pp. 246-48.

Janzen, D. H. 1983h. *Hymenaea courbaril*. In: Janzen '83c. Pp. 253-56.

Janzen, D. H. 1983i. *Odocoileus virginianus*. In: Janzen '83c. Pp. 81-3.

Janzen, D. H. 1983j. *Swallenochloa subtessellata*. In: Janzen '83c. Pp. 330-31.

Janzen, D. H. 1983k. *Tapirus bairdii*. In: Janzen '83c. Pp. 496-7.

Janzen, D. H. 1986. El venado en la biología del bosque seco tropical. In: Solís et al. '86. Pp. 9-11.

Janzen, D. H., and W. Hallwachs. 1982. The hooded skunk, *Mephitis macroura*, in lowland northwestern Costa Rica. Brenesia 19/20: 549-52.

Janzen, D. H., and P. S. Martin. 1982. Neotropical anachronisms: the fruits the mastodons left behind. Science.

Janzen, D. H., and D. E. Wilson. 1983. Mammals: Introduction. In: Janzen '83c. Pp. 426-42.

Jenkins, F. A., Jr., and D. McClearn. 1984. Mechanisms of hindfoot reversal in climbing mammals. J. Morphol. 182: 197-219.

Jiménez, I. 1995. Presencia de contaminantes en el Parque Nacional Tortuguero y su posible impacto sobre el manatí (*Trichechus manatus*). Unpublished manuscript. Univ. Nacional, Heredia.

Jiménez, I. 1998. Ecología y conservación del manatí antillano (*Trichechus manatus*) en el noreste de Costa Rica. Base de datos de los humedales del noreste de Costa Rica asociada a un sistema de información geográfica. Master's thesis. Univ. Nacional, Heredia, Costa Rica.

Jiménez, I. 2000. Los manatíes del río San Juan y los canales de Tortuguero. Friends of the Earth.

Jiménez, I., and M. Altrichter. 1998. Estado de conservación del manatí (*Trichechus manatus*) en la gran Reserva Biológica Indio Maíz, Nicaragua. Informe técnico presentado a Amigos de la Tierra y Idea Wild. Univ. Nacional, Heredia, Costa Rica.

Jiménez, J. 1970. Condición económica de los monos en Costa Rica. O'Bios 2:21-40.

Jiménez, J. 1973. Daño ocasionado al banano por el murciélago *Glossophaga soricina* en el Valle de La Estrella, Limón, Costa Rica. Rev. Biol. Trop. 21(1): 69-81.

Jones, C. B. 1978. Aspects of reproductive behavior in the mantled howler monkey, *Alouatta palliata* Gray. Ph.D Thesis. Cornell Univ.

Jones, C. B. 1980a. The functions of status in the mantled howler monkey, *Alouatta palliata* Gray: intraspecific competition for group membership in a folivorous Neotropical primate. Primates 21: 389-405.

Jones, C. B. 1980b. Seasonal parturition mortality and dispersal in the mantled howler monkey, *Alouatta palliata* Gray. Brenesia 17: 1-10.

Jones, C. B. 1985. Reproductive patterns in mantled howler monkeys: estrus, mate choice and competition. Primates 26: 130-42.

Jones, J. K., Jr. 1966. Bats from Guatemala. Univ. Kansas Publ., Mus. Nat. Hist. 16: 439-72.

Jones, J. K., Jr., J. Arroyo-Cabrales, and R. D. Owen. 1988. Revised checklist of bats (Chiroptera) of Mexico and Central America. Occas. Pap. Mus. Texas Tech Univ. 120.

Jones, M. L. 1982. Longevity of captive mammals. Zool. Garten 52: 113-28.

Julien-Laferrière. D. 1993. Radio-tracking observations on ranging and foraging patterns by kinkajous (*Potos flavus*) in French Guiana. J. Trop. Ecol. 9: 19-32.

Julien-Laferrière. D. 1997. The influence of moonlight on activity of woolly opossums (*Caluromys philander*). J. Mamm. 78(1): 251-55.

Kalko, E. K. V., C. O. Handley, Jr., and D. Handley. 1996. Organization, diversity, and long-term dynamics of a neotropical bat community. In: Cody & Smallwood '96. Pp. 503-53.

Kalko, E. K. V., H.-U. Schnitzler, I. Kaipf, and A. D. Grinnell. 1998. Echolocation and foraging behavior of the lesser bulldog bat, *Noctilio albiventris*: preadaptations for piscivory? Behav. Ecol. Sociobiol. 42: 305-19.

Kalmbach, E. R. 1943. The armadillo: its relation to agriculture and game. Texas Game, Fish and Oyster Comm., Austin, Texas.

Kaplan, J. B., and R. A. Mead. 1991. Conservation status of the eastern spotted skunk. Mustelid and Viverrid Conservation Newsletter 4: 15.

Kaufmann, J. H. 1962. Ecology and social behavior of the coati, *Nasua narica*, on Barro Colorado Island, Panama. Univ. Calif. Publ. Zool. 60(3): 95-222.

Kaufmann, J. H. 1983. *Nasua narica*. In: Janzen '83c. Pp. 478-80.

Kaufmann, J. H. 1987. Ringtail and coati. In: Novak et al. '87. Pp. 501-08.

Kaufmann, J. H., and A. Kaufmann. 1965. Observations on the behavior of tayras and grisons. Z. Säugetierk. 30: 146-55.

Kays, R. W. 1999. Food preferences of kinkajous (*Potos flavus*): a frugivorous carnivore. J. Mamm. 80(2): 589-99.

Kays, R. W., and J. L. Gittleman. 1995. Home range size and social behavior of kinkajous (*Potos flavus*) in the republic of Panama. Biotropica 27(4): 530-34.

Keast, A., F. C. Erk, and B. Glass (eds.). 1972. Evolution, mammals, and southern continents. Univ. New York Press, Albany.

Keleman, G., and J. Sade. 1960. The vocal organ of the howling monkey (*Alouatta palliata*). J. Morph. 107: 123-40.

Kerby, G. 1984. Other big cats. In: Macdonald '84a. Pp. 48-9.

Kile, T. L., and R. L. Marchinton. 1977. White-tailed deer rubs and scrapes: spatial, temporal and physical characteristics and social role. Am. Midl. Nat. 97: 257-66.

Kiltie, R. A. 1981a. Distribution of palm fruits on a rain forest floor: why white-lipped peccaries forage near objects. Biotropica 13: 141-45.

Kiltie, R. A. 1981b. Stomach contents of rain forest peccaries (*Tayassu tajacu* and *T. pecari*). Biotropica 13: 234-36.

Kiltie, R. A. 1981c. The function of interlocking canines in rain forest peccaries (*Tayassuidae*). J. Mamm. 62: 459-69.

Kiltie, R. A. 1982. Bite force as a basis for niche differentiation between rain forest peccaries (*Tayassu tajacu* and *T. pecari*). Biotropica 14(3): 188-95.

Kiltie, R. A. 1984. Size ratios among sympatric Neotropical cats. Oecologia 61: 411-16.

Kiltie, R. A., and J. Terborgh. 1983. Observation on the behavior of rain forst peccaries in Peru: why do white-lipped peccaries form herds? Z. Tierpsychol. 62: 241-55.

Kimbal, W. H., and L. B. Martin (eds.). 1993. Species, species concepts, and primate evolution. Plenum Press, New York.

King, C. M. 1977. The effects of the nematode parasite *Skrajabingylus nasicola* on British weasels (*Mustela nivalis*). J. Zool. Lond. 182: 225-49.

King, C. M. 1989. The natural history of weasels. Comstock-Cornell Univ. Press.

Kinlaw, A. 1995. *Spilogale putorius*. Mamm. Species 511. Am. Soc. Mammalogists.

Kinzey, W. 1997a. *Alouatta*. In: Kinzey '97b. Pp. 174-85.

Kinzey, W. 1997b. New World primates: ecology, evolution and behavior. Adline, New York.

Kirksey, E. R., J. F. Pagels, and C. R. Blem. 1975. The role of the tail in temperature regulation of the cotton rat, *Sigmodon hispidus*. Comp. Biochem. Physiol. 52A: 707-11.

Kirsch, J. A. W. 1977. The classification of marsupials. In: Hunsaker '77a. Pp. 1-50.

Kitchener, A. 1991. The natural history of the wild cats. Comstock-Cornell Univ. Press.

Kleiman, D. G. 1974. Patterns of behavior in hystricomorph rodents. In: Rowlands & Weir '74. Pp. 171-209.

Kleiman, D. G., J. F. Eisenberg, and E. Maliniak. 1979. Reproductive parameters and productivity of Caviomorph rodents. In: Eisenberg '79. Pp. 173-83.

Klein, L. L., and D. J. Klein. 1976. Neotropical primates: aspects of habitat usage, population density and regional distribution in La Macarena, Colombia. In: Thorington & Heltne '76. Pp. 70-79.

Klein, L. L., and D. J. Klein. 1977. Feeding behavior of the Colombian spider monkey. In: Clutton-Brock '77. Pp. 153-81.

Klingel, H. 1977. Communication in Perissodactyla. In: Sebeok '77. Pp. 715-27.

Koford, C. B. 1976. Latin American cats: economic values and future prospects. Proc. Third Intl. Symp. World's Cats 3(1): 79-88.

Koford, C. B. 1983a. *Felis onca*. In: Janzen '83c. Pp. 470-71.

Koford, C. B. 1983b. *Felis wiedii*. In: Janzen '83c. Pp. 471-2.

Konecny, M. J. 1989. Movement patterns and food habits of four sympatric carnivore species in Belize, Central America. In: Redford & Eisenberg '89. Pp. 243-64.

Kraatz, W. C. 1930. Mouse opossums stowaways on bananas. Science 71: 288.

Krantz, G. S. 1970. Human activities and megafaunal extinctions. Am. Sci. 58: 164-70.

Kunz, T. H. 1987. Post-natal growth and energetics of suckling bats. In: Fenton et al. '87. Pp. 394-420.

Kunz, T. H., and C. A. Diaz. 1995. Folivory in fruit-eating bats, with new evidence from *Artibeus jamaicensis* (Chiroptera: Phyllostomidae). Biotropica 27(1): 106-120.

Kunz, T. H., M. S. Fujita, A. P. Brooke, and G. F. McCracken. 1994. Convergence in tent architecture and tent-making behavior among Neotropical and Paleotropical bats. J. Mamm. Evol. 2: 57-78.

Kunz, T. H., and E. D. Pierson. 1994. Bats of the world: an introduction. In: Nowak '94. Pp 1-46.

Kurten, B. 1973a. Pleistocene jaguars in North America. Comment. Biologic. 62: 1-23.

Kurten, B. 1973b. Geographic variation in size in the puma. Comment. Biologic. 63: 1-8.

Laack, L. L. 1991. Ecology of the ocelot (*Felis pardalis*) in South Texas. M.S. Thesis. Texas A & I Univ., Kingsville.

Langguth, A. 1975. Ecology and evolution in the South American Canids. In: Fox '75. Pp. 192-206.

Langtimm, C. A. 2000. Singing mice. In: Nadkarni & Wheelwright 2000. Pp. 236-38.

Langtimm, C. A., and R. Unnasch. 2000. Mice, birds, and pollination of *Blakea clorantha*. In: Nadkarni & Wheelwright 2000. Pp. 241.

Larson, D., and H. F. Howe. 1987. Dispersal and destruction of *Virola surinamensis* seeds by agoutis: appearance and reality. J. Mamm. 68: 859-60.

Lassieur, S., and D. E. Wilson. 1989. *Lonchorhina aurita*. Mamm. Species 347. Am. Soc. Mammalogists.

Laundré, J. W., and B. L. Keller. 1984. Home range size of coyotes: a critical review. J. Wildl. Mgmt. 48: 127-39.

LaVal, R. K. 1973a. A revision of the Neotropical bats of the genus *Myotis*. Los Angeles County Nat. Hist. Mus. Sci. Bull. 15.

LaVal, R. K. 1973b. Observations on the biology of *Tadarida brasiliensis cynocephala* in southeatern Louisiana. Am. Midl. Nat. 89: 112-20.

LaVal, R. K. 1977. Notes on some Costa Rican bats. Brenesia 10/11: 77-83.

LaVal, R. K., and H. S. Fitch. 1977. Structure, movements and reproduction in three Costa Rican bat communities. Occas. Pap. Mus. Nat. Hist. Univ. Kansas 69.

Lawrence, G. A. 1991. The non-insect pests of cocoa. J. Agric. Soc. Trinidad & Tobago 1: 16-19.

Lawton, R. O. 2000. Tapir. In: Nadkarni & Wheelwright 2000. Pp. 242-43.

Layne, J. N. 1958. Observations on freshwater dolphins in the upper Amazon. J. Mamm. 39: 1-22.

Layne, J. N., and D. Glover. 1977. Home range of the armadillo in Florida. J. Mamm. 58: 411-13.

Layne, J. N., and D. Glover. 1985. Activity patterns of the common long-nosed armadillo *Dasypus novemcinctus* in south-central Florida. In: Montgomery '85a. Pp. 407-17.

Leatherwood, S., and R. R. Reeves. 1983. The Sierra Club handbook of whales and dolphins. Sierra Club Books, San Francisco.

Lefebvre, L. W., T. J. O'Shea, G. B. Rathbun, and R. C. Best. 1989. Distribution, staus, and biogeography of the West Indian Manatee. In: Woods ' 89. Pp. 567-610.

Leigh, E. G., Jr., A. S. Rand, and D. M. Windsor (eds.). 1996. The ecology of a tropical forest: seasonal rythms and long-term changes. Second edition. Smithsonian Institution Press, Washington.

Leigh, E. G., Jr., and D. M. Windsor. 1996. Forest production and regulation of primary consumers on Barro Colorado Island. In: Leigh et al. '96. Pp. 111-22.

Lemke, T. O. 1984. Foraging ecology of the long-nosed bat, *Glossophaga soricina*, with respect to resource availability. Ecology 65: 538-48.

Lemoine, J. 1977. Some aspects of ecology and behavior of ringtails (*Bassariscus astutus*) in Sta. Helena. California Press, Berkeley.

Leopold, A. S. 1959. Wildlife of Mexico. Univ. Calif. Press, Berkeley.

Levings, S. C., and D. M. Windsor. 1996. Seasonal and annual variation in litter arthropod populations. In: Leigh et al. '96. Pp. 355-87.

Leyhausen, P., and M. Falkena. 1966. Breeding the Brazilian ocelot-cat *Leopardus tigrinus* in captivity. Intl. Zoo Yearb. 6: 176-82.

Ligon, S. 1983. *Trichechus manatus*. In: Janzen '83c. Pp. 498-500.

Lilly, J. C. 1977. The cetacean brain. Oceans 10(4): 4-7.

Linares, O. J. 1981. Tres nuevos carnívoros prociónidos fósiles del Mioceno de Norte y Sudamérica. Ameghiniana 18: 113-21.

Linduska, J. P. 1947. Longevity of some Michigan farm game mammals. J. Mamm. 28: 129-29.

Lindzey, F. 1987. Mountain lion. In: Novak et al. '87. Pp. 657-68.

Lippold, L. K. 1988. A census of primates in Cabo Blanco Absolute Nature Reserve, Costa Rica. Brenesia 29: 101-05.

Locket, N. A. 1977. Adaptations to the deep-sea environment. In: Cresticelli '77. Pp. 67-192.

Löhmer, R. 1976. Zur Verhaltensontogenese bei *Procyon cancrivorus*. Z. Säugetierk. 41: 42-58.

Long, A., and P. S. Martin. 1974. Death of American ground sloths. Science 186: 638-40.

Longino, J. T. 1984. True anting by the capuchin, *Cebus capucinus*. Primates 25: 243-45.

López, E. 1985. Análisis de población del venado colablanca (*Odocoileus virginianus*) en la Isla San Lucas, Golfo de Nicoya, Costa Rica. In: Subdirección General... '85. Pp. 47-50.

López, J. E. 1996. Hábitos alimentarios de murciélagos frugívoros y su participación en la dispersión de semillas, en bosques secundarios húmedos, de Costa Rica. Masters Thesis. Univ. Nacional, Heredia.

López, M. T. 1999. Aspectos nuticionales de la dieta del chancho de monte (*Tayassu pecari*) en un bosque húmedo tropical de Costa Rica. Masters Thesis. Univ. Nacional, Heredia.

López-Forment, W. 1980. Longevity of wild *Desmodus rotundus* in Mexico. Proc. 5th Internatl. Bat Res. Conf. Pp. 143-44.

Lord, R. D. 1988. Control of vampire bats. In: Greenhall & Schmidt '88. Pp. 215-26.

Lotze, J. H., and S. Anderson. 1979. *Procyon lotor*. Mamm. Species 119. Am. Soc. Mammalogists.

Lowery, G. H. Jr. 1974. The mammals of Louisiana and its adjacent waters. Louisiana State Univ. Press.

Lowry, J. B. 1989. Green-leaf fractionation by fruit bats: is this feeding behaviour a unique nutritional strategy for herbivores? Aust. Wild. Res. 16: 203-06.

Lubin, Y. D. 1983a. *Nasutitermes*. In: Janzen '83c. Pp. 743-45.

Lubin, Y. D. 1983b. *Tamandua mexicana*. In: Janzen '83c. Pp. 494-96.

Lubin, Y. D., and G. G. Montgomery. 1981. Defenses of *Nasutitermes* termites (Isoptera: Termitidae) against *Tamandua* anteaters (Edentata: Myrmecophagidae). Biotropica 13: 66-76.

Lubin, Y. D., G. G. Montgomery, and O. P. Young. 1977. Food resources of anteaters (Edentata: Myrmecophagidae). 1. A year's census of arboreal nests of ants and termites on Barro Colorado Island, Panama Canal Zone. Biotropica 9(1): 26-34.

Luckett, W. P., and J. -L. Hartenberger (eds.). 1985. Evolutionary relationships among rodents: a multidisciplinary analysis. Plenum Press, New York.

Luckett, W. P., and F. S. Szalay (eds.). 1975. Phylogeny of the primates: a multidisciplinary approach. Plenum Press, New York.

Ludlow, M. E., and M. E. Sunquist. 1987. Ecology and behavior of ocelots in Venezuela. Nat. Geog. Res. 3: 447-61.

Lumer. C. 1983. *Blakea*. In: Janzen '83c. Pp. 194-5.

Lumer, C., and R. D. Schoer. 1986. Pollination of *Blakea austin-smithii* and *B. penduliflora* (Melastomaceae) by small rodents in Costa Rica. Biotropica 18: 363-4.

Lyall-Watson, M. 1963. A critical re-examination of food "washing" in the raccoon (*Procyon lotor* Linn.). Proc. Zool. Soc. Lond. 141: 371-93.

Lynch, G. M. 1967. Long-range movement of a raccoon in Manitoba. J. Mamm. 48: 659-60.

MacClintock, D. 1985. Phoebe the kinkajou. Scribners and Sons, New York.

Macdonald, D. W. 1984a. Encyclopedia of mammals. Facts on File, New York.

Macdonald, D. W. 1984b. The carnivores. In: Macdonald '84a. Pp. 16-25.

Macdonald, D. W. 1984c. The dog family. In: Macdonald '84a. Pp. 56-7.

Macdonald, D. W. 1995. European mammals: evolution and behaviour. Harper Collins, London.

MacDonald, S., and S. Mason. 1992. A note on *Lutra longicaudis* in Costa Rica. IUCN Otter Specialist Group Bull. 7: 37-38.

MacLaren, J. P. 1967. Manatees as a naturalistic biological mosquito control method. Mosquito News 27(3): 387-93.

MacNamara, M., and W. Eldridge. 1987. Behavior and reproduction in captive pudu (*Pudu pudu*) and red brocket (*Mazama americana*), a descriptive and comparative analysis. In: Wemmer '87. Pp. 371-86.

Magnusson, W. E., R. C. Best, and V. M. F. da Silva. 1980. Numbers and behavior of Amazonian dolphins, *Inia geoffrensis* and *Sotalia fluviatalis fluviatalis*, in the river Solimoes, Brasil. Aquatic Mammals 8(1): 27-32.

Malavassi, L. 1991. I. Captura del grisón (*Gallictis vittata*). II. Actividad del grisón (*Gallictis vittata*) en cautiverio. III. Posible ciclo reproductivo del grisón (*Gallictis vittata*) en Costa Rica. Masters Thesis. Univ. Nacional, Heredia.

Maliniak, E., and J. F. Eisenberg. 1971. The breeding of *Proechimys semispinosus* in captivity. Intl. Zoo Yearb. 11: 93-8.

Manaro, A. J. 1961. Observations on the behavior of the spotted skunk in Florida. Quat. J. Florida Acad. Sci. 24: 59-63.

Manzani, P. R., and E. L. A. Monteiro-Filho. 1989. Notes on the food habits of the jaguarundi *Felis yagouaroundi* (Mammalia: Carnivora). Mammalia 53: 659-60.

Marchinton, R. L., and D. H. Hirth. 1984. Behavior. In: Halls '84. Pp. 129-68.

Mares, M., and H. Genoways. 1982. Mammalian biology in South America. Pymatuning Symposia in Ecology 6. Special Pub. Ser. Pymatuning Lab. of Ecol., Univ. of Pittsburg.

Marsden, H. M., and N. R. Holler. 1964. Social behavior in confined populations of the cottontail and swamp rabbit. Wildl. monogr. 13. Wildlife Society.

Marshall, L. G. 1978. *Chironectes minimus*. Mamm. Species 109. Am. Soc. Mammalogists.

Martin, I. G. 1981. Venom of the short-tailed shrew (*Blarina brevicauda*) as an insect immobilizing agent. J. Mamm. 62: 189-92.

Martin P. G. 1977. Marsupial biogeography and plate tectonics. In: Stonehouse & Gilmore '77. Pp. 97-115.

Maser, C., J. M. Trappe, and R. A. Nussbaum. 1978. Fungal-small mammal interrrelationships with emphasis on Oregon coniferous forests. Ecology 59: 799-809.

Mason, C. F., and S. M. Macdonald. 1986. Otters: ecology and conservation. Cambridge Univ. Press, London.

Matamoros, Y. H. 1980. Investigaciones preliminares sobre la reproducción, comportamiento, alimentación y manejo del tepezcuintle (*Cuniculus paca*) en cautiverio. In: Salinas '80. Pp. 961-94.

Matamoros, Y. H. 1981. Anatomía e histología del sistema reproductivo del tepezcuintle (*Cuniculus paca*). Rev. Biol. Trop. 29: 155-64.

Matamoros, Y. H. 1982. Notas sobre la biología del tepezcuintle, *Cuniculus paca,* Brisson (Rodentia: Dasyproctidae) en cautiverio. Brenesia 19/20: 71-82.

Mayer, J. J., and R. M. Wetzel. 1987. *Tayassu pecari*. Mamm. Species 293. Am. Soc. Mammalogists.

McBee, K., and R. J. Baker. 1982. *Dasypus novemcinctus*. Mamm. Species 162. Am. Soc. Mammalogists.

McCarthy, T. J. 1987a. Additional mammalian prey of the carnivorous bats *Chrotopterus auritus* and *Vampyrum spectrum*. Bat. Res. News 28: 1-3.

McCarthy, T. J. 1987b. Distributional records of bats from the Caribbean lowlands of Belize and adjacent Guatemala and Mexico. Fieldiana Zool., n.s. 39: 137-62.

McCarthy, T. J. 1992. Notes concerning the jaguarundi cat (*Herpailurus yagouaroundi*) in Caribbean lowlands of Belize and Guatemala. Mammalia 56: 302-06.

McClearn, D. 1990. Limb proportions of raccoons (*Procyon*) and coatis (*Nasua*). Am. Zool. 30: 25(A).

McClearn, D. 1992a. Locomotion, posture, and feeding behavior of kinkajous, coatis, and raccoons. J. Mamm. 73(2): 245-61.

McClearn, D. 1992b. The rise and fall of a mutualism? Coatis, tapirs, and ticks on Barro Colorado Island, Panama. Biotropica 24(2a): 220-22.

McCoy, M. B. 1985. Seasonal movement, home range, activity and diet of collared peccaries in Costa Rican dry forest. Masters thesis. Humboldt State Univ., California.

McCoy, M. B., C. S. Vaughan, M. A. Rodríguez, and D. Kitchen. 1990. Seasonal movement, home range, activity and diet of collared peccaries (*Tayassu tajacu*) in Costa Rican dry forest. Vida Silv. Neotrop. 2: 6-20.

McCoy, M. B., C. S. Vaughan, and V. Villalobos. 1983. An interesting feeding habitat for the collared peccary (*Tayassu tajacu*) in Costa Rica. Brenesia 21:456-57.

McCracken, G. F., and M. K. Gustin. 1987. Batmom's daily nightmare. Nat. Hist. 96(10): 66-73.

McCullough, C. R., and E. K. Fritzell. 1984. Ecological observations of eastern spotted skunks on the Ozark Plateau. Transactions Missouri Acad. Sci. 18: 25-32.

McDade, L. A., K. S. Bawa, H. A. Hespenheide, and G. S. Hartshorn (eds.). 1994. La Selva: ecology and natural history of a neotropical rainforest. Univ. Chicago Press, Chicago.

McDonnell, M. J., and E. W. Stiles. 1983. The structural complexity of old field vegetation and the recruitment of bird-dispersed plant species. Oecologia 56: 109-16.

McDonough, C. M., and W. J. Loughry. 1997. Influences on activity patterns in a population of nine-banded armadillos. J. Mamm. 78(3): 932-41.

McHargue, L. A., and G. S. Hartshorn. 1983a. *Carapa guianensis*. In: Janzen '83c. Pp. 206-7.

McHargue, L. A., and G. S. Hartshorn. 1983b. Seed and seedling ecology of *Carapa guianensis*. Turrialba 33: 399-404.

McKenna, M. C. 1975. Toward a phylogenetic classification of the Mammalia. In: Luckett & Szalay '75. Pp. 21-46.

McMahan, L. R. 1986. The international cat trade. In: Miller & Everett '86. Pp. 461-87.

McManus, J. J. 1967. Observations on sexual behavior of the opossum, *Didelphis marsupialis*. J. Mamm. 48(3): 486-7.

McManus, J. J. 1970. Behavior of captive opossums, *Didelphis marsupialis virginiana*. Am. Midl. Nat. 84: 144-69.

McManus, J. J. 1974. *Didelphis marsupialis*. Mamm. Species 40. Am. Soc. Mammologists.

McNab, B. 1978. Energetics of arboreal folivores: physiological problems and ecological consequences of feeding on an ubiquitous food supply. In: Montgomery '78. Pp. 153-62.

McNab, B. 1980. Energetics and the limits to a temperate distribution in armadillos. J. Mamm. 61: 606-22.

McNab, B. 1982. The physiological ecology of South American mammals. In: Mares & Genoways '82. Pp. 187-207.

McNab, B. 1983. Ecological and behavioral consequences of adaptation to various food resources. In: Eisenberg & Kleiman '83. Pp. 664-97.

McNab, B. 1985. Energetics, population biology, and distribution of xenarthrans, living and extinct. In: Montgomery '85a. Pp. 219-32.

McNab, B. 1989. Basal rate of metabolism, body size, and food habits in the order Carnivora. In: Gittleman '89. Pp 335-54.

McPherson, A. B. 1985. The biogeography of Costa Rican rodents: an ecological, geological, and evolutionary approach. Brenesia 25-26: 229-44.

McPherson, A. B. 1986. A biogeographical analysis of factors influencing the distribution of Costa Rican rodents. Brenesia 23: 97-273.

McPherson, A. B., R. Zeledón, and S. Shelton. 1985. Comments on the status of *Metachirus nudicaudatus dentatus* (Goldman, 1912) in Costa Rica. Brenesia 24: 375-7.

Mead, R. A. 1968a. Reproduction in eastern forms of the spotted skunk (genus *Spilogale*). J. Zool. 156: 119-36.

Mead, R. A. 1968b. Reproduction in western forms of the spotted skunk (genus *Spilogale*). J. Mamm. 49: 373-90.

Mead, R. A. 1981. Delayed implantation in mustelids, with special emphasis on the spotted skunk. J. Rep. Fert. Suppl. 29: 11-24.

Mead, R. A. 1989. The physiology and evolution of delayed implantation in carnivores. In: Gittleman '89. Pp. 437-64.

Medellín, R. A. 1994. Seed dispersal of Cecropia obtusifolia by two species of opossums in the Selva Lacandona, Chiapas, Mexico. Biotropica 26(4): 400-07.

Medellín, R. A., G. Cancino Z., A. Clemente M., and R. O. Guerrero V. 1992. Noteworthy records of three mammals from Mexico. Southwest. Nat. 37: 427-29.

Medellín, R. A., and G. Ceballos (eds.). 1993. Avances en el estudio de mamíferos de México.

Mellen, J. D. 1993. A comparative analysis of scent marking, social and reproductive behavior of small cats (*Felis*). Am. Zool. 33: 151-66.

Melquist, W. E. 1984. Status of otters (Lutrinae) and spotted cats (Felidae) in Latin America. Report to IUCN. College of Forestry, Wildlife, and Range Sciences, Univ. Idaho.

Melquist, W. E., and A. E. Dronkert. 1987. River otter. In: Novak et al. '87. Pp. 626-41.

Melquist, W. E., and M. G. Hornocker. 1983. Ecology of river otters in west central Idaho. Will. Monogr. 83.

Mena, R. A. 1978. Fauna y caza en Costa Rica. Litografía e Imprenta LIL, S.A, San José.

Meritt, D. A., Jr. 1976. The nutrition of edentates. Int. Zoo Yearb. 16: 38-46.

Meritt, D. A., Jr. 1985a. Naked-tailed armadillos, *Cabassous* sp. In: Montgomery '85a. Pp. 389-91.

Meritt, D. A., Jr. 1985b. The two-toed hoffmann's sloth, *Chloepus hoffmanni* Peters. In: Montgomery '85. Pp. 333-41.

Meritt, D. A., Jr., and G. F. Meritt. 1976. Sex ratios in Hoffmann's sloth, *Chloepus hoffmanni* Peters and three-toed sloth, *Bradypus infuscatus* Wagler in Panama. Am. Midl. Nat. 96: 472-73.

Merrett, P. K. 1983. Edentates. Zoological Trust of Guernsey.

Michels, K. M., B. E. Fischer, and J. I. Johnson. 1960. Raccoon performance on color discrimination problems. J. Comp. Physiol. Psychol. 53: 379-80.

Miles, M. A., A. A. de Souza, and M. M. Póvoa. 1981. Mammal tracking and nest location in Brazilian forest with an improved spool-and-line device. J. Zool. Lond. 195: 331-47.

Miller, L. A. 1975. The behavior of flying green lacewings, *Chrysopa carnea*, in the presence of ultrasound. J. Insect Physiol. 21: 205-19.

Miller, L. A. 1991. Arctiid moth clicks can degrade the accuracy of range difference discrimination in echolocating big brown bats, *Eptesicus fuscus*. J. Comp. Physiol. 168: 571-79.

Miller, S. D., and D. D. Everett (eds.). 1986. Cats of the world: biology, conservation, and management. Nat. Wildl. Fed., Washington.

Milton, K. 1978. Behavioral adaptations to leaf-eating in the mantled howler monkey. In: Montgomery '78. Pp. 335-50.

Milton, K. 1980. The foraging strategy of howler monkeys. Columbia Univ. Press, New York.

Milton, K. 1984. Leaf-eaters of the New World: diet and energy conservation in the mantled howler monkey. In: Macdonald '84a. Pp. 368-9.

Milton, K. 1996. Dietary quality and demographic regulation in a howler monkey population. In: Leigh et al. '96. Pp. 273-89.

Milton, K., T. M. Casey, and K. K. Casey. 1979. The basal metabolism of mantled howler monkeys (*Alouatta palliata*). J. Mamm. 60: 373-76.

Mitchell, C. L., S. Boinski, and C. P. van Schaik. 1991. Competitive regimes and female bonding in two species of squirrel monkey (*Saimiri oerstedi* and *S. sciureus*). Behav. Ecol. Sociobiol. 28: 55-60.

Mittermeier, R. A. 1978. Locomotion and posture in *Ateles geoffroyi* and *Ateles paniscus*. Folia Primatol. 30: 161-93.

Mittermeier, R. A. 1988. Primate diversity and the tropical forest. In: Wilson '88. Pp. 145-54.

Mittermeier, R. A., A. B. Rylands, A. F. Coimbra-Filho, and G. A. B. Fonseca (eds.). 1988. Ecology and behavior of Neotropical primates. Vol. 2. World Wildlife Fund, Washington D.C.

Mock, O. B., and C. H. Conaway. 1975. Reproduction of the least shrew (*Cryptotis parva*) in captivity. In: Antikatzides et al. '75. Pp. 59-74.

Moiseff, A., and R. R. Hoy. 1983. Sensitivity to ultrasound in an identified auditory interneuron in the cricket: a possible neural link to phonotactic behavior. J. Comp. Physiol. 152: 155-67.

Molina, H., C. Roldán, A. Sáenz, and S. Torres. 1986. Hallazgo de *Bradypus griseus* y *Choloepus hoffmanni* (Edentata: Bradypodidae) en tierras altas de Costa Rica. Rev. Biol. Trop. 34: 165-6.

Molina, I., and S. Palmer. 1998. The history of Costa Rica. Editorial de la Universidad de Costa Rica.

Mondolfi, E. 1986. Notes on the biology and status of the small wild cats in Venezuela. In: Miller & Everett '86. Pp. 125-46.

Mondolfi, E., and R. Hoogesteijn. 1986. Notes on the biology and status of the jaguar in Venezuela. In: Miller & Everett '86. Pp. 85-123.

Mondolfi, E., and G. Medina P. 1957. Contribución al conocimiento del "perrito de agua" (*Chironectes minimus* Zimmermann). Mem. Soc. Cient. Nat. La Salle 17: 140-55.

Monge, J. I. 1989. Ciclo reproductivo y dieta de la ardilla *Sciurus variegatoides* (Sciuridae, Rodentia) en la Península de Nicoya, Costa Rica. Thesis: Universidad Nacional de Heredia.

Monge, J. I. 1992. Características poblacionales y uso del hábitat de la rata de la caña (*Sigmodon hispidus*) en Cañas, Guanacaste, Costa Rica. Masters Thesis. Univ. Nacional, Heredia.

Monge-Nájera, J., and B. Morera. 1986. La dispersión del coyote (*Canis latrans*) y la evidencia de los antiguos cronistas. Brenesia 25/26: 251-60.

Monge-Nájera, J., and B. Morera. 1987. Why is the coyote (*Canis latrans*) expanding its range? A critique of the deforestation hypothesis. Rev. Biol. Trop. 35: 169-71.

Montgomery, G. G. (ed.). 1978. The ecology of arboreal folivores. Smithsonian Institution Press, Washington.

Montgomery, G. G. 1979. El grupo aliméntico (feeding guild) del oso hormiguero. Convivencia y especialización de las presas de sustento de los osos hormigueros neotropicales (Edentata, Myrmecophagidae). ConCiencia 6(1): 3-6.

Montgomery, G. G. 1983a. *Bradypus variegatus*. In: Janzen '83c. Pp. 453-6.

Montgomery, G. G. 1983b. *Cyclopes didactylus*. In: Janzen '83c. Pp.461-3.

Montgomery, G. G. (ed.) 1985a. The evolution and ecology of armadillos, sloths, and vermilinguas. Smithsonian Institution Press, Washington.

Montgomery, G. G. 1985b. Impact of vermilinguas (*Cyclopes*, *Tamandua*: Xenarthra = Edentata) on arboreal ant populations. In: Montgomery '85a. Pp. 351-63.

Montgomery, G. G. 1985c. Movements, foraging and food habits of the four extant species of Neoptropical vermilinguas (Mammalia: Myrmecophagidae). In: Montgomery '85a. Pp. 365-77.

Montgomery, G. G., W. W. Cochran, and M. E. Sunquist. 1973. Radio-locating arboreal vertebrates in tropical forest. J. Wildl. Mgmt. 37: 426-28.

Montgomery, G. G., and Y. D. Lubin. 1977. Prey influences on movements of neotropical anteaters. In: Phillips & Jonkel '77. Pp. 103-31.

Montgomery, G. G., and M. E. Sunquist. 1974. Contact-distress calls of young sloths. J. Mamm. 55: 211-13.

Montgomery, G. G., and M. E. Sunquist. 1975. Impact of sloths on Neotropical forest energy flow and nutrient cycling. In: Golley & Medina '75. Pp. 69-98.

Montgomery, G. G., and M. E. Sunquist. 1978. Habitat selection and use by two-toed and three-toed sloths. In: Montgomery '78. Pp. 329-59.

Moojen, J. 1943. Algunos mammiferos coleccionados no nordeste do Brasil. Bol. Mus. Nac., Nova Ser. No. 5.

Moore, J. C. 1961. Geographic variation in some reproductive characteristics of diurnal squirrels. Bull. Am. Mus. Nat. Hist. 122.

Moore, W. G., and R. L. Marchinton. 1974. Marking behavior and its social function in white-tailed deer. In: Geist & Walther '74. Pp. 447-56.

Mora, J. M., and I. Moreira. 1984. Mamíferos de Costa Rica. Editorial Univ. Estatal a Distancia, San José.

Morrison, D. W. 1978. Lunar phobia in a Neotropical fruit bat, *Artibeus jamaicensis* (Chiroptera: Phyllostomidae). Anim. Behav. 26: 852-55.

Morrison, D. W. 1983. *Artibeus jamaicensis*. In: Janzen '83c. Pp. 449-51.

Morúa, A. P. 1986. Lista anotada y observaciones de los mamíferos del Refugio Nacional de Vida Silvestre Tapantí, Costa Rica. Thesis: Universidad de Costa Rica.

Moscow, D., and C. Vaughan. 1987. Troop movements and food habits of white-faced monkeys in a tropical dry forest. Rev. Biol. Trop. 35(2): 287-97.

Müller, E., and E. Kulzer. 1977. Body temperature and oxygen uptake in the kinkajou (*Potos flavus* Schreber), a nocturnal tropical carnivore. Archs. Internat. Physiol. Biochim. 86: 153-63.

Müller, E., and H. Rost. 1983. Respiratory frequency, total evaporative water loss and heart rate in the kinkajou (*Potos flavus* Schreber). Z. Säugetierk. 48: 217-26.

Murie, O. J. 1974. A field guide to animal tracks. Second edition. Houghton Mifflin Co., Boston.

Murray, J. L., and G. L. Gardner. 1997. *Leopardus pardalis*. Mamm. Species 548. Am. Soc. Mammalogists.

Murray, P. F., and T. Strickler. 1975. Notes on the structure and function of cheek pouches within the Chiroptera. J. Mamm. 56: 673-76.

Nadkarni, N. M., and N. T. Wheelwright. 2000. Monteverde: ecology and conservation of a tropical cloud forest. Oxford University Press.

Naranjo, E. J. 1995a. Abundancia y uso de habitat del tapir (*Tapirus bairdii*) en un bosque tropical húmedo de Costa Rica. Vida Silv. Neotrop. 4(1): 20-30.

Naranjo, E. J. 1995b. Hábitos de alimentación del tapir (*Tapirus bairdii*) en un bosque tropical húmedo de Costa Rica. Vida Silv. Neotrop. 4(1): 30-37.

Navarro, L. D., and D. E. Wilson. 1982. *Vampyrum spectrum*. Mamm. Species 184. Am. Soc. Mammalogists.

Nelson, C. E. 1965. *Lonchorhina aurita* and other bats from Costa Rica. Texas J. Sci. 17: 303-06.

Neville, M. K., K. E. Glander, F. Braza, and A. B. Rylands. 1988. The howling monkeys, genus *Alouatta*. In: Mittermeier et al. '88. Pp. 349-453.

Newcomer, M. W., and D. D. DeFarcy. 1985. White-faced capuchin (*Cebus capucinus*) predation on a nesting coati (*Nasua narica*). J. Mamm. 66: 185-86.

Nitikman, L. Z. 1985. *Sciurus granatensis*. Mamm. Species 246. Am. Soc. Mammalogists.

Norberg, U. M., and J. M. V. Rayner. 1987. Ecological morphology and flight in bats (Mammalia; Chiroptera): wing adaptations, flight performance, foraging strategy, and echolocation. Phil. Trans. R. Soc. Lond. B, Biol. Sci. 316: 335-427.

Novacek, M. J. 1985. Evidence for echolocation in the oldest known bat. Nature 315: 140-41.

Novak, M., J. A. Baker, M. E. Obbard, and B. Malloch (eds). 1987. Wild furbearer management and conservation in North America. Ontario Ministry of Natural Resources, Toronto.

Novaro, A. J., R. S. Walker, and M. Suarez. 1995. Dry season habits of the gray fox (*Urocyon cinereoargenteus fraterculus*) in the Belizean Peten. Mammalia 59: 19-24.

Nowak, R. M. 1991. Walker's mammals of the world: fifth edition. Volumes I and II. Johns Hopkins Univ. Press, Baltimore.

Nowak, R. M. 1994. Walker's bats of the world. Johns Hopkins Univ. Press, Baltimore.

Nowell, K., and P. Jackson. 1996. Wild cats: status survey and conservation action plan. IUCN/SSC Cat Specialist Group, IUCN, Gland, Switzerland.

O'Brien, S. J., G. E. Collier, R. E. Benveniste, W. G. Nash, A. K. Newman, J. M. Simonson, J. A. Eichelberger, U. S. Seal, D. Janssen, M. Bush, and D. E. Wildt. 1987. Setting the molecular clock in the Felidae: The great cats, *Panthera*. In: Tilson & Seal '87. Pp. 10-27.

O'Connell, M. A. 1979. Ecology of didelphid marsupials from northern Venezuela. In: Eisenberg '79. Pp. 73-87.

O'Connell, M. A. 1983. *Marmosa robinsoni*. Mamm. Species 203. Am. Soc. Mammalogists.

O'Connell, M. A. 1984. American opossums. In: Macdonald '84a. Pp. 830-37.

Odell, D. K., D. Forrester, and E. Asper. 1978. Growth and sexual maturation in the West Indian manatee. Amer. Soc. Mamm. Abstr. Tech. Paper, 58th Annual Meeting. Pp. 7-8.

O'Donnell, D. J. 1981. Manatees and man in Central America. Ph.D. thesis, Univ. California, UCLA.

O'Farrell, M. J., and W. L. Gannon. 1999. A comparison of acoustic versus capture techniques for the inventory of bats. J. Mamm. 80(1): 24-30.

Ojeda C., M., and L. B. Keith. 1982. Sex and age composition and breeding biology of cottontail rabbit populations in Venezuela. Biotropica 14(1): 99-107.

Oldenburg, P. W., P. J. Ettestad, W. E. Grant, and E. Davis. 1985. Structure of collared peccary herds in south Texas: spatial and temporal dispersion of herd members. J. Mamm. 66: 764-70.

Olimpio, J. 1992. Consideraçoes preliminares sobre hábitos alimentares de *Lutra longicaudis* (Olfers, 1818) (Carnivora: Mustelidae), na lagoa do Peri, Ilha de Santa Catarina. Anales III Reunión de Especialistas en Mamíferos Acuáticos de Am. del Sur. Brasil.

Oliveira, T. G. de. 1994. Neotropical cats: ecology and conservation. Edufma, Sao Luís, Brazil.

Oliveira, T. G. de. 1998a. *Herpailurus yagouaroundi*. Mamm. Species 578. Am. Soc. Mammalogists.

Oliveira, T. G. de. 1998b. *Leopardus wiedii*. Mamm. Species 579. Am. Soc. Mammalogists.

Oliver, W. L. R. 1976. The management of yapoks (*Chironectes minimus*) at Jersey Zoo, with observations on their behavior. Ann. Rept. Jersey Wildl. Pres. Trust 13: 32-36.

Olmos, F. 1993. Diet of sympatric Brazilian caatinga peccaries (*Tayassu tajacu* and *T. pecari*). J. Trop. Ecol. 9: 255-58.

Oppenheimer, J. R. 1968. Behavior and ecology of the white-faced monkey, *Cebus capucinus*, on Barro Colorado Island, C. Z. Ph.D. diss., Univ. of Illinois.

Oppenheimer, J. R. 1977. Communication in New World monkeys. In: Sebeok '77. Pp. 851-89.

Oppenheimer, J. R. 1996. *Cebus capucinus*: home range, population dynamics, and interspecific relationships. In: Leigh et al. '96. Pp. 253-72.

Oppenheimer, J. R., and G. E. Lang. 1969. *Cebus* monkeys: effect on branching of *Gustavia* trees. Science 165: 187-88.

Osgood, W. H. 1943. The mammals of Chile. Field Mus. Nat. Hist. Zool. Ser. 30: 1-268.

O'Shea, T. J., and C. A. Salisbury. 1991. Belize - a last stronghold for manatees in the Caribbean. Oryx 25: 156-64.

Overall, K. L. 1980. Coatis, tapirs, and ticks: a case of mammalian interspecific grooming. Biotropica 12(2): 158.

Pardini, R. 1998. Feeding ecology of the neotropical river otter (*Lontra longicaudis*) in an Atlantic Forest stream, southeastern Brazil. J. Zool. London 245: 385-91.

Pardini, R., and E. Trajano. 1999. Use of shelters by the neotropical river otter (*Lontra longicaudis*) in an Atlantic Forest stream, southeastern Brazil. J. Mamm. 80(2): 600-610.

Parera, A. 1992. Dieta de *Lutra longicaudis* en la laguna Iberá, provincia de Corrientes, Argentina. Libro de Resúmenes, V Reunión de Especialistas en Mamíferos Acuáticos de América del Sur. Buenos Aires.

Parera, A. 1993. Dieta de *Lutra longicaudis* en el río Iguazú, Misiones, Argentina. Libro de Resúmenes, XVI Reunión Argentina de Ecología. Puerto Madryn, Argentina.

Parfit, M. 2000. Hunt for the first Americans. National Geographic. Dec. 2000. Pp. 41-67.

Parra, R. 1978. Comparison of foregut and hindgut fermentation in herbivores. In: Montgomery '78. Pp. 205-30.

Patterson, B., and R. R. Pascual. 1972. The fossil mammal fauna of South America. In: Keast et al. '72. Pp. 247-309.

Patterson, D., and R. M. Timm (eds.). 1987. Studies in Neotropical mammalogy: essays in honor of Philip Hershkovitz. Fieldiana Zool., n.s. 39.

Patton, J. L. 1984. Pocket gophers. In: Macdonald '84a. Pp. 628-31.

Pearson, O. P. 1942. On the cause and nature of a poisonous action produced by the bite of the shrew. J. Mamm. 23: 159-66.

Pearson, O. P. 1946. Scent glands of the short-tailed shrew. Anat. Rec. 94: 615-29.

Peck, S. B., and A. Forsyth. 1982. Composition, structure and competetive behavior in a guild of Ecuadorian rain forest dung beetles (Coleoptera, Scarabaeidae). Canadian J. Zool. 60: 1624-34.

Perry, S., and L. Rose. 1994. Begging and transfer of coati meat by white-faced capuchin monkeys, *Cebus capucinus*. Primates 35(4): 409-15.

Peters, G., and M. H. Hast. 1994. Hyoid structure, laryngeal anatomy, and vocalization in felids (Mammalia: Carnivora: Felidae). Zeitschrift für Säugetierk. 59: 87-104.

Peterson, R. L., and P. Kirmse. 1969. Notes on *Vampyrum spectrum*, the false vampire bat, in Panama. Can. J. Zool. 47: 140-2.

Peres, C. A. 1996. Population status of white-lipped *Tayassu pecari* and collared peccaries *T. tajacu* in hunted and unhunted Amazonian forests. Biological Conservation 77: 115-23.

Pérez, E. M. 1992. *Agouti paca*. Mamm. Species 404. Am. Soc. Mammalogists.

Perrin, W. F., R. L. Brownell, Jr., and D. P. DeMaster (eds.). 1984. Reproduction in whales, dolphins and porpoises. Reports of the International Whaling Commission. Special Issue 6.

Perrin, W. F., and S. B. Reilly. 1984. Reproductive parameters of dolphins and small whales of the family Delphinidae. In: Perrin et al '84.

Perry, S. 1996. Female-female social relationships in wild white-faced capuchin monkeys, *Cebus capucinus*. Am. J. Primatol. 40: 167-82.

Perry, S., and L. Rose. 1994. Begging and transfer of coati meat by white-faced capuchin monkeys, *Cebus capucinus*. Primates 35: 409-15.

Petersen, M. K. 1979. Behavior of the margay. Carnivore 2: 69-76.

Petersen, M. K., and M. K. Petersen. 1978. Growth rate and other postnatal developmental changes in margays. Carnivore 1: 87-92.

Peterson, R. L., and P. Kirmse. 1969. Notes on *Vampyrum spectrum*, the false vampire bat, in Panama. Can. J. Zool. 47: 140-42.

Pettigrew, J. D. 1986. Flying primates? Megabats have the advanced pathway from eye to midbrain. Science 231: 1304-6.

Pettigrew, J. D., B. G. M. Jamieson, S. K. Robson, L. S. Hall, I. I. McAnally, and J. M. Cooper. 1989. Phylogenetic relations between microbats (Mammalia: Chiroptera and Primates). Phil. Trans. R. Soc. Lond. B, Biol. Sci. 325: 489-559.

Phillips, C. J., and J. K. Jones, Jr. 1968. Additional comments on reproduction in the woolly opossum, (*Caluromys derbianus*) in Nicaragua. J. Mamm. 49: 320-21.

Phillips, C. J., and J. K. Jones, Jr. 1969. Notes on reproduction and development in the four-eyed opossum, *Philander opossum*, in Nicaragua. J. Mamm. 50: 345-48.

Phillips, R. L., and C. Jonkel (eds.). 1977. Proceedings of the 1975 predator symposium. Montana Forest and Conservation Experiment Station, Univ. Montana, Missoula.

Pine, R. H. 1972. The bats of the genus *Carollia*. Texas Agric. Exp. Stn., Tech. Monogr., Texas A & M Univ. 8: 1-125.

Pine, R. H. 1993. A new species of *Thyroptera* Spix (Mammalia: Chiroptera: Thyropteridae) from the Amazon basin of northeastern Peru. Mammalia 57: 213-25.

Plumpton, D. L., and J. K. Jones, Jr. 1992. *Rhynchonycteris naso*. Mamm. Species 413. Am. Soc. Mammalogists.

Poduschka, W. 1977. Insectivore communication. In: Sebeok '77. Pp. 600-33.

Poglayen-Neuwall, I. 1962. Beitrage zu einem Ethogramm des Wickelbären (*Potos flavus* Schreber). Z. Säugetierk. 27: 1-44.

Poglayen-Neuwall, I. 1965. Notes on care, display and breeding of olingos *Bassaricyon*. Intl. Zoo Yearb. 6: 169-71.

Poglayen-Neuwall, I. 1966. On the marking behavior of the kinkajou (*Potos flavus* Schreber). Zoologica 51: 137-41.

Poglayen-Neuwall, I. 1973. Preliminary notes on the maintenance and behaviour of the Central American cacomistle *Bassariscus sumichrasti*. Intl. Zoo Yearb. 13: 207-11.

Poglayen-Neuwall, I. 1975. Copulatory behavior, gestation, and parturition of the tayra *Eira barbara*. Z. Säugetierk. 40: 176-89.

Poglayen-Neuwall, I. 1978. Breeding, rearing and notes on the behavior of tayras *Eira barbara* in captivity. Intl. Zoo Yearb. 18: 134-40.

Poglayen-Neuwall, I. 1992. Report on a little known procyonid, *Bassariscus (Jentinkia) sumichrasti* (de Saussure, 1860). Small Carnivore Conservation 7: 1-3.

Poglayen-Neuwall, I., and I. Poglayen-Neuwall. 1965. Gefangenschaftsbeobachtungen an Makibären (*Bassaricyon gabbii* Allen, 1876). Z. Säugetierk. 30: 321-66.

Poglayen-Neuwall, I., and I. Poglayen-Neuwall. 1976. Postnatal development of tayras (Carnivora: *Eira barbara* L., 1758). Zool. Beitr. 22: 345-405.

Poglayen-Neuwall, I., and I. Poglayen-Neuwall. 1980. Gestation period and parturition of the ringtail *Bassariscus astutus*. Z. Säugetierk. 45: 73-81.

Poglayen-Neuwall, I., and I. Poglayen-Neuwall. 1995. Observations on the ethology and biology of the Central American cacomistle, *Bassariscus sumichrasti* (Saussure, 1860), in captivity, with notes on its ecology. Zoological Garten 65(1): 11-49.

Poglayen-Neuwall, I., and D. E. Toweill. 1988. *Bassariscus astutus*. Mamm. Species 327. Am. Soc. Mammalogists.

Pounds, J. A., M. P. L. Fogden, and J. H. Campbell. 1999. Biological response to climate change on a tropical mountain. Nature 398: 611-15.

Porter, F. L. 1978. Roosting patterns and social behavior in captive *Carollia perspicillata*. J. Mamm. 59: 627-30.

Powell, J. A. 1978. Evidence of carnivory in manatees (*Trichechus manatus*). J. Mamm. 59: 442.

Prator, T., W. D. Thomas, M. Jones, and M. Dee. 1988. A twenty-year overview of selected rare carnivores in captivity. In: Dresser et al. '96. Pp. 191-229.

Pratt, H. D., B. F. Bjornson, and K. S. Littig. 1977. Control of domestic rats and mice. U.S. Pub. Health Serv., Center for Disease Control, Atlanta.

Price, P. W., T. M. Lewinson, G. W. Fernandes, and W. W. Benson (eds.). 1990. Plant-animal interactions: evolutionary ecology in tropical and temperate regions. John Wiley & Sons, New York.

Priewert, F. W. 1961. Record of an extensive movement by a raccoon. J. Mamm. 42: 113.

Pruitt, C. H., and G. M. Burghardt. 1977. Communication in terrestrial carnivores: Mustelidae, Procyonidae, and Ursidae. In: Sebeok '77. Pp. 767-93.

Quesada, R., and G. Hanan. 1988. Osteología del tepezcuintle (*Cuniculus paca*), cinturas y miembros. Anatomía, Histología, y Embriología 17: 60-71.

Quigley, H. B. 1988. Ecology and conservation of the jaguar in the Pantanal region of Brazil. Ph.D. Thesis. Univ. Idaho.

Quinn, T. H., and J. J. Baumel. 1993. Chiropteran tendon locking mechanism. J. Morph. 216: 197-208.

Rabinowitz, A. R. 1986a. Jaguar: one man's battle to establish the world's first jaguar reserve. Anchor Books.

Rabinowitz, A. R. 1986b. Jaguar predation on domestic livestock. Wildl. Soc. Bull. 14(2).

Rabinowitz, A. R., and B. G. Nottingham. 1986. Ecology and behavior of the jaguar (*Panthera onca*) in Belize, Central America. J. Zool. Lond. 210: 149-59.

Radinsky, L. B. 1975. Evolution of the felid brain. Brain Behav. Evol. 11: 214-54.

Radinsky, L. B. 1987. The evolution of vertebrate design. Univ. Chicago Press, Chicago.

Rasmussen, D. T. 1990. Primate origins: lessons from a neotropical marsupial. Am. J. Primatol. 22: 263-77.

Rasmussen, D. T., and A. M. Rasmussen. 1984. The behavior and ecology of *Caluromys derbianus* (Marsupialia: Didelphidae) in a dry tropical forest of Costa Rica.

Rathbun, G. B., and T. J. O'Shea. 1984. The manatee's simple social life. In: Macdonald '84a. Pp. 300-301.

Redford, K. H. 1985a. Feeding and food preference in captive and wild giant anteaters (*Myrmecophaga tridactyla*). J. Zool. Lond. 205: 559-72.

Redford, K. H. 1985b. Food habits of armadillos (Xenarthra: Dasypodidae). In: Montgomery '85a. Pp. 429-37.

Redford, K. H. 1987. Ants and termites as food. Patterns of mammalian myrmecophagy. In: Genoways '87. Pp. 349-400.

Redford, K. H. 1989. The kinkajou (*Potos flavus*) as a myrmecophage. Mammalia 53(1): 132-4.

Redford, K. H., and J. F. Eisenberg (eds.). 1989. Advances in neotropical mammology. Sandhill Crane Press, Gainseville.

Redford, K. H., and J. F. Eisenberg. 1992. Mammals of the Neotropics, Volume 2. The southern cone: Chile, Argentina, Uruguay, Paraguay. Univ. Chicago Press, Chicago.

Reid, F. A. 1997. A field guide to the mammals of Central America and southeast Mexico. Oxford Univ. Press, New York.

Reid, F. A., and C. A. Langtimm. 1993. Distributional and natural history notes for selected mammals from Costa Rica. Southwest. Nat. 38: 299-302.

Reig, O. 1981. Teoría del origen y desarrollo de la fauna de mamíferos de América del Sur. Monografie Naturae. Museo Municipal de Ciencias Naturales Lorenzo Scaglia, Mar del Plata, Argentina.

Rettig, N. L. 1978. Breeding behavior of the harpy eagle (*Harpia harpyja*). Auk 95: 629-43.

Reynolds, J. E. III. 1979. The semisocial manatee. Nat. Hist. 88(2): 44-53.

Reynolds, J. E. III, and D. K. Odell. 1991. Manatees and dugongs. Facts on File, New York.

Reynolds, J. E. III, W. A. Szelistowski, and M. A. León. 1995. Status and conservation of manatees (*Trichechus manatus manatus*) in Costa Rica. Biol. Cons. 71: 193-96.

Rice, D. W. 1984. Cetaceans. In: Anderson & Jones '84. Pp. 447-90.

Rich, P. V. and T. H. Rich. 1983. The Central American dispersal route: biotic history and paleogeography. In: Janzen '83c. Pp. 12-34.

Richard, A. 1970. A comparative study of the activity patterns and behavior of *Alouatta villosa* and *Ateles geoffroyi*. Folia Primatol. 12: 241-63.

Ridgeway, S. H., and R. Harrison (eds.). 1985. Handbook of Marine Mammals, vol. 3: the sirenians and baleen whales. Academic Press, London.

Rink, R., and J. Miller. 1967. Temperature, weight (=age), and resistance to asphixia in pouch-young opossums. Cryobiology 4: 24-29.

Roberts, M., S. Brand, and E. Maliniak. 1985. The biology of captive prehensile-tailed porcupines, *Coendou prehensilis*. J. Mamm. 66: 476-82.

Robinette, W. J., J. S. Gashwiler, and O. W. Morris. 1959. Food habits of the cougar in Utah and Nevada. J. Wildl. Mgmt. 23: 261-73.

Robinson, J. G., and J. F. Eisenberg. 1985. Group size and foraging habits of the collared peccary *Tayassu tajacu*. J. Mamm. 66: 153-5.

Robinson, J. G., and K. H. Redford (eds.). 1991. Neotropical wildlife use and conservation. Univ. Chicago Press.

Robinson, R. 1970. Homologous mutants in mammalian coat colour variation. Symp. Zool. Soc. Lond. 26: 251-69.

Robinson, R. 1976. Homologous genetic varaition in the Felidae. Genetica 46: 1-31.

Rodríguez, B., and D. E. Wilson. 1999. Lista y distribución de las especies de murciélagos de Costa Rica. Occ. Pap. Cons. Biol. Conservation International.

Rodríguez, J. 1985. Ecología de la guatusa (*Dasyprocta punctata punctata*, Gray). In: Subdirección General... '85. Pp. 9-22.

Rodríguez, J. 1993. *Thyroptera discifera* en Costa Rica. Rev. Biol. Trop. 41(3).

Rodríguez, J., and F. A. Chinchilla. 1996. Lista de mamíferos de Costa Rica. Rev. Biol. Trop. 44(2): 877-90.

Rodríguez, M. 1985. Algunos aspectos sobre comportamiento, alimentación y nivel de población de los monos (Primates: Cebidae) en el Refugio de Fauna Silvestre Palo Verde (Guanacaste, Costa Rica). In: Subdirección General... '85. Pp. 53-71.

Rodríguez, M., and V. Solís. 1994. Ciclo de vida del venado cola blanca en la Isla San Lucas, Costa Rica. In: Vaughan & Rodríguez '94. Pp. 63-71.

Rodríguez, M., C. Vaughan, and M. McCoy. 1985a. Composición elemental de algunos afloramientos minerales utilizados por los ungulados silvestres en Palo Verde. In: Subdirección General... '85. Pp. 75-79.

Rodríguez, M., C. Vaughan, V. Villalobos, and M. McCoy. 1985b. Notas sobre los movimientos del venado colablanca (*Odocoileus virginianus Rafinesque*) en un bosque tropical seco de Costa Rica. In: Subdirección General... '85. Pp. 37-46.

Roeder, K. D., and A. E. Treat. 1961. The detection and evasion of bats by moths. Am. Sci. 49: 135-48.

Rogers, D. S. 1990. Genic evolution, historical biogeography and systematic relationships among spiny pocket mice (subfamily Heteromyinae). J. Mamm. 71: 668-85.

Rogers, D. S., and J. E. Rogers. 1992. *Heteromys oresterus*. Mamm. Species 396. Am. Soc. Mammalogists.

Rood, J. P. 1970. Ecology and social behavior of the desert cavy (*Microcavia australis*). Am. Midl. Nat. 83: 415-54.

Rosatte, R. C. 1987. Striped, spotted, hooded, and hog-nosed skunks. In: Novak et al. '87. Pp. 599-613.

Rose, L. M. 1994a. Benefits and costs of resident males to females in white-faced capuchins, *Cebus capucinus*. Am. J. Primatol. 32: 235-48.

Rose, L. M. 1994b. Sex differences in diet and foraging behavior in white-faced capuchins (*Cebus capucinus*). Intl. J. Primatol. 15(1): 95-114.

Rosenblatt, J. S., C. Beer, M. -C. Busnel, and P. J. B. Slater (eds.). 1986. Advances in the study of behavior. Academic Press, Orlando, Florida.

Rosenthal, M. A. 1975. Observations on the water opossum or yapok *Chironectes minimus* in captivity. Intl. Zoo Yearb. 15: 4-6.

Rowlands, I. W. (ed.). 1966. Comparative biology of reproduction in mammals. Academic Press, New York.

Rowlands, I. W. and B. J. Weir. 1974. The biology of hystricomorph rodents. Academic Press, London.

Rue, L. L. III. 1968. Sportsman's guide to game animals. Outdoor Life Books. Popular Science Pub. Comp., Inc.

Ruiz Loaiciga, A. M. 1984. Observaciones ecológicas de *Sigmodon hispidus* en áreas de cultivo de caña de azúcar del Ingenio Taboga, S. A., Cañas, Guanacaste. Thesis: Universidad de Costa Rica.

Russell, E. M. 1984. Marsupials. In: Macdonald '84a. Pp. 824-9.

Russell, J. K. 1979. Reciprocity in the social behavior of coatis, *Nasua narica*. Ph. D. diss. Univ. of North Carolina, Chapel Hill.

Russell, J. K. 1981. Exclusion of adult male coatis from social groups: protection from predation. J. Mamm. 62(1): 206-8.

Russell, J. K. 1983. Altruism in coati bands: nepotism or reciprocity? In: Wasser '83. Pp. 263-90.

Russell, J. K. 1984. Coatis. In: Macdonald '84a. Pp. 102-3.

Russell, J. K. 1996. Timing of reproduction by coatis (*Nasua narica*) in relation to fluctuations in food resources. In: Leigh et al. '96. Pp. 413-31.

Ryan, M. J., and M. D. Tuttle. 1984. To eat or be eaten: predatory habits of a frog-eating bat. In: Macdonald '84a. Pp. 810-11.

Ryan, M. J., M. D. Tuttle, and R. M. R. Barclay. 1983. Behavioral responses of the frog-eating bat, *Trachops cirrhosus*, to sonic frequencies. J. Comp. Physiol., ser. A, 150: 413-18.

Rydell, J. 1989. Food habits of northern (*Eptescius nilssoni*) and brown long-eared (*Plecotus auritus*) bats in Sweden. Holarctic Ecology 12: 16-20.

Sader, S. A., and A. T. Joyce. 1988. Deforestation rates and trends in Costa Rica, 1940 to 1983. Biotropica 20: 11-19.

Sáenz, J. 1994a. Ecología del pizote (*Nasua narica*) y su papel como dispersador de semillas en el bosque seco tropical, Costa Rica. Masters Thesis. Univ. Nacional, Heredia.

Sáenz, J. 1994b. Reintroducción del venado cola blanca en el noroeste de Costa Rica. In: Vaughan & Rodríguez '94. Pp. 383-416.

Salas, D. S. 1974. Algunas observaciones sobre el hábito de vida del "zorro de balsa" *Caluromys derbianus* (Marsupialia: Didelphidae) en la vertiente del Pacífico de Costa Rica. O'Bios. 11(7): 11-15.

Salinas, P. (ed.). 1980. Zoología Neotropical: actas del VIII congreso latinoamericano de zoología, Mérida.

Sánchez, C., and C. Chávez. 1985. Observaciones del murciélago de cápsula *Diclidurus virgo* Thomas. II. Reunión Iberoamericana de Conservación y Zoología de Vertebrados 1: 411-16.

Sánchez, R. E. 1991. Utilización de habitat, comportamiento y dieta del mono congo (*Alouatta palliatta*) en un bosque premontano húmedo, Costa Rica. Masters Thesis. Univ. Nacional, Heredia.

Sanderson, G. C. 1983. *Procyon lotor*. In: Janzen '83c. Pp. 485-8.

Sarich, V. M. 1985. Xenarthran systematics: albumin immunological evidence. In: Montgomery '85a. Pp. 77-81.

Savage, R. J. G. 1977. Evolution in carnivorous mammals. Palaeontology 20: 237-71.

Schaik, C. P. van. 1983. Why are diurnal primates living in groups? Behaviour 88: 120-43.

Schaller, G. B. 1980. Epitaph for a jaguar Anim. Kingdom 83(2): 4-11.

Schaller, G. B., and P. G. Crawshaw, Jr. 1980. Movement patterns of jaguar. Biotropica 12(3): 161-68.

Schaller, G. B., Jinchu, H., Wenski, P., and Jing, Z. 1985. The giant pandas of Wolong. Chap. 8: The giant panda - bear or raccoon? Univ. Chicago Press. Pp. 225-49.

Schaller, G. B., and J. M. C. Vasconcelos. 1978. Jaguar predation on capybara. Z. Saugetierk. 43: 296-301.

Scheffer, V. B. 1973. The last days of the sea cow. Smithsonian 3: 64-67.

Schevill, W. E. and W. A. Watkins. 1965. Underwater calls of *Trichechus*. Nature 205: 373-74.

Schmitt, D. M. 1966. How to prepare skeletons. Ward's Natural Science Establishment, Inc.

Schnitzler, H. U., E. K. V. Kalko, I. Kaipf, and A. D. Grinell. 1994. Fishing and echolocation behavior in the greater bulldog bat, *Noctilio leporinus*. Behav. Ecol. Sociobiol. 35: 327-45.

Schutt, W. A., Jr. 1993. Digital morphology in the Chiroptera: the passive digital lock. Acta Anatomica 148: 219-27.

Schutt, W. A., Jr. 1995. The chiropteran hindlimb: functional, behavioral, and evolutionary correlates of morphology. Ph.D. diss., Cornell Univ., Ithaca, New York.

Schutt, W. A., Jr., F. Muradali, N. Mondol, K. Joseph, and K. Brockman. 1999. Behavior and maintenance of captive white-winged vampire bats, *Diaemus youngi*. J. Mamm. 80(1): 71-81.

Seamon, J. O., and G. H. Adler. 1999. Short-term use of space by a neotropical forest rodent, *Proechimys semispinosus*. J. Mamm. 80(3): 899-904.

Searfoss, G. 1995. Skulls and bones: a guide to the skeletal structures and behavior of North American mammals. Stackpole Books.

Sebeok, T. (ed). 1977. How animals communicate. Indiana Univ. Press.

Seidensticker, J. C., IV, M. G. Hornocker, W. V. Wiles, and J. P. Messick. 1973. Mountain lion social organization in the Idaho Primitive Area. Wildl. Monogr. no. 35.

Seidensticker, J. C., IV, and S. Lumpkin. 1992. Mountain lions don't stalk people. True or false? Smithsonian 22 (February issue): 113-22.

Sekulic, R. 1982. The function of howling in red howler monkeys (*Alouatta seniculus*). Behaviour 81: 38-54.

Servin, J., and C. Huxley. 1993. Biología del coyote (*Canis latrans*) en la Reserva de la Biósfera "La Michilía", Durango. In: Medellín & Ceballos '93. Pp. 197-204.

Setz, E. Z. F. and I. Sazima. 1987. Bats eaten by Nambiquara indians in western Brazil. Biotropica 19(2): 190.

Seymour, K. 1989. *Panthera onca*. Mamm. Species 340. Am. Soc. Mammalogists.

Sharman, G. B. 1970. Reproductive physiology of marsupials. Science 167: 1221-28.

Shaw, J. H., T. S. Carter, and J. C. Machado-Neto. 1985. Ecology of the giant anteater *Myrmecophaga tridactyla* in Serra de Canastra, Minas Gerais, Brasil: a pilot study. In: Montgomery '85b. Pp. 379-84.

Shaw, J. J. 1985. The hemoflagellates of sloths, vermilinguas (anteaters), and armadillos. In: Montgomery '85a. Pp. 279-84.

Shaw, J. H., J. C. Machado-Neto, and T. S. Carter. 1987. Behavior of free-living anteaters (*Myrmecophaga tridactyla*). Biotropica 19: 255-59.

Sheffield, S. R., and H. H. Thomas. 1997. *Mustela frenata*. Mamm. Species 570. Am. Soc. Mammalogists.

Sheldon, J. W. 1992. Wild dogs: the natural history of the non-domestic Canidae. Academic Press.

Shump, K. A., Jr. and A. U. Shump. 1982. *Lasiurus borealis*. Mamm. Species 183. Am. Soc. Mammalogists.

Silva, G., J. Benítez, and J. Jiménez. 1993. Uso del hábitat por monos araña (*Ateles geoffroyi*) y aullador (*Alouatta palliata*) en áreas perturbadas. In: Medellín & Ceballos '93. Pp. 421-35.

Simmons, N. 1994. The case for chiropteran monophyly. American Museum Novitates 3103: 1-54.

Sisk, T. and C. Vaughan. 1984. Notes on some aspects of the natural history of the giant pocket gopher (*Orthogeomys* Merriam) in Costa Rica. Brenesia 22: 233-47.

Skutch, A. F. 1960. A forest view of kinkajous. Anim. Kingdom 63: 25-28.

Skutch, A. F. 1980. Naturalist on a tropical farm. Univ. California Press.

Skutch, A. F. 1992. A naturalist in Costa Rica. University Press of Florida.

Smith, C. C. 1977. Feeding behavour and social organization in howling monkeys. In: Clutton-Brock '77. Pp. 96-126.

Smith, J. D. 1972. Systematics of the chiropteran family Mormoopidae. Univ. Kansas Mus. Nat. Hist. Misc. Publ. 56.

Smith, N. 1974. Agouti and babassu. Oryx 12: 581-82.

Smith, T. E., A. Tomlinson, J. Mlotkiewicz, and D. H. Abbott. 1994. Unique ratios of highly volatile chemicals in circumgenital scent marks may provide a basis for discerning individual identity in female common marmosets. Abstracts presented at Chemical Signals in Vertebrates VII Conference, Tubingen, Germany.

Smith, W. P. 1982. Status and habitat use of Columbian white-tailed deer in Douglas County, Oregon. Ph.D. diss. Oregon State Univ., Corvallis.

Smuts, B. B. 1987. Gender, aggression, and influence. In: Smuts et al. '87. Pp. 400-412.

Smuts, B. B., D. L. Cheney, R. M. Seyfarth, R. W. Wrangham, and T. T. Struhsaker (eds.). 1987. Primate societies. Univ. Chicago Press.

Smythe, N. 1970a. Ecology and behavior of the agouti (*Dasyprocta punctata*) and related species on Barro Colorado Island, Panama. Ph.D. Thesis, Univ. Maryland.

Smythe, N. 1970b. The adaptive value of the social organization of the coati (*Nasua narica*). J. Mamm. 51: 818-20.

Smythe, N. 1978. The natural history of the Central American agouti (*Dasyprocta punctata*). Smithsonian Contrib. Zool. 257: 1-52.

Smythe, N. 1983. *Dasyprocta punctata* and *Agouti paca*. In: Janzen '83c. Pp. 463-5.

Smythe, N. 1987. The paca (*Cuniculus paca*) as a domestic source of protein for the neotropical, humid lowlands. Appl. Anim. Behav. Sci. 17: 155-70.

Smythe, N., W. E. Glanz, and E. G. Leigh, Jr. 1996. Population regulation in some terrestrial frugivores. In: Leigh et al. '96. Pp. 227-38.

Snow, J. L., J. K. Jones, Jr., and W. D. Webster. 1980. *Centurio senex*. Mamm. Species 138. Am. Soc. Mammalogists.

Solís, V., M. Rodríguez, and C. Vaughan (eds.). 1986. Actas del primer taller nacional sobre el venado cola blanca (*Odocoileus virginianus*) del pacífico seco, Costa Rica. Univ. Nacional, Heredia.

Solís, V. 1994. Uso tradicional y conservación del venado cola blanca en Costa Rica. In: Vaughan & Rodríguez '94. Pp. 351-57.

Sowls, L. K. 1983. *Tayassu tajacu*. In: Janzen '83c. Pp. 497-8.

Sowls, L. K. 1997. Javelinas and other peccaries: their biology, management, and use. Texas A & M Univ. Press.

Spínola, R. M. 1994. Dieta, abundancia relativa y actividad de marcaje de la nutria neotropical (*Lutra longicaudis*) en la Estación Biológica La Selva, Costa Rica. Masters Thesis. Univ. Nacional, Heredia.

Spínola, R. M., and C. Vaughan. 1995a. Abundancia relativa y actividad de marcaje de la nutria neotropical (*Lutra longicaudis*) en Costa Rica. Vida Silv. Neotrop. 4(1): 38-45.

Spínola, R. M., and C. Vaughan. 1995b. Dieta de la nutria neotropical (*Lutra longicaudis*) en la Estación Biológica La Selva, Costa Rica. Vida Silv. Neotrop. 4(2): 125-32.

Springer, J. T. 1980. Fishing behavior of coyotes on the Columbia River, southcentral Washington. J. Mamm. 61: 373-74.

Stains, H. J. 1984. Carnivores. In: Anderson & Jones '84. Pp. 491-522.

Starrett, A., and R. S. Casebeer. 1968. Records of bats from Costa Rica. Los Angeles Co. Mus. Contrib. Sci. 148: 1-21.

Stebbings, R. E. 1984. Bats. In: Macdonald '84a. Pp. 786-803.

Steiner, K. E. 1981. Nectivory and potential pollination by a Neotropical marsupial. Annals Missouri Bot. Gard. 68: 505-13.

Stejneger, L. 1887. How the great northern sea-cow (*Rytina*) became exterminated. The American Naturalist 21(12): 1047-54.

Stewart, R. 1995. Bribris, semillas de Sibö. Editorial Costarricense de Enseñanza Radiofónica.

Stiles, F. G., and A. F. Skutch. 1989. A guide to the birds of Costa Rica. Cornell Univ. Press.

Stoddart, D. M. 1984. Rodents. In: Macdonald '84a. Pp. 594-600.

Stokes, D., and L. Stokes. 1986. A guide to animal tracking and behavior. Little, Brown and Co.

Stonehouse, B., and D. P. Gilmore (eds.). 1977. The biology of marsupials. Univ. Park Press, Baltimore.

Stoner, K. E. 1993. Habitat preferences, foraging patterns, intestinal parasitic infections, and diseases in mantled howler monkeys, *Alouatta palliata* (Mammalia: Primates: Cebidae), in a rainforest in northeastern Costa Rica. Ph.D. diss., Univ. of Kansas, Lawrence.

Storrs, E. E., and H. P. Burchfield. 1985. Leprosy in wild common long-nosed armadillos *Dasypus novemcinctus*. In: Montgomery '85a. Pp. 265-68.

Storrs, E. E., H. P. Burchfield, and R. J. W. Rees. 1989. Reproduction delay in the common long-nosed armadillo *Dasypus novemcinctus*. In: Redford & Eisenberg '89. Pp. 535-48.

Stott, K., and C. J. Selsor. 1961. Association of trogons and monkeys on Barro Colorado. Condor 63: 508.

Subdirección General de Vida Silvestre, Dirección General Forestal del Ministerio de Agricultura y Ganadería, y la Universidad Nacional de Heredia. 1985. Investigaciones sobre fauna silvestre de Costa Rica. Editorial Universidad Estatal a la Distancia, San José.

Sunquist, M. E. 1981. The social organization of tigers (*Panthera tigris*) in Royal Chitawan National Park, Nepal. Smith. Contrib. Zool. 336.

Sunquist, M. E. 1992. The ecology of the ocelot: the importance of incorporating life history traits into conservation plans. Memorias del Simposio Orgaganizado por Fundeci, 1991. Pp. 117-28.

Sunquist, M. E., S. N. Austad, and F. Sunquist. 1987. Movement patterns and home range in the common opossum, *Didelphis marsupialis*. J. Mamm. 68: 173-76.

Sunquist, M. E., and G. G. Montgomery. 1973a. Activity patterns and rate of movement of toe-toed and three-toed sloths. J. Mamm. 54: 946-54.

Sunquist, M. E., and G. G. Montgomery. 1973b. Activity pattern of a translocated silky anteater (*Cyclopes didactylus*). J. Mamm. 54: 782.

Sunquist, M. E., and G. G. Montgomery. 1973c. Arboreal copulation by coatimundi (*Nasua narica*). Mammalia 37: 517-18.

Sunquist, M. E., and F. Sunquist. 1989. Ecological constraints on predation by large felids. In: Gittleman '89. Pp. 283-301.

Sunquist, M. E., F. Sunquist, and D. E. Daneke. 1989. Ecological separation in a Venezuelan llanos carnivore community. In: Redford & Eisenberg '89. Pp. 197-232.

Suriykke, A., and L. A. Miller. 1985. The influence of Arctiid moth clicks on bat echolocation: jamming or warning? J. Comp. Physiol. 156: 831-43.

Suthers, R. and J. Fattu. 1973. Fishing behavior and acoustic orientation by the bat *Noctilio labialis*. Anim. Behav. 21: 61-66.

Svendsen, G. E. 1976. Vocalizations of the longtailed weasel (*Mustela frenata*). J. Mamm. 57: 398-99.

Svendsen, G. E. 1982. Weasels, *Mustela* species. In: Chapman & Feldhamer '82.

Swank, W. G., and J. G. Teer. 1989. Status of the jaguar - 1987. Oryx 23: 14-21.

Symington, M. 1988a. Demography, ranging patterns, and activity budgets of black spider monkeys (*Ateles paniscus chamek*) in Manu National Park, Peru. Am. J. Primatol. 15: 45-67.

Symington, M. 1988b. Food competition and foraging party size in the black spider monkey (*Ateles paniscus chamek*). Behaviour 105: 117-34.

Szalay, F. S., M. J. Novacek, and M. C. McKenna (eds.). 1993. Mammalian phylogeny: placentals. Springer-Verlag, New York.

Taber, A. B., A. J. Novaro, N. Neris, and F. H. Colman. 1997. The food habits of sympatric jaguar and puma in the Paraguayan Chaco. Biotropica 29(2): 204-13.

Taylor, B. K. 1985. Functional anatomy of the forelimb in vermilinguas (anteaters). In: Montgomery '85a. Pp. 163-71.

Teer, J. 1994. El venado cola blanca: historia natural y principios de manejo. In: Vaughan & Rodríguez '94. Pp. 33-47.

Telford, S. R., Jr., R. J. Tonn, J. J. Gonzalez, and P. Betancourt. 1979. Densidad, área de distribución y movimiento de poblaciones de *Didelphis marsupialis* en los llanos altos de Venezuela. Bol. Dirección Malariología Saneamiento Ambiental 19(3-4): 119-28.

Terry, R. P. 1983. Observations on the captive behaviour of *Sotalia fluviatalis guianensis*. Aquat. Mamm. 10: 95-105.

Terwilliger, V. J. 1978. Natural history of Baird's tapir on Barro Colorado Island, Panama Canal Zone. Biotropica 10: 211-20.

Tesh, R. B. 1970. Notes on the reproduction, growth, and development of echimyid rodents in Panama. J. Mamm. 51: 199-202.

Tewes, M. E., and D. D. Everett. 1986. Status and distribution of the endangered ocelot and jaguarundi in Texas. In: Miller & Everett '86. Pp. 147-56.

Tewes, M. E., and D. J. Schmidly. 1987. The neotropical felids: jaguar, ocelot, margay and jaguarundi. In: Novak et al. '87. Pp. 697-711.

Thies, W., E. K. V. Kalko, and H.-U. Schnitzler. 1998. The roles of echolocation and olfaction in two Neotropical fruit-eating bats, *Carollia perspicillata* and *C. castanea*, feeding on *Piper*. Behav. Ecol. Sociobiol. 42: 397-409.

Thomas, D. W., B. Crawford, S. Eastman, R. Glofscheskie, and M. Heir. 1984. A reappraisal of the feeding adaptations in the hairs of nectar-feeding bats. J. Mamm. 65: 481-84.

Thomas, W. 1975. Observations on captive brockets, *Mazama americana* and *M. gouazoubira*. Int. Zoo Yearb.15:77-8.

Thorington, R. W., Jr., and E. M. Thorington. Postcranial proportions of *Microsciurus* and *Sciurillus*, the American pygmy tree squirrels. In: Redford & Eisenberg '89. Pp. 125-33.

Thorington, R. W., Jr., and P. G. Heltne (eds.). 1976. Neotropical primates: field studies and conservation. Natl. Acad. Sci., Washington, D. C.

Thornback, J., and M. Jenkins. 1982. The IUCN mammal red data book. Part I: Threatened mammalian taxa of the Americas and the Australasian zoogeographic region (excluding Cetacea). Internatl. Union Conserv. Nat., Gland, Switzerland.

Tilson, R. L., and U. S. Seal (eds.). 1987. Tigers of the world. Noyes, New Jersey.

Timm, R. M. 1982. *Ectophylla alba*. Mamm. Species 166. Am. Soc. Mammalogists.

Timm, R. M. 1984. Tent construction by *Vampyressa* in Costa Rica. J. Mamm. 65(1): 166-67.

Timm, R. M. 1987. Tent construction by bats of the genera *Artibeus* and *Uroderma*. In: Patterson & Timm '87. Pp. 187-212.

Timm, R. M. 1989. A review and reappraisal of the night monkey, *Aotus lemurinus* (Primates: Cebidae), in Costa Rica. Rev. Biol. Trop. 36(2B): 537-40.

Timm, R. M. 1994a. Appendix 8. Mammals. In: McDade et al. '94. Pp. 394-98.

Timm, R. M. 1994b. The mammal fauna. In: McDade et al. '94. Pp. 229-37.

Timm, R. M., and R. K. LaVal. 1998. A field key to the bats of Costa Rica. Occas. Pub. Ser. Center of Lat. Am. Studies. Univ. Kansas, Lawrence, Kansas.

Timm, R. M., and R. K. LaVal. 2000a. Appendix 10. Mammals of Monteverde. In: Nadkarni 2000. Pp. 553-57.

Timm, R. M., and R. K. LaVal. 2000b. Mammals. In: Nadkarni 2000. Pp. 223-36.

Timm, R. M., R. K. LaVal, and B. Rodríguez. 1999. Clave de campo para los murciélagos de Costa Rica.

Timm, R. M., and J. Mortimer. 1976. Selection of roost sites by Honduran white bats, *Ectophylla alba* (Chiroptera: Phyllostomatidae). Ecology 57: 385-89.

Timm, R. M., D. E. Wilson, B. L. Clauson, R. K. LaVal, and C. S. Vaughan. 1989. Mammals of the La Selva-Braulio Carrillo complex, Costa Rica. North American Fauna. U.S. Fish and Wildlife Service Publ. 75: 1-162.

Tomasi, T. E. 1979. Echolocation by the short-tailed shrew. J. Mamm. 60: 751-59.

Tomblin, D. C., and G. H. Adler. 1998. Differences in habitat use between two morphologically similar tropical forest rodents. J. Mamm. 79(3): 953-61.

Torrealba, I. M. 1993. Ecología de los grupos de saínos (*Tayassu tajacu*) y daños que ocasionan en los cultivos vecinos a la estación biológica La Selva, Costa Rica. Masters thesis. Univ. Nacional, Heredia, Costa Rica.

Toxopeus, H. 1985. Botany, types and populations. In: Wood & Lass '85. Pp. 11-37.

Trapp, G. R. 1972. Some anatomical and behavioral adaptations of ringtails *Bassariscus astutus*. J. Mamm. 53: 549-57.

Trapp, G. R. 1978. Comparative behavioral ecology of the ringtail (*Bassariscus astutus*) and gray fox (*Urocyon cinereoargenteus*) in southwestern Utah. Carnivore 1: 3-32.

Trapp, G. R., and D. L. Hallberg. 1975. Ecology of the gray fox (*Urocyon cinereoargenteus*): a review. In: Fox '75. Pp. 164-78.

Tschapka, M., L. T. Brooke, and A. P. Wesserthal. 2000. *Thyroptera discifera* (Chiroptera: Thyroptidae). A new record for Costa Rica and observations on echolocation. Z. Säugetierk. 65: 193-98.

Turkowski, F. J. 1971. The tree fox. Animal Kingdom 74: 18-21.

Turner, A., and M. Antón. 1997. The big cats and their fossil relatives: an illustrated guide to their evolution and natural history. Columbia Univ. Press.

Turner, D. C. 1975. The vampire bat: a field study in behavior and ecology. Johns Hopkins Univ. Press, Baltimore.

Turner, D. C. 1983. *Desmodus rotundus*. In: Janzen '83c. Pp. 467-8.

Turner, D. C. 1984. A myth explored: hunting behavior of vampire bats. In: Macdonald '84a. Pp. 812-13.

Tuttle, M. D. 1970. Distribution and zoogeography of Peruvian bats, with comments on natural history. Univ. Kansas Sci. Bull. 49: 45-86.

Tuttle, M. D. 1982. The amazing frog-eating bat. National Geographic 161 (1). January issue. Pp. 78-91.

Tuttle, M. D. 1995. Saving North America's beleaguered bats. Natl. Geog. 188 (2), August.

Tuttle, M. D., and M. J. Ryan. 1981. Bat predation and the evolution of frog vocalizations in the neotropics. Science 214: 677-78.

Tuttle, M. D., L. K. Taft, and M. J. Ryan. 1981. Accoustical location of calling frogs by *Philander* opossums. Biotropica 13(3): 233-34.

Uieda, W. 1994. Comportamento alimentar de morcegos hematófagos ao atacar aves, caprinos e suínos, em condicoes de cativeiro. Ph.D. Thesis. Universidade Estadual de Campinas, Brazil.

Uieda, W., S. Buck, and I. Sazima. 1992. Feeding behavior of the vampire bats *Diaemus youngi* and *Diphylla ecaudata* on smaller birds in captivity. J. Brazil. Assoc. Advcmt. Sci. 44: 410-12.

Uieda, W., I. Sazima, and A. Storti Filho. 1980. Aspectos da biologia do morcego *Furipterus horrens* (Mammalia, Chiroptera, Furipteridae). Rev. Brasil. Biol. 40: 59-66.

Valerio, C. E. 1999. Costa Rica: ambiente y biodiversidad. Instituto Navional de Biodiversidad (INBio), Heredia.

Vandermeer, J. H. 1979. Hoarding behavior of captive *Heteromys demarestianus*, (Rodentia) on the fruits of *Welfia georgii*, a rainforest dominant palm in Costa Rica. Brenesia 16: 107-16.

Vandermeer, J. 1983a. Pejibaye Palm. In: Janzen '83c. Pp. 98-101.

Vandermeer, J. 1983b. *Welfia georgii*. In: Janzen '83c. Pp. 346-49.

Vandermeer, J. H., J. Stout, and S. Risch. 1979. Seed dispersal of a common Costa Rican rain forest palm (*Welfia georgii*). Trop. Ecol. 20: 17-26.

Van Gelder, R. 1953. The egg-opening technique of a spotted skunk. J. Mamm. 34: 255-6.

Van Gelder, R. 1959. A taxonomic revision of the spotted skunk (genus *Spilogale*). Bull. Am. Mus. Nat. Hist. 117: 229-392.

Van Roosmalen, M. G. M. 1980. Habitat preferences, diet, feeding strategy and social organization of the black spider monkey (*Ateles paniscus paniscus*) in Surinam. PhD thesis. Agricultural Univ. of Wageningen.

Van Roosmalen, M. G. M., and L. L. Klein. 1988. The spider monkeys, genus *Ateles*. In: Mittermeier et al. '88. Pp. 455-537.

Van Valkenburgh, B., and C. B. Ruff. 1987. Canine tooth strength and killing behaviour in large carnivores. J. Zool. Lond. 212: 1-19.

Van Zyll de Jong, C. G. 1972. A systematic review of the nearctic and neotropical river otters (genus *Lutra*, Mustelidae, Carnivora). Royal Ontario Mus. Life Sci. Contrib. no. 80.

Van Zyll de Jong, C. G. 1987. A phylogenetic study of the Lutrinae (Carnivora: Mustelidae) using morphological data. Can. J. Zool. 65: 2536-44.

Vaughan, C. 1978. Pilot study on the population status of Baird's tapir, *Tapirus bairdii*, a Costa Rican endangered species. Mimeograph. Organization for Tropical Studies, San José, Costa Rica.

Vaughan, C. 1980. Predation of *Coendou mexicanus* by large felids. Brenesia 18: 368.

Vaughan, C. 1983a. A report on dense forest habitat for endangered wildlife species in Costa Rica. Universidad Nacional, Heredia.

Vaughan, C. 1983b. Coyote range expansion in Costa Rica and Panama. Brenesia 21: 27-32.

Vaughan, C., T. Kotowski, and L. Saénz. 1994. Ecology of the Central American cacomistle, *Bassariscus sumichrasti*, in Costa Rica. Small Carnivore Conservation 11: 4-7.

Vaughan, C., and M. McCoy. 1984. Estimación de las poblaciones de algunos mamíferos en el Parque Nacional Manuel Antonio, Costa Rica. Brenesia 22: 207-17.

Vaughan, C., and M. A. Rodríguez. 1986. Comparación de los hábitos alimentarios del coyote (*Canis latrans*) en dos localidades en Costa Rica. Vida Silv. Neotrop. 1: 6-11.

Vaughan, C., and M. A. Rodríguez (eds). 1994. Ecología y manejo del venado cola blanca en México y Costa Rica. Editorial de la Universidad Nacional, Heredia.

Vaughan, T. A. 1972. Mammalogy. W. B. Saunders Company.

Vaughan, T. A., and G. C. Bateman. 1970. Functional morphology of the forelimb of mormoopid bats. J. Mamm. 51: 217-35.

Vaughn, R. 1974. Breeding the tayra *Eira barbara* at Antelope Zoo, Lincoln. Intl. Zoo Yearb. 14: 120-22.

Vehrencamp, S. L., F. G. Stiles, and J. W. Bradbury. 1977. Observations on the foraging behavior and avian prey of the Neotropical carnivorous bat *Vampyrum spectrum*. J. Mamm. 58: 469-78.

Verts, B. J. 1967. The biology of the striped skunk. Univ. Chicago Press.

Voigt, D. R. 1984. Skunks. In: Macdonald '84a. Pp. 122-3.

Waage, J. K., and R. C. Best. 1985. Arthropod associates of sloths. In: Montgomery '85a. Pp. 297-311.

Waage, J. K., and G. G. Montgomery. 1976. *Cryptoses cholopei*: A coprophagous moth that lives on a sloth. Science 193: 157-58.

Wace, N. M. 1986. The rat problem on oceanic islands - research is needed. Oryx 20: 79-86.

Walker, P. L., and J. G. H. Cant. 1977. A population survey of kinkajous (*Potos flavus*) in a seasonally dry tropical forest. J. Mamm. 58: 100-02.

Walther, F. R. 1977. Artiodactyla. In: Sebeok '77. Pp. 655-714.

Washabaugh, K., and C. T. Snowdon. 1998. Chemical communication of reproductive status in female cotton-top tamarins (*Saguinus oedipus oedipus*). Am. J. Primatol. 45: 337-49.

Wasser, S. K. (ed.). 1983. Social behavior of female vertebrates. John Wiley, New York.

Watkins, L. C., J. K. Jones, Jr., and H. H. Genoways. 1972. Bats of Jalisco, Mexico. Spec. Publ. Mus.Texas Tech Univ.1.

Watt, E. M. 1987. A scatological analysis of parasites and food habits of jaguar (*Panthera onca*) in the Cockscombe Basin of Belize. Masters Thesis. Univ. Toronto.

Wayne, R. K., and S. J. O'Brien. 1987. Allozyme divergence within the Canidae. Syst. Zool. 36(4): 339-355.

Webb, S. D., and L. G. Marshall. 1982. Historical biogeography of Recent South American land mammals. In: Mares & Genoways '82. Pp. 39-52.

Weber, J-M., and C. Mermod. 1983. Experimental transmission of *Skrjabingylus nasicola*, parasitic nematode of mustelids. Acta. Zool. Fenn. 174: 237-38.

Weber, J-M., and C. Mermod. 1985. Quantitative aspects of the life cycle of *Skrjabingylus nasicola*, a parasitic nematode of the frontal sinuses of mustelids. Z. Parasit. 71: 631-38.

Webster, W. D. 1993. Systematics and evolution of bats of the genus *Glossophaga*. Spec. Publ. Mus. Texas Tech Univ. 36.

Weir, B. J. 1974. Reproductive characteristics of hystricomorph rodents. In: Rowlands & Weir '74. Pp. 265-99.

Wells, N. M., and J. Giacalone. 1985. *Syntheosciurus brochus*. Mamm. Species 249. Am. Soc. Mammalogists.

Wemmer, C. (ed.). 1987. Biology and management of the Cervidae. Smithsonian Inst. Press, Washington D. C.

Wemmer, C., and K. Scow. 1977. Communication in the Felidae with emphasis on scent marking and contact patterns. In: Sebeok '77. Pp. 749-66.

Wenstrup, J. J., and R. A. Suthers. 1984. Echolocation of moving targets by the fish-catching bat, *Noctilio leporinus*. J. Comp. Physiol., ser. A 155: 75-89.

Werdelin, L. 1985. Small Pleistocene felines of North America. J. Vert. Paleont. 5: 194-210.

Westergaard, G. C., and D. M. Fragaszy. 1987. The manufacture and use of tools by capuchin monkeys (*Cebus apella*). J. Comp. Psychol. 102: 152-59.

Wetzel, R. M. 1977. The Chacoan peccary, *Catagonus wagneri* (Rusconi). Bull. Carneg. Mus. Nat. Hist. No. 3, Pittsburgh.

Wetzel, R. M. 1980. A revision of the naked-tailed armadillos, genus *Cabassous* McMurtrie. Ann. Carnegie Mus. Nat. Hist. 49: 323-57.

Wetzel, R. M. 1982. Systematics, distribution, ecology and conservation of South American edentates. In: Mares & Genoways '82. Pp. 345-75.

Wetzel, R. M. 1983. *Dasypus novemcinctus*. In: Janzen '83c. Pp. 465-7.

Wetzel, R. M. 1985a. Taxonomy and distribution of armadillos, Dasypodidae. In: Montgomery '85a. Pp. 23-46.

Wetzel, R. M. 1985b. The identification and distribution of recent Xenarthra. In: Montgomery '85a. Pp. 5-21.

Wetzel, R. M., and E. Mondolfi. 1979. The subgenera and species of long-nosed armadillos, genus *Dasypus* L.. In: Eisenberg '79. Pp. 43-63.

Wheelwright, N. T. 1983. Fruits and the ecology of resplendent quetzals. Auk 100: 286-301.

Whitaker, J. O., Jr. 1974. *Cryptotis parva*. Mamm. Species 43. Am. Soc. Mammalogists.

Whitaker, J. O. Jr., and J. S. Findley. 1980. Foods eaten by some bats from Costa Rica and Panama. J. Mamm. 61: 540-44.

Wilkins, K. T. 1989. *Tadarida brasiliensis*. Mamm. Species 331. Am. Soc. Mammalogists.

Wilkinson, G. S. 1985a. The social organization of the common vampire bat. I. Pattern and cause of association. Behav. Ecol. Sociobiol. 17: 111-21.

Wilkinson, G. S. 1985b. The social organization of the common vampire bat. II. Mating system, genetic structure, and relatedness. Behav. Ecol. Sociobiol. 17: 123-34.

Wilkinson, G. S. 1988. Social organization and behavior. In: Greenhall & Schmidt '88. Pp. 85-97.

Wille T., A. 1987. Corcovado: meditaciones de un biólogo - un estudio ecológico. Editorial Univ. Estatal a la Distancia.

Williams, K. D. 1984. The Central American tapir (*Tapirus bairdii* Gill) in northwestern Costa Rica. Ph.D. dissertation. Michigan State Univ.

Williams, T. C., L. C. Ireland, and J. M. Williams. 1973. High-altitude flights of the free-tailed bat, *Tadarida brasiliensis*, observed with radar. J. Mamm. 54: 807-21.

Willig, M. R. 1986. Bat community structure in South America: a tenacious chimera. Rev. Chilena de Hist. Nat. 59: 151-68.

Wilson, D. E. 1971. Ecology of *Myotis nigricans* (Mammalia: Chiroptera) on Barro Colorado Island, Panama Canal Zone. J. Zool. Lond. 163: 1-13.

Wilson, D. E. 1978. *Thyroptera discifera*. Mamm. Species 104. Am. Soc. Mammalogists.

Wilson, D. E. 1983. *Myotis nigricans*. In: Janzen '83c. Pp. 477-8.

Wilson, D. E. 1997. Bats in question: the Smithsonian answer book. Smithsonian Institution Press.

Wilson, D. E., and J. S. Findley. 1972. Randomness in bat homing. Am. Nat. 106: 418-24.

Wilson, D. E., and J. S. Findley. 1977. *Thyroptera tricolor*. Mamm. Species 71. Am. Soc. Mammalogists.

Wilson, D. E., and R. K. LaVal. 1974. *Myotis nigricans*. Mamm. Species 39. Am. Soc. Mammalogists.

Wilson, E. O. 1988. Biodiversity. National Academy Press, Washington, D.C.

Wohlgenant, T. J. 1994. Roost interactions between the common vampire bat (*Desmodus rotundus*) and two frugivorous bats (*Phyllostomus discolor* and *Sturnira lilium*) in Guanacaste, Costa Rica. Biotropica 26(3): 344-48.

Wolff, A. 1981. The use of olfaction in food location and discrimination of food by *Carollia perspicillata*. M. S. Thesis, Univ. of Wisconsin, Milwaukee.

Wong, G. 1990. Uso del hábitat, estimación de la composición y densidad poblacional del mono tití (*Saimiri oerstedi citrinellus*) en la zona de Manuel Antonio, Quepos, Costa Rica. Masters Thesis. Univ. Nacional, Heredia.

Wong, G., Y. Matamoros, and U. Seal (eds.). 1994. Population and habitat viability assessment workshop for *Saimiri oerstedi citrinellus*. Simón Bolívar Zoo.

Wood, G. A. R., and R. A. Lass (eds.). 1985. Cocoa. Longman Group UK Ltd., Harlow, England.

Woodburne, M. O. 1968. The cranial myology and osteology of *Dicotyles tajacu*, the collared peccary, and its bearing on classification. Mem. S. California Acad. Sci. 7: 1-48.

Woodburne, M. O., and W. J. Zinmeister. 1982. Fossil land mammal from Antarctica. Science 218: 284-86.

Woodburne, M. O., and W. J. Zinmeister. 1984. The first land mammal from Antarctica and its biogeographic implications. J. Paleont. 58: 913-48.

Woodman, N. 1992. Biogeographical and evolutionary relationships among Central American small-eared shrews of the genus *Cryptotis* (Mammalia: Insectivora: Soricidae). Ph.D diss., Univ. Kansas, Lawrence.

Woodman, N., and R. M. Timm. 1993. Intraspecific and interspecific variation in the *Cryptotis nigrescens* species complex of small-eared shrews (Insectivora: Soricidae), with the description of a new species from Colombia. Fieldiana Zool., n.s., 74: 1-30.

Woods, C. A. (ed.). 1989. Biogeography of the West Indies. Sandhill Crane Press, Gainesville, Fla.

Wrangham, R. W. 1980. An ecological model of female-bonded primate groups. Behaviour 75: 262-300.

Wrangham, R. W. 1987. Evolution of social structure. In Smuts et al. '87. Pp 282-96.

Wroot, A. 1984. Insectivores. In: Macdonald '84a. Pp. 738-43.

Wujek, D. E. and J. M. Cocuzza. 1986. Morphology of hair of two- and three-toed sloths (Edentata: Bradypodidae). Rev. Biol. Trop. 34(2): 243-6.

Yager, D. D., and R. R. Hoy. 1986. The cyclopean ear: a new sense for the praying mantis. Science 231: 727-29.

Yager, R. H., and C. B. Frank. 1972. The nine-banded armadillo for medical research. Inst. Lab. Anim. Res. News 15(2): 4-5.

Yancey, F. D., II, J. R. Goetz, and C. Jones. 1998. *Saccopteryx bilineata*. Mamm. Species 581. Am. Soc. Mammalogists.

Yanosky, A. A., and C. Mercolli. 1993. Activity pattern of *Procyon cancrivorus* (Carnivora: Procyonidae) in Argentina. Rev. Biol. Trop. 41(1): 157-59.

Yates, T. L. 1984. Insectivores, elephant shrews, tree shrews, and dermopterans. In: Anderson & Jones '84. Pp. 117-44.

Young, A. M. 1971. Foraging of vampire bats (*Desmodus rotundus*) in Atlantic wet lowland Costa Rica. Rev. Biol. Trop. 18: 73-88.

Young B. E. 1996. An experimental analysis of small clutch size in tropical house wrens. Ecology 77: 472-88.

Young, O. P. 1982. Agressive interaction between howler monkeys and turkey vultures: the need to thermoregulate behaviorally. Biotropica 14(3): 228-31.

Zeiner, H. G. 1975. Behavior of striped and spotted skunks. Ph.D. diss. Univ. California, Berkeley.

Zervanos, S. M., and N. F. Hadley. 1973. Adaptational biology and energy relationships of the collared peccary (*Tayassu tajacu*). Ecology 54: 759-74.

Zetek, J. 1930. The water opossum *Chironectes panamensis*. J. Mamm. 11: 470-71.

Zúñiga, T. 1994. Abundancia relativa y uso tradicional del tepezcuintle (*Agouti paca*) en el Refugio de Vida Silvestre Barra del Colorado, Costa Rica. Masters Thesis. Univ. Nacional, Heredia.

INDEX

A

B

C

D

E

F

G

H

I

J

K

L

M

N

O

P

R

S

T

U

V

W

X

Y

Z

ABOUT THE AUTHOR

British naturalist Mark Wainwright has lived and worked in Monteverde, Costa Rica, since 1991. In addition to writing and illustrating, he works as a guide and as an instructor for tropical ecology courses; he has also participated in a number of research projects. His previous publications include five titles in the Costa Rica Field Guide series—*Mammals*, *Reptiles*, *Amphibians*, *Animal Tracks*, and *Cloud Forest Birds*—and the children's book *Jungle Jumble*.

COLOR PLATES

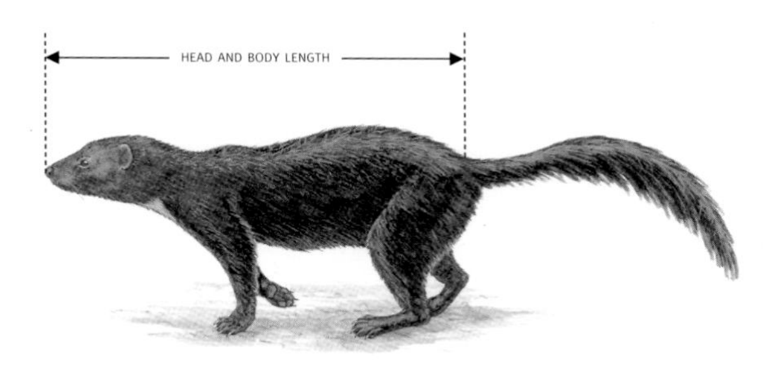

Mammals are measured from the tip of the snout to the base of the tail, a measurement referred to as the head and body length.

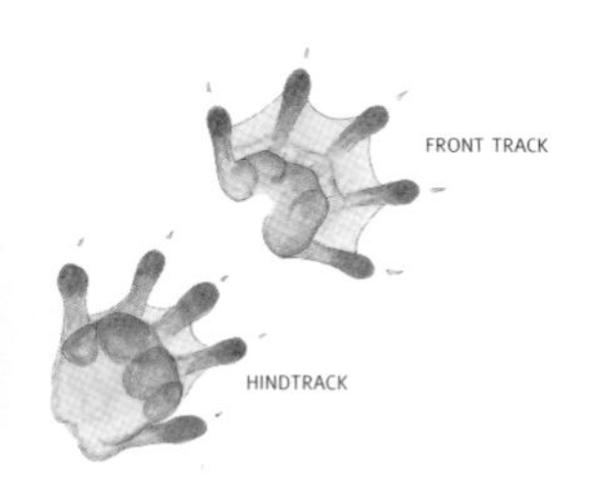

When tracks of both the front and hind feet are illustrated, the front tracks always appear above the hindtracks.

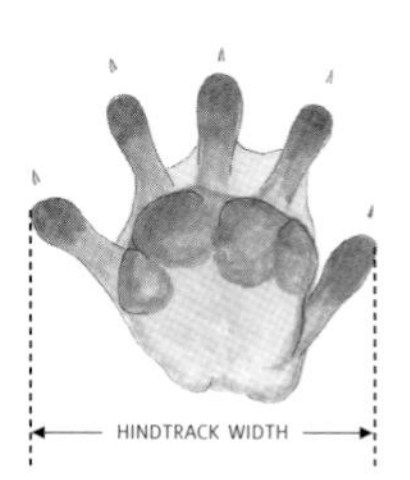

The track is measured across its widest point.

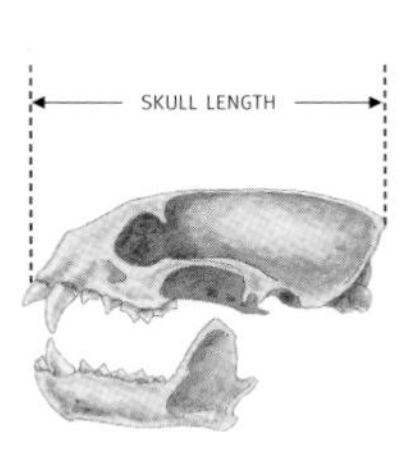

Skulls are measured lengthwise.

PLATE 1: OPOSSUMS

1 Common Opossum (*Didelphis marsupialis*), p. 32.

Nocturnal; terrestrial and arboreal; solitary. Shaggy, unkempt fur; pale underfur visible beneath long gray or black guard hairs; cheeks dirty yellow; facial whiskers black; white portion of tail usually longer than black portion; 40 cm (16 in.).

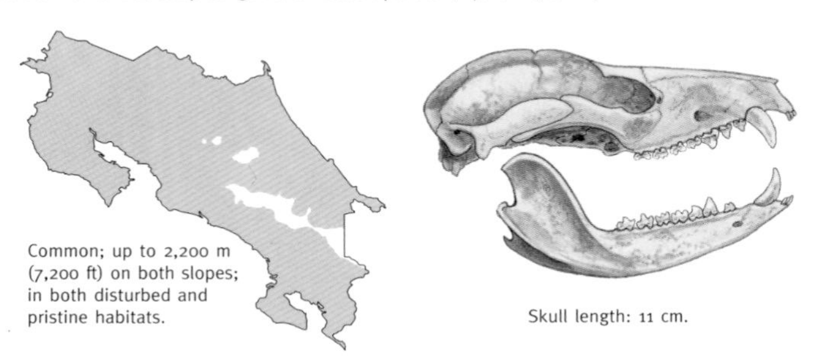

Common; up to 2,200 m (7,200 ft) on both slopes; in both disturbed and pristine habitats.

Skull length: 11 cm.

2 Virginia Opossum (*Didelphis virginiana*), p. 35.

Nocturnal; mostly terrestrial; solitary. Cheeks clean white; facial whiskers white; white portion of tail usually shorter than black portion; smaller and darker than Virginia opossums in North America; 43 cm (17 in.).

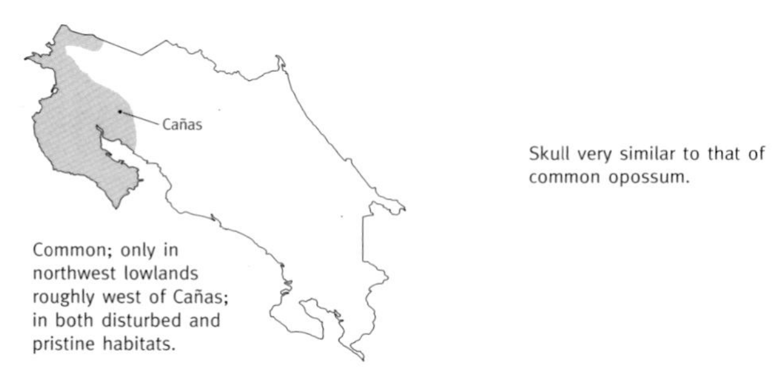

Common; only in northwest lowlands roughly west of Cañas; in both disturbed and pristine habitats.

Skull very similar to that of common opossum.

3 Common Gray Four-Eyed Opossum (*Philander opossum*), p. 41.

Nocturnal; terrestrial and arboreal; solitary. Tail furred at base; naked portion of tail black at base, changing abruptly to white on outer half; compare with brown four-eyed opossum (pl. 2); 30 cm (12 in.).

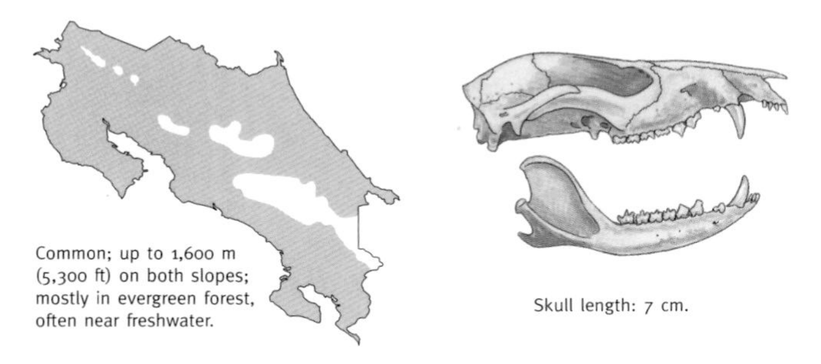

Common; up to 1,600 m (5,300 ft) on both slopes; mostly in evergreen forest, often near freshwater.

Skull length: 7 cm.

1
2
3
Common or Virginia opossum tracks. Tracks of these species are indistinguishable. Hindtrack is about 6.5 cm wide.
A tail mark is often visible in opossum tracks.
Common gray four-eyed opossum tracks. Hindtrack is about 5 cm wide.

1 Central American Woolly Opossum (*Caluromys derbianus*), p. 37.

Nocturnal; arboreal; solitary. Orangey fur, pink ears, and half-furred, half-white tail unique; adult males have conspicuous blue scrotums; 25 cm (10 in.).

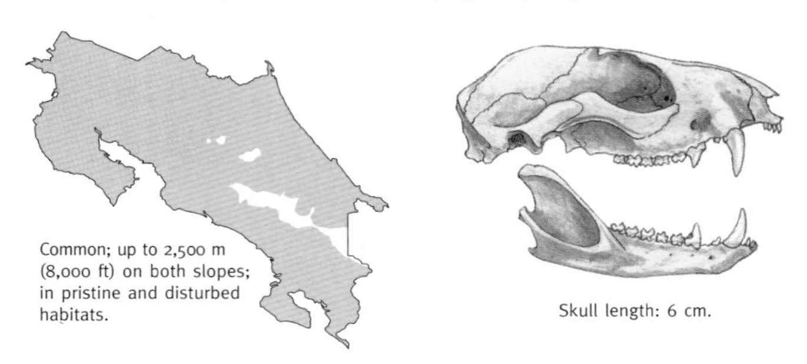

Common; up to 2,500 m (8,000 ft) on both slopes; in pristine and disturbed habitats.
Skull length: 6 cm.

2 Brown Four-eyed Opossum (*Metachirus nudicaudatus*), p. 39.

Nocturnal; terrestrial and arboreal; solitary. Distinguished from common gray four-eyed opossum (pl. 1) by brown back and more uniform tail coloration; 26 cm (10 in.).

Exceedingly rare; up to 1,200 m (4,000 ft) on Atlantic slope only; in forested habitats.

3 Yapok or Water Opossum (*Chironectes minimus*), p. 43.

Mostly nocturnal; only aquatic opossum; solitary. Banded coloration consistent and unique, although yapok appears entirely black when wet; fur short and dense; tail white only at tip; 28 cm (11 in.).

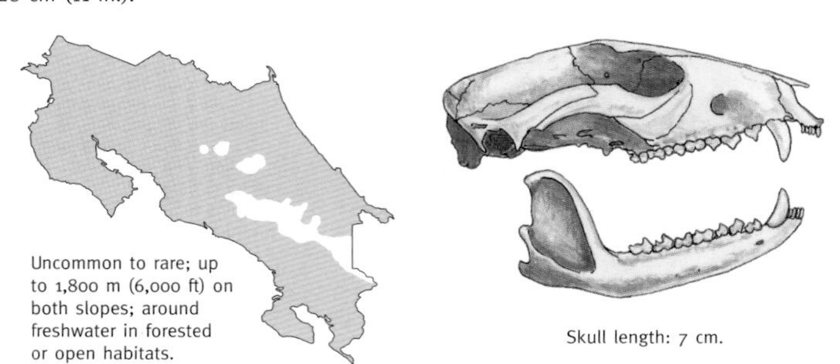

Uncommon to rare; up to 1,800 m (6,000 ft) on both slopes; around freshwater in forested or open habitats.
Skull length: 7 cm.

Woolly opossum tracks.
Hindtrack is about 5 cm wide.

Yapok tracks.
Hindtrack is about 5 cm wide.

PLATE 3: MOUSE OPOSSUMS

1 Alston's Mouse Opossum (*Micoureus alstoni*), p. 44.

Nocturnal; arboreal; solitary. Fur long, woolly, and grayish; tail usually white on outer half, sometimes only at tip; 18 cm (7 in.).

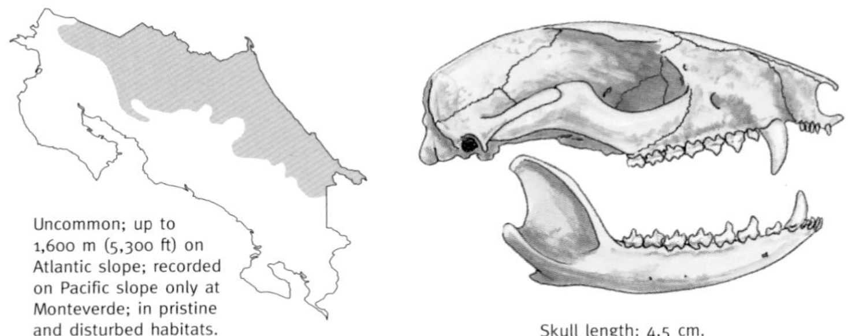

Uncommon; up to 1,600 m (5,300 ft) on Atlantic slope; recorded on Pacific slope only at Monteverde; in pristine and disturbed habitats.

Skull length: 4.5 cm.

2 Mexican Mouse Opossum (*Marmosa mexicana*), p. 44.

Nocturnal; arboreal; solitary. Fur short, smooth, and brown or orangey brown; tail uniform in color; 15 cm (6 in.).

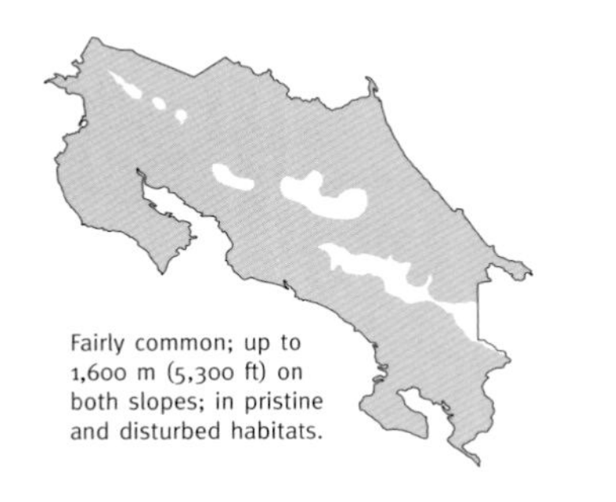

Fairly common; up to 1,600 m (5,300 ft) on both slopes; in pristine and disturbed habitats.

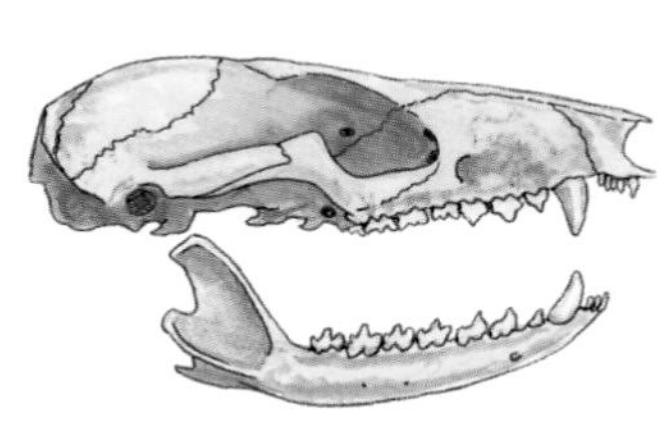

Skull length: 3.5 cm.

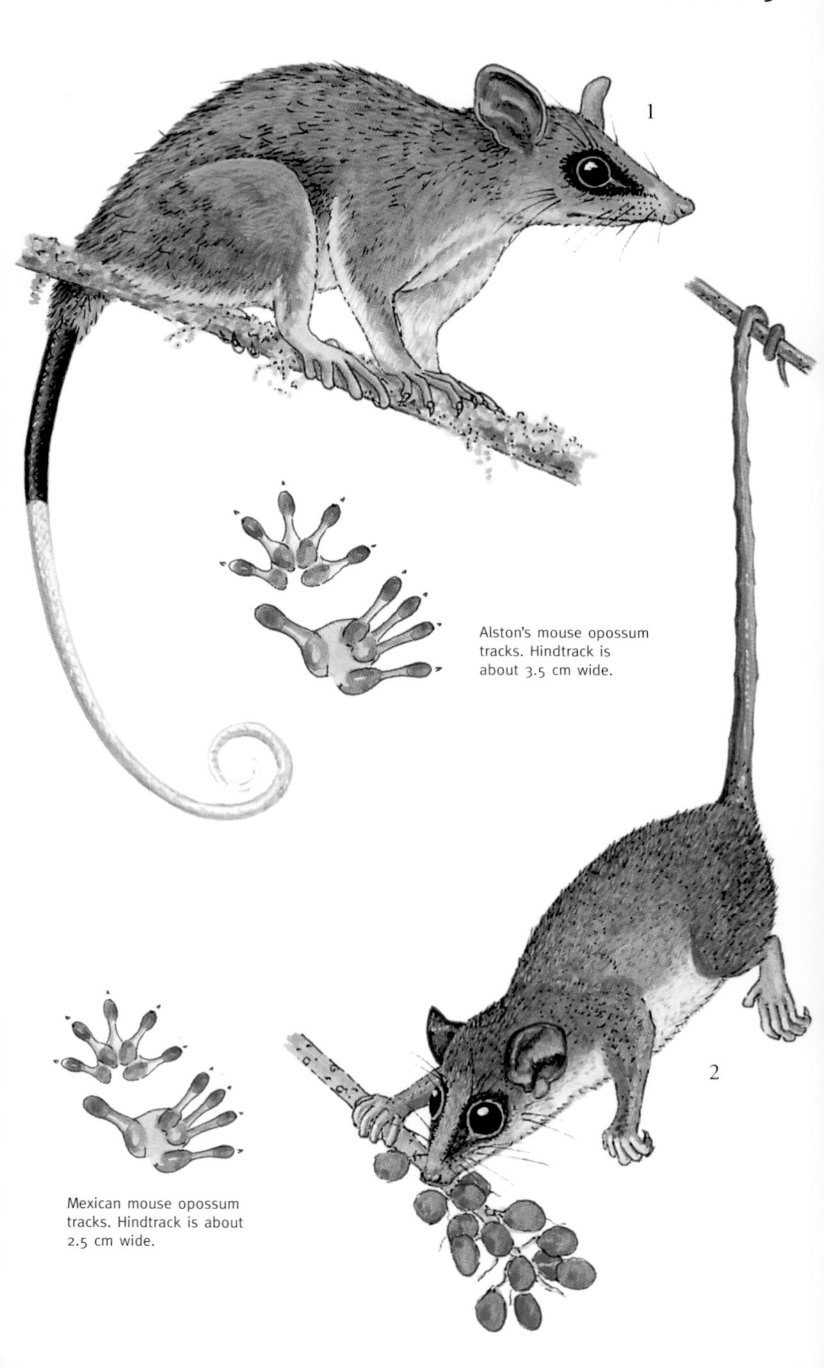

Alston's mouse opossum tracks. Hindtrack is about 3.5 cm wide.

Mexican mouse opossum tracks. Hindtrack is about 2.5 cm wide.

1 **Silky Anteater** (*Cyclopes didactylus*), p. 58.

Nocturnal; arboreal; solitary. Fur with metallic sheen; usually with dark line down belly; tail furred to tip; each forefoot with just one large claw and one small claw; 17 cm (7 in.).

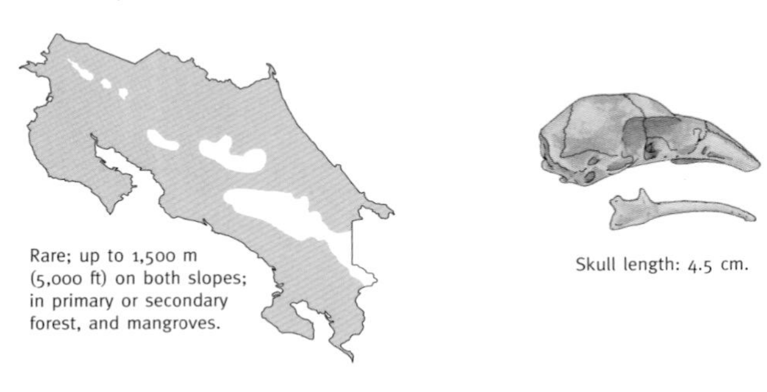

Rare; up to 1,500 m (5,000 ft) on both slopes; in primary or secondary forest, and mangroves.

Skull length: 4.5 cm.

2 **Northern Tamandua** (*Tamandua mexicana*), p. 55.

Diurnal and nocturnal; arboreal and terrestrial; solitary. Distinguished from giant anteater by smaller size and prehensile tail; folds claws of forefeet inwards and places weight on knuckles when walking on ground; 60 cm (24 in.).

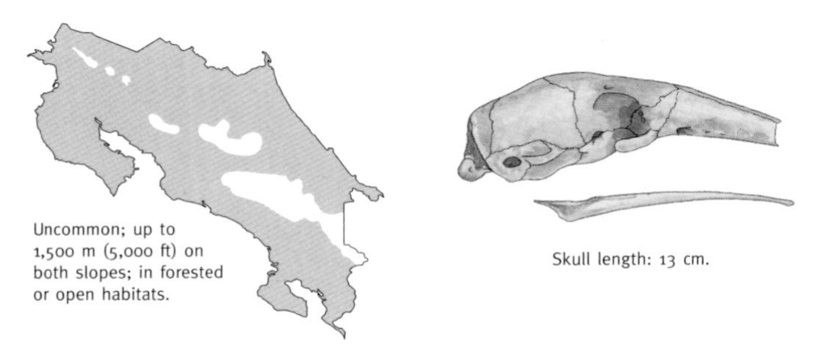

Uncommon; up to 1,500 m (5,000 ft) on both slopes; in forested or open habitats.

Skull length: 13 cm.

3 **Giant Anteater** (*Myrmecophaga tridactyla*), p. 53.

Diurnal and nocturnal; terrestrial; solitary. Distinguished from northern tamandua by larger size and bushy tail; folds claws of forefeet inwards and places weight on knuckles when walking; 130 cm (50 in.).

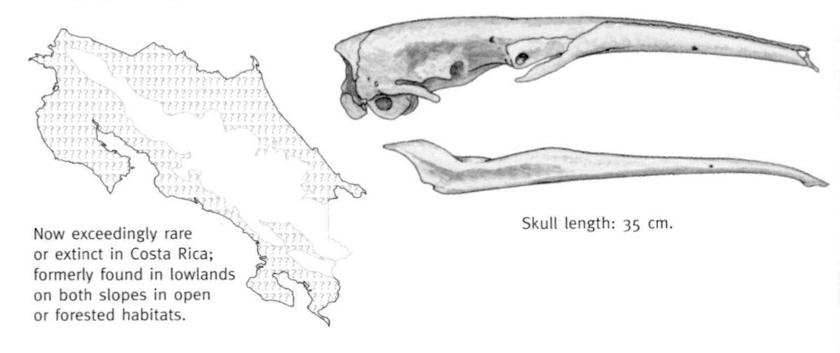

Now exceedingly rare or extinct in Costa Rica; formerly found in lowlands on both slopes in open or forested habitats.

Skull length: 35 cm.

Tamandua tracks; close-up and entire track pattern. Hindtrack is about 4.5 cm wide.

Giant anteater tracks. Hindtrack is about 7 cm wide.

1 **Brown-Throated Three Toed Sloth** (*Bradypus variegatus*), p. 61.

Nocturnal and diurnal; arboreal; solitary. Fur wirey; face with black mask, short snout, and "smiling" mouth; possesses stubby tail; males with patch of short orange and brown fur in center of back; three toes on each hand and foot; 60 cm (24 in.).

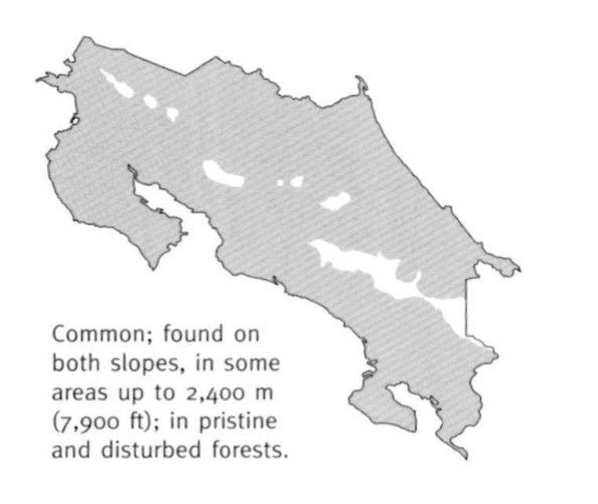

Common; found on both slopes, in some areas up to 2,400 m (7,900 ft); in pristine and disturbed forests.

Skull length: 10 cm.

2 **Hoffmann's Two-Toed Sloth** (*Choloepus hoffmanni*), p. 65.

Mostly nocturnal; arboreal; solitary. Face surrounded by pale fur; snout longer than that of three-toed sloth; no tail; three toes on each foot, two on each hand; common; 60 cm (24 in.).

Common; found on both slopes up to the tree line; in pristine and disturbed forests.

Skull length: 11 cm.

1

2

PLATE 6: ARMADILLOS

1 **Northern Naked-Tailed Armadillo** (*Cabassous centralis*), p. 67.

Mostly nocturnal; terrestrial; solitary. Proportionately broader and flatter body than nine-banded long-nosed armadillo; ears broad and widely separated; huge central claw on each forefoot; almost no scales on tail; 40 cm (16 in.).

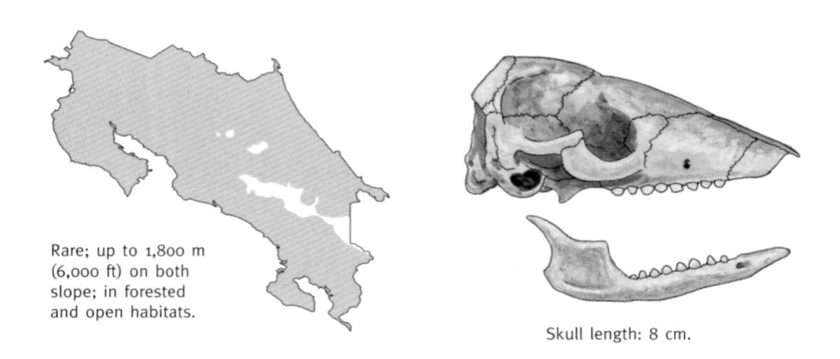

Rare; up to 1,800 m (6,000 ft) on both slope; in forested and open habitats.

Skull length: 8 cm.

2 **Nine-Banded Long-Nosed Armadillo** (*Dasypus novemcinctus*), p. 69.

Mostly nocturnal; terrestrial; solitary. Distinguished from northern naked-tailed armadillo by more arched carapace and longer, thinner snout, more upright and narrower ears that are set close together, and by armored tail; 45 cm (18 in.).

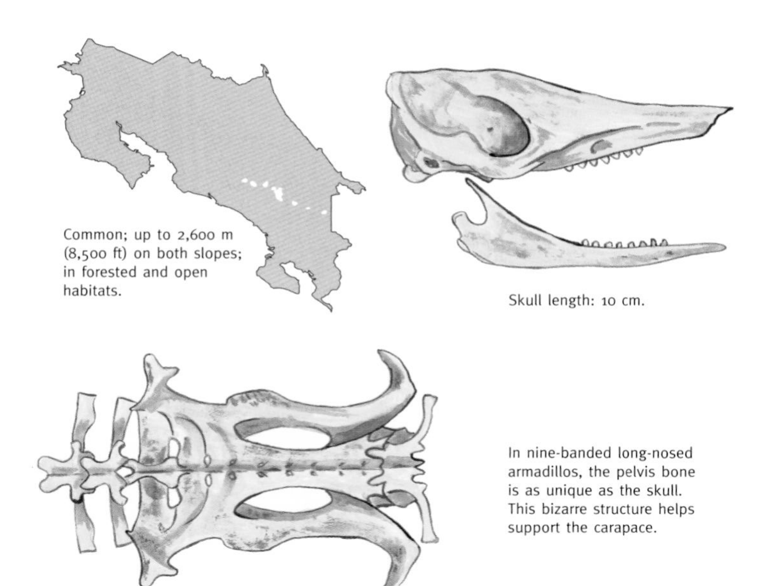

Common; up to 2,600 m (8,500 ft) on both slopes; in forested and open habitats.

Skull length: 10 cm.

In nine-banded long-nosed armadillos, the pelvis bone is as unique as the skull. This bizarre structure helps support the carapace.

A tail mark is often visible in armadillo tracks.

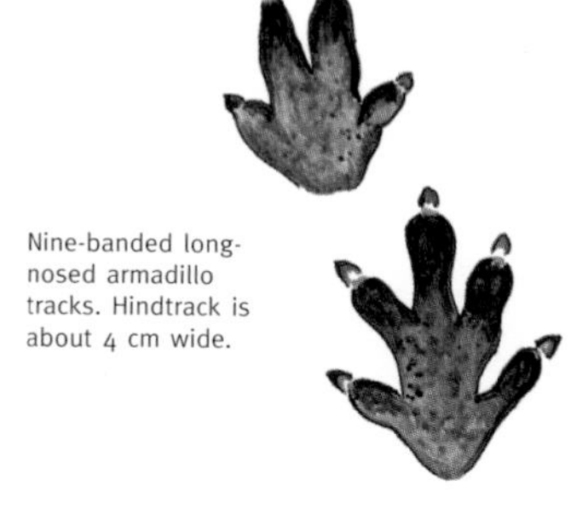

Nine-banded long-nosed armadillo tracks. Hindtrack is about 4 cm wide.

PLATE 7: SAC-WINGED BATS

1 **Long-Nosed Bat** (*Rhynchonycteris naso*), p. 91.

One of three bat species in Costa Rica with wavy pale lines on back; lines sometimes inconspicuous in this species; of the three, this is the only one with a grizzled gray coloration and tufts of pale fur on forearms; 4 cm (1 ½ in.).

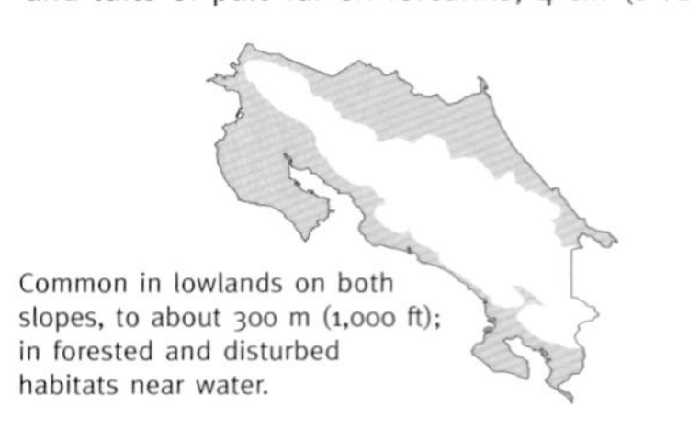
Common in lowlands on both slopes, to about 300 m (1,000 ft); in forested and disturbed habitats near water.

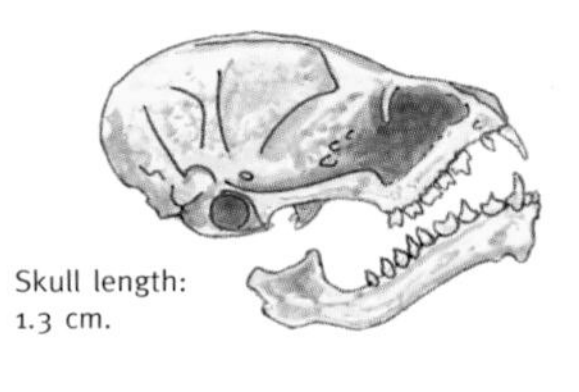
Skull length: 1.3 cm.

2 **Greater White-Lined Bat** (*Saccopteryx bilineata*), p. 91.

Only other bats in Costa Rica with wavy pale lines on back are the long-nosed bat (above) and the lesser white-lined bat (*Saccopteryx leptura*, not illustrated). *S. leptura* is smaller than *S. bilineata* and has dark brown rather than black fur with less pronounced wavy lines; 5 cm (2 in.).

Common in wet lowlands of Atlantic and south Pacific, to about 500 m (1650 ft); rare in dry lowlands of north Pacific; in primary and secondary forests.

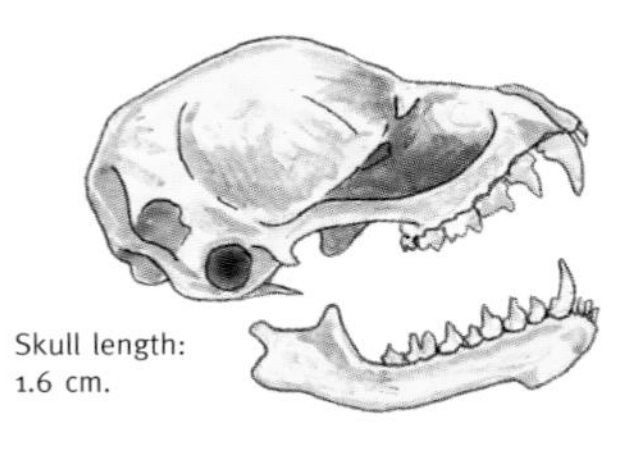
Skull length: 1.6 cm.

3 **Northern Ghost Bat** (*Diclidurus albus*), p. 91.

No other bat has a scent gland on the tail. The only other whitish bat in Costa Rica, *Ectophylla* alba (pl. 12), is much smaller and has a noseleaf. 7 cm (3 in.).

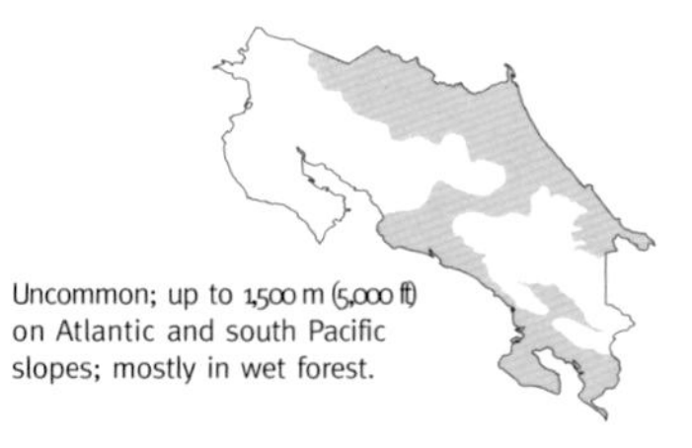
Uncommon; up to 1,500 m (5,000 ft) on Atlantic and south Pacific slopes; mostly in wet forest.

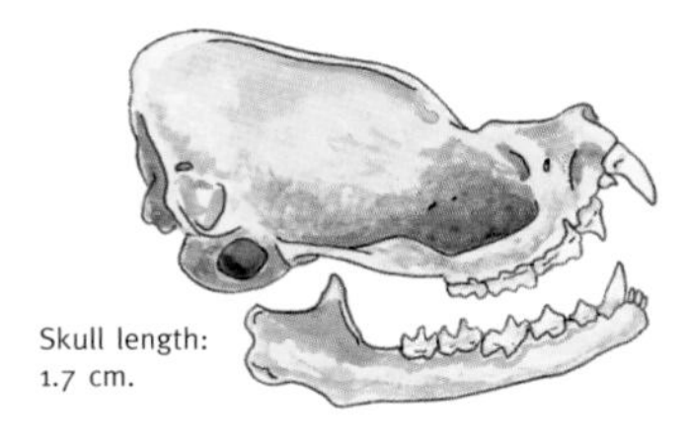
Skull length: 1.7 cm.

4 **Gray Sac-Winged Bat** (*Balantiopteryx plicata*), p. 91.

The only uniform gray-colored bat with "wing sacs" (see p. 92); wings brownish; 5 cm (2 in.).

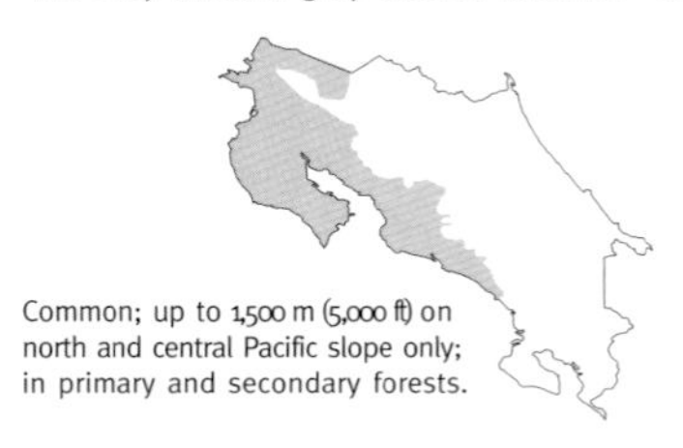
Common; up to 1,500 m (5,000 ft) on north and central Pacific slope only; in primary and secondary forests.

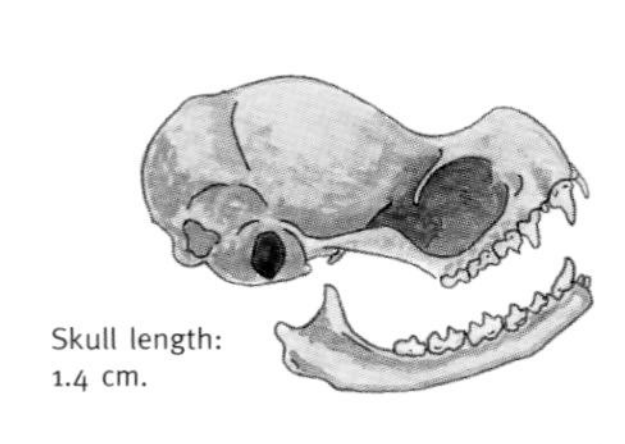
Skull length: 1.4 cm.

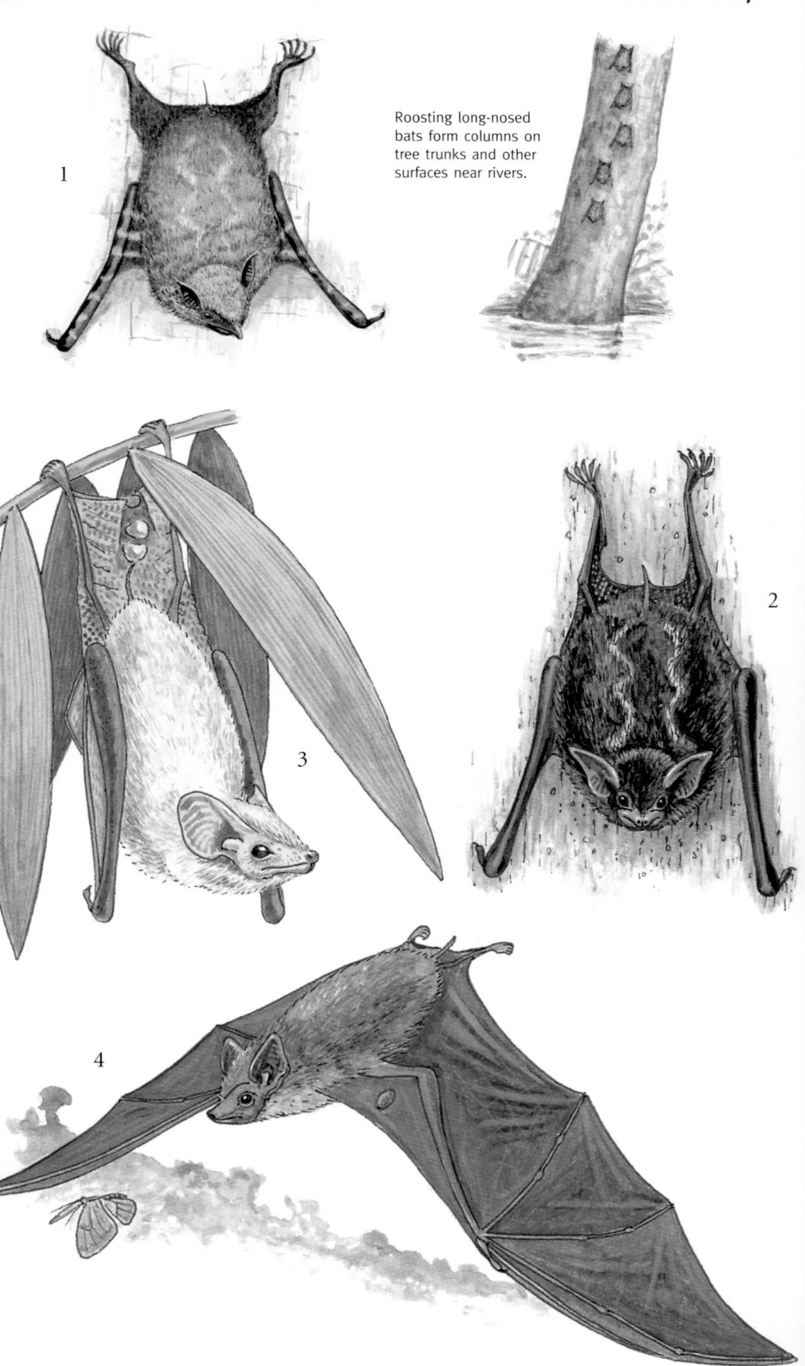
1
Roosting long-nosed bats form columns on tree trunks and other surfaces near rivers.
2
3
4

PLATE 8: FISHING AND LEAF-CHINNED BATS

1 **Greater Fishing Bat** (*Noctilio leporinus*), p. 94.

Bulldog-like mouth shape; upperparts vary from orange to brown to gray; proportionately huge, rakelike feet; 12 cm (5 in.).

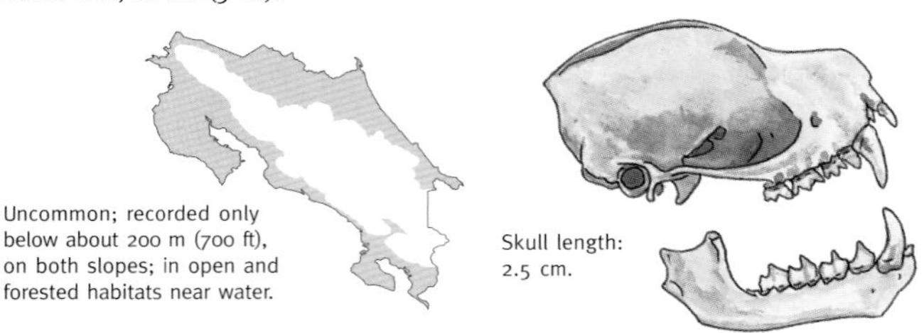

Uncommon; recorded only below about 200 m (700 ft), on both slopes; in open and forested habitats near water.

Skull length: 2.5 cm.

2 **Lesser Fishing Bat** (*Noctilio albiventris*), p. 94.

Similar to greater fishing bat in coloration and head shape but smaller in size and with proportionately much smaller feet; 7 cm (3 in.).

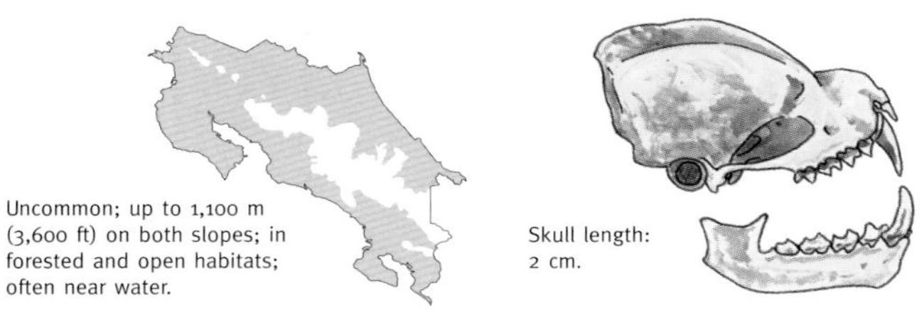

Uncommon; up to 1,100 m (3,600 ft) on both slopes; in forested and open habitats; often near water.

Skull length: 2 cm.

3 **Davy's Naked-Backed Bat** (*Pteronotus davyi*), p. 96.

Like Parnell's mustached bat but wing membranes attach along middle of back, covering fur beneath. 5 cm (2 in.).

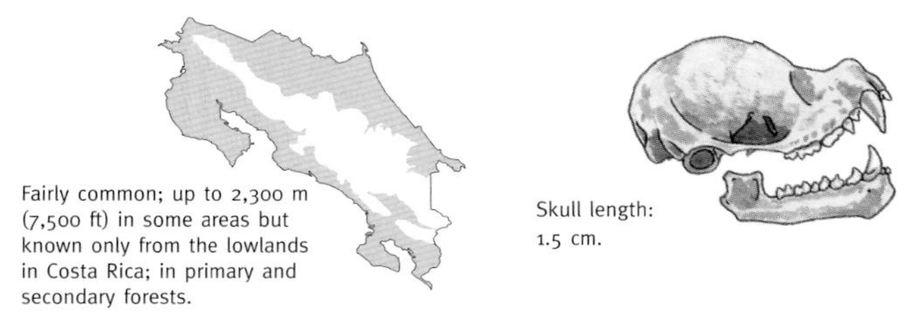

Fairly common; up to 2,300 m (7,500 ft) in some areas but known only from the lowlands in Costa Rica; in primary and secondary forests.

Skull length: 1.5 cm.

4 **Parnell's Mustached Bat** (*Pteronotus parnellii*), p. 96.

Upperparts vary from brown to orange; pointed ears that slant forwards; mustache of bristles above thick, flared lips; protruding flaps of skin on chin; tiny eyes; tail protrudes from tail membrane; 7 cm (3 in.).

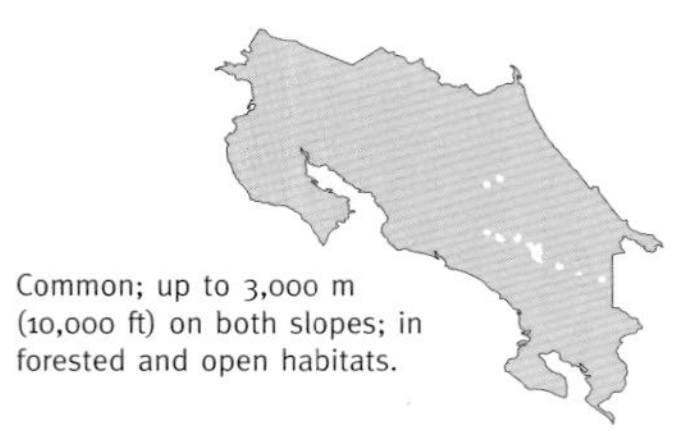

Common; up to 3,000 m (10,000 ft) on both slopes; in forested and open habitats.

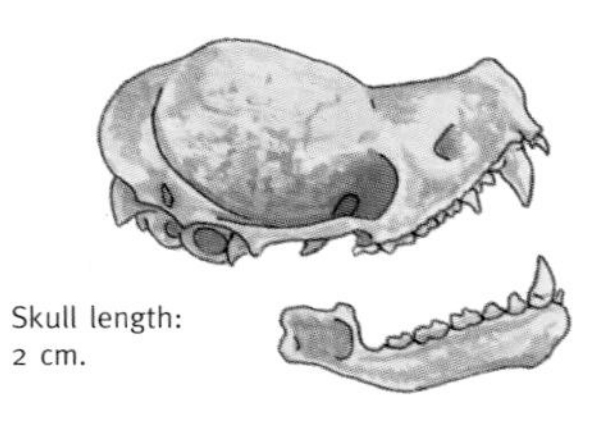

Skull length: 2 cm.

PLATE 8

PLATE 9: SPEAR-NOSED BATS

1 Frog-Eating Bat (*Trachops cirrhosus*), p. 100, catching mud puddle frog (*Physalaemus pustulosus*).

Upperparts vary from gray-brown to orangey-brown; only leaf-nosed bat with pronounced warts around mouth; 9 cm (3 ½ in.).

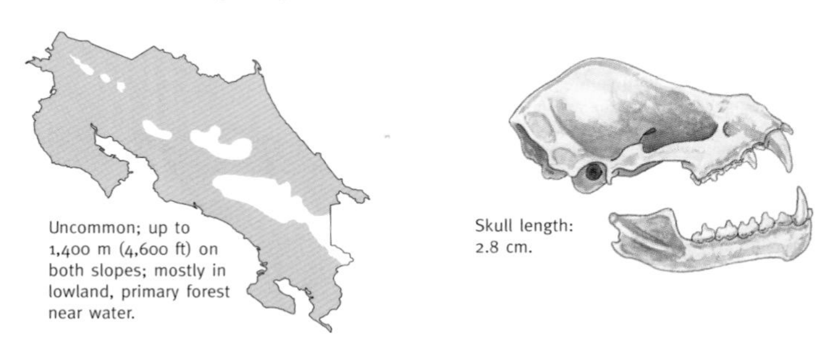

Uncommon; up to 1,400 m (4,600 ft) on both slopes; mostly in lowland, primary forest near water.
Skull length: 2.8 cm.

2 Tome's Long-Eared Bat (*Lonchorhina aurita*), p. 100.

Upperparts vary from brown to reddish; long, pointed ears; no other bat has such a proportionately long noseleaf; tail memebrane extends to tip of long tail; 6 cm (2 ½ in.).

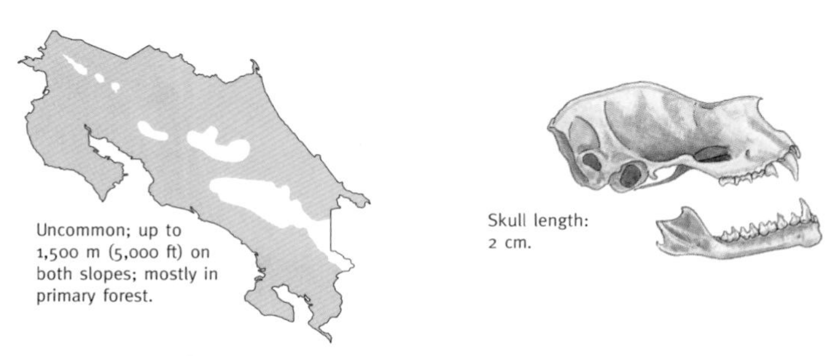

Uncommon; up to 1,500 m (5,000 ft) on both slopes; mostly in primary forest.
Skull length: 2 cm.

3 False Vampire Bat (*Vampyrum spectrum*), p. 100, snatching sleeping turquoise-browed motmot (*Eumomota superciliosa*).

Upperparts vary from brown to orangish to reddish, with indistinct pale stripe down center of back; long muzzle, pale noseleaf; large, rounded ears; much larger than any other New World bat; 15 cm (6 in.); wingspan 80 cm.

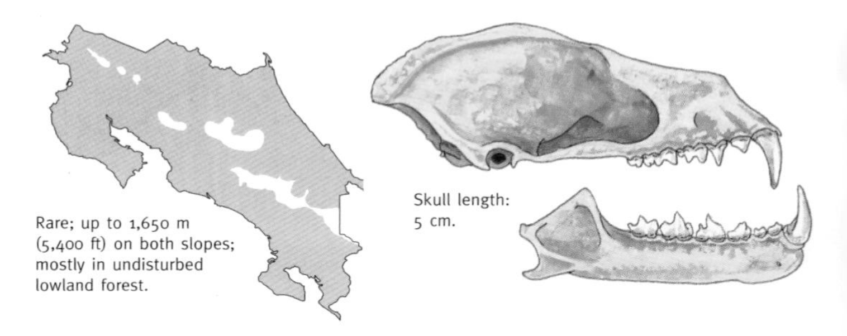

Rare; up to 1,650 m (5,400 ft) on both slopes; mostly in undisturbed lowland forest.
Skull length: 5 cm.

1
2
3

1 Pallas' Long-Tongued Bat (*Glossophaga soricina*), p. 104, visiting flower of bull's eye vine (*Mucuna urens*).

Upperparts vary from grayish-brown to reddish-brown; fur paler at base, darker at tip; long, thin snout; small noseleaf; groove in lower lip; tail very short; tail membrane naked; 5 cm (2 in.).

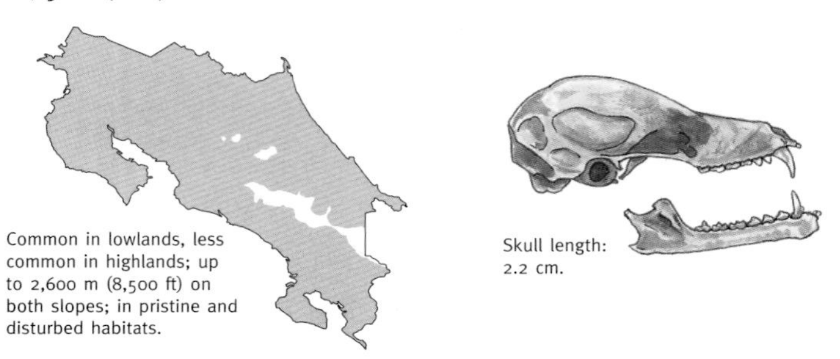

Common in lowlands, less common in highlands; up to 2,600 m (8,500 ft) on both slopes; in pristine and disturbed habitats.

Skull length: 2.2 cm.

2 Geoffroy's Tailless or **Hairy-Legged Bat** (*Anoura geoffroyi*), p. 104, licking pollen from fur.

Fur paler at base, darker at tip; long, thin snout; small noseleaf; groove in lower lip; legs very hairy; no tail; very short, hairy tail membrane; 7 cm (3 in.).

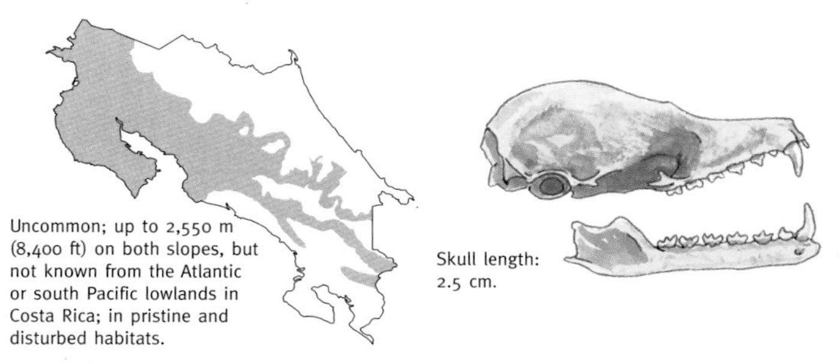

Uncommon; up to 2,550 m (8,400 ft) on both slopes, but not known from the Atlantic or south Pacific lowlands in Costa Rica; in pristine and disturbed habitats.

Skull length: 2.5 cm.

3 Orange Nectar Bat (*Lonchophylla robusta*), p. 104, visiting flower of kapok tree (*Ceiba pentandra*).

Long, thin snout; small noseleaf; groove in lower lip; tail very short; tail membrane naked; only nectar-feeding bat with orange fur; 7 cm (3 in.).

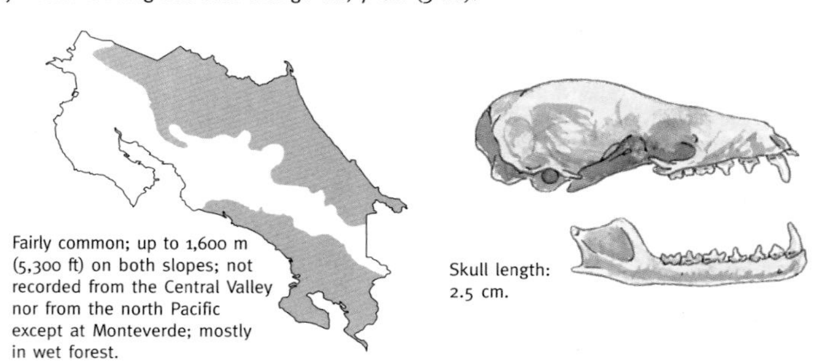

Fairly common; up to 1,600 m (5,300 ft) on both slopes; not recorded from the Central Valley nor from the north Pacific except at Monteverde; mostly in wet forest.

Skull length: 2.5 cm.

1

2

3

1 Seba's Short-Tailed Fruit Bat (*Carollia perspicillata*), p. 106, plucking fruit of black pepper relative (*Piper* sp.).

Upperparts vary from gray-brown to orange; one of four short-tailed fruit bats in Costa Rica; four species distinguished from one another only by subtle differences in size and in banding pattern of fur; as a group, distinguished from nectar-feeding bats by shorter snout and lack of groove in lower lip; from tailless fruit bats by presence of short tail; all short-tailed fruit bats have U-shaped row of warts around larger central wart on chin; 6 cm (2 ½ in.).

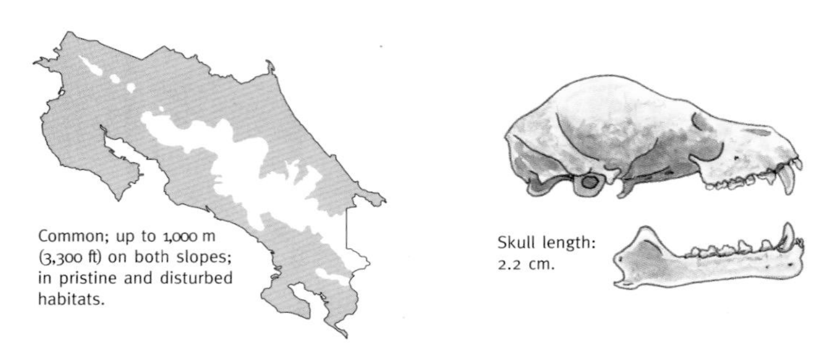

Common; up to 1,000 m (3,300 ft) on both slopes; in pristine and disturbed habitats.
Skull length: 2.2 cm.

2 Highland Epauleted Bat (*Sturnira ludovici*), p. 109, plucking fruit of tomato relative (*Solanum* sp.).

One of four very similar bats in this genus in Costa Rica, distinguished only by subtle differences in size, color, hairiness, and dentition. 7 cm (3 in.).

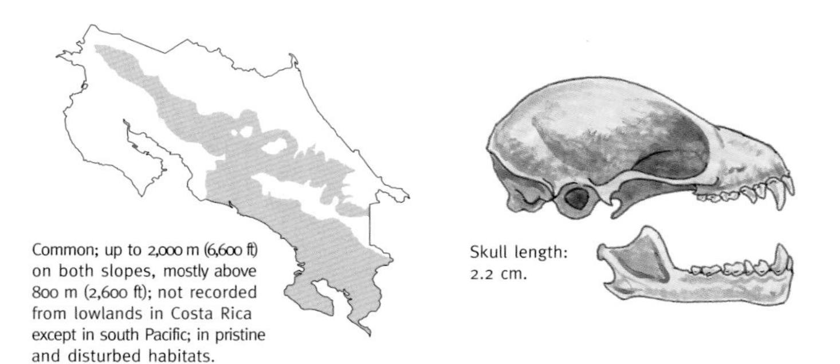

Common; up to 2,000 m (6,600 ft) on both slopes, mostly above 800 m (2,600 ft); not recorded from lowlands in Costa Rica except in south Pacific; in pristine and disturbed habitats.
Skull length: 2.2 cm.

1
2

PLATE 12: TAILLESS FRUIT BATS

1 **Common Tent-Making Bat** (*Uroderma bilobatum*), p. 109.

One of several tent-making bat species with similar coloration. 6 cm (2 ½ in.).

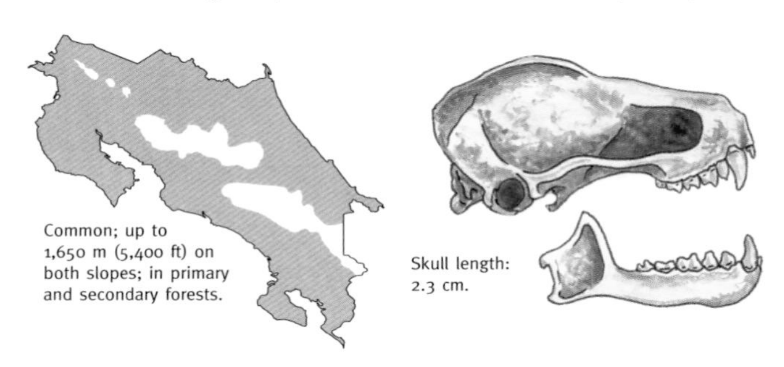
Common; up to
1,650 m (5,400 ft) on
both slopes; in primary
and secondary forests.
Skull length:
2.3 cm.

2 **White Tent Bat** (*Ectophylla alba*), p. 109

The only other whitish bat in Costa Rica, the northern ghost bat (*Diclidurus albus*, pl. 7), is larger, lacks a noseleaf, and does not build tents in heliconia leaves; 4 cm (1 ½ in.).

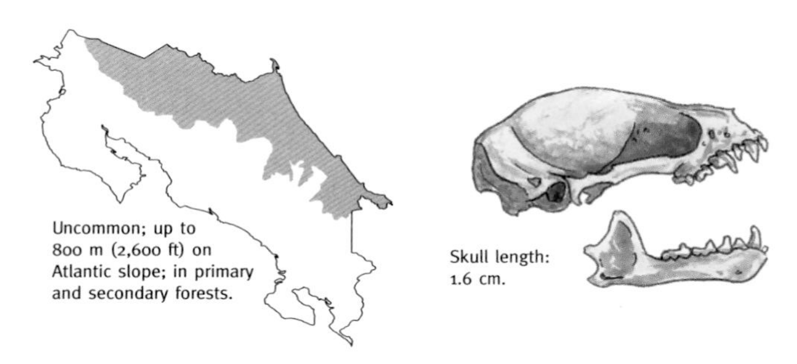
Uncommon; up to
800 m (2,600 ft) on
Atlantic slope; in primary
and secondary forests.
Skull length:
1.6 cm.

3 **Wrinkle-Faced** or **Lattice-Winged Bat** (*Centurio senex*), p. 109, eating mango.

Bizarre face and barred wings unique; 6 cm (2 ½ in.).

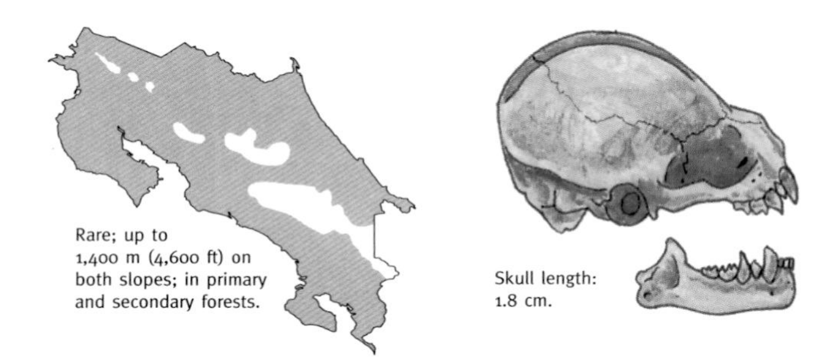
Rare; up to
1,400 m (4,600 ft) on
both slopes; in primary
and secondary forests.
Skull length:
1.8 cm.

1
2
3

PLATE 13: VAMPIRE BATS

1 **Common Vampire Bat** (*Desmodus rotundus*), p. 113

Upperparts sometimes orangey; M-shaped noseleaf; triangular ears, with length greater than width; small eyes; wingtips sometimes pale but not pure white; 8 cm (3 in.).

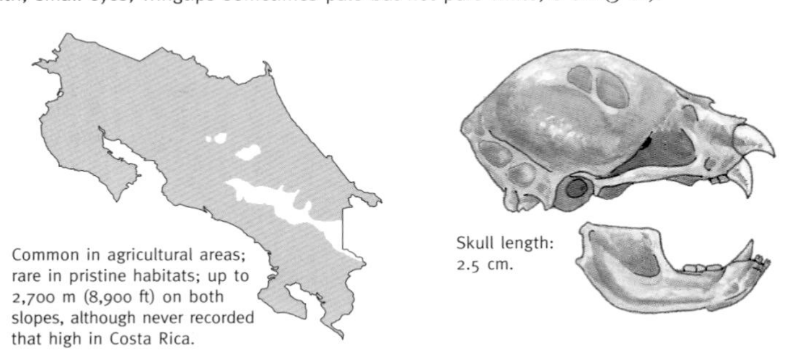

Common in agricultural areas; rare in pristine habitats; up to 2,700 m (8,900 ft) on both slopes, although never recorded that high in Costa Rica.
Skull length: 2.5 cm.

2 **White-Winged Vampire Bat** (*Diaemus youngi*), p. 113.

M-shaped nosepad; triangular ears, with length greater than width; small eyes; pure white on wingtips; note scent glands in corners of mouth; 8 cm (3 in.).

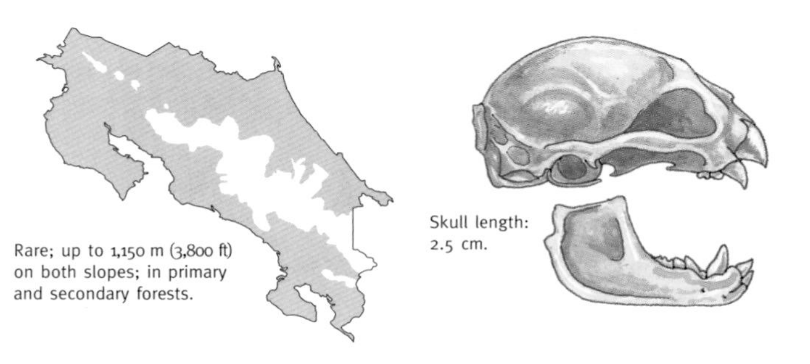

Rare; up to 1,150 m (3,800 ft) on both slopes; in primary and secondary forests.
Skull length: 2.5 cm.

3 **Hairy-Legged Vampire Bat** (*Diphylla ecaudata*), p. 113.

M-shaped nosepad; round ears, with width greater than length; large eyes; very hairy legs; 8 cm (3 in.).

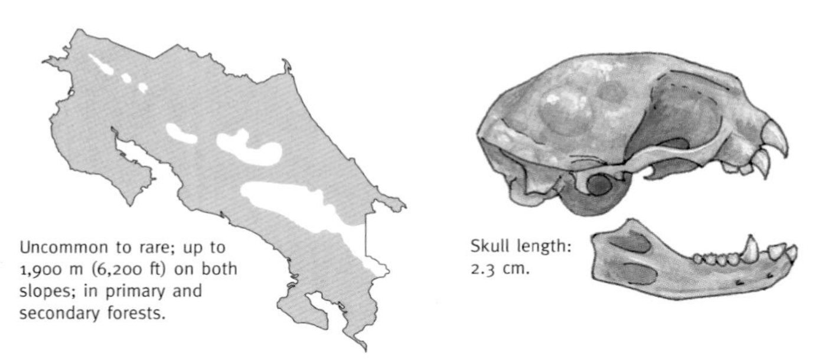

Uncommon to rare; up to 1,900 m (6,200 ft) on both slopes; in primary and secondary forests.
Skull length: 2.3 cm.

Common vampire tracks. Hindtrack is about 1 cm wide.

PLATE 14: FUNNEL-EARED, THUMBLESS & DISK-WINGED BATS

1 **Mexican Funnel-Eared Bat** (*Natalus stramineus*), p. 117

Tiny, V-shaped body; upperparts pale brown, orangey, or yellowish; brown membranes; tiny eyes; pale, pointed ears; mustache of bristles above mouth; very long legs and tail; tail membrane extends to tip of tail; 5 cm (2 in.).

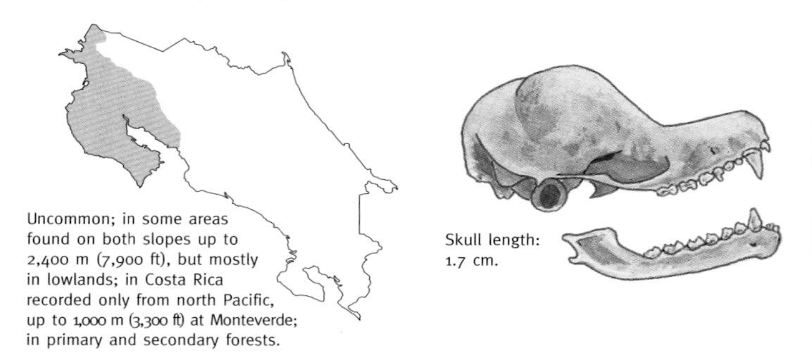

Uncommon; in some areas found on both slopes up to 2,400 m (7,900 ft), but mostly in lowlands; in Costa Rica recorded only from north Pacific, up to 1,000 m (3,300 ft) at Monteverde; in primary and secondary forests.

Skull length: 1.7 cm.

2 **Thumbless Bat** (*Furipterus horrens*), p. 117.

Smallest bat in Costa Rica; unusual, smoky-gray coloration; tiny eyes; short, broad ears; mustache of bristles above mouth; thumbs clawless and barely visible; very long legs; tail membrane extends beyond tip of tail; 3.5 cm (1 ⅓ in.).

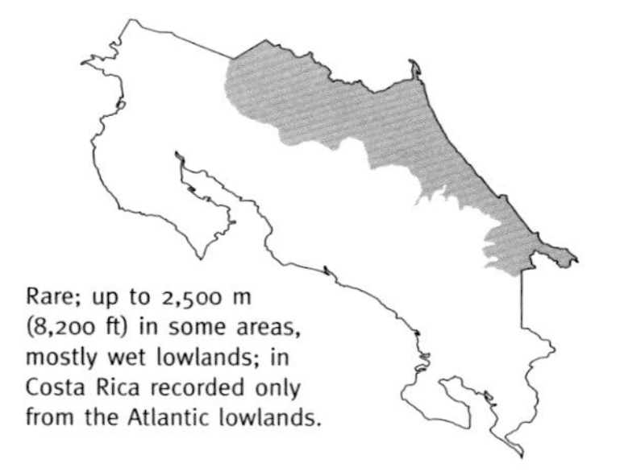

Rare; up to 2,500 m (8,200 ft) in some areas, mostly wet lowlands; in Costa Rica recorded only from the Atlantic lowlands.

3 **Spix's Disk-Winged Bat** (*Thyroptera tricolor*), p. 117.

Upperparts vary from dark to reddish brown; pointy ears; narrow snout; tail extends just beyond tail membrane; suction cups on thumbs and heels, and habit of roosting in young, rolled heliconia leaves unique; 4 cm (1 ½ in.).

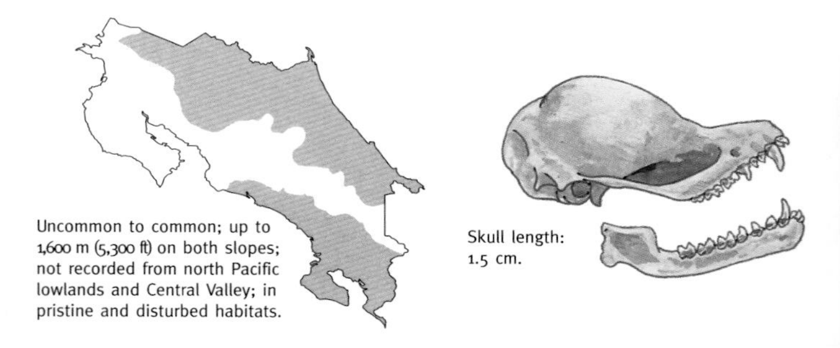

Uncommon to common; up to 1,600 m (5,300 ft) on both slopes; not recorded from north Pacific lowlands and Central Valley; in pristine and disturbed habitats.

Skull length: 1.5 cm.

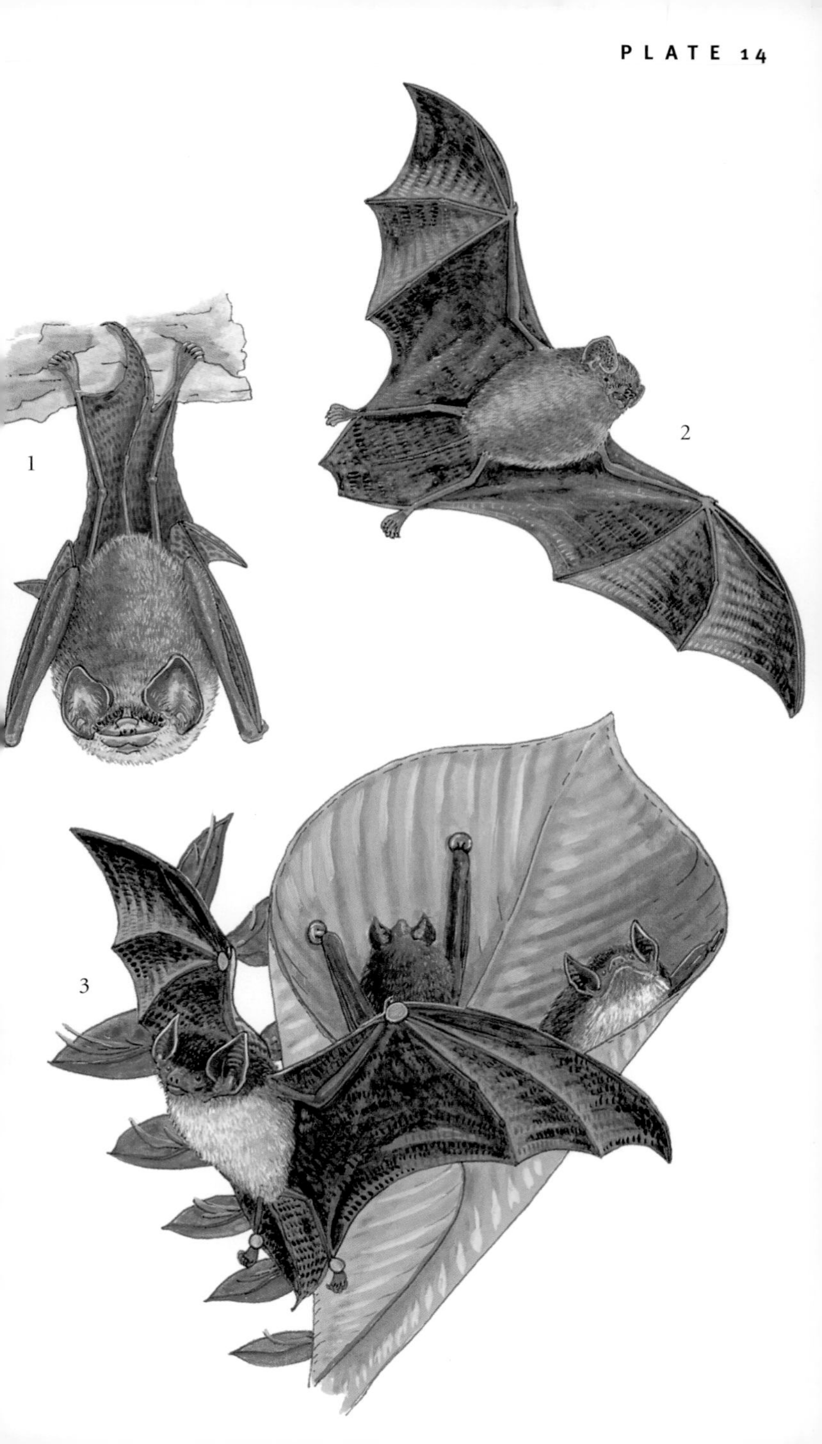
1
2
3

PLATE 15: PLAIN-NOSED BATS

1 **Black Myotis** (*Myotis nigricans*), p. 119.

Upperparts vary from blackish to dark brown or reddish brown; long, narrow ears and tragus; large, V-shaped tail membrane; one of several similar-looking bats in this genus; 4 cm (1 ½ in.).

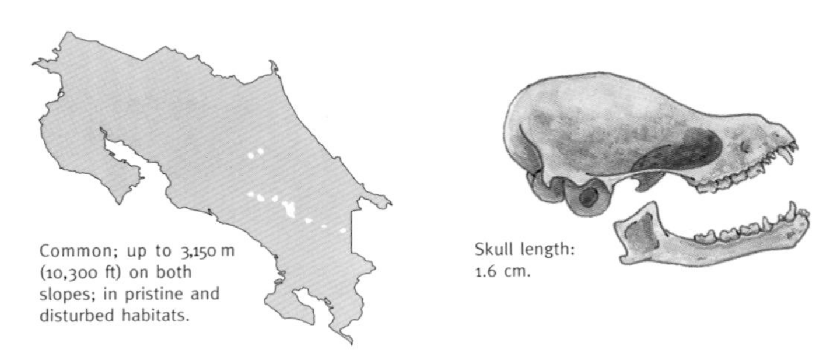

Common; up to 3,150 m (10,300 ft) on both slopes; in pristine and disturbed habitats.
Skull length: 1.6 cm.

2 **Southern Yellow Bat** (*Lasiurus ega*), p. 119.

Large, V-shaped tail membrane thickly furred at base; upperparts vary from yellow to grayish tan, becoming brighter yellow or rusty on tail membrane; broad ears; fairly long snout; only yellowish bat in Costa Rica; 6 cm (2 ½ in.).

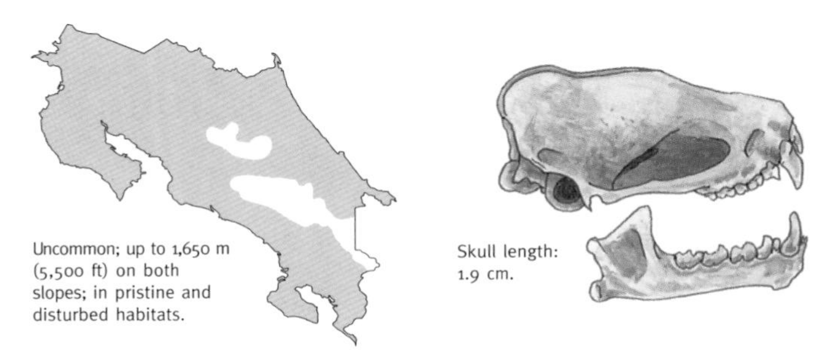

Uncommon; up to 1,650 m (5,500 ft) on both slopes; in pristine and disturbed habitats.
Skull length: 1.9 cm.

3 **Western Red Bat** (*Lasiurus blossevillii*), p. 119.

Large, V-shaped tail membrane thickly furred almost to tip; broad ears; fairly long snout; only other similar-looking bat, the tacaruna bat (*Lasiurus castaneus*), has upperparts dark brick red rather than light orangey red; 5 cm (2 in.).

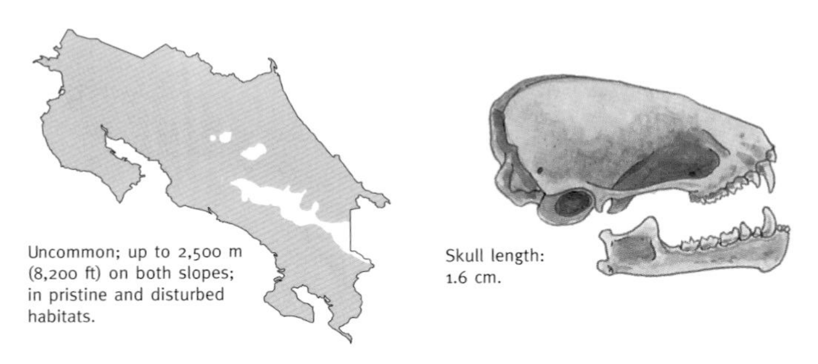

Uncommon; up to 2,500 m (8,200 ft) on both slopes; in pristine and disturbed habitats.
Skull length: 1.6 cm.

1
2

PLATE 18: MONKEYS

1 **Mantled Howler Monkey** (*Alouatta palliata*), p. 139.

Diurnal; arboreal; usually in troops, of up to 45 individuals. All dark except for a mantle of light yellowish to reddish fur on sides, and conspicuous white testicles in males; beard of fur on chin, longer in males; prehensile tail; 50 cm (20 in.).

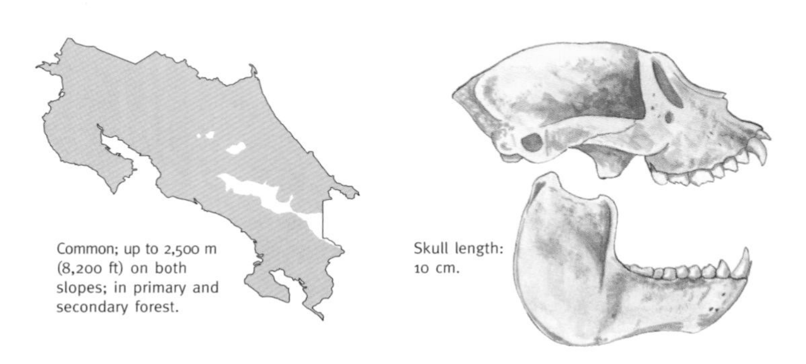

Common; up to 2,500 m (8,200 ft) on both slopes; in primary and secondary forest.
Skull length: 10 cm.

2 **Central American Spider Monkey** (*Ateles geoffroyi*), p. 146.

Diurnal; arboreal; in loose communities of up to 40 individuals, often split into small subgroups. Pale tan (especially in northwest lowlands) to reddish body with dark, very long limbs and tail; proportionately small head with conspicuous, pinkish mask of bare skin around eyes and muzzle; prehensile tail; 50 cm (20 in.).

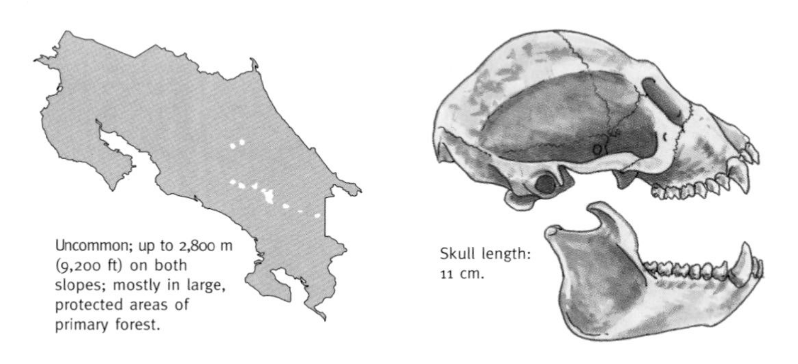

Uncommon; up to 2,800 m (9,200 ft) on both slopes; mostly in large, protected areas of primary forest.
Skull length: 11 cm.

1

2

PLATE 19: SQUIRRELS

1 Alfaro's Pygmy Squirrel (*Microsciurus alfari*), p. 155.

Diurnal; mostly arboreal; solitary, in pairs, or in small, loose groups. Much smaller than any other Costa Rican squirrel; upperparts dark brown; underparts gray to tan; small, low-set ears that do not protrude above the top of the head when seen in profile; slightly reddish tail, relatively narrower than in other squirrels, and tapering toward tip; 14 cm (5 1/2 in.).

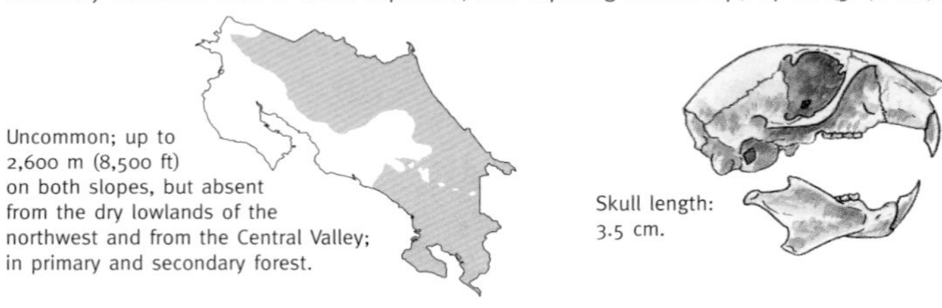

Uncommon; up to 2,600 m (8,500 ft) on both slopes, but absent from the dry lowlands of the northwest and from the Central Valley; in primary and secondary forest.

Skull length: 3.5 cm.

2 Deppe's Squirrel (*Sciurus deppei*), p. 157.

Diurnal; mostly arboreal; usually solitary. Dark brown upperparts; pale grayish underparts; ears protrude above top of head when seen in profile; tail not very bushy, with inconspicuous grayish frosting; 20 cm (8 in.).

Previously, perhaps erroneously, listed for many parts of the country; known to occur for certain only in the Guanacaste mountain range, where it is fairly common; in pristine and disturbed habitats.

3 Red-Tailed Squirrel (*Sciurus granatensis*), p. 159.

Diurnal, mostly arboreal; usually solitary. Dark brown upperparts; orange-tan underparts; ears project above top of head; bushy tail with conspicuous orange-tan frosting; 24 cm (9 in.).

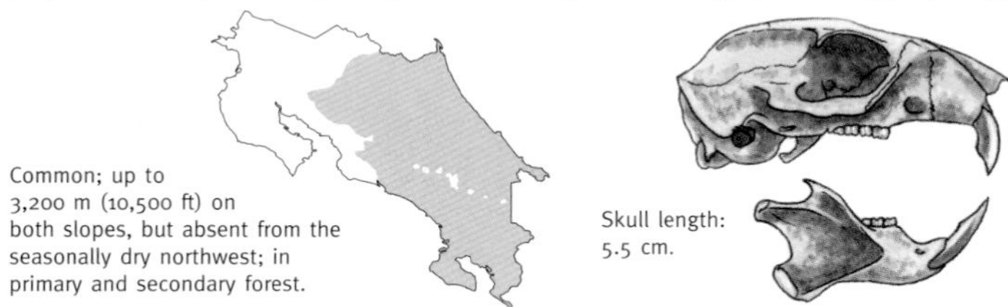

Common; up to 3,200 m (10,500 ft) on both slopes, but absent from the seasonally dry northwest; in primary and secondary forest.

Skull length: 5.5 cm.

4 Montane or Poás Squirrel (*Syntheosciurus brochus*), p. 164.

Diurnal; mostly arboreal; solitary, in pairs, or in small, loose groups. Dark orangey-brown upperparts; dark orange underparts; short, bushy, tail with orange frosting; tiny ears do not protrude above crown of head when seen in profile; red-tailed squirrel is larger, with larger ears that protrude above crown; 17 cm (7 in.).

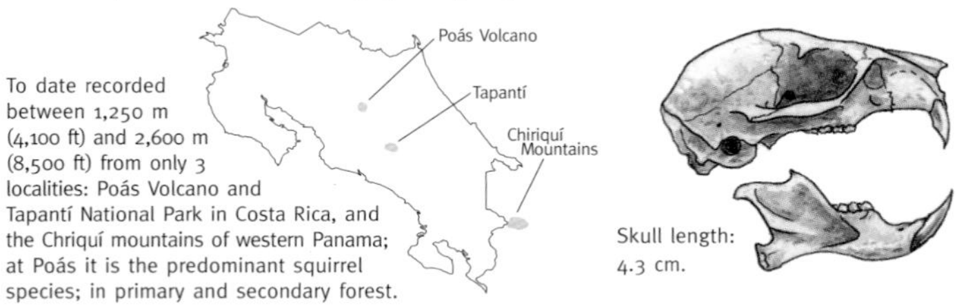

To date recorded between 1,250 m (4,100 ft) and 2,600 m (8,500 ft) from only 3 localities: Poás Volcano and Tapantí National Park in Costa Rica, and the Chiriquí mountains of western Panama; at Poás it is the predominant squirrel species; in primary and secondary forest.

Skull length: 4.3 cm.

1
2
3
4

Variegated Squirrel (*Sciurus variegatoides*), p. 161.

Diurnal; mostly arboreal; usually solitary. Highly variable between areas, depending on the subspecies; somewhat larger than the red-tailed squirrel (pl. 19) and considerably larger than all other Costa Rican squirrels; heavy white frosting on tail, present in all but the black subspecies, distinguish this from all other Costa Rican squirrels; 28 cm (11 in.).

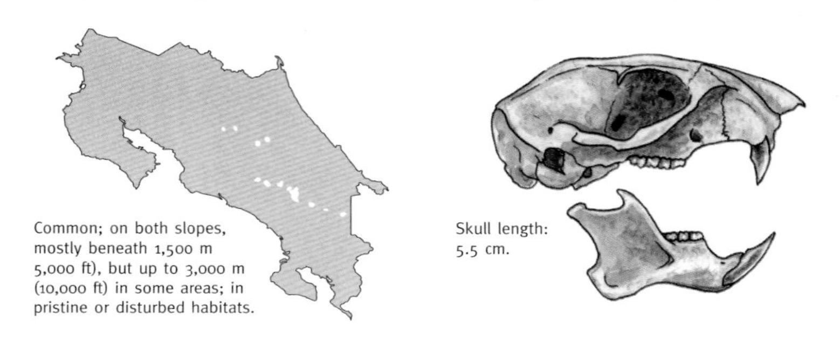

Common; on both slopes, mostly beneath 1,500 m 5,000 ft), but up to 3,000 m (10,000 ft) in some areas; in pristine or disturbed habitats.

Skull length: 5.5 cm.

DISTRIBUTION OF VARIEGATED SQUIRREL SUBSPECIES IN COSTA RICA

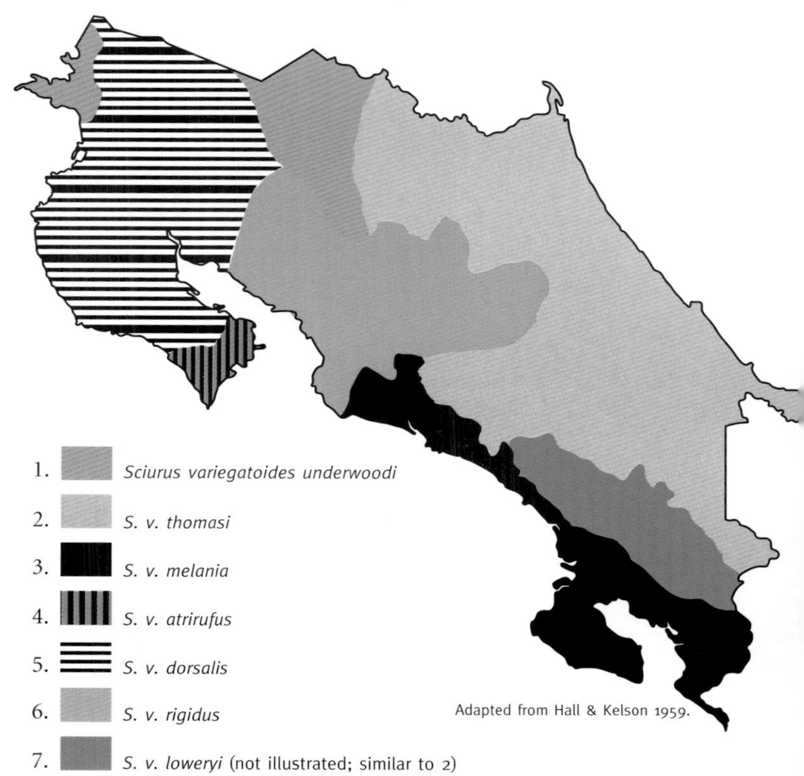

1
2
3
4
6
5
Typical footfall pattern
of squirrels.
uirrel tracks. Hindfoot
th varies from 1 to
cm, depending on
cies.

PLATE 21: POCKET GOPHERS

1 **Chiriquí Pocket Gopher** (*Orthogeomys cavator*), p. 166.

Fossorial; solitary. Dark blackish brown upperparts; no white markings; 25 cm (10 in.).

Endemic to southern Costa Rica and western Panama, from sea level up to 2,400 m (7,900 ft); rarely seen but fairly common in this small region; mostly in agricultural areas.

2 **Variable Pocket Gopher** (*Orthogeomys heterodus*), p. 166.

Diurnal and nocturnal; fossorial; solitary. Upperparts sometimes darker than illustrated, but not as dark as in other Costa Rican gophers; muzzle usually pale; a few individuals have white markings on head or rump; 25 cm (10 in.).

Endemic to Costa Rica's Central Valley and the surrounding mountains, from 1,500 to 2,400 m (5,000 to 7,900 ft); rarely seen but fairly common in this small region; mostly in agricultural areas.

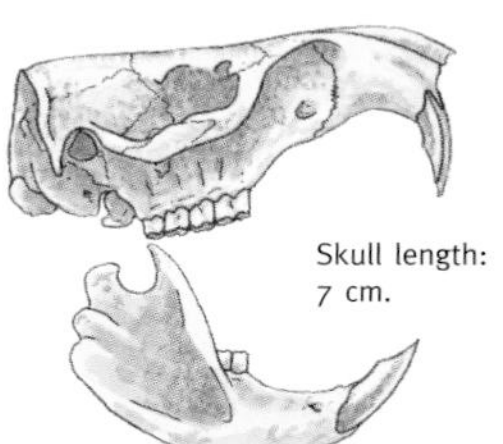

Skull length: 7 cm.

3 **Cherrie's Pocket Gopher** (*Orthogeomys cherriei*), p. 166.

Fossorial; solitary. Dark blackish brown upperparts with conspicuous white patch on top of head; 20 cm (8 in.).

Endemic to Costa Rica; found in the Tilarán and Central mountain ranges and Caribbean lowlands, from sea level to 1450 m (4,800 ft); rarely seen but fairly common in this small region; in agricultural areas and forest.

4 **Underwood's Pocket Gopher** (*Orthogeomys underwoodi*), p. 166.

Fossorial; solitary. Dark blackish brown upperparts with conspicuous white band around rump; shape and width of band variable; 20 cm (8 in.).

Endemic to southwestern Costa Rica, from sea level to 1450 m (4,800 ft); rarely seen but fairly common in this small region; mostly in agricultural areas.

1

2

3

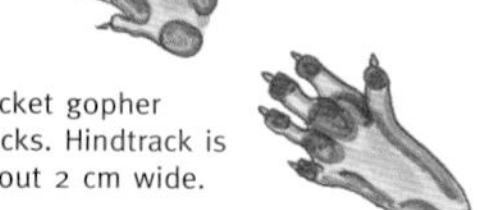

Pocket gopher tracks. Hindtrack is about 2 cm wide.

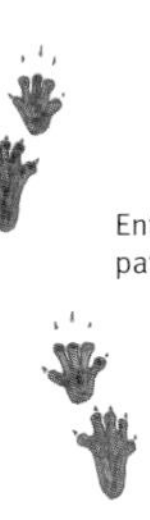

Entire track pattern.

4

1 Salvin's Spiny Pocket Mouse (*Liomys salvini*), p. 169, eating seeds of th Guanacaste tree (*Enterolobium cyclocarpum*).

Nocturnal; terrestrial; solitary. Pouch hidden in fur on outside of each cheek; thin, dark spine hidden in fur on back and sides; dark to pale gray or gray-brown upperparts; forelegs entirel whitish; 13 cm (5 in.).

Common; in Costa Rica found only in the northwest and in the Central Valley, in some areas up to 1,500 m (5,000 ft); in forested and open habitats.

2 Forest Spiny Pocket Mouse (*Heteromys demarestianus*), p. 169, eating seeds of wild nutmeg (*Virola sebifera*).

Nocturnal; terrestrial; solitary. Pouch hidden in fur on outside of each cheek; thin, dark spines hidden in fur on back and sides; dark gray-brown upperparts, usually grading to orangey brown on sides; dark fur extends down front of forelegs; 14 cm (5 1/2 in.).

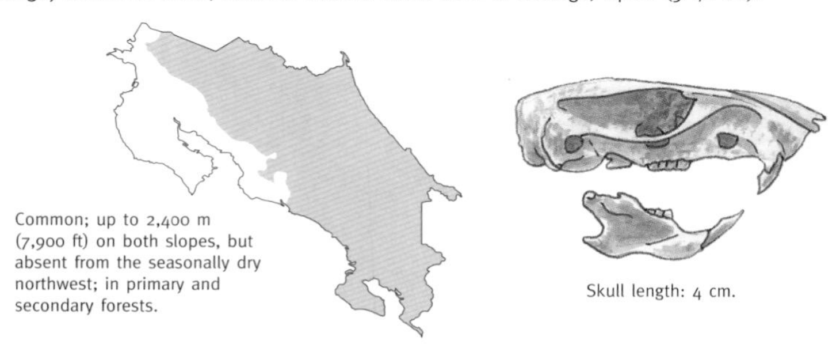

Common; up to 2,400 m (7,900 ft) on both slopes, but absent from the seasonally dry northwest; in primary and secondary forests.

Skull length: 4 cm.

3 Mountain Spiny Pocket Mouse (*Heteromys oresterus*), p. 169.

Nocturnal; terrestrial; solitary. Pouch hidden in fur on outside of each cheek; not spiny, despite name; dark gray upperparts; forelegs entirely whitish; 15 cm (6 in.).

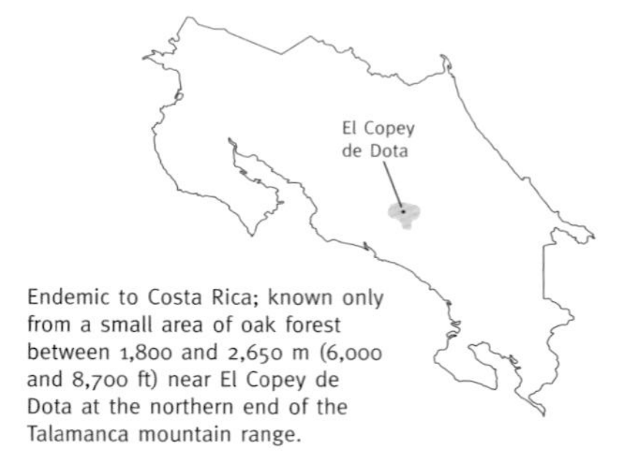

Endemic to Costa Rica; known only from a small area of oak forest between 1,800 and 2,650 m (6,000 and 8,700 ft) near El Copey de Dota at the northern end of the Talamanca mountain range.

Small rodent tracks. Rodents have four digits on each forefoot, while shrews (pl. 23) have five.

1 Least Shrew (*Cryptotis parva*), p. 77.

Nocturnal and diurnal; terrestrial; may nest in groups but forages alone. Distinguished from rodents by long snout and five toes on forefeet; *C. parva* is the only brown shrew in Costa Rica; the other four species are all blackish; 6 cm (2 ½ in.).

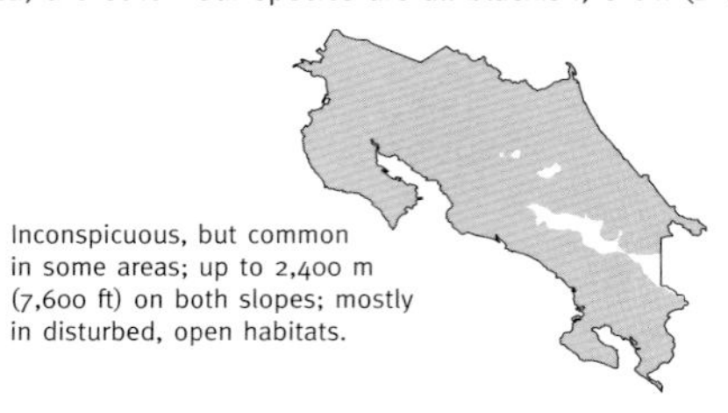

Inconspicuous, but common in some areas; up to 2,400 m (7,600 ft) on both slopes; mostly in disturbed, open habitats.

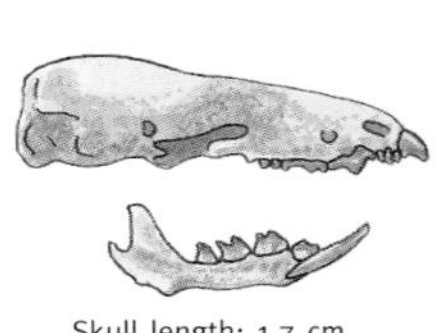

Skull length: 1.7 cm.

2 Vesper Rat (*Nyctomys sumichrasti*), p. 171.

Nocturnal; arboreal; solitary. Upperparts vary from bright orangey in dry lowlands to browner and darker in wet highlands; no other Costa Rican rodent has this coloration and a black mask; tail tufted, not prehensile; mouse opossums (pl. 3) are similar in color but have a naked, prehensile tail; 12 cm (5 in.).

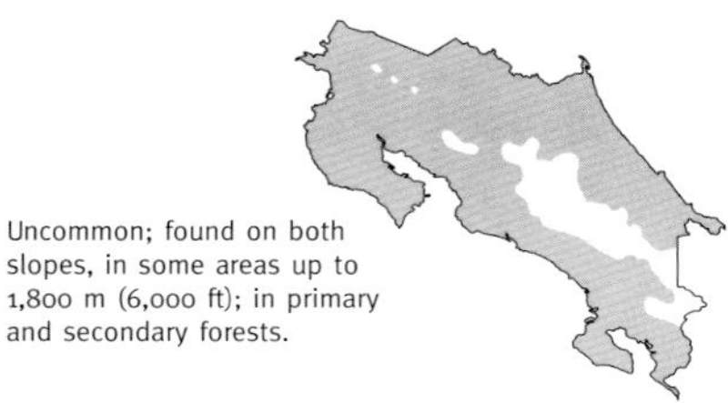

Uncommon; found on both slopes, in some areas up to 1,800 m (6,000 ft); in primary and secondary forests.

3 Alston's Singing Mouse (*Scotinomys teguina*), p. 171.

Mostly diurnal; mostly terrestrial; solitary. Dark blackish-brown upperparts, with subtle tawny tinge on sides; 8 cm (3 in.).

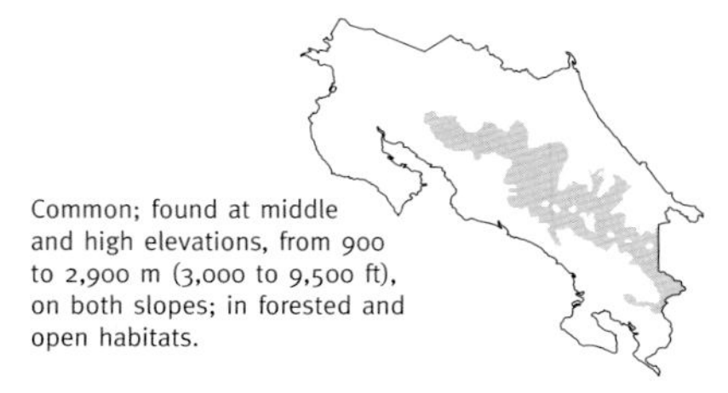

Common; found at middle and high elevations, from 900 to 2,900 m (3,000 to 9,500 ft), on both slopes; in forested and open habitats.

4 Hispid Cotton Rat (*Sigmodon hispidus*), p. 171.

Mostly diurnal; terrestrial; solitary. Shaggy, grizzled fur; conspicuous tan-colored eye ring; 15 cm (6 in.).

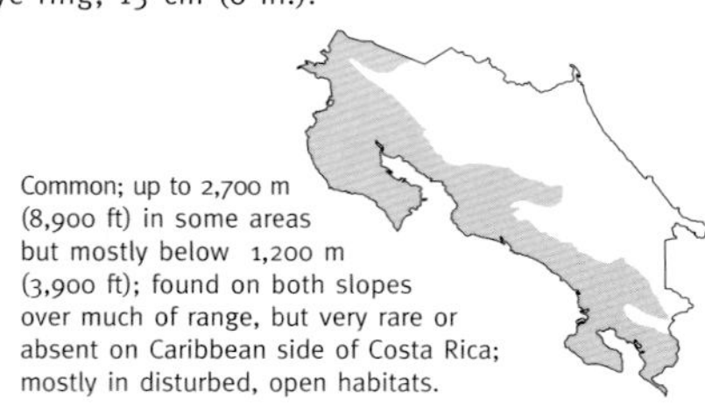

Common; up to 2,700 m (8,900 ft) in some areas but mostly below 1,200 m (3,900 ft); found on both slopes over much of range, but very rare or absent on Caribbean side of Costa Rica; mostly in disturbed, open habitats.

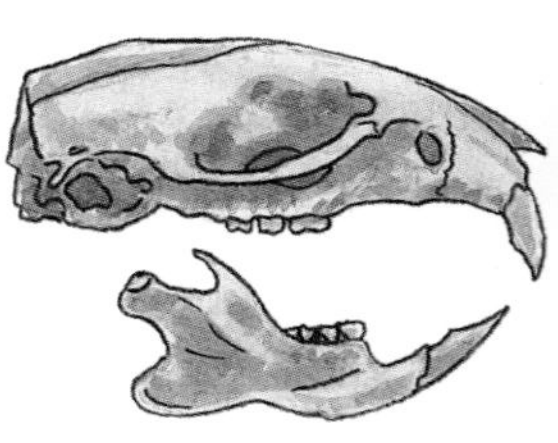

Skull length: 3.2 cm.

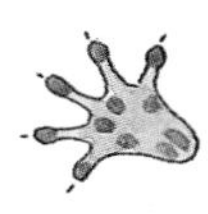

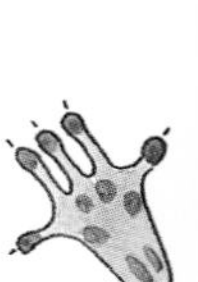

Shrew tracks. Shrews have five digits on each forefoot, while mice have only four.

PLATE 24: NATIVE MURID RODENTS

1 Goldman's Water Mouse (*Rheomys raptor*), p. 171.

Nocturnal and diurnal; terrestrial and aquatic; solitary. Dense, smooth fur; upperparts smoky gray or dark brown, with a few scattered white hairs on sides and rump; tiny ears; densely-whiskered snout; disproportionately large hindfeet; 12 cm (5 in.).

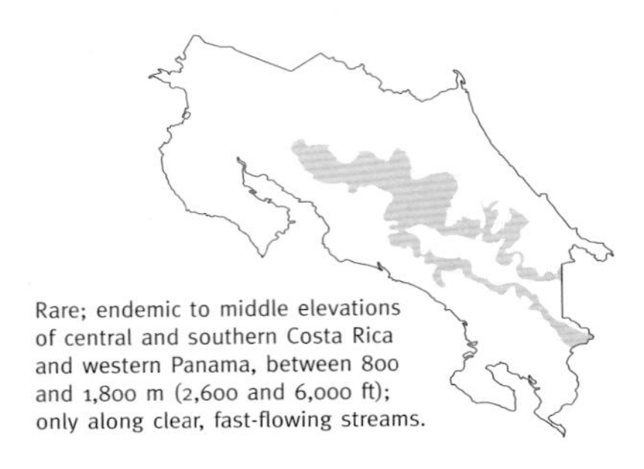

Rare; endemic to middle elevations of central and southern Costa Rica and western Panama, between 800 and 1,800 m (2,600 and 6,000 ft); only along clear, fast-flowing streams.

2 Dusky Rice Rat (*Melanomys caliginosus*), p. 171.

Mostly diurnal; mostly terrestrial; solitary. Fur short and shiny, almost black on top of back, grading to rusty brown on sides; feet and tail entirely black; 12 cm (5 in.).

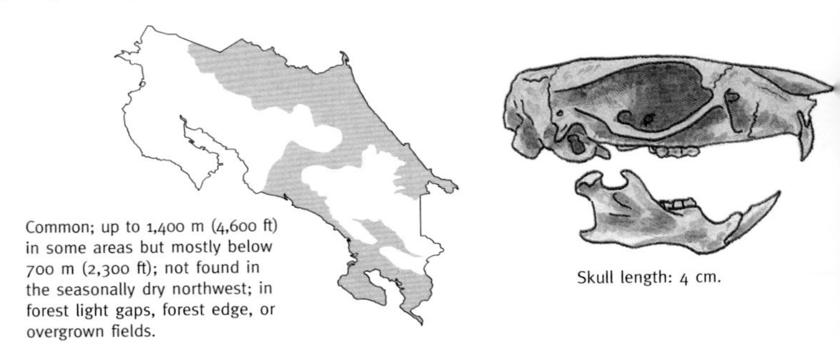

Common; up to 1,400 m (4,600 ft) in some areas but mostly below 700 m (2,300 ft); not found in the seasonally dry northwest; in forest light gaps, forest edge, or overgrown fields.

Skull length: 4 cm.

3 Watson's Climbing Rat (*Tylomys watsoni*), p. 171.

Nocturnal; arboreal and terrestrial; solitary. Long, dense, fine fur; upperparts bluish to brownish gray; tail half black, half white, resembling that of some opossums; large size and two-toned tail make this species unmistakeable; 24 cm (10 in.).

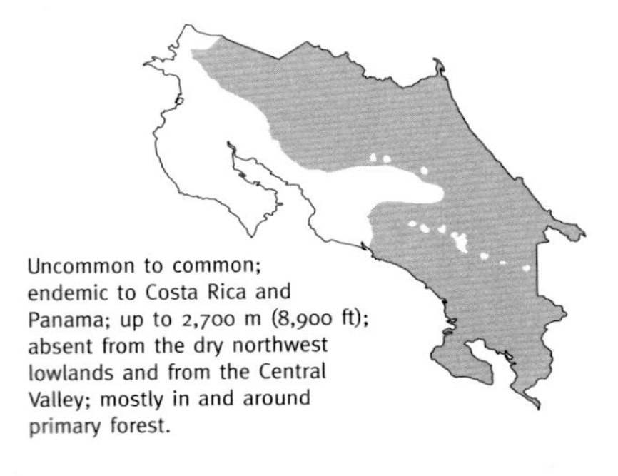

Uncommon to common; endemic to Costa Rica and Panama; up to 2,700 m (8,900 ft); absent from the dry northwest lowlands and from the Central Valley; mostly in and around primary forest.

1
2
3

1 **Roof or Black Rat** (*Rattus rattus*), p. 173.

Mostly nocturnal; terrestrial and arboreal; lives in colonies but usually forages alone. Upperparts either dark gray or tawny; fur long, coarse, and unkempt; tail longer than head and body length; 17 cm (7 in.).

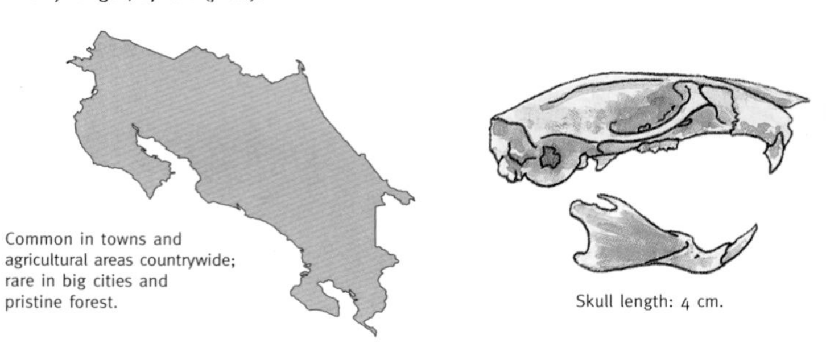

2 **Norway or Brown Rat** (*Rattus norvegicus*), p. 173.

Mostly nocturnal; terrestrial and fossorial; lives in sometimes huge colonies, but often forages alone. Upperparts tawny; fur long, coarse, and unkempt; tail shorter than head and body length; 22 cm (9 in.).

Common in large towns and
cities countrywide; rare in
rural areas, and absent from
Common in towns and
rare in big cities and
pristine forest.

away from buildings.

3 **House Mouse** (*Mus musculus*), p. 173.

Nocturnal; mostly terrestrial; can live in small colonies but usually forages alone. Upperparts vary from pale tawny-brown to dark gray-brown; 8 cm (3 in.).

Common around buildings in

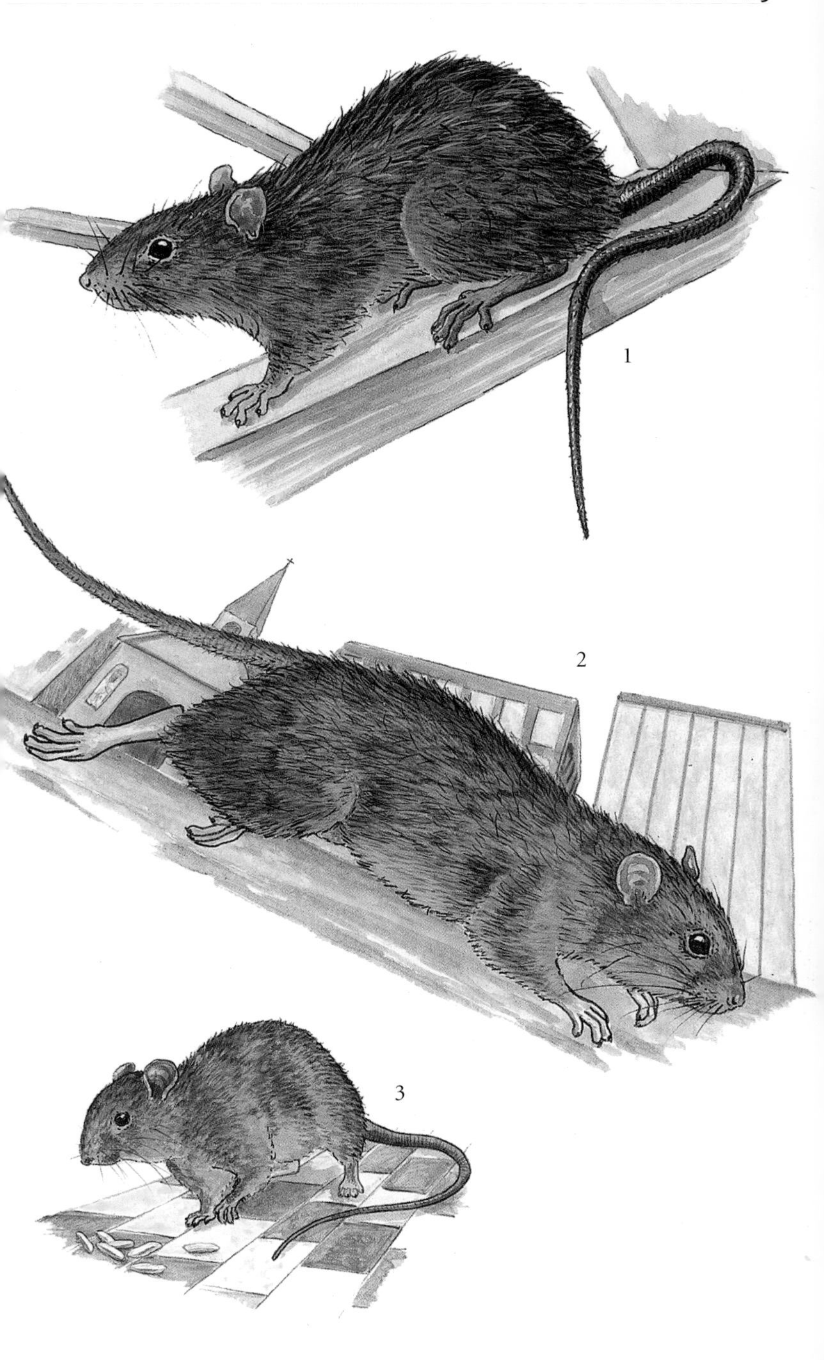
1
2
3

1 **Mexican Hairy Porcupine** (*Coendou mexicanus*), p. 175.

Nocturnal; arboreal; solitary. Covered with pale yellow spines largely concealed by long, black fur, but conspicuous around face; prehensile tail. Costa Rica´s only porcupine; 45 cm (18 in.).

Absent from south Pacific lowlands and uncommon in Atlantic lowlands; common everywhere else; up to 3,200 m (10,500 ft); in primary and secondary forest.

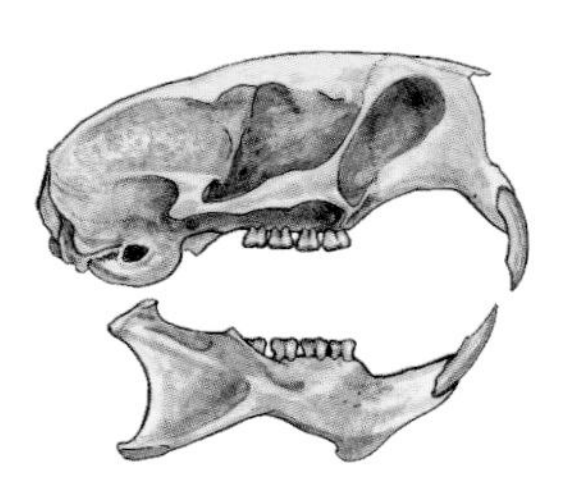

Skull length: 9 cm.

2 **Central American Agouti** (*Dasyprocta punctata*), p. 178.

Diurnal; terrestrial; shares territory with mate, but pair often far apart. Fur finely grizzled; upperparts vary from reddish to yellowish brown; ears much shorter than those of rabbits; when moving resembles a tiny ungulate; 50 cm (20 in.).

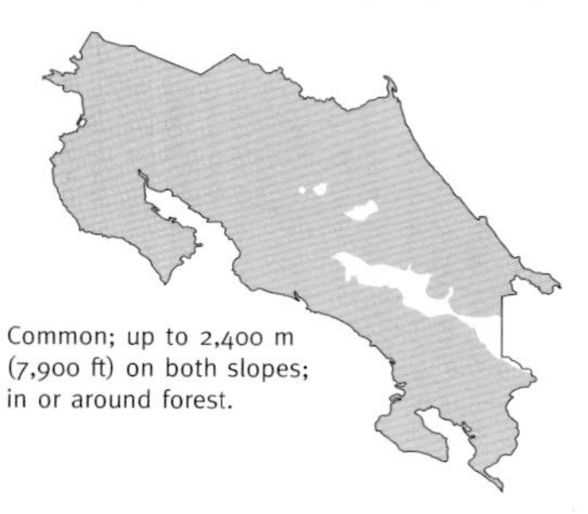

Common; up to 2,400 m (7,900 ft) on both slopes; in or around forest.

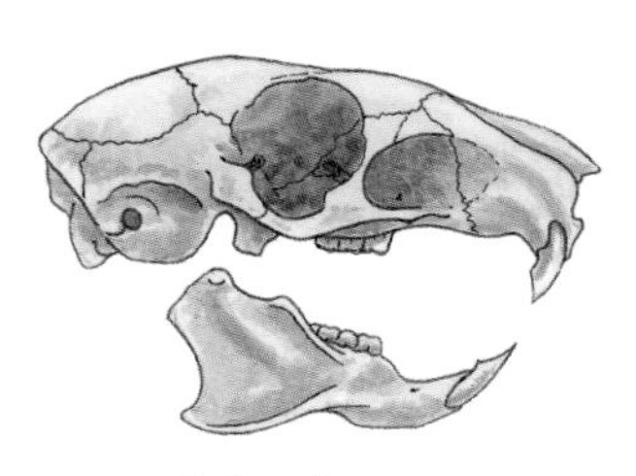

Skull length: 11 cm.

3 **Paca** (*Agouti paca*), p. 180.

Nocturnal; terrestrial; shares territory with mate, but pair often far apart. Similar to agouti in shape, but larger, and with rows of white spots along dark reddish brown upperparts; 70 cm (28 in.).

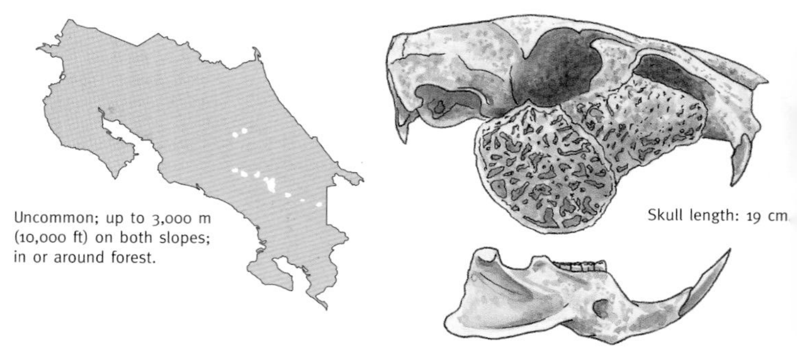

Uncommon; up to 3,000 m (10,000 ft) on both slopes; in or around forest.

Skull length: 19 cm

Mexican hairy
porcupine tracks.
Hindtrack is about
3 cm wide.
1
Agouti tracks.
Hindtrack is
about 3 cm wide.
2
Entire agouti
track pattern;
often, as here,
tips of toes only
portion visible.
Paca tracks.
Hindtrack is about
5 cm wide.
3

1 Tome's Spiny Rat (*Proechimys semispinosus*), p. 183, amidst *Welfia* seeds.

Nocturnal; terrestrial; solitary. Chestnut red upperparts contrast with pure white underparts; back and sides covered with thin spines, although spines are largely hidden in fur and inconspicuous; narrow head with large eyes; M-shaped rear edge to ear; tail often missing; 25 cm (10 in.).

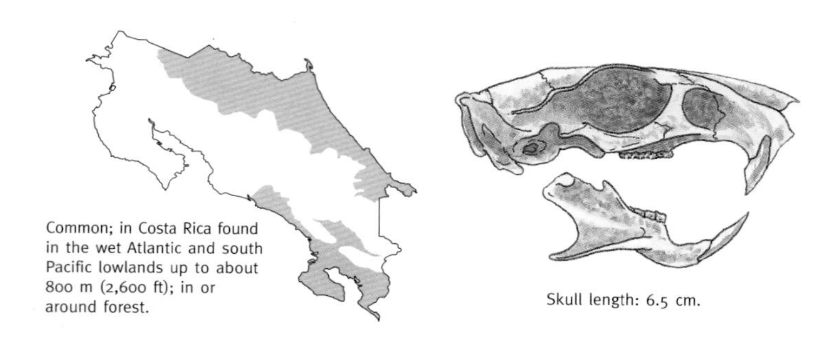

Common; in Costa Rica found in the wet Atlantic and south Pacific lowlands up to about 800 m (2,600 ft); in or around forest.

Skull length: 6.5 cm.

2 Armored Rat (*Hoplomys gymnurus*), p. 183, feeding on cedro macho seeds (*Carapa guianensis*).

Nocturnal; terrestrial; solitary. Cinnamon brown to almost black on back; slightly paler on sides; contrasting pure white underparts; back and sides covered with thick spines that are arranged in evenly-spaced rows and protrude above fur conspicuously; spines sometimes have pale tips, especially on sides; narrow head with large eyes; M-shaped rear edge to ear; tail often missing; 25 cm (10 in.).

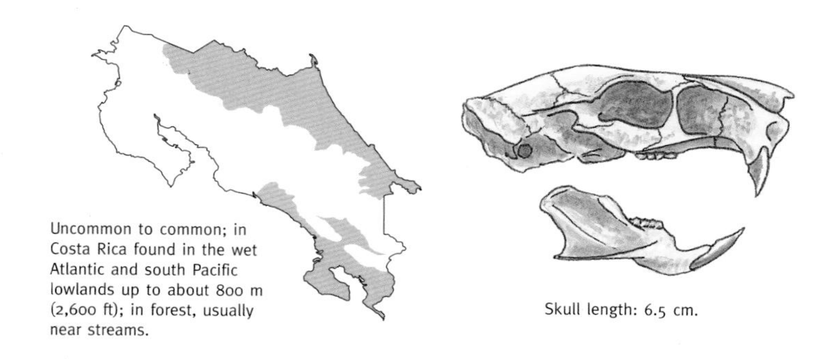

Uncommon to common; in Costa Rica found in the wet Atlantic and south Pacific lowlands up to about 800 m (2,600 ft); in forest, usually near streams.

Skull length: 6.5 cm.

The width of the hindtrack of Tome's spiny rat and of the armored rat is about 2.5 cm. The only other species that leaves tracks this size and shape is Watson's climbing rat (pl. 24).

1 **Tapiti** (*Sylvilagus brasiliensis*) and **Dice's Cottontail** (*S. dicei*), p. 191.

Mostly nocturnal; terrestrial; solitary. The two species are practically indistinguishable (the only difference is that Dice's cottontail is slightly larger and usually slightly darker than the tapiti), but they occur in different areas; *S. brasiliensis*: 38 cm (15 in.); *S. dicei*: 42 cm (17 in.).

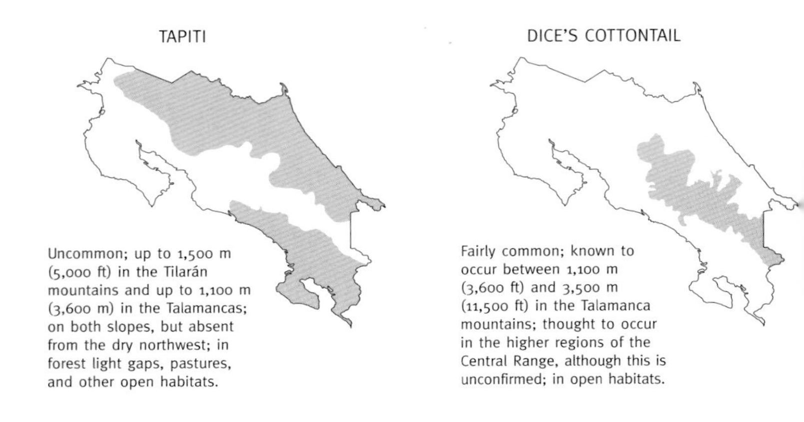

Uncommon; up to 1,500 m (5,000 ft) in the Tilarán mountains and up to 1,100 m (3,600 m) in the Talamancas; on both slopes, but absent from the dry northwest; in forest light gaps, pastures, and other open habitats.

Fairly common; known to occur between 1,100 m (3,600 ft) and 3,500 m (11,500 ft) in the Talamanca mountains; thought to occur in the higher regions of the Central Range, although this is unconfirmed; in open habitats.

2 **Eastern Cottontail** (*Sylvilagus floridanus*), p.193.

Mostly nocturnal; terrestrial; solitary. Differs from tapiti and Dice's cottontail in having a larger, more conspicuous tail, longer ears, and more orange on the nape, although the three species' ranges do not overlap; 38 cm (15 in.).

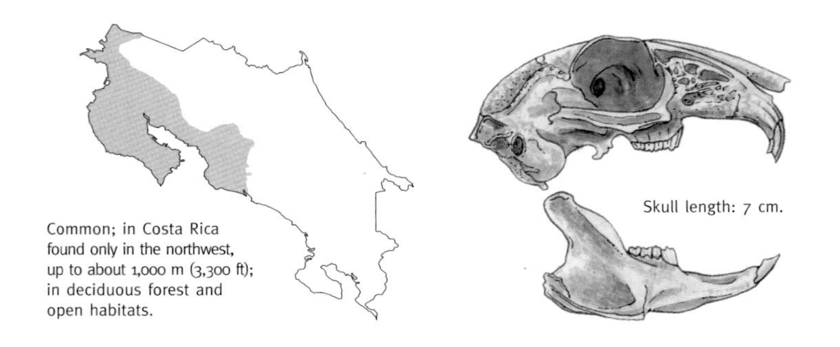

Common; in Costa Rica found only in the northwest, up to about 1,000 m (3,300 ft); in deciduous forest and open habitats.

Skull length: 7 cm.

Close-up and entire footfall pattern of rabbit tracks; the width of the hindtrack is about 2 cm in all three Costa Rican species.

1 **Gray Fox** (*Urocyon cinereoargenteus*), p. 208.

Nocturnal and diurnal; mostly terrestrial; solitary or in pairs. Upperparts vary from brownish to bluish gray. Only fox species in Costa Rica. Cacomistle (pl. 31) has black and white rings on tail, lacks fox's orangey markings, and is mostly arboreal; 50 cm (20 in.).

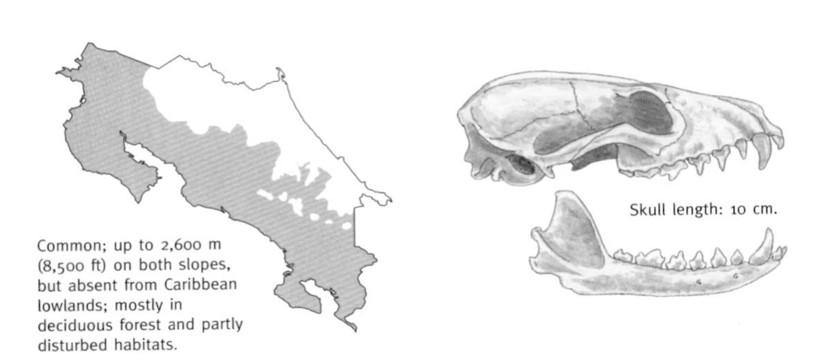

Common; up to 2,600 m (8,500 ft) on both slopes, but absent from Caribbean lowlands; mostly in deciduous forest and partly disturbed habitats.
Skull length: 10 cm.

2 **Coyote** (*Canis latrans*), p. 210.

Nocturnal and diurnal; terrestrial; solitary or in small groups. Distinguished from most domestic dogs by long, slender legs, large, pointed ears, long muzzle, and bushy, black-tipped tail; 80 cm (31 in.).

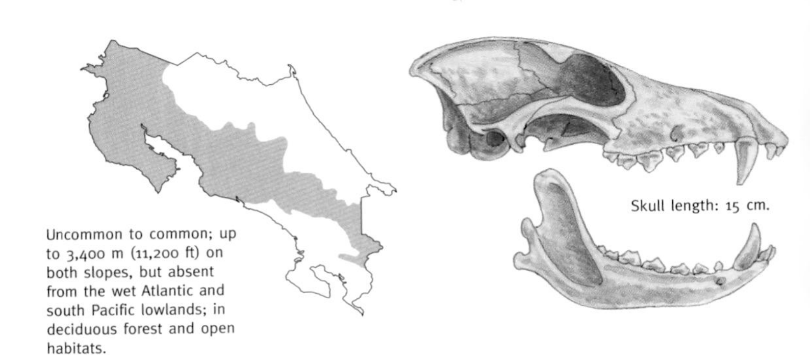

Uncommon to common; up to 3,400 m (11,200 ft) on both slopes, but absent from the wet Atlantic and south Pacific lowlands; in deciduous forest and open habitats.
Skull length: 15 cm.

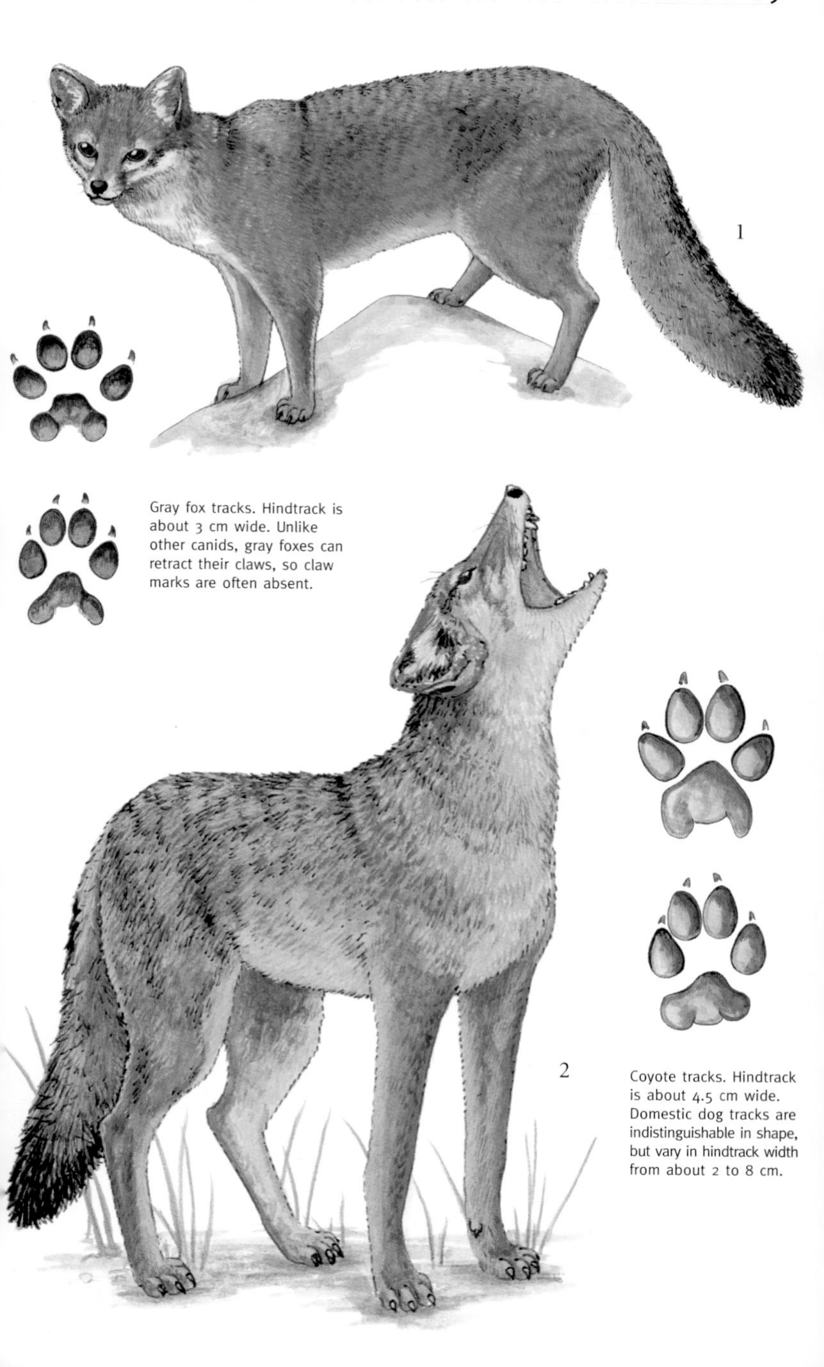

Gray fox tracks. Hindtrack is about 3 cm wide. Unlike other canids, gray foxes can retract their claws, so claw marks are often absent.

Coyote tracks. Hindtrack is about 4.5 cm wide. Domestic dog tracks are indistinguishable in shape, but vary in hindtrack width from about 2 to 8 cm.

1 **Crab-Eating Raccoon** (*Procyon cancrivorus*), p. 220.

Mostly nocturnal; terrestrial and arboreal; mostly solitary. Hard to distinguish from the northern raccoon, which also eats plenty of crabs (see p. 221), but the two species occur together only in southwest; fur grows forwards on nape; legs dark; 60 cm (24 in.).

Uncommon; up to 1,200 m (4,000 ft), but mostly in lowlands; southwest only, up to Orotina and the Central Valley; mostly in relatively undisturbed habitats.

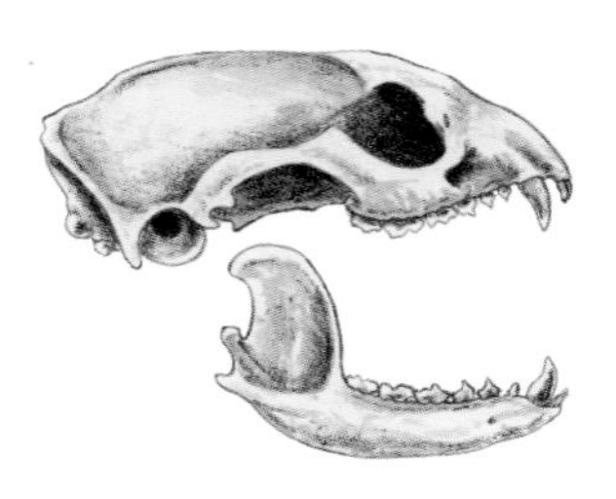

Skull length: 12 cm.

2 **Northern Raccoon** (*Procyon lotor*), p. 222.

Mostly nocturnal; terrestrial and arboreal; mostly solitary. Hard to distinguish from the slightly larger crab-eating raccoon (see p. 221), but the two species occur together only in southwest; fur grows backwards on nape; legs pale; 50 cm (20 in.).

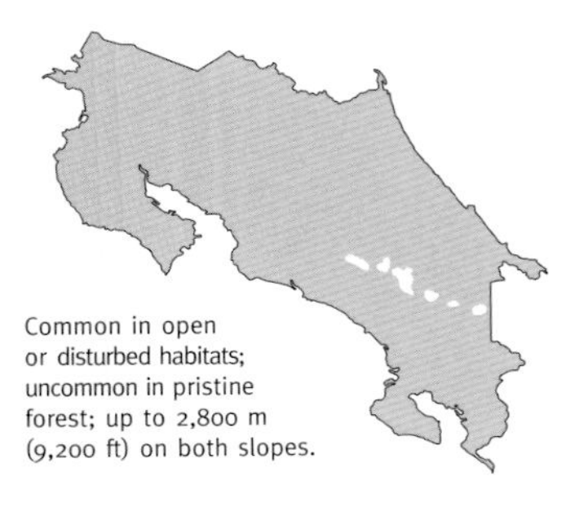

Common in open or disturbed habitats; uncommon in pristine forest; up to 2,800 m (9,200 ft) on both slopes.

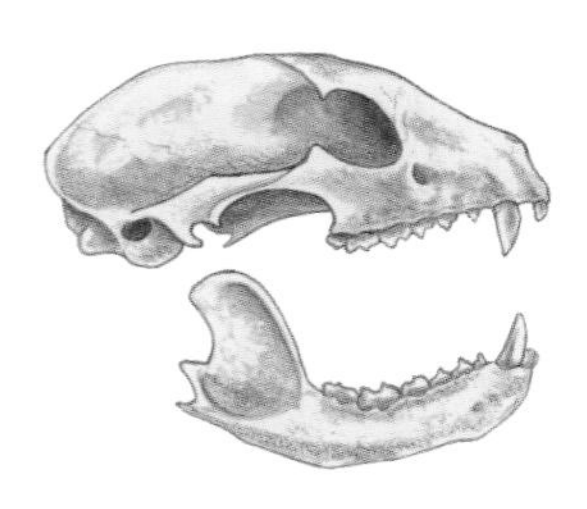

Skull length: 11 cm.

3 **White-Nosed Coati** (*Nasua narica*), p. 225.

Diurnal; terrestrial and arboreal; males solitary; females and young social, travelling in bands of up to 25 individuals or more. Darkness of overall coloration, extent of frosting on shoulders, and distinctness of rings on tail all variable; long muzzle; tail not bushy, often held vertically; 55 cm (22 in.).

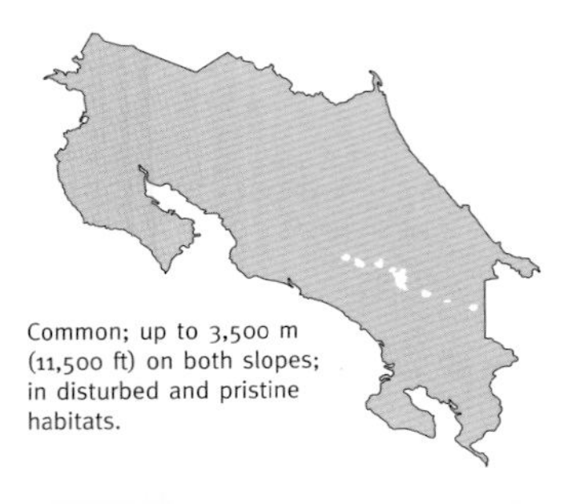

Common; up to 3,500 m (11,500 ft) on both slopes; in disturbed and pristine habitats.

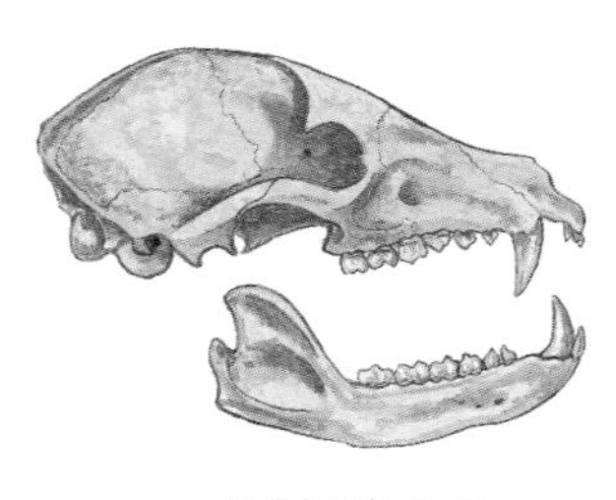

Skull length: 12 cm.

1

coon tracks.
dtrack width in
h species is
ut 6 cm.

2

Coati tracks.
Hindtrack width
is about 4.5 cm.

3

1 **Central American Cacomistle** (*Bassariscus sumichrasti*), p. 218.

Mostly nocturnal; mostly arboreal; mostly solitary. Only other mammals with such conspicuosly banded tails are raccoons, which are larger and have black masks and shorter tails; olingo is similar in size and shape, but has only faint banding on tail; 45 cm (18 in.).

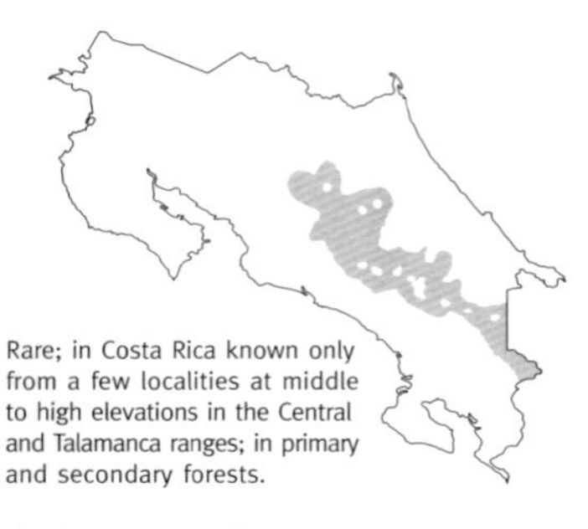

Rare; in Costa Rica known only from a few localities at middle to high elevations in the Central and Talamanca ranges; in primary and secondary forests.

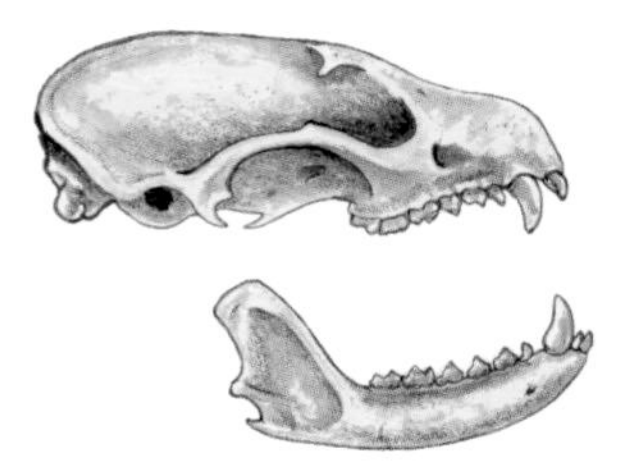

Skull length: 8.5 cm.

2 **Kinkajou** (*Potos flavus*), p. 229.

Mostly nocturnal; arboreal; mostly solitary. Upperparts vary from orangey to olingolike gray-brown, the latter mostly in the lowlands; tail prehensile; 50 cm (20 in.).

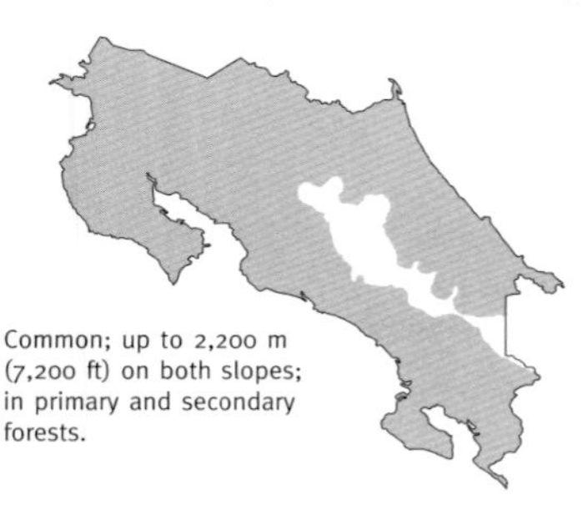

Common; up to 2,200 m (7,200 ft) on both slopes; in primary and secondary forests.

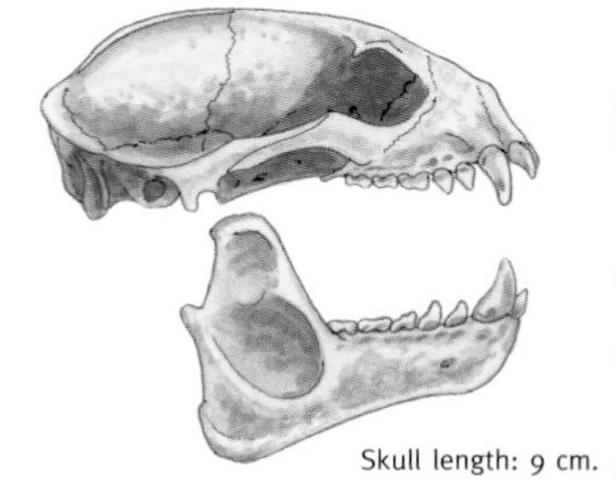

Skull length: 9 cm.

3 **Olingo** (*Bassaricyon gabbii*), p. 232.

Mostly nocturnal; arboreal; solitary. Upperparts usually drab gray-brown, rarely paler or more golden; smaller than kinkajou, with a more bushy, non-prehensile tail; 43 cm (17 in.).

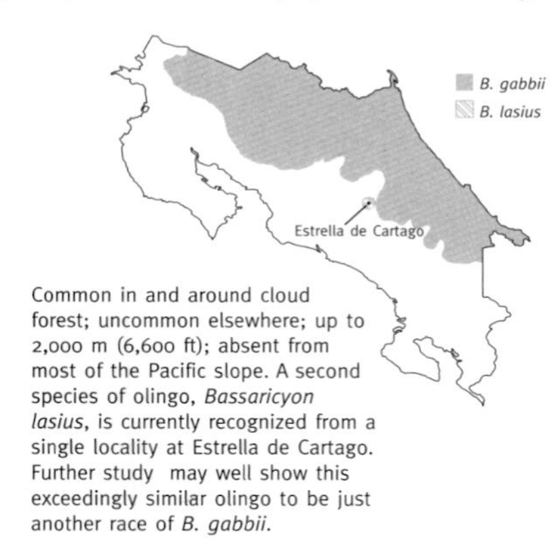

Common in and around cloud forest; uncommon elsewhere; up to 2,000 m (6,600 ft); absent from most of the Pacific slope. A second species of olingo, *Bassaricyon lasius*, is currently recognized from a single locality at Estrella de Cartago. Further study may well show this exceedingly similar olingo to be just another race of *B. gabbii*.

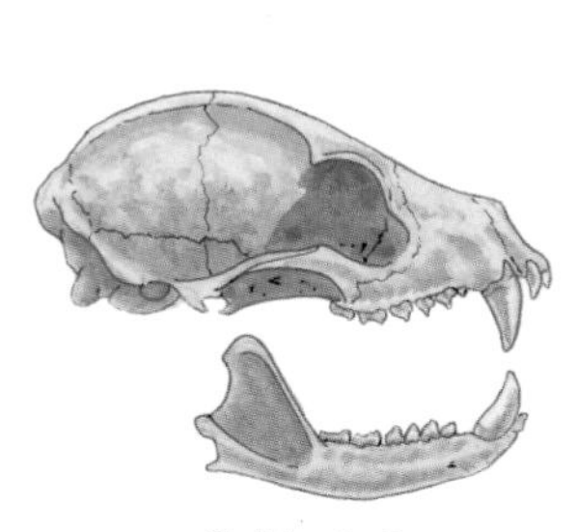

Skull length: 8 cm.

1
2
3
Cacomistle tracks.
Hindtrack width
is about 3 cm.
Kinkajou tracks. Hindtrack width
is about 4 cm. Olingo tracks are
identical in shape but a little
smaller, with the hindtrack about
3 cm wide. Olingos have webbing
between their toes, although the
webbing seldom appears in tracks.

PLATE 32: WEASEL FAMILY

1 **Greater Grison** (*Grison vittata*), p. 242.

Nocturnal and diurnal; terrestrial and aquatic; solitary or in small groups. Distinctive black, white, and gray coloration unlike that of any other Costa Rican mammal; 50 cm (20 in.).

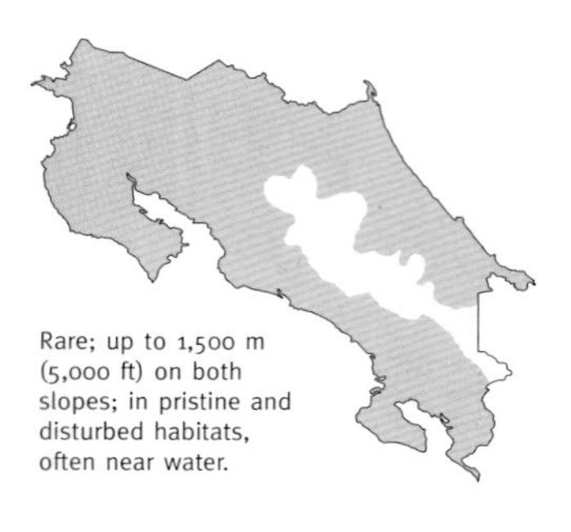

Rare; up to 1,500 m (5,000 ft) on both slopes; in pristine and disturbed habitats, often near water.

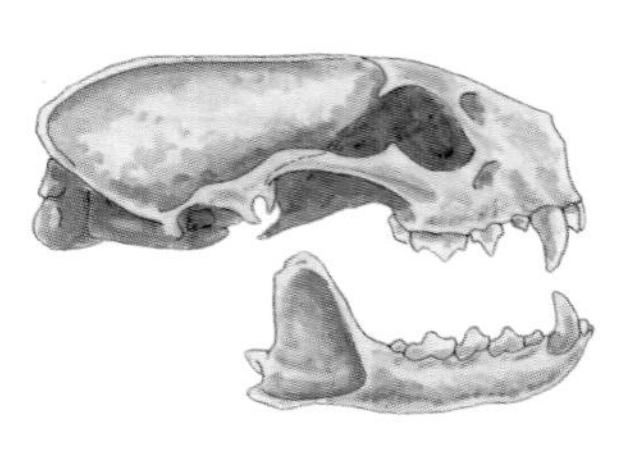

Skull length: 9 cm.

2 **Tayra** (*Eira barbara*), p. 245.

Diurnal; terrestrial and arboreal; solitary or in pairs. All dark, except for a slightly paler head and, usually, a cream-colored patch on the throat; distinguished from both the jaguarundi (pl. 35) and the Neotropical river otter (below) by bushy tail; 65 cm (26 in.).

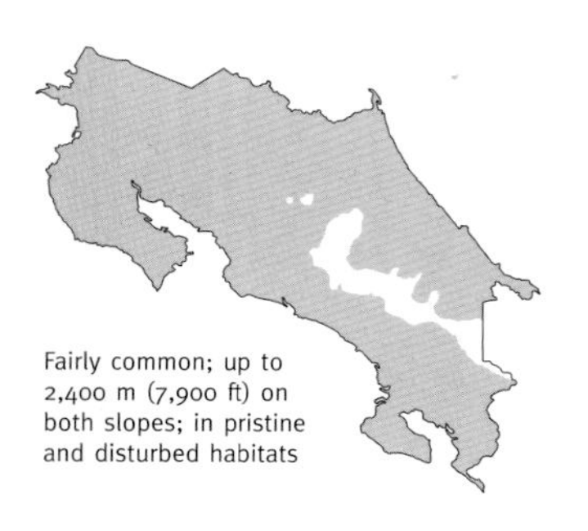

Fairly common; up to 2,400 m (7,900 ft) on both slopes; in pristine and disturbed habitats

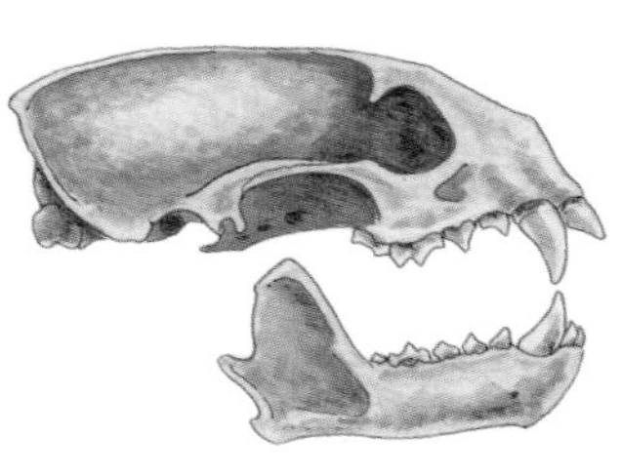

Skull length: 12 cm.

3 **Neotropical River Otter** (*Lutra longicaudis*), p. 253.

Diurnal and nocturnal; aquatic and terrestrial; solitary or small family groups. Fur short and sleek, including on tail; tail very broad at base, tapering toward tip; 60 cm (24 in.).

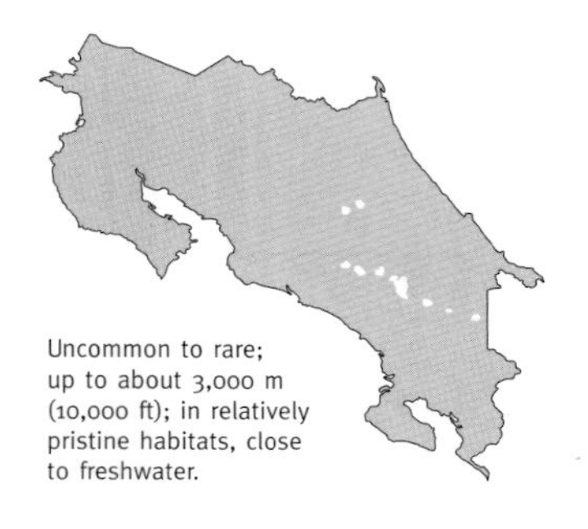

Uncommon to rare; up to about 3,000 m (10,000 ft); in relatively pristine habitats, close to freshwater.

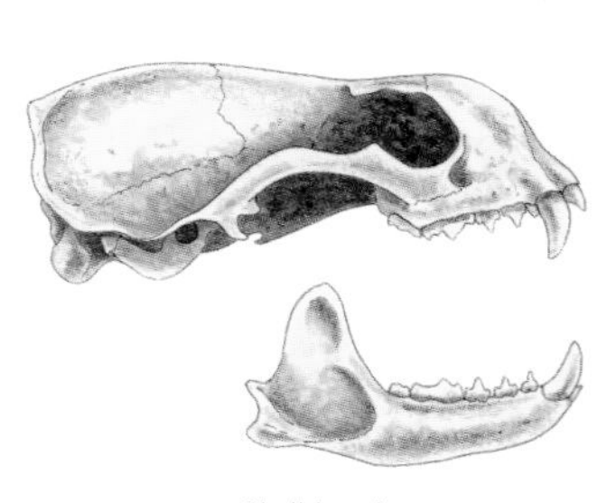

Skull length: 12 cm.

Grison tracks. Width of hindtrack is about 5 cm.

Tayra tracks. Width of hindtrack is about 8 cm.

Neotropical river otter tracks. Width of hindtrack is about 8 cm.

PLATE 33: WEASEL FAMILY

1 Long-Tailed Weasel (*Mustela frenata*), p. 239.

Mostly diurnal; mostly terrestrial and fossorial, but climbs to get at bird nests; solitary. Squirrellike, but proportionately longer and thinner, and with a less bushy tail; very quick and restless, and hard to observe for long; white markings on head variable, sometimes absent; 20 cm (8 in.).

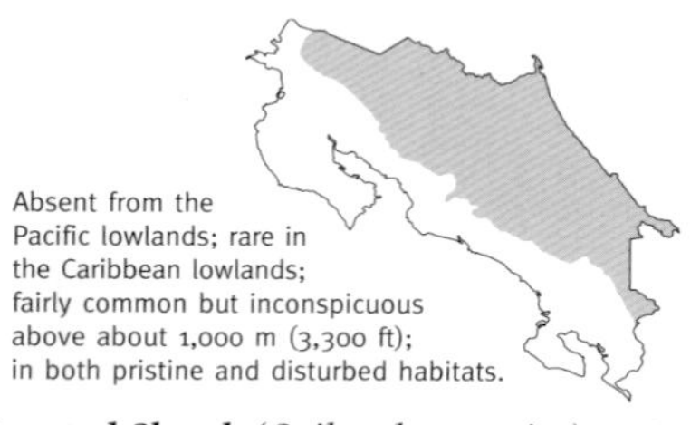

Absent from the Pacific lowlands; rare in the Caribbean lowlands; fairly common but inconspicuous above about 1,000 m (3,300 ft); in both pristine and disturbed habitats.

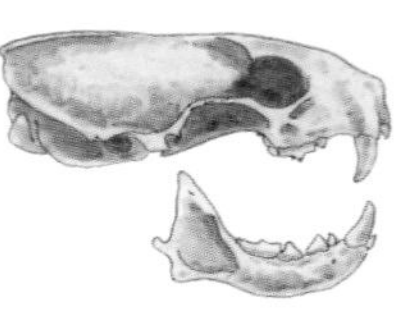

Skull length: 5 cm.

2 Spotted Skunk (*Spilogale putorius*), p. 248.

Nocturnal; mostly terrestrial; solitary. Pattern of white stripes and spots variable and unique to each individual, but usually similar to that illustrated; bushy tail, black at base and white at tip; notably smaller than other two Costa Rican skunks; 25 cm (10 in.).

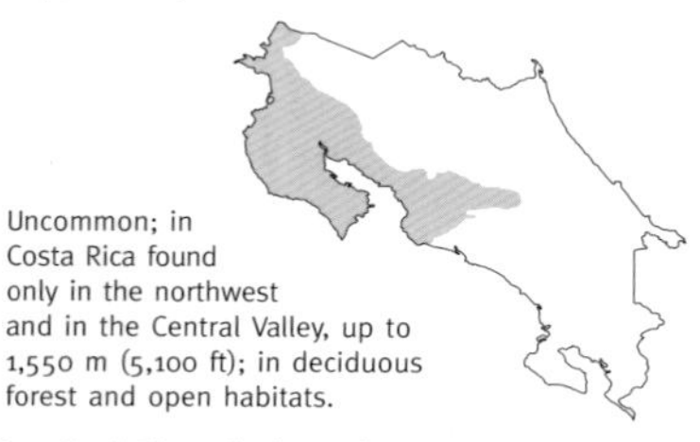

Uncommon; in Costa Rica found only in the northwest and in the Central Valley, up to 1,550 m (5,100 ft); in deciduous forest and open habitats.

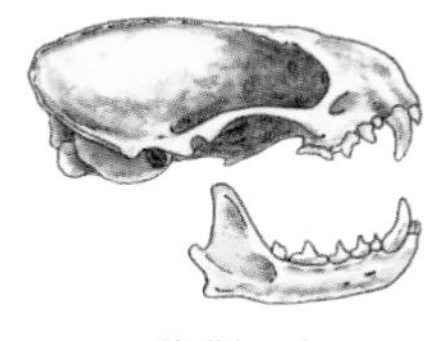

Skull length: 5 cm.

3 Hooded Skunk (*Mephitis macroura*), p. 251.

Nocturnal; terrestrial; solitary. Black with a thin white stripe along each side of the body; sometimes with a thin, white, vertical stripe on forehead; tail bushy, mostly black with white in the center of the tail near its base; 35 cm (14 in.).

Common in the northwest lowlands; not found anywhere else; in deciduous forest and open habitats.

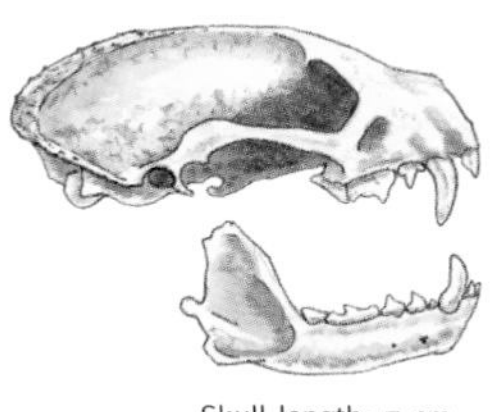

Skull length: 7 cm.

4 Striped Hog-Nosed Skunk (*Conepatus semistriatus*), p. 247.

Nocturnal; terrestrial; solitary. Black with two thick white stripes along the top of the back that meet on the top of the head; bushy white tail with a little black near the base, often held erect; long, conical snout; 40 cm (16 in.).

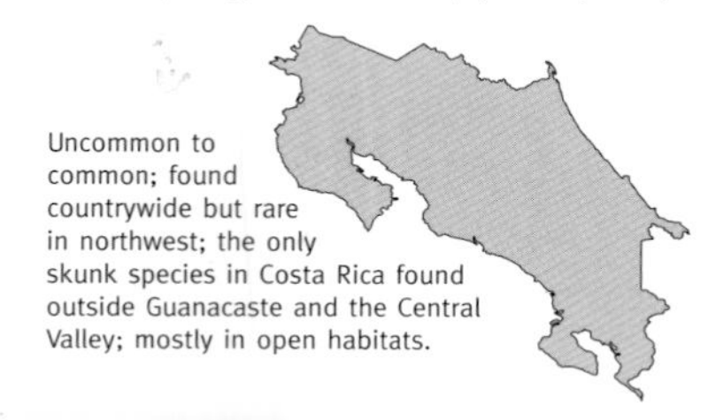

Uncommon to common; found countrywide but rare in northwest; the only skunk species in Costa Rica found outside Guanacaste and the Central Valley; mostly in open habitats.

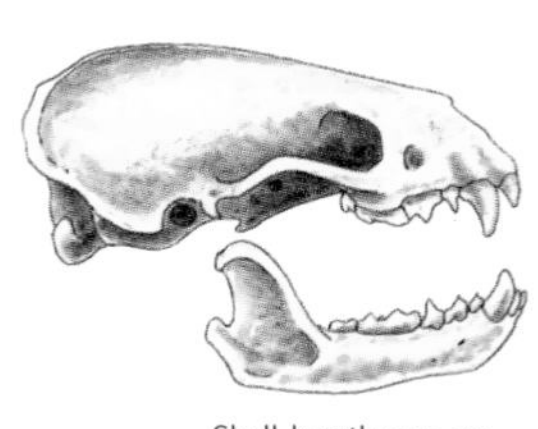

Skull length: 7.5 cm.

Long-tailed
weasel tracks.
Hindtrack width
is about 1.2 cm
1
2
3
Skunk tracks.
Hindtrack width
is about 2 cm in
the spotted skunk,
2.5 cm in the hooded
skunk, and 3.5 cm
in the striped hog-
nosed skunk.
4

PLATE 34: CAT FAMILY

1 **Oncilla** (*Leopardus tigrina*), p. 262.

Probably mostly nocturnal; terrestrial and arboreal; solitary. Difficult to distinguish from margay (see p. 262), but smaller (size of house cat or little larger), spots and streaks usually smaller and more numerous, proportionately smaller eyes and forefeet, and proportionately shorter tail (about 55% of head and body length); 50 cm (20 in.).

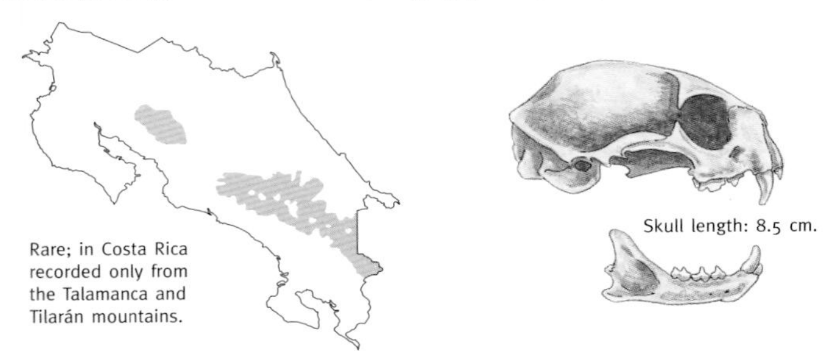

Rare; in Costa Rica recorded only from the Talamanca and Tilarán mountains.

Skull length: 8.5 cm.

2 **Margay** (*Leopardus wiedii*), p. 264.

Mostly nocturnal; terrestrial and arboreal; solitary. Difficult to distinguish from oncilla (see p. 262), but larger, with spots and streaks usually larger, fewer, and enclosed rather than open, proportionately larger eyes and forefeet, broader muzzle, and longer tail (about 70% of head and body length); 55 cm (22 in.).

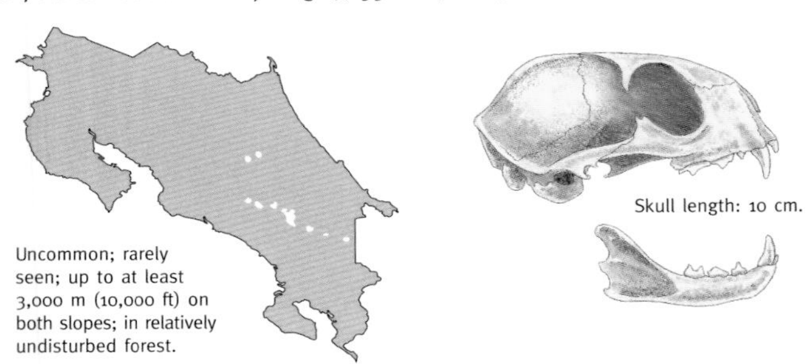

Uncommon; rarely seen; up to at least 3,000 m (10,000 ft) on both slopes; in relatively undisturbed forest.

Skull length: 10 cm.

3 **Ocelot** (*Leopardus pardalis*), p. 266.

Nocturnal and diurnal; mostly terrestrial; solitary. Larger than margay, with proportionately shorter tail (45% of head and body length), broader front paws, and rosettes usually fused together more than in margay, forming more continuous bands along side of body; 75 cm (30 in.).

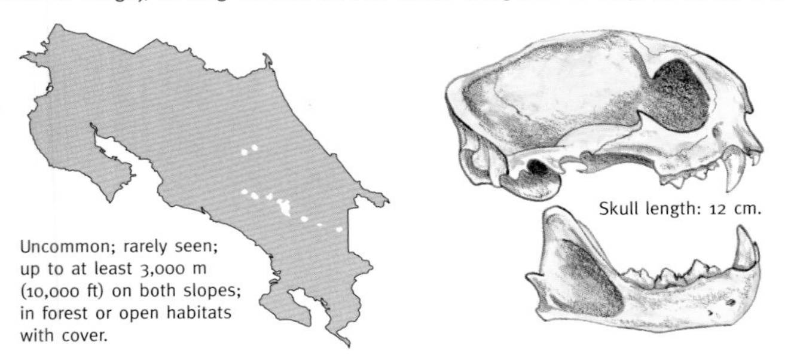

Uncommon; rarely seen; up to at least 3,000 m (10,000 ft) on both slopes; in forest or open habitats with cover.

Skull length: 12 cm.

:illa tracks. Hindtrack
ith is about 2.2 cm.
istinguishable from
ıse cat tracks.
1
2
Margay tracks.
Hindtrack width
is about 3 cm.
Ocelot tracks. Hindtrack
width is about 5 cm.
Forepaws are considerably
broader than hindpaws.
3

PLATE 35: CAT FAMILY

1 **Jaguarundi** (*Herpailurus yagouaroundi*), p. 269.

Mostly diurnal; mostly terrestrial; solitary. Coloration varies considerably, from black, to brownish gray, to sandy or reddish brown; fur often heavily grizzled; black form distinguished from tayra (pl. 32) by longer, much thinner tail; 65 cm (26 in.).

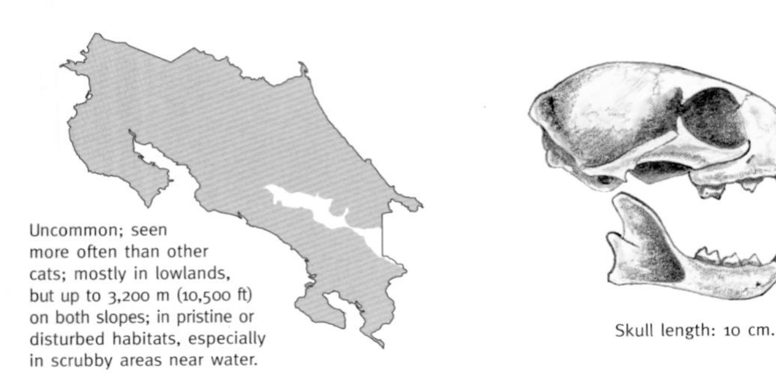

Uncommon; seen more often than other cats; mostly in lowlands, but up to 3,200 m (10,500 ft) on both slopes; in pristine or disturbed habitats, especially in scrubby areas near water.
Skull length: 10 cm.

2 **Puma** (*Puma concolor*), p. 272.

Nocturnal and diurnal; mostly terrestrial; solitary. Upperparts vary from pale to reddish brown; black and white facial markings conspicuous; tail usually with black tip; young spotted; 110 cm (43 in.).

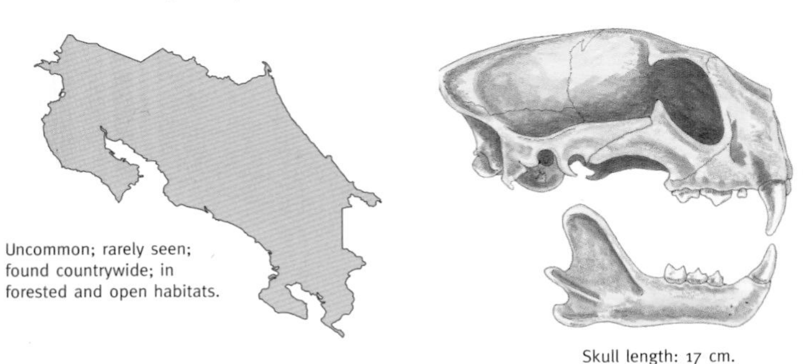

Uncommon; rarely seen; found countrywide; in forested and open habitats.
Skull length: 17 cm.

3 **Jaguar** (*Panthera onca*), p. 275.

Nocturnal and diurnal; mostly terrestrial; solitary. Much larger than other spotted cats; 150 cm (60 in.).

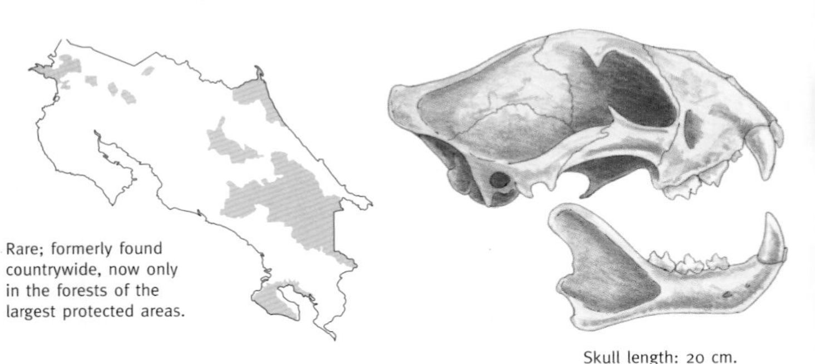

Rare; formerly found countrywide, now only in the forests of the largest protected areas.
Skull length: 20 cm.

1
Jaguarundi tracks. Hindtrack width is about 3 cm. Toes of jaguarundi are slightly more elongate than those of the margay, but the two species' tracks are very hard to distinguish.
Puma tracks. Hindtrack width is about 8 cm.
2
Jaguar tracks. Hindtrack width is about 9.5 cm. Jaguars have round toes, while pumas have egg-shaped toes. Also, the rear edge of the jaguar heel pad is less indented than that of the puma.
3

PLATE 36: TAPIR & PECCARIES

1 Baird's Tapir (*Tapirus bairdii*), p. 297.

Nocturnal and diurnal; terrestrial; solitary. Upperparts vary from reddish to grayish brown; fur very short and smooth; long, curved, prehensile upper lip; young reddish brown with white spots and streaks; much larger than any other Costa Rican mammal; 200 cm (80 in.).

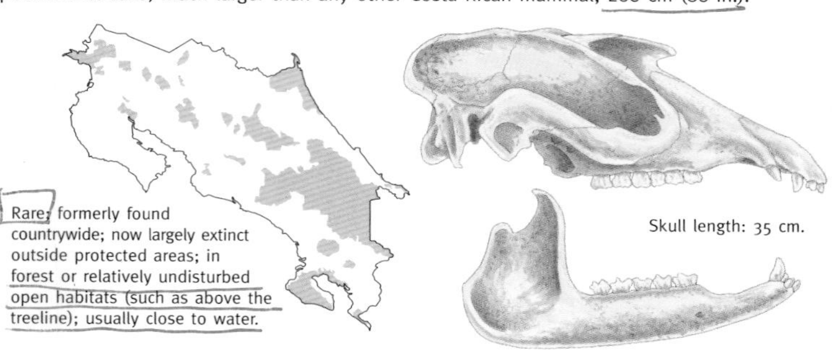

Rare; formerly found countrywide; now largely extinct outside protected areas; in forest or relatively undisturbed open habitats (such as above the treeline); usually close to water.

Skull length: 35 cm.

2 Collared Peccary (*Tayassu tajacu*), p. 309.

Diurnal and nocturnal; terrestrial; travel in herds of a few to 30 individuals. Upperparts dark gray grizzled with either light gray or pale tan, with an inconspicuous pale tan collar from top of shoulder to back of cheek; some individuals with more pale tan fur around mouth and cheeks; fur along top of back long and wiry, raised in alarm; head proprtionately huge, triangular in shape; snout piglike; legs short and thin; young reddish brown; 90 cm (35 in.).

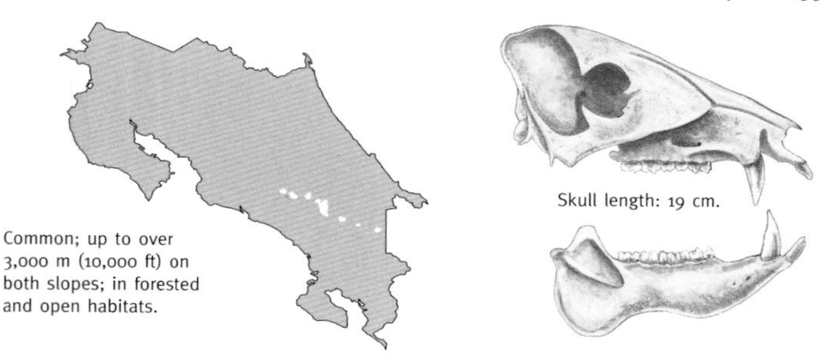

Common; up to over 3,000 m (10,000 ft) on both slopes; in forested and open habitats.

Skull length: 19 cm.

3 White-Lipped Peccary (*Tayassu pecari*), p. 313.

Diurnal and nocturnal; terrestrial; travels in herds of 50 to 300 individuals. Upperparts dark gray; patch of pure white fur on throat usually conspicuous, but small and inconspicuous in a few individuals; fur longer than that of collared peccary; fur along top of back especially long and wiry, raised in alarm; head proportionately huge, triangular in shape; snout piglike; legs short and thin; young reddish brown; 100 cm (40 in.).

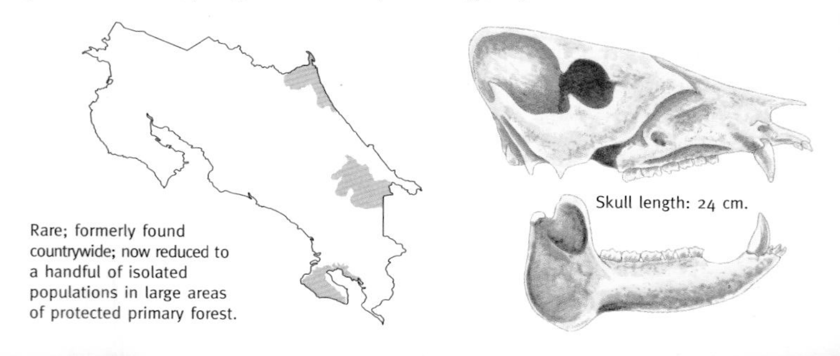

Rare; formerly found countrywide; now reduced to a handful of isolated populations in large areas of protected primary forest.

Skull length: 24 cm.

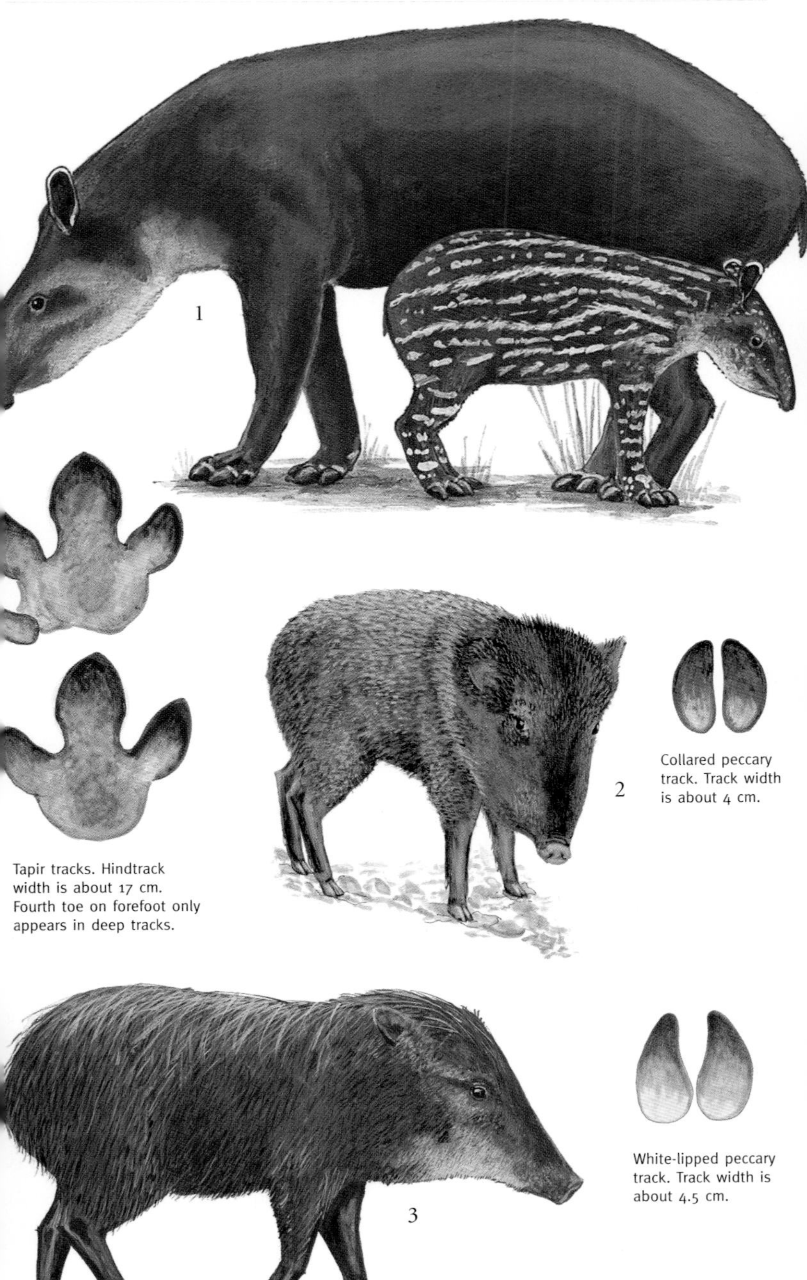

Tapir tracks. Hindtrack width is about 17 cm. Fourth toe on forefoot only appears in deep tracks.

Collared peccary track. Track width is about 4 cm.

White-lipped peccary track. Track width is about 4.5 cm.

1 Red Brocket Deer (*Mazama americana*), p. 318.

Diurnal and nocturnal; terrestrial; solitary. Upperparts reddish brown; face reddish brown or blackish, without markings; usually carries head low when moving; males' antlers short and unbranched; young spotted; 110 cm (43 in.).

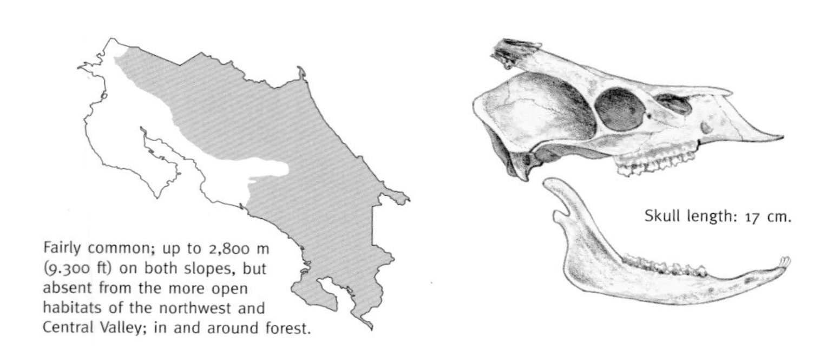

Fairly common; up to 2,800 m (9.300 ft) on both slopes, but absent from the more open habitats of the northwest and Central Valley; in and around forest.
Skull length: 17 cm.

2 White-Tailed Deer (*Odocoileus virginianus*), p. 321.

Nocturnal and diurnal; terrestrial; solitary or in small groups. Upperparts pale to orange brown, paler than those of red brocket; face with conspicuous white and black markings; usually carries head high when moving; males' antlers branched and longer than those of brocket deer; young spotted; 130 cm (51 in.).

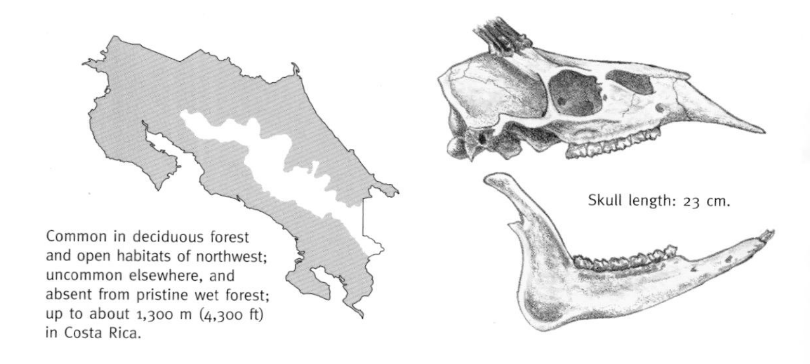

Common in deciduous forest and open habitats of northwest; uncommon elsewhere, and absent from pristine wet forest; up to about 1,300 m (4,300 ft) in Costa Rica.
Skull length: 23 cm.

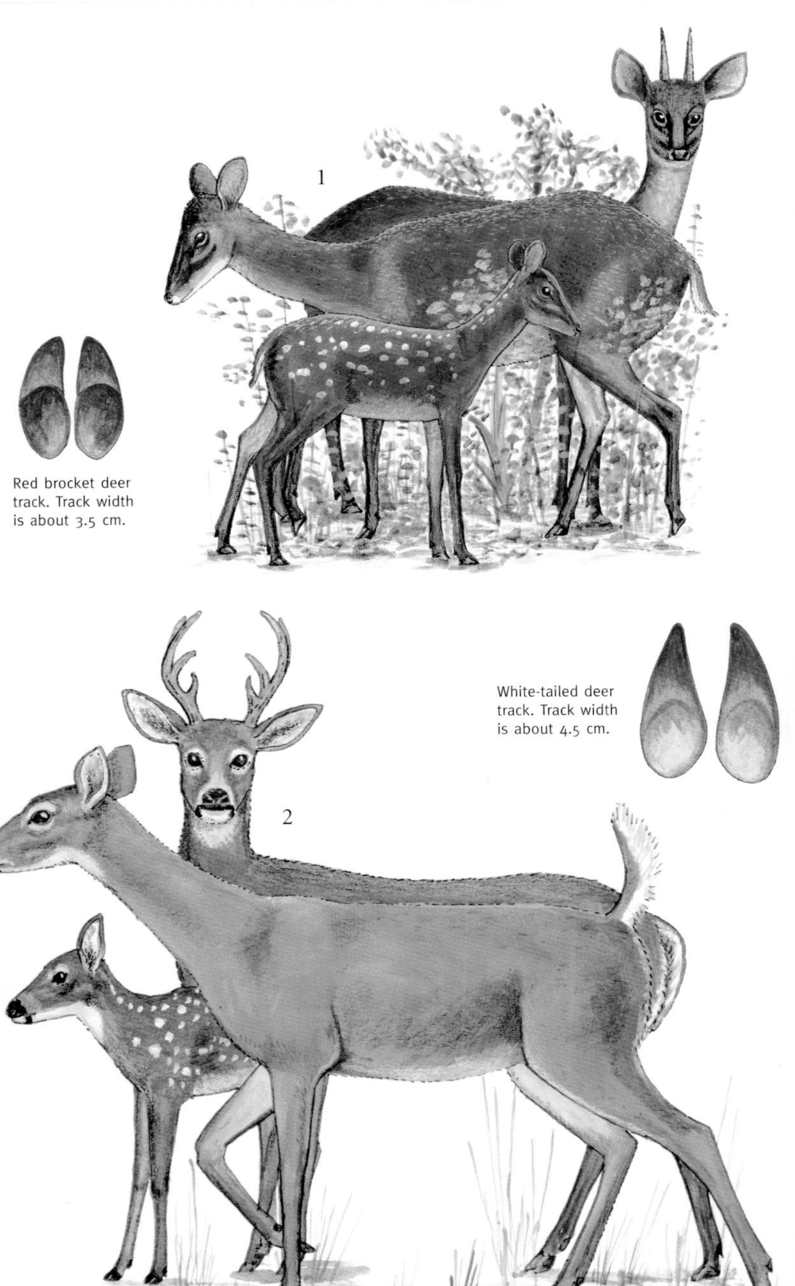

Red brocket deer track. Track width is about 3.5 cm.

White-tailed deer track. Track width is about 4.5 cm.

1 **West Indian Manatee** (*Trichechus manatus*), p. 286.

Nocturnal and diurnal; aquatic; mostly solitary. Almost no fur; large, bulbous snout; round, flattened tail; 300 cm (120 in.).

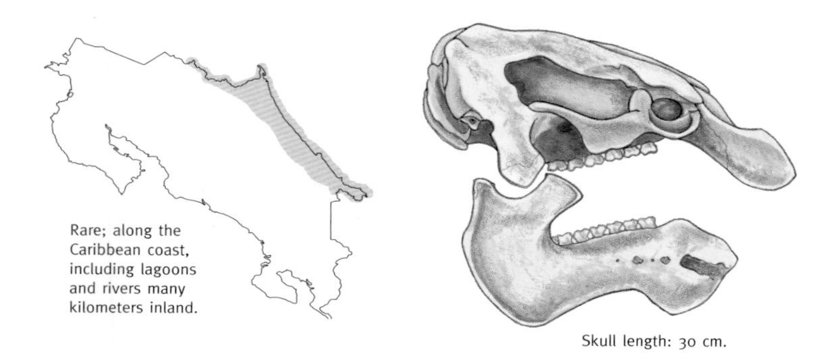

Rare; along the Caribbean coast, including lagoons and rivers many kilometers inland.

Skull length: 30 cm.

2 **Tucuxi** (*Sotalia fluviatalis*), p. 331.

Most active in early morning and late evening; aquatic; travels in groups of up to 30, but usually just 2 to 4. The only Costa Rican dolphin found in freshwater well beyond river mouths; 170 cm (67 in.).

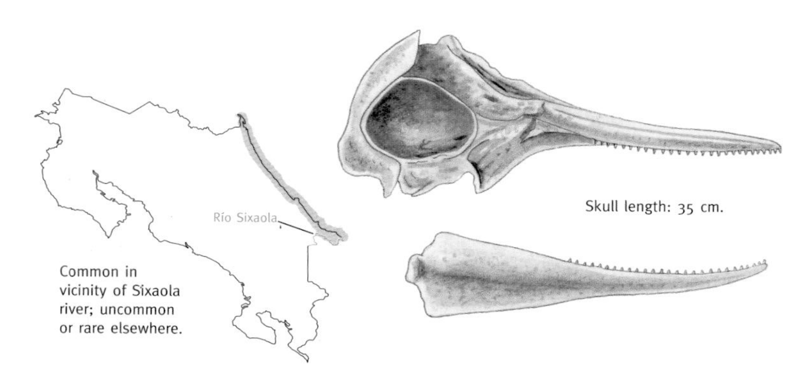

Río Sixaola
Common in vicinity of Sixaola river; uncommon or rare elsewhere.

Skull length: 35 cm.

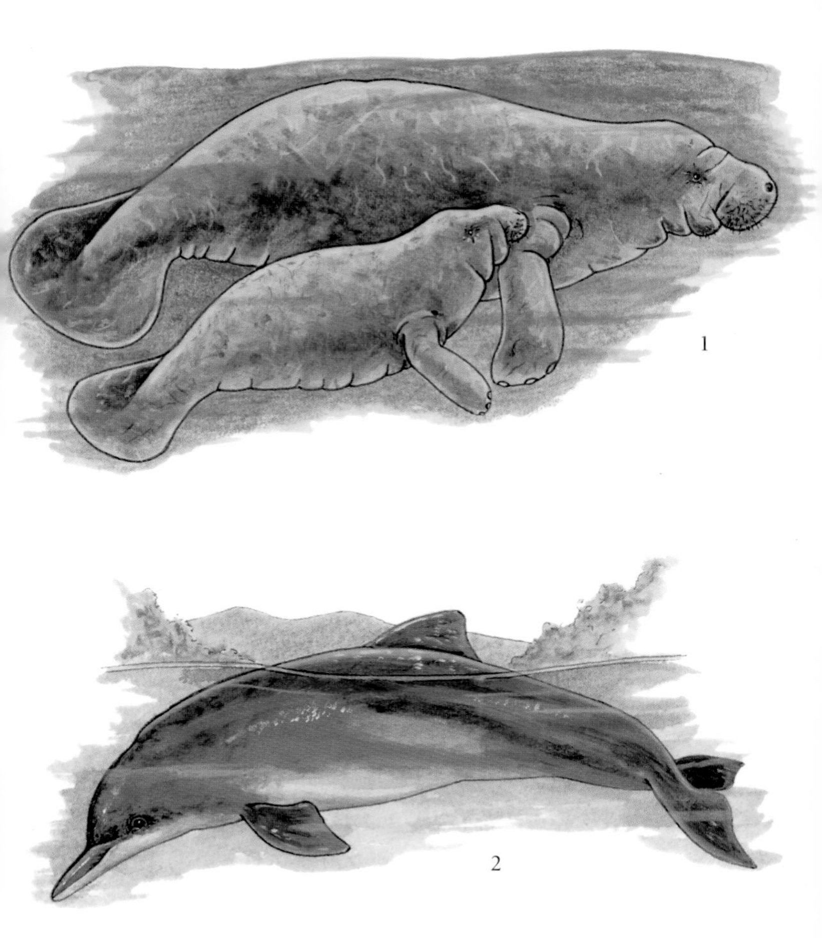
1
2